AN INDEX OF

COMPILED BY

JOHN B. CORBIN

STATE GEOLOGICAL SURVEY PUBLICATIONS

ISSUED IN SERIES

SUPPLEMENT, 1963-1980

THE SCARECROW PRESS
METUCHEN, NJ, & LONDON • 1982

Library of Congress Cataloging in Publication Data

Corbin, John Boyd.
    An index of state geological survey publications
issued in series : supplement, 1963-1980.

    Includes indexes.
    1. Geology--Indexes. I. Title.
Z6031.C6 Suppl. [QE26.2]   016.5573   81-18501
ISBN 0-8108-1501-X   AACR2

TABLE OF CONTENTS

PREFACE	v
LIST OF SURVEYS	vii
Alabama	3
Alaska	12
Arizona	18
Arkansas	20
California	22
Colorado	31
Connecticut	34
Delaware	37
Florida	39
Georgia	46
Hawaii	50
Idaho	56
Illinois	60
Indiana	74
Iowa	80
Kansas	83
Kentucky	93
Louisiana	103
Maine	107
Maryland	108
Massachusetts	112
Michigan	113
Minnesota	116

Mississippi	120
Missouri	124
Montana	130
Nebraska	137
Nevada	138
New Hampshire	142
New Jersey	143
New Mexico	149
New York	158
North Carolina	159
North Dakota	161
Ohio	169
Oklahoma	174
Oregon	178
Pennsylvania	182
Rhode Island	190
South Carolina	191
South Dakota	193
Tennessee	198
Texas	201
Utah	210
Vermont	216
Virginia	218
Washington	223
West Virginia	226
Wisconsin	231
Wyoming	235
AUTHOR INDEX	239
SUBJECT INDEX	273

PREFACE

The purpose of this supplement to <u>An Index of State
Geological Survey Publications Issued in Series</u>
(Scarecrow Press, 1965) is to provide a single-
source checklist of all monographic publications issued in numbered series by the state geological surveys or their designated equivalents from 1962
through 1980, for collection development, interli-
brary loan, and general reference. I have included
in this supplement Alaska, Hawaii, and South Carolina, as well as some older series omitted from the
base index. Omitted are monographic materials not
in numbered series; periodical-type publications,
such as monthly mineral notes; and a few series of
highly ephemeral information; as are maps and annual or biennial reports of a strictly administrative
nature. Materials are arranged alphabetically by
series within each state, with author and subject
indexes to the individual listings within the series.

Grateful acknowledgement must be made to Don
Richter, Stephen F. Austin State University Library,
Nacogdoches, Texas; Craig Likness, Trinity Univer-
sity Library, San Antonio, Texas; and Mary Pound,
University of Texas General Libraries, Austin,
Texas. Their assistance in the compilation of this
supplement is appreciated.

LIST OF SURVEYS

Geological Survey of Alabama
Drawer O
University, AL 35486

Alaska Division of Geological and Geophysical Surveys
3001 Porcupine Drive
Anchorage, AK 99501

Arizona Bureau of Geology and Mineral Technology
University of Arizona
Tucson, AZ 85721

Arkansas Geological Commission
Vardelle Parham Geology Center
3815 West Roosevelt Road
Little Rock, AR 72204

California Division of Mines and Geology
1416 Ninth Street, Room 1341
Sacramento, CA 95814

Colorado Geological Survey
1313 Sherman Street, Room 715
Denver, CO 80203

Connecticut Natural Resources Center
Room 553, State Office Building
165 Capitol Avenue
Hartford, CT 06115

Delaware Geological Survey
University of Delaware
Newark, DE 19711

Florida Bureau of Geology
903 West Tennessee Street
Tallahassee, FL 32304

Georgia Department of Natural Resources
Earth and Water Division
19 Hunter Street, S.W., Room 400
Atlanta, GA 30334

Hawaii Division of Water and Land Development
Box 373
Honolulu, HI 96809

Idaho Bureau of Mines and Geology
Moscow, ID 83843

Illinois Geological Survey
121 Natural Resources Building
Urbana, IL 61801

Indiana Geological Survey
611 North Walnut Grove
Bloomington, IN 47401

Iowa Geological Survey
123 North Capitol Street
Iowa City, IA 52242

Kansas Geological Survey
1930 Avenue A, Campus West
University of Kansas
Lawrence, KS 66044

Kentucky Geological Survey
311 Breckinridge Hall
University of Kentucky
Lexington, KY 40506

Louisiana Geological Survey
Box G, University Station
Baton Rouge, LA 70803

Maine Bureau of Geology
State Office Building, Room 211
Augusta, ME 04330

Maryland Geological Survey
Merryman Hall
Johns Hopkins University
Baltimore, MD 21218

Michigan Geological Survey Division
Box 30028
Lansing, MI 48909

Minnesota Geological Survey
1633 Eustis Street
St. Paul, MN 55108

Mississippi Geologic, Economic, and Topographic Survey
Box 4915
Jackson, MS 39216

Missouri Division of Geological and Land Survey
Box 250
Rolla, MO 65401

Montana Bureau of Mines and Geology
Montana College of Mineral Science and Technology
Butte, MT 59701

Nebraska Conservation and Survey Division
University of Nebraska
Lincoln, NE 68508

Nevada Bureau of Mines and Geology
University of Nevada
Reno, NV 89557

New Hampshire Department of Resources and Economic Development
James Hall, University of New Hampshire
Durham, NH 03824

New Jersey Bureau of Geology and Topography
Box 1390
Trenton, NJ 08625

New Mexico Bureau of Mines and Mineral Resources
Socorro, NM 87801

New York State Geological Survey
New York State Education Building, Room 973
Albany, NY 12224
(Indexed in Clapp's Museum Publications;[1] not indexed herein)

---

[1]Clapp, Jane. Museum Publications (New York: Scarecrow Press, 1962). Pts. 1-2.

North Carolina Department of Natural Resources and Community Development
Geological Survey Section
Box 27687
Raleigh, NC 27611

North Dakota Geological Survey
University Station
Grand Forks, ND 58202

Ohio Division of Geological Survey
Fountain Square, Building 6
Columbus, OH 43224

Oklahoma Geological Survey
830 Van Vleet Oval, Room 163
Norman, OK 73069

Oregon Department of Geology and Mineral Industries
1069 State Office Building
1400 SW Fifth Avenue
Portland, OR 97201

Pennsylvania Bureau of Topographic and Geologic Survey
Department of Environmental Resources
Box 2357
Harrisburg, PA 17120

South Carolina Geological Survey
State Development Board
Harbison Forest Road
Columbia, SC 29210

South Dakota Geological Survey
Science Center, University of South Dakota
Vermillion, SD 57069

Tennessee Division of Geology
G-5 State Office Building
Nashville, TN 37219

Texas Bureau of Economic Geology
University of Texas
Box X, University Station
Austin, TX 78712

Utah Geological and Mineral Survey
606 Black Hawk Way
Salt Lake City, UT 84108

Vermont Geological Survey
Agency of Environmental Conservation
Montpelier, VT 05602

Virginia Division of Mineral Resources
Box 3667
Charlottesville, VA 22903

Washington Division of Geology and Earth Resources
Olympia, WA 98504

West Virginia Geological and Economic Survey
Box 879
Morgantown, WV 26505

Wisconsin Geological and Natural History Survey
1815 University Avenue
Madison, WI 53706

Wyoming Geological Survey
Box 3008, University Station
University of Wyoming
Laramie, WY 82071

AN INDEX OF
STATE GEOLOGICAL SURVEY PUBLICATIONS
ISSUED IN SERIES

Supplement, 1963-1980

ALABAMA

Atlas Series. 1971-

AS. 1 Environmental geology and hydrology, Madison County, Alabama, Meridianville quadrangle, by P. E. LaMoreaux and others. 1971.   1

AS. 2 Metropolitan Birmingham and Jefferson County, Alabama: Flood, March 19-20, 1970, by K. W. Causseaux. 1973.   2

AS. 3 Environmental geology and hydrology, Madison County, Alabama, Maysville area, by P. H. Moser and others. 1972.   3

AS. 4 Environmental geology and hydrology, Madison County, Alabama, Triana area, by P. H. Moser and others. 1973.   4

AS. 5 Environmental geology and hydrology, Madison County, Alabama, Madison area, by P. H. Moser and others. 1973.   5

AS. 6 Environmental geology: An aid to growth and development in Lauderdale, Colbert, and Franklin Counties, Alabama, by P. H. Moser and L. W. Hyde. 1975.   6

AS. 7 Environmental geology of an area in west-central Alabama, by P. H. Moser and M. J. Keener. 1975.   7

AS. 8 Environmental geology and hydrology, Huntsville and Madison County, Alabama, by F. L. Doyle and others. 1975.   8

AS. 9 Not yet published.   9

AS. 10 Preliminary areal geologic maps of the Valley and Ridge Province, Jefferson County, Alabama,

by J. T. Kidd and S. W. Shannon. 1977.   10

AS. 11 Drainage areas for Jefferson County, Alabama, by J. C. Scott. 1977.   11

AS. 12 Drainage areas for the Upper Black Warrior River Basin, Alabama, by J. C. Scott. 1978.   12

AS. 13 Drainage areas for the Choctawhatchee River Basin in Alabama.   13

AS. 14 Engineering geology in Jefferson County, Alabama.   14

AS. 15 Areal geology of Jefferson County, Alabama.   15

AS. 16 Summary of streamflow in Jefferson County, Alabama.   16

Bulletin. 1886-

B. 68A Geology and ground-water resources of Montgomery County, Alabama, with special reference to the Montgomery area, by D. B. Knowles, H. L. Reade, and J. C. Scott. 1963.   17

(B. 69-70 in base volume)

B. 71 Geology and ground-water resources of the Athens area, Alabama, by W. M. McMaster. 1963.   18

B. 72 Geology and ground-water resources of Franklin County, Alabama, a reconnaissance report, by R. R. Peace. 1963.   19

B. 73 Geology and ground-water resources of St. Clair County, Alabama, a reconnaissance report, by L. V. Causey. 1963.   20

B. 74 Geology and ground-water resources of Escambia County, Alabama, by J. W. Cagle and J. G. Newton. 1963.   21

B. 75 Ground-water resources of Russell County, Alabama, a reconnaissance, by J. C. Scott. 1964.   22

B. 76 Geology and ground-water resources of Morgan County, Alabama, by C. L. Dodson and W. F. Harris. 1965.   23

B. 77 Geology and ground-water resources of the Russellville area, Alabama, by R. R. Peace. 1964.   24

B. 78 Geology and ground-water resources of Lawrence County, Alabama, by W. F. Harris and W. M. McMaster. 1965.   25

B. 79 Geology and ground-water resources of Cherokee County, Alabama, a reconnaissance, by L. V. Causey. 1965.   26

B. 80 Biostratigraphic correlation of the type Shubuta member of the Yazoo clay and Red Bluff clay and their equivalents in southwestern Alabama, by P. B. Deboo. 1965.   27

B. 81 Availability of ground water in Talladega County, Alabama, a reconnaissance, by L. V. Causey. 1965.   28

B. 82 Reservoir temperatures in north-central Alabama, by L. B. Peirce. 1964.   29

B. 83 Ground-water resources of Pickens County, Alabama, a reconnaissance, by K. D. Wahl. 1965.   30

B. 84 Surface water in southwestern Alabama, by L. B. Peirce, with a section on chemical quality of surface water, by S. M. Rogers. 1966.   31

B. 85 Ground water in Marshall County, Alabama, a reconnaissance, by T. H. Sanford. 1966.   32

B. 86 Geology and ground-water resources of Greene County, Alabama, by K. D. Wahl. 1966.   33

B. 87 7-Day low flows and flow duration of Alabama streams, by L. B. Peirce. 1967.   34

B. 88 A subsurface study of southeastern Alabama, by D. B. Moore and T. J. Joiner. 1969.   35

B. 89 Water laws of Alabama, by A. J. Harris and W. T. Watson. 1968.   36

B. 89 (Revised) Water laws of Alabama, by W. T. Watson and S. F. Harrison. 1974.   37

B. 90 Talc and anthophyllite deposits in Tallapoosa and Chambers Counties, Alabama, by T. L. Neathery. 1968.   38

B. 91 Pecten perplanus stock (Oligocene) of the southeastern United States, by L. N. Glawe. 1969.   39

B. 92A Effects of water and soil on drainage structures in Alabama, Phase I, by L. W. Hyde. 1971.   40

B. 92B Effects of water and soil on drainage structures in Alabama, Phase II, by V. M. Shamburger, J. S. Ellard, and L. W. Hyde. 1970.   41

B. 93 An oil and gas evaluation of North Alabama, by J. W. Jewell. 1969.   42

B. 94A Hydrology of limestone terranes, annotated bibliography of carbonate rocks, by P. E. LaMoreaux, D. Raymond, T. J. Joiner, and others. 1970.   43

B. 94B Hydrology of limestone terranes, geologic investigations, by J. L. Sonderegger and J. C. Kelly. 1970.   44

B. 94C Hydrology of limestone terranes, photogeologic investigations,

by J. L. Sonderegger. 1970.   45

B. 94D Hydrology of limestone terranes, geophysical investigations,
by T. J. Joiner and W. L. Scarbrough. 1969.   46

B. 94E Hydrology of limestone terranes; progress of knowledge about
hydrology of carbonate terranes, by
P. E. LaMoreaux, H. E. LeGrand,
V. T. Stringfield, and J. S. Tolson,
with an annotated bibliography of
carbonate rocks, by W. M. Warren
and J. D. Moore. 1975.   47

B. 94F Hydrology of limestone ter-
ranes; quantitative studies, by J. D.
Moore, G. F. Moravec, and P. E.
LaMoreaux. 1975.   48

B. 94G Hydrology of limestone terranes; development of karst and its
effects on the permeability and circulation of water in carbonate rocks,
with special reference to the southwestern states, by V. T. Stringfield,
H. E. LeGrand, and P. E. LaMoreaux. 1977.   49

B. 95 Ostracoda and the Silurian
stratigraphy of northwestern Alabama, by R. F. Lundin and G. D.
Newton. 1969.   50

B. 96 Strippable coal in the Fabius
area, Jackson County, Alabama,
by T. W. Daniel and M. H. Fies.
1971.   51

B. 97 Water resources and geology
of Winston County, Alabama, by
K. D. Wahl, W. F. Harris, and P. O.
Jefferson. 1971.   52

B. 98 Nonfenestrate Ectoprocta
(Bryozoa) of the Bangor Limestone
(Chester) of Alabama, by F. K.
McKinney. 1971.   53

B. 99 Subsurface geology of southwest Alabama, by D. B. Moore.
1971.   54

B. 100 Bauxite and kaolin in the
Eufaula bauxite district, Alabama,
by O. M. Clarke. 1973.   55

B. 101 A strippable lignite bed in

south Alabama, by T. W. Daniel.
1973.   56

B. 102 Exploring Alabama caves, by
T. W. Daniel and W. D. Coe. 1973.   57

B. 103 Lower Mississippian (Kinderhookian) arenaceous foraminifera from
the Maury Formation at Gipsy Limestone County, Alabama, by J. E. Con-
kin and P. F. Ciesielski. 1973.   58

B. 104 Deep-well disposal in Alabama,
by W. E. Tucker and R. E. Kidd.
1973.   59

B. 105 Karst and paleohydrology of
carbonate rock terranes in semiarid
and arid regions with a comparison
to humid karst of Alabama, by V. T.
Stringfield, P. E. LaMoreaux, and
H. E. LeGrand. 1974.   60

B. 106 Development of a hydrologic
concept for the greater Mobile metro-
politan-urban environment, by J. F.
Riccio, J. D. Hardin, and G. M. Lamb,
with a section on electrical resistivity
by L. Scarbrough and K. P. Hanby.
1973.   61

B. 107 Not yet published.   62

B. 108 A bibliography of coastal Alabama with selected annotations, by
R. L. Lipp and R. L. Chermock. 1975.   63

B. 109 Geology of the Lineville East,
Ofelia, Wadley North and Mellow Val-
ley Quadrangles, Alabama, by T. L.
Neathery and J. W. Reynolds. 1975.   64

B. 110 Not yet published.   65

B. 111 Not yet published.   66

B. 112 Not yet published.   67

B. 113 7-Day low flows and flow duration of Alabama streams through 1973,
by E. C. Hayes. 1978.   68

B. 114 Not yet published.   69

B. 115 The fishes of the Birmingham-
Jefferson County region of Alabama

with ecologic and taxonomic notes, by M. F. Mettee. 1978. 70

B. 116 Not yet published. 71

B. 117 Low-flow characteristics of Alabama streams, by R. H. Bingham. 1979. 72

Circular. 1895-

C. 19 Curious creatures in Alabama rocks, a guidebook for amateur fossil collectors, by C. W. Copeland. 1963. 73

C. 20A Residual clays of the Piedmont province in Alabama, by O. M. Clarke. 1963. 74

C. 20B Clay and shale of northwestern Alabama, by O. M. Clarke. 1966. 75

C. 20C Clay and shale of northeastern Alabama, by O. M. Clarke. 1968. 76

C. 20D Clay and shale of southeastern Alabama, by O. M. Clarke. 1968. 77

C. 20E Clays of southwestern Alabama, by O. M. Clarke. 1971. 78

C. 21 Structural geology of the Birmingham red iron district, by T. A. Simpson. 1963. 79

C. 22 Water problems associated with oil production in Alabama, by W. J. Powell, L. E. Carroon, and J. R. Avrett. 1964. 80

C. 23 Ground-water levels in Alabama in 1959 and 1960, by D. M. O'Rear. 1964. 81

C. 24 Ground-water conditions in the Huntsville area, Alabama, January 1960 through June 1961, by T. H. Sanford. 1965. 82

C. 25 Ground water in the vicinity of Bryce State Hospital, Negro Colony, Tuscaloosa County, Alabama, by K. D. Wahl. 1965. 83

C. 26 Clay and shale for lightweight aggregate in Alabama, by H. D. Pallister, E. L. Hastings, and T. W. Daniel. 1964. 84

C. 27 Resources and beneficiation studies of copper-bearing pyrite ore, Pyriton, Clay County, Alabama, by W. E. Lamont and E. L. Hastings. 1964 85

C. 28 Clay mineral analysis of some Warrior Basin underclays, by W. J. Metzger, 1964. 86

C. 29 Some ostracods from the Pennington Formation of Alabama, by R. Ehrlich. 1964. 87

C. 30 Pennsylvanian stratigraphy of the Warrior Basin, Alabama, by W. J. Metzger. 1965. 88

C. 31 Reconnaissance radiation survey of Marion County, Alabama, by R. C. MacElvain. 1965. 89

C. 32 Flow characteristics of Alabama streams, a basic data report, by C. F. Hains. 1968. 90

C. 33 Surface water resources of Calhoun County, Alabama, by J. R. Harkins, with a section on quality of water, by R. G. Grantham. 1965. 91

C. 34 Reduction roasting, sodium hydroxide leaching, and magnetic concentration of brown iron ore, Crenshaw County, Alabama, by J. P. Hansen and C. H. T. Wilkins. 1965. 92

C. 35 Subsurface geology of the Gilbertown oil field, by D. B. Moore. 1967. 93

C. 36 A compilation of surface water quality data in Alabama, by J. R. Avrett. 1966. 94

C. 37 A compilation of ground water quality data in Alabama, by J. R. Avrett. 1968. 95

C. 38 Rocks and minerals of Alabama, a guidebook for Alabama rockhounds, by T. W. Daniel, T. L. Neathery, and T. A. Simpson. 1966. 96

C. 39 Alabama's mineral industry, by M. W. Szabo. 1967. 97

C. 40 Water, minerals, and oil in the South--a frontier of our nation's resources, by P. E. LaMoreaux. 1967. 98

C. 41 Paragenesis of the Turkey Heaven Mountain kyanite deposits, Cleburne County, Alabama, by T. L. Neathery. 1967. 99

C. 42 Geophysical prospecting for ground water in the Piedmont area, Alabama, by T. J. Joiner, J. C. Warman, W. L. Scarbrough, and D. B. Moore. 1967. 100

C. 43 Oil and gas wells in Alabama, January 1966, by W. E. Tucker, T. J. Joiner, and H. G. White. 1967. 101

C. 44 Sand and gravel in Alabama, a summary report, by M. W. Szabo. 1967. 102

C. 45 The petroleum industry in Alabama, by H. G. White. 1967. 103

C. 46 The geology and mineral resources of Limestone County, Alabama, a reconnaissance survey, by T. W. Daniel and E. L. Hastings. 1968. 104

C. 47 Geology of the Alabama Coastal Plain, by C. W. Copeland. 1968. 105

C. 48 The petroleum industry in Alabama, 1965-1966, by H. G. White. 1968. 106

C. 49 Product research on anthophyllite from Tallapoosa County, Alabama, by C. H. T. Wilkins, J. P. Hansen, and R. E. Moyer. 1968. 107

C. 50 Sulfide mineralization in the Cragford district, Clay and Randolph Counties, Alabama, by T. L. Neathery. 1968. 108

C. 51 Gravity survey of Monroe County, Alabama, by R. C. MacElvain. 1968. 109

C. 52 Caves of Madison County, Alabama, by W. B. Jones and W. W. Varnedoe. 1968. 110

C. 53 The petroleum industry in Alabama, 1967, by H. G. White. 1969. 111

C. 54 Gravity survey of Clarke County, Alabama, by R. G. McWilliams and T. J. Joiner. 1970. 112

C. 55 Base- and precious-metal and related deposits of Alabama, by S. I. Spaine. 1969. 113

C. 56 A problem of subsidence in a limestone terrane at Columbiana, Alabama, by W. J. Powell and P. E. LaMoreaux. 1969. 114

C. 57 Electrical resistivity survey in the Piedmont area, by W. L. Scarbrough, T. J. Joiner, and J. C. Warman. 1969. 115

C. 58 Deep-well disposal study for Baldwin, Escambia, and Mobile Counties, Alabama, by R. M. Alverson. 1970. 116

C. 59 Gravity survey of Marengo County, Alabama, by R. G. McWilliams. 1970. 117

C. 60 Gravity survey of Baldwin County, Alabama, by R. G. McWilliams. 1970. 118

C. 61 Gravity survey of Marion County, Alabama, by T. H. Clements and J. Sandy. 1970. 119

C. 62 Gravity survey of Wilcox County, Alabama, by R. G. McWilliams. 1970. 120

C. 63 Gravity survey of Winston County, Alabama, by T. H. Clements and J. Sandy. 1970. 121

C. 64 Acid mine drainage in Cane Creek basin near Oakman, Walker County, Alabama, by L. W. Hyde. 1971. 122

C. 65 Evaluation, by test drilling of geophysical methods used for ground-water development in the Piedmont area, Alabama, by G. V. Wilson, T. J. Joiner, and J. C. Warman. 1970. 123

C. 66 Mineral resources of Wilcox County, Alabama, by O. M. Clarke, T. W. Daniel, and M. W. Szabo. 1970. 124

C. 67 Gold values of samples from the Devils Backbone area, Talla-poosa County, Alabama, by J. F. O'Neill and J. L. Vallely. 1971. 125

C. 68 Sinkhole problem in and near Roberts Industrial Subdivision, Birmingham, Alabama, by J. G. Newton and L. W. Hyde. 1971. 126

C. 69 Bibliography of offshore and estuarine areas of Alabama with selected annotations, by J. H. Cammack, T. J. Joiner, and R. D. Schneeflock. 1971. 127

C. 70 Oil and gas exploration in Alabama in 1970, by D. B. Moore. 1971. 128

C. 71 Hydrogeology of sanitary land-fill sites in Alabama, by J. F. Riccio and L. W. Hyde. 1971. 129

C. 72 Coal reserves of Bibb County, Alabama, by R. Q. Shotts. 1971. 130

C. 73 A summary of the composition and technology of the coals of Bibb County, Alabama, compiled from the published literature, by R. W. Shotts. 1971. 131

C. 74 Gravity survey of the Mobile Delta, Alabama, by R. E. Kidd and G. V. Wilson. 1971. 132

C. 75 Techniques of evaluating effects of water on drainage structures in Alabama, by J. S. Ellard. 1971. 133

C. 76 Contamination of ground water in a limestone aquifer in the Stevenson area, Alabama, by B. L. Bailey and A. M. Malatino. 1972. 134

C. 77 Bibliography of Alabama coals, by O. E. Gilbert and W. E. Smith. 1972. 135

C. 78 Mineral resources of Marion County, Alabama, by M. W. Szabo, T. W. Daniel, and O. M. Clarke. 1972. 136

C. 79 Mercury concentrations in sediment samples from the Tennessee, Mobile, Warrior, and Tombigbee Rivers, Alabama, by N. A. Lloyd, H. S. Chaffin, and J. T. McLendon. 1972. 137

C. 80 Aquatic-biotic community structure as an indicator of pollution, by G. G. Dills and D. T. Rogers. 1972. 138

C. 81 Mineral resources of Calhoun County, Alabama, by T. L. Neathery, O. M. Clarke, W. E. Smith, M. W. Szabo, and D. H. White. 1972. 139

C. 82 Distribution of Holocene Foraminiferida in Mobile Bay and the effect of salinity changes, by G. M. Lamb. 1972. 140

C. 83 Sinkhole problem along proposed route of Interstate Highway 459 near Greenwood, Alabama, by J. G. Newton, C. W. Copeland, and L. W. Scarbrough. 1973. 141

C. 84 Evaluation of techniques for selecting sites for induced infiltration along the Alabama River, by L. Scarbrough, with a section of photogeology, by R. E. Kidd. 1972. 142

C. 85 Barite in Alabama, by T. H. Hughes and R. E. Lynch. 1973. 143

C. 86 Phosphate resources of Perry County, Alabama, by M. W. Szabo. 1973. 144

C. 87 Gravity survey of Lamar County, Alabama, by G. V. Wilson. 1973. 145

C. 88 Gravity survey of Limestone

County, Alabama, by T. H. Clements and R. E. Kidd. 1973. 146

C. 89 Engineering evaluation of Alabama tar sands, by I. Moftah. 1973. 147

C. 90 A field guide to mineral deposits in south Alabama, by M. W. Szabo. 1973. 148

C. 91 Mineral resources of Marengo County, Alabama, by O. M. Clarke, M. W. Szabo, and W. E. Smith, with a report on lignite, by T. W. Daniel. 1975. 149

C. 92 History of water supply of the Mobile area, Alabama, by J. F. Riccio and C. A. Gazzier. 1974. 150

C. 93 Bibliography of the mineral resources of Alabama, exclusive of coal, iron, and petroleum, by O. E. Gilbert. 1974. 151

C. 94 Ceramic investigation in the Eufaula bauxite district, Alabama, by M. E. Tyrrell and O. M. Clarke. 1974. 152

C. 95 Pennsylvanian sediment-fossil relationships in part of the Black Warrior Basin of Alabama, by J. E. McKee. 1975. 153

C. 96 Occurrence of iron bacteria in ground-water supplies of Alabama, by N. Valkenburg, R. Christian, and M. Green. 1975. 154

C. 97 Regional flood depth-frequency relations for Alabama, by C. F. Hains. 1977. 155

C. 98 Energy distribution systems of Alabama, by H. S. Chaffin. 1976. 156

C. 99 Not yet published. 157

C. 100 Porters Creek lightweight aggregate, by O. M. Clarke and M. E. Tyrrell. 1976. 158

C. 101 Sinkhole occurrence in western Shelby County, Alabama, by W. M. Warren. 1976. 159

C. 102 Chalk resources of west-central Alabama by M. W. Szabo and M. A. Beg. 1977. 160

C. 103 Ground-water resources of the Birmingham and Cahaba valleys of Jefferson County, Alabama, by T. B. Moffett and P. H. Moser. 1978. 161

County Report. 1923-1963.

CR. 8 Geology and ground-water resources of Lauderdale County, Alabama, by H. B. Harris, R. R. Peace, and W. F. Harris. 1963. 162

(CR. 9 in base index)

CR. 10 Geology and ground-water resources of Colbert County, Alabama, by H. B. Harris, G. K. Moore, and L. R. West. 1963. 163

CR. 11 General geology and ground-water resources of Limestone County, Alabama, a reconnaissance report, by W. M. McMaster and W. F. Harris. 1963. 164

Information Series. 1955-

IS. 27 Chemical quality of water of Alabama streams, 1960, a reconnaissance study, by R. N. Cherry. 1963. 165

(IS. 28-29 in base index)

IS. 30 Brown iron ore deposits in Barbour County, Alabama, an interim report, by E. L. Hastings. 1963. 166

IS. 31 Gamma radiation survey of the Gilbertown area, Alabama, by W. B. Collins. 1963. 167

IS. 32 Alabama's mineral industry, by T. E. Cook, M. P. Turner, and T. A. Simpson. 1963. 168

IS. 33 Oil and gas activities in Alabama, by H. G. White and W. E. Tucker. 1964. 169

IS. 34 Brown iron ore deposits in Pike, Crenshaw, and Butler Counties,

Alabama, an interim report, by
E. L. Hastings and E. Smith.
1964. 170

IS. 35 Temperature of Alabama
streams, by J. R. Avrett and L. E.
Carroon. 1964. 171

IS. 36 Mineral and water resources,
Butler County, Alabama. 1967.
172

IS. 37 Mineral and water resources
of Barbour County, Alabama. 1968.
173

IS. 38 Mineral and water resources
of Houston County, Alabama. 1969.
174

IS. 39 Mineral and water resources
of Geneva County, Alabama. 1969.
175

IS. 40 Mineral, water, and energy
resources of Wilcox County, Alabama. 1969. 176

IS. 41 Geologic and hydrologic research through space-acquired data
for Alabama, delineation of linear
features and application to reservoir
engineering using Apollo 9 multispectral photography, by W. J. Powell, C. W. Copeland, and J. A. Drahovzal. 1970. 177

IS. 42 Use of water in Alabama,
1970, with projections to 2020, by
L. B. Peirce. 1972. 178

IS. 43 Identification of user cate-
gories and users of ERTS-acquired
data in Alabama, by T. J. Joiner,
J. V. Walters, E. T. Miller, and
C. C. Wielchowsky, with research
contributions by H. R. Henry, J. F.
Vallery, C. H. Mullis, C. E. Brett,
and W. Bondarenko. 1973. 179

IS. 44 Water resources monitoring
and evaluation: a key to environmental protection in Alabama oil
fields, by W. J. Powell, M. E.
Davis, B. L. Bailey, and E. R. German. 1973. 180

IS. 45 Mineral, water, and energy

resources of Winston County, Alabama, by W. E. Smith. 1972. 181

IS. 46 Hurricanes and tornadoes in
Alabama, by R. L. Chermock. 1976.
182

IS. 47 Mining and minerals in Alabama, by W. E. Smith and O. E. Gilbert. 1975. 183

IS. 48 Use of water in Alabama, 1975,
with projections to 2020, by M. F.
Mettee, P. H. Moser, and L. Dean.
1978. 184

IS. 49 General order prescribing rules
and regulations governing the conservation of oil and gas in Alabama (as
amended) and oil and gas laws of
Alabama (as amended) with oil and gas
forms, by W. T. Watson. 1974. 185

IS. 50 Shoreline and bathymetric
changes in the coastal area of Alabama, a remote-sensing approach, by
J. D. Hardin, C. D. Sapp, J. L. Emplaincourt, and K. E. Richter. 1976.
186

IS. 51 The environment of offshore
and estuarine Alabama, by R. L. Chermock, with contributions by P. A.
Boone and R. L. Lipp. 1974. 187

IS. 52 Not yet published. 188

IS. 53 Coal--its importance to Alabama, by W. E. Ward and F. E. Evans.
1975. 189

IS. 54 Mineral laws of Alabama, by
S. V. Hammond. 1975. 190

IS. 55 Urbanization and flooding in
Shades Creek basin, Jefferson County,
Alabama, by A. L. Knight. 1976.
191

IS. 56 The forest resources of the
Birmingham, Jefferson County region
of Alabama, by R. L. Chermock.
1977. 192

Map Abstract. 1972-

MA. 1 The map abastract of population

and housing: Alabama, 1970, by
N. G. Lineback, C. T. Traylor,
and N. E. Turnage. 1972.   193

MA. 2 The map abstract of water
resources: Alabama, by N. G.
Lineback, L. B. Peirce, and N. E.
Turnage. 1974.   194

Monograph. 1883-

M. 12 Mississippian stratigraphy
of Alabama, by W. A. Thomas.
1972.   195

M. 13 Stratigraphic distribution of
Paleocene and Eocene fossils in the
eastern Gulf Coast region, by L. D.
Toulmin. 1977.   196

Oil and Gas Report. 1976-

OGR. 1 General order prescribing
rules and regulations governing the
conservation of oil and gas in Alabama and oil and gas laws of Alabama with oil and gas board forms,
by A. C. Freeman, J. H. Griggs,
and K. P. Hanby. 1976.   197

OGR. 2 Oil and gas wells in Alabama, by R. M. Poe, G. V. Wilson,
and J. S. Tolson. 1979.   198

OGR. 3 The petroleum industry in
Alabama, 1977, by J. H. Masingill,
P. F. Hall, and J. S. Tolson. 1977.
                               199

OGR. 4 Remote surveillance of oil-
and gas-field activities in Alabama,
by C. D. Sapp and K. E. Richter.
1975.   200

OGR. 5 Evaluation of water quality
data from oil field areas in south
Alabama: Specific conductance-
chloride relationship, by J. S. Ellard and R. L. Lipp. 1978.   201

Quadrangle Series. 1975-

QS. 1 Geology and mineral resources of the Moontown quadrangle,
Madison Co. , Alabama, by H. S.

Chaffin and M. W. Szabo. 1975. 202

QS. 2 Geology and mineral resources
of the Toney quadrangle, Madison Co. ,
Alabama, by H. S. Chaffin and M. W.
Szabo. 1975.   203

QS. 3 Geology and mineral resources
of New Hope quadrangle, Madison Co. ,
Alabama, by H. S. Chaffin and M. W.
Szabo. 1975.   204

QS. 4 Geology and mineral resources
of the Pride quadrangle, Colbert Co. ,
Alabama, by M. W. Szabo. 1975.
                                205

QS. 5 Geology and mineral resources
of Fisk quadrangle, Madison Co. ,
Alabama, by H. S. Chaffin and M. W.
Szabo. 1975.   206

QS. 6 Geology and mineral resources
of the Tuscumbia quadrangle, Colbert
Co. , Alabama, by M. W. Szabo. 1975.
                                207

QS. 7 Geology and mineral resources
of the Leighton quadrangle, Colbert
Co. , Alabama, by M. W. Szabo. 1975.
                                208

ALASKA

Geochemical Report. 1964-1973.

GR. 1 Geochemical investigations of selected areas in southeastern Alaska, by C. F. Herbert and W. H. Race. 1964. 209

GR. 2 Geochemical investigation of the Slana District, south-central Alaska, 1963 and 1964, by D. H. Richter. 1965. 210

GR. 3 A geochemical investigation in the Richardson area, Big Delta Quadrangle, by R. H. Saunders. 1965. 211

GR. 4 Geochemical investigations of selected areas in south-central Alaska, 1964, by M. W. Jasper. 1965. 212

GR. 5 A geochemical investigation between Chatanika and Circle Hot Springs, Alaska, by W. M. Burand. 1965. 213

GR. 6 Geochemical investigations of selected areas in southeastern Alaska, 1964 and 1965, by C. F. Herbert and W. H. Race. 1965. 214

GR. 7 Geochemical investigations along highway and secondary roads in south-central Alaska, 1965, by M. W. Jasper. 1965. 215

GR. 8 Geochemical and geological investigations of Admiralty Island, Alaska, by W. H. Race and A. W. Rose. 1967. 216

GR. 9 A geochemical investigation along the Taylor Highway, east-central Alaska, by R. H. Saunders. 1966. 217

GR. 10 A geochemical investigation of the Nenana Highway area, central Alaska, by W. M. Burand. 1966. 218

GR. 11 A geochemical investigation of stream sediments in the Elliott Highway area, by W. M. Burand. 1966. 219

GR. 12 A geochemical investigation of Minook Creek, Rampart District, by W. M. Burand and R. H. Saunders. 1966. 220

GR. 13 Geochemical investigations of selected areas in the Yukon-Tanana region of Alaska, 1965 and 1966, by W. M. Burand. 1968. 221

GR. 14 Geochemical investigations, Willow Creek southerly to Kenai Lake region, south-central Alaska, by M. W. Jasper. 1967. 222

GR. 15 Geochemical investigations along the Valdez to Chitina Highway in southcentral Alaska, 1966, by M. W. Jasper. 1967. 223

GR. 16 Geochemical investigation of a portion of the Fortymile District, east-central Alaska, by W. H. Smith. 1968. 224

GR. 17 A geochemical investigation of the Wood River-Tikchik Lakes area, southwestern Alaska, by G. R. Eakins. 1968. Supplement, 1969. 225

GR. 18 Geology and geochemistry of part of the Iron Creek area, Solomon D-6 Quadrangle, Seward Peninsula, Alaska, by R. R. Asher. 1969. 226

GR. 19 Preliminary geochemistry and geology, Little Falls Creek area, Talkeetna Mountains Quadrangle, Alaska, by R. E. Anderson. 1969. 227

GR. 20 Geology and geochemistry at Kontrashibuna Lake, Lake Clark Region, southwestern Alaska, by G. R. Eakins. 1970.   228

GR. 21 Geochemical survey and geological reconnaissance of the White River area, south-central Alaska, by J. Knaebel. 1970.   229

GR. 22 Geology and geochemistry of the Belt Creek-Libby River area, Seward Peninsula, by R. R. Asher. 1970.   230

GR. 23 Geochemistry and geology, Boundary area, Fortymile District, Eagle A-1 Quadrangle, Alaska, by R. R. Asher. 1970.   231

GR. 24 Analyses of rock and stream-sediment samples, Hetta Inlet area, Prince of Wales Island, Craig Quadrangle, by G. Herreid. 1971.   232

GR. 25 Analyses of rock and stream-sediment samples, Wild Lake area, Wiseman Quadrangle, Arctic Alaska, by E. R. Chipp. 1972.   233

GR. 26 Analyses of rock and stream-sediment samples, Mt. Hayes A-6 Quadrangle, south-central Alaska, by T. E. Smith, T. C. Trible, and D. R. Stein. 1973.   234

GR. 27 Analyses of stream sediment samples, Craig A-2 and vicinity, Prince of Wales Island, southeastern Alaska, by G. Herreid and T. C. Trible. 1973. 235

Geologic Report. 1962-

RE. 1 Preliminary report on geologic mapping in the Coast Range Mineral Belt, Alaska, by G. Herreid. 1962.   236

RE. 2 Bedrock geology of the Rainbow Mountain area, Alaska Range, by L. G. Hanson. 1963.   237

RE. 3 Geology of the Portage Creek-Susitna River area, Alaska, by

D. H. Richter. 1963.   238

RE. 4 Geology and mineral deposits of the Denali-MacLaren River area, Alaska, by M. A. Kaufman. 1964.   239

RE. 5 Geology of the Niblack Anchorage area, southeastern Alaska, by G. Herreid. 1964.   240

RE. 6 Geology and mineral deposits of the Ahtell Creek area, Slana District, south-central Alaska, by D. H. Richter. 1964.   241

RE. 7 Geology of the Dry Pass area, southeastern Alaska, by G. Herreid and M. A. Kaufman. 1964.   242

RE. 8 Geology of the Paint River area, Iliamna Quadrangle, Alaska, by D. H. Richter and G. Herreid. 1965.   243

RE. 9 A geologic and geochemical traverse along the Nellie Juan River, Kenai Peninsula, Alaska, by G. Herreid. 1965.   244

RE. 10 Geology of the Bluff area, Solomon Quadrangle, Seward Peninsula, Alaska, by G. Herreid. 1965.   245

RE. 11 Geology of the Omikak-Otter Creek area, Bendeleben Quadrangle, Seward Peninsula, Alaska, by G. Herreid. 1965.   246

RE. 12 Geology of the Bear Creek area, Seward Peninsula, Candle Quadrangle, Alaska, by G. Herreid. 1965.   247

RE. 13 Geology and geochemical investigations near Paxson, northern Copper River Basin, Alaska, by A. W. Rose and R. H. Saunders. 1965.   248

RE. 14 Geology and mineral deposits of the Rainy Creek area, Mt. Hayes Quadrangle, Alaska, by A. W. Rose. 1965.   249

RE. 15 Geology and mineralization of the Midas Mine and Sulphide Gulch

areas near Valdez, Alaska, by
A. W. Rose. 1965. 250

RE. 16 Geology and mineral deposits of central Knight Island, Prince William Sound, Alaska, by D. H. Richter. 1965. 251

RE. 17 Geology and geochemistry of the Hollis and Twelve-Mile Creek areas, Prince of Wales Island, southeastern Alaska, by G. Herreid and A. W. Rose. 1966. 252

RE. 18 Geology of chromite-bearing ultramafic rocks near Eklutna, Anchorage Quadrangle, Alaska, by A. W. Rose. 1966. 253

RE. 19 Geology of part of the Amphitheatre Mountains, Mt. Hayes Quadrangle, Alaska, by A. W. Rose. 1966. 254

RE. 20 Geological and geochemical investigations in the Eureka Creek and Rainy Creek areas, Mt. Hayes Quadrangle, Alaska, by A. W. Rose. 1966. 255

RE. 21 Geology of the Slana District, south-central Alaska, by R. H. Richter. 1966. 256

RE. 22 Geology and geochemistry of the Nixon Fork area, Medfra Quadrangle, Alaska, by G. Herreid. 1966. 257

RE. 23 The geology and geochemistry of the Inmachuk River map area, Seward Peninsula, Alaska, by G. Herreid. 1966. 258

RE. 24 Preliminary geology and geochemistry of the Sinuk River area, Seward Peninsula, Alaska, by G. Herreid. 1966. 259

RE. 25 Geological and geochemical investigations in the Metal Creek area, Chugach Mountains, Alaska, by D. H. Richter. 1967. 260

RE. 26 Geological and geochemical investigations southwest of Farewell, Alaska, by G. Herreid. 1968. 261

RE. 27 Geology and mineral deposits of the Dolomi area, Prince of Wales Island, Alaska, by G. Herreid. 1967. 262

RE. 28 Geology of the Upper Chistochina River area, Mt. Hayes Quadrangle, Alaska, by A. W. Rose. 1967. 263

RE. 29 Progress report on the geology and geochemistry of the Sinuk area, Seward Peninsula, Alaska, by G. Herreid. 1968. 264

RE. 30 Geology of the Upper Slana-Mentasta Pass area, by D. H. Richter. 1967. 265

RE. 31 Geology and stream sediment geochemistry of Anton Carsen Bay and vicinity, Kodiak Island, Alaska, by A. W. Rose and D. H. Richter. 1967. 266

RE. 32 Geology of an area on the Upper Talkeetna River, Talkeetna Mountains Quadrangle, Alaska, by A. W. Rose. 1967. 267

RE. 33 Geological and geochemical study, Solomon C-5 Quadrangle, Seward Peninsula, Alaska, by R. R. Asher. 1969. 268

RE. 34 Geology and geochemistry, Diana Lakes area, Western Talkeetna Mountains, Alaska, by R. E. Anderson. . 1969. 269

RE. 35 Geology and geochemistry, Sithylemenkat Lake area, Beetles Quadrangle, Alaska, by G. Herreid. 1969. 270

RE. 36 Geology and geochemistry of the Sinuk area, Seward Peninsula, Alaska, by G. Herreid. 1970. 271

RE. 37 Geology and geochemistry in the southeastern part of the Cosmos Hills, Shungnak D-2 Quadrangle, Alaska, by C. E. Fritts. 1969. 272

RE. 38 Uranium in Alaska, by G. R. Eakins. 1969. 273

RE. 39 Geology and geochemistry of

the Cosmos Hills, Ambler River
and Shungnak Quadrangles, Alaska,
by C. E. Fritts. 1970. 274

RE. 40 Geology of the Spirit Mountain nickel-copper prospect and surrounding area, by G. Herreid.
1970. 275

RE. 41 An experiment in geobotanical prospecting for uranium, Bokan Mountain area, southeastern
Alaska, by G. R. Eakins. 1970.
276

RE. 42 Geology and geochemistry
of the Chandalar area, Brooks
Range, Alaska, by E. R. Chipp.
1970. 277

RE. 43 Withdrawn (not published).
278

RE. 44 Uranium investigations in
southeastern Alaska, by G. R. Eakins. 1975. 279

RE. 45 Geology of the Rainbow
Mountain-Gulkana Glacier area,
eastern Alaska Range, with an emphasis on Upper Paleozoic strata;
stratigraphy, structure, petrology,
and sedimentology of the Late Paleozoic and Tertiary rocks in the
Rainbow Mountain-Gulkana Glacier
area, by G. C. Bond. 1976. 280

RE. 46 Geology of the Eureka Creek
area, east-central Alaska, by J. H.
Stout. 1976. 281

RE. 47 The Teklanika formation,
a new Paleocene volcanic formation
in the central Alaska Range, by W.
G. Gilbert, V. M. Ferrell, and D.
L. Turner. 1976. 282

RE. 48 Geology and geochemistry
of the Craig A-2 Quadrangle and
vicinity, Prince of Wales Island,
southeastern Alaska, by G. Herreid,
T. K. Bundtzen, and D. L. Turner.
1978. 283

RE. 49 Gravity survey of Beluga
Basin and adjacent area, Cook Inlet
region, south-central Alaska, by
S. W. Hackett. 1977. 284

RE. 50 Metamorphic rocks of the
Toklat-Teklanika Rivers area, Alaska,
by W. G. Gilbert and E. Redman.
1977. 285

RE. 51 Short notes on Alaskan geology;
recent research on Alaskan geology.
1976. 286

RE. 52 Clay mineralogy and petrology
of the coal-bearing group near Healy,
Alaska, by D. M. Triplehorn. 1976.
287

RE. 53 Not yet published. 288

RE. 54 Salinity study, Cook Inlet Basin, Alaska, by D. L. McGee. 1977.
289

RE. 55 Short notes on Alaskan geology;
recent research in Alaskan geology.
1977. 290

RE. 56 Aeromagnetic map of southwestern Brooks Range, Alaska, by
S. W. Hackett. 1977. 291

RE. 57 Not yet published. 292

RE. 58 Geology of Ruby Ridge, southwestern Brooks Range, Alaska, by
W. G. Gilbert and others. 1977. 293

RE. 59 The Mount Galen volcanics--
a new Middle Tertiary volcanic formation in the central Alaska Range, by
J. E. Decker and W. G. Gilbert. 1978.
294

RE. 60 Not yet published. 295

RE. 61 Short notes on Alaskan geology;
recent research on Alaskan geology.
1979. 296

RE. 62 Tertiary formations and associated Mesozoic rocks in the Alaska
Peninsula area, Alaska, and their
petroleum-reservoir and source-rock
potential, by W. M. Lyle and others.
1979. 297

Information Circular. 196?-

IC. 1 Proper claim staking in Alaska.
Rev. 1980. 298

IC. 2 Mineral rights of aliens in the State of Alaska. Rev. 1968. 299

IC. 3 Hand placer mining methods. Rev. 1980. 300

IC. 4 Uranium prospecting in Alaska. Rev. 1968. 301

IC. 5 General Alaskan mineral and energy information. Rev. 1978. 302

IC. 6 Alaska prospecting information. Rev. 1978. 303

IC. 7 Alaskan companies and prospectors. Rev. 1979. 304

IC. 8 Consultants available for work in Alaska. Rev. 1979. 305

IC. 9 Alaska rockhound information. Rev. 1978. 306

IC. 10 Skin diving for gold in Alaska. Rev. 1968. 307

IC. 11 List of reports issued by the Division of Geological and Geophysical Surveys. Rev. 1980. 308

IC. 12 Services of the Division of Geological Survey. Rev. 1975. 309

IC. 13 Dangers in old mine openings. 1962. 310

IC. 14 Mining laws applicable in Alaska. Rev. 1979. 311

IC. 15 A prospector's guide to the sale and lease of mineral properties. 1969. 312

IC. 16 Alaska map information. Rev. 1977. 313

IC. 17 Companies interested in Alaskan mining possibilities. Rev. 1978. 314

IC. 18 Amateur gold prospecting in Alaska. 1980. 315

IC. 19 The status of mining claims on native lands, by J. W. Katz. 1974. 316

IC. 20 State of Alaska aeromagnetic surveys. Rev. 1979. 317

IC. 21 Open-file reports. Rev. often. 318

IC. 22 Progress and prospects in marine mining in Alaska, by C. N. Conwell. 1976. 319

IC. 23 Index map of DGGS publications. 1979. 320

IC. 24 Mining-claim recording districts in Alaska. 1979. 321

IC. 25 Information on water and water rights in Alaska. 1979. 322

Miscellaneous Paper. 1964-

MP. 1 The great Alaska earthquake, Mar. 27, 1964. 1964. 323

MP. 2 Preliminary results of stream sediment sampling, upper MacLaren River area, south-central Alaska, by T. E. Smith, T. C. Trible, and D. R. Stein. n. d. 324

MP. 3 Stratabound copper-gold occurrence, northern Talkeetna Mountains, Alaska, by T. E. Smith, T. K. Bundtzen, and T. C. Trible. 1975. 325

MP. 4 Reconnaissance geology of the Nelchina Glacier gold-silver-copper-lead deposit, north-central Chugach Mountains, Alaska, by M. W. Henning and G. H. Pessel. n. d. 326

Special Report. 1967-

SR. 1 History of mines and prospects, Ketchikan District, prior to 1952, by J. Bufvers. 1967. 327

SR. 2 Mineral occurrences in the Yukon-Tanana region, by R. H. Saunders. 1967. 328

SR. 3 A petrified forest on Unga Island, by G. R. Eakins. 1970. 329

SR. 4 Mineralization near Stepovak Bay, Alaska Peninsula, by G. R. Eakins. 1970. 330

SR. 5 Preliminary report on stratigraphy of Kenai Group, upper Cook Inlet, by D. C. Hartman, G. H. Pessel, and D. L. McGee. 1972. 331

SR. 6 Sedimentary zeolite deposits of the upper Mantanuska Valley, Alaska, by D. B. Hawkins. 1973. 332

SR. 7 Mineral preparation of ores from Friday Creek, Kantishna Mining District, Alaska, by C. N. Conwell. 1974. 333

SR. 8 Herendeen Bay-Chignik coals, southern Alaska Peninsula, by C. N. Conwell and D. M. Triplehorn. 1978. 334

SR. 9 Mordenite deposits and zeolite zonation in the Horn Mountains area, south-central Alaska, by D. B. Hawkins. 1976. 335

SR. 10 Radiometric dates from Alaska--a 1975 compilation, by D. L. Turner, D. Grybeck, and F. H. Wilson. 1975.. 336

SR. 11 Commercial-grade mordenite deposits of the Horn Mountains, south-central Alaska, by D. B. Hawkins. 1976. 337

SR. 12 Investigation of Alaska's uranium potential, by G. R. Eakins and R. B. Forbes. 1976.

SR. 13 Mineral occurrences near Cantwell, south-central Alaska, by R. G. Hickman and C. Craddock. 1976. 339

SR. 14 Mineral occurrences in the upper Wood River, Edgar Creek, and West Fork Glacier areas, central Alaska, by K. W. Sherwood, C. Craddock, and T. E. Smith. 1976. 340

SR. 15 Not yet published. 341

SR. 16 Surficial geology and processes, Prudhoe Bay Oil Field, Alaska, with hydrologic implications, by R. G. Updike and M. D. Howland. 1979. 342

Bulletin. 1915-

B. 172 The use of compressed air in small mines, by H. E. Krumlauf. 1963. 343

B. 173 Bibliography of the geology and mineral resources of Arizona, 1848-1964, by R. T. Moore and E. D. Wilson. 1965. 344

B. 174 Guidebook 1--Highways of Arizona; U. S. Highway 666, by E. D. Wilson. 1965. 345

B. 175 Field tests for the common mineral elements, by G. H. Roseveare. 1966. 346

B. 176 Geologic guidebook 2--Highways of Arizona; Arizona highways 77 and 177, by H. W. Peirce. 1967. 347

B. 177 Mineral deposits of the Fort Apache Indian Reservation, Arizona, by R. T. Moore. 1968. 348

B. 178 A brief history of the United Verde Open Pit, Jerome, Arizona, by E. M. J. Alenius. 1968. 349

B. 179 Mineral deposits of the Gila River Indian Reservation, Arizona, by E. D. Wilson. 1969. 350

B. 180 Mineral and water resources of Arizona, prepared by the U. S. Geological Survey, the Arizona Bureau of Mines, and the U. S. Bureau of Reclamation. 1969. 351

B. 181 Minerals of Arizona, by F. W. Galbraith and D. J. Brennan. 1970. 352

B. 182 Coal, oil, natural gas, helium, and uranium in Arizona, by H. W. Peirce, S. B. Keith, and J. C. Wilt. 1970. 353

B. 183 Geologic guidebook--Highways of Arizona; Arizona highways 85, 86, and 386, by S. B. Keith. 1971. 354

B. 184 Geologic guidebook 4--Highways of Arizona; Arizona highways 87, 88, and 188, by C. F. Royse, M. F. Sheridan, and H. W. Peirce. 1971. 355

B. 185 Arizona well information, by H. W. Peirce and J. R. Scurlock. 1972. 356

B. 186 Geology of the Virgin and Beaverdam Mountains, Arizona, by R. T. Moore. 1972. 357

B. 187 Index of mining properties in Cochise County, Arizona, by S. B. Keith. 1973. 358

B. 188 The mineral industry of Arizona in 1971, by L. Moore. n. d. 359

B. 189 Index of mining properties in Pima County, Arizona, by S. B. Keith. 1974. 360

B. 190 Bibliography of the geology and mineral resources of Arizona, 1965-1970, by J. S. Vuich and J. C. Wilt. 1974. 361

B. 191 Index of mining properties in Santa Cruz County, Arizona, by S. B. Keith. 1975. 362

B. 192 Index of mining properties in Yuma County, Arizona, by S. B. Keith. 1978. 363

Circular. 1941?-

C. 1 Diatomaceous earth, by E. D. Wilson. n. d. 364

C. 2 Alunite, by R. E. Heineman.
n. d.   365

C. 3 Mica, by E. D. Wilson. n. d.
                                          366

C. 4 Vermiculite, by E. D. Wilson.
n. d.   367

C. 5 Gypsum, by F. W. Galbraith.
n. d.   368

C. 6 Manganese, by E. D. Wilson.
n. d.   369

C. 7 Quartz, by R. E. Heineman.
n. d.   370

C. 8 The use and limitations of the
spectograph in examining prospectors' samples, by F. S. Wartman
and A. J. Thompson. n. d.   371

C. 9 Quicksilver (Mercury), by
E. D. Wilson. 1941.   372

C. 10 Not published?   373

C. 11 Magnesium, by E. D. Wilson.
1943.   374

C. 12 Arizona perlite, by E. D.
Wilson and G. H. Roseveare.
1945.   375

C. 13 Radioactive uranium and
thorium, by J. W. Anthony. 1948.
                                          376

C. 14 Fifty common Arizona minerals, by F. W. Galbraith and J. W.
Anthony. 1952.   377

C. 15 Fluorspar in Arizona, by
E. D. Wilson. 1950.   378

C. 16 Strata-bound sulfide deposits
and suggestions for exploration in
Arizona, by J. S. Vuich. 1974.
                                          379

C. 17 Utilization of municipal
waste water for froth flotation of
copper and molybdenum sulfides,
by W. W. Fisher and S. Rudy.
1976.   380

C. 18 Chemical analyses of coal
samples from the Black Mesa
field, Arizona, by R. T. Moore and
others. 1977.   381

C. 19 A survey of uranium favorabil-
ity of Paleozoic rocks in the Mongol-
lon Rim and slope region, east-central
Arizona, by H. W. Peirce and others.
1977.   382

C. 20 Geology of the Socorro Peak
area, western Harquahala Mountains,
by R. J. Varga. 1977.   383

Special Paper. 1977-

SP. 1 Late Cenozoic geology of the
White Mountains, Arizona, by R. K.
Merrill and T. L. Pewe. 1977. 384

SP. 2 Guidebook to the geology of
central Arizona, by D. M. Burt and
T. L. Pewe. 1978.   385

ARKANSAS

Bulletin. 1929-

B. 21 Quartz, rectorite, and cookeite from the Jeffery Quarry, near North Little Rock, Pulaski County, Arkansas, by H. D. Miller and C. Milton. 1964. 386

B. 22 Fossils of Arkansas, by T. Freeman. 1966. 387

Information Circular. 1932-

IC. 20B Geology of Paris Quadrangle, Logan County, Arkansas, by B. R. Haley. 1961. 388

IC. 20C Geology of the Barber Quadrangle, Sebastian County and vicinity, Arkansas, by B. R. Haley. 1966. 389

IC. 20D Geologic formations penetrated by the Shell Oil Company No. 1 Western Coal and Mining Company well on the Backbone Anticline, Sebastian County, Arkansas, by B. R. Haley and S. E. Frezon. 1965. 390

IC. 20E Geology of Knoxville Quadrangle, Johnson and Pope Counties, Arkansas, by E. A. Merewether. 1967. 391

IC. 20F Geology of the Greenwood Quadrangle, Arkansas-Oklahoma, by B. R. Haley and T. A. Hendricks. 1968. 392

IC. 20G Geology of the Scranton and New Blaine Quadrangles, Logan and Johnson Counties, Arkansas, by B. R. Haley. 1968. 393

IC. 20H Geology of the Coal Hill, Hartman, and Clarksville Quadrangles, Johnson County and vicinity, Arkansas, by E. A. Merewether and B. R. Haley. 1969. 394

IC. 20I Geology of the Van Buren and Lavaca Quadrangles, Arkansas-Oklahoma, by B. R. Haley and T. A. Hendricks. 1972. 395

IC. 20J Geology of the Knoxville and Delaware Quadrangles, Johnson and Logan Counties and vicinity, Arkansas, by E. A. Merewether. 1972. 396

IC. 20K Low-volatile bituminous coal and semianthracite in the Arkansas Valley Coal Field, by B. R. Haley. 1978. 397

IC. 20L Inventory of surface and underground coal mines in the Arkansas Valley Coal Field, by W. V. Bush and L. B. Gilbreath. 1978. 398

(IC. 21 in base index)

IC. 22 Geology of a uranium-bearing black shale of Late Devonian age in north-central Arkansas, by V. E. Swanson and E. R. Landis. 1962. 399

IC. 23 Mercury district of southwest Arkansas, by B. F. Clardy and W. V. Bush. 1975. 400

Water Resources Circular. 1955-

WRC. 9 Water supply characteristics of selected Arkansas streams, by M. S. Hines. 1965. 401

WRC. 10 Storage requirements for Arkansas streams, by J. L. Patterson. 1967. 402

Water Resources Summary. 1962-

WRS. 1 Resume of the ground-water

resources of Bradley, Calhoun,
and Ouachita Counties, Arkansas,
by D. R. Albin. 1962.   403

WRS. 2   Ground-water temperatures
in the coastal plain of Arkansas, by
R. O. Plebuch.   1962.   404

WRS. 3   Changes in ground-water
levels in deposits of Quaternary
Age in northeastern Arkansas, by
R. O. Plebuch.   1962.   405

CALIFORNIA

Bulletin. 1888-

B. 183 Franciscan and related rocks and their significance in the geology of western California, by E. H. Bailey, W. P. Irwin, D. L. Jones. 1964. 406

B. 184 Geology of the Oroville (15') Quadrangle, Butte County, California, by R. S. Creely. 1965. 407

B. 185 Geology of the E½ Mt. Hamilton (15') Quadrangle, Alameda and Santa Clara Counties, California, by S. M. Soliman. 1965. 408

B. 186 Geology of the central Santa Ynez Mountains, Santa Barbara County, California, by T. W. Dibblee. 1966. 409

B. 187 Quartzite in California, by W. E. Ver Planck. 1966. 410

B. 188 Geology of the Fremont Peak and Opal Mountain (15') Quadrangles, San Bernardino County, California, by T. W. Dibblee. 1968. 411

B. 189 Minerals of California, by J. Murdoch, R. W. Webb, I. Campbell, and E. M. Learned. 1966. 412

B. 190 Geology of northern California, ed. by E. H. Bailey. 1966. 413

Chapter I, Introduction, by C. W. Jennings. 414

Chapter II, Klamath Mountains Province, by W. P. Irwin, G. A. Davis, and J. P. Albers 415

Chapter III, Cascade Range, Modoc

Plateau, and Great Basin Provinces, by G. A. Macdonald and T. E. Gay. 416

Chapter IV, Sierra Nevada Province, by C. Wahrhaftig, V. E. McMath, C. Durrell, D. B. Siemmons, and W. B. Clark. 417

Chapter V, Great Valley Province, by O. Hackel, J. F. Poland, and R. E. Evenson, and E. W. Hart. 418

Chapter VI, Coast Ranges Province, by B. M. Page, R. R. Compton, P. F. Dickert, M. N. Christensen, and F. F. Davis. 419

Chapter VII, Offshore area, by G. A. Rusnak, J. R. Curray, T. A. Wilson, and J. L. Mero. 420

Chapter VIII, San Andreas Fault, by G. B. Oakeshott, T. W. Dibblee, B. K. Meade, and J. B. Small. 421

Chapter IX, Subcrustal structure, by R. H. Chapman, A. Griscom, and J. P. Eaton. 422

Chapter X, Field trip guides, by A. J. Galloway, M. G. Bonilla, J. Schlocker, E. E. Brabb, M. E. Maddock, R. E. Wallace, S. N. Davis, D. O. Emerson, E. I. Rich, D. L. Peck, C. Wahrhaftig, L. D. Clark, and C. W. Chesterman. 423

B. 191 Mineral resources of California, by the U. S. Geological Survey and J. P. Albers, I. Campbell, P. B. King, G. B. Oakeshott, W. P. Irwin, T. E. Gay, C. A. Repenning, D. L. Jones, W. O. Addicott, P. C. Bateman, J. H. Stewart, D. C. Ross, T. W. Dibblee, D. F. Hewett, G. L. Peterson, R. G. Gastil, E. C. Allison, W. B. Hamilton, G. B. Cleveland, Q. A. Aune, S. J. Rice, M. B. Smith, F. H. Weber,

22

E.B. Gross, W.C. Smith, G.I. Smith, P.K. Morton, C.D. Edgerton, T.P. Thayer, F.R. Kelley, E.R. Landis, J.T. Alfors, A.R. Kinkel, F.G. Lesure, C.W. Chesterman, D.M. Lemmon, J.R. McNitt, D.E. White, W.B. Clark, C.F. Withington, L. Moore, G.H. Espenshade, R.M. Stewart, O.E. Bowen, W.P. Irwin, A.R. Smith, F.F. Davis, D.F. Hewett, E.H. Bailey, M.C. Stinson, R.U. King, P.E. Hotz, R.L. Parker, J.W. Padan, C.W. Jennings, F.J. Schambeck, H.D. Gower, L.A. Wright, C. Durrell, J.W. Adams, H.B. Goldman, J.L. Burnett, C.T. Weiler, H.K. Stager, G.N. Broderick, L.A. Wright, J.R. Evans, C.H. Gray, N. Herz, G.W. Walker, A.P. Butler, R.P. Fischer, and B.W. Troxel. 1966. 424

B.192 Geology of the Ono (15') Quadrangle, Shasta and Tehama Counties, California, by M.A. Murphy, P.U. Rodda, and D.M. Morton. 1969. 425

B.193 Gold districts of California, by W.B. Clark. 1970. 426

B.194 The mineral economics of the carbonate rocks--limestone and dolomite resources of California, by O.E. Bowen, C.H. Gray, and J.R. Evans. 1973. 427

B.195 Geology of the San Andreas 15-minute quadrangle, Calaveras County, California, by L.D. Clark. 1970. 428

B.196 San Fernando, California, earthquake of 9 February 1971, ed. by G.B. Oakeshott. 1974. 429

Part I, Geology and geophysics, by A.K. Baird, D.M. Morton, P.L. Ehig, G.B. Oakeshott, G.A. Brown, R.B. Saul, F.H. Weber, A.G. Barrows, J.E. Kahle, J.R. Evans, California Division of Mines and Geology, F. Beach, Leighton & Associates, J.L. Smith, R.B. Fallgren, R.W. Greensfelder, J.C. Savage, R.O. Burford, W.T. Kinoshita, R.V. Sharp, H.W. Oliver, S.L. Robbins, R.B. Grannell,

R.W. Alewine, S. Biehler, R.H. Chapman, G.W. Chase, A.G. Sylvester, D.D. Pollard, and J.E. Slosson. 430

Part II, Seismology, by C.R. Allen, T.C. Hanks, J.H. Whitcomb, B.A. Bolt, B.S. Gopalakrishnan, W.K. Cloud, D.E. Hudson, L.L. Davis, L.R. West, and B.A. Bolt. 431

Part III, Damage, by K.V. Steinbrugge, E.E. Schader, D.F. Moran, J.F. Meehan, California Division of Highways, J. Penzien, R.W. Clough, C.J. Cartright, D.F. Moran, and C.M. Duke. 432

Part IV, Disaster response, by Los Angeles County Earthquake Commission and F.I. Ross. 433

Part V, Mimimizing losses, by G.B. Oakeshott and W.G. Bruer. 434

B.197 Limestone, dolomite, and shell resources of the Coast Range Province, California, by E.W. Hart. 1978. 435

B.198 Urban geology master plan for California, by J.T. Alford, J.L. Burnett, and T.E. Gay. 1973. 436

B.199 Basic geology of the Santa Margarita area, San Luis Obispo County, California, by E.W. Hart. 1976. 437

B.200 Geology of the San Diego Metropolitan Area, California, by M.P. Kennedy and G.L. Peterson. 1975. 438

B.201 Not yet published. 439

B.202 Geology of the Point Reyes Peninsula, Marin County, California, by A.J. Galloway. 1977. 440

B.203 Vertical deformation, stress accumulation, and secondary faulting in the vicinity of the Transverse Ranges of southern California, by D.A. Rodgers. 1979. 441

County Report (New Series). 1962-

CR.3 San Diego County, California,

mines and mineral resources, by
F. H. Weber. 1963.   442

CR. 4 Trinity County, California,
mines and mineral resources, by
J. C. O'Brien. 1965.   443

CR. 5 Monterey County, California,
mines and mineral resources, by
E. W. Hart. 1966.   444

CR. 6 Shasta County, California,
mines and mineral resources, by
P. A. Lydon and J. C. O'Brien.
1974.   445

CR. 7 Imperial County, California,
mines and mineral resources, by
P. K. Morton. 1977.   446

CR. 8 Alpine County, California,
mines and mineral resources, by
W. B. Clark. 1977.   447

Preliminary Report. 1914-

PR. 9 Not yet published.   448

PR. 10 Geology of the NE$\frac{1}{4}$ and
NW$\frac{1}{4}$ Canada Gobernadora ($7\frac{1}{2}$')
Quadrangle, Orange County, Cal-
ifornia, by P. K. Morton. 1970.
                                  449

PR. 11 Map of surface breaks re-
sulting from the San Fernando,
California, earthquake of February
9, 1971, by A. G. Barrows, J. E.
Kahle, F. H. Weber, and R. B. Saul.
1971.   450

PR. 12 Geology of the Shadow Moun-
tains and north part of the Shadow
Mountains-SE $7\frac{1}{2}$' Quadrangles,
western San Bernardino County,
California, by B. Troxel and J. N.
Gundersen. 1970.   451

PR. 13 Preliminary fault and geologic map of California, by C. W.
Jennings. 1974.   452

PR. 14 Geology and mineral re-
sources study of southern Ventura
County, California, by F. H. Weber,
with G. B. Cleveland, J. E. Kahle,
E. F. Kiessling, R. V. Miller, M. F.
Mills. D. M. Morton, and B. A. Cil-

weck. 1973.   453

PR. 15 Preliminary geo-environmental
maps of Orange County, California,
by P. K. Morton, R. V. Miller, and
D. L. Fife. 1973.   454

PR. 16 Geology for planning on the
Sonoma County Coast between the Rus-
sian and Gualala Rivers, by M. E.
Huffman. 1972.   455

PR. 17 Environmental geologic analysis of the Monte Bello Ridge Mountain
Study Area, Santa Clara County, by
T. H. Rogers and C. F. Armstrong.
1973.   456

PR. 18 Environmental geological analysis of the South County Study Area,
Santa Clara County, by J. W. Williams, with C. F. Armstrong, E. W.
Hart, and T. H. Rogers. 1973.   457

PR. 19 Geological and geophysical investigations for tri-cities seismic
safety and environmental resources
study, by C. C. Bishop and R. D. Knox.
1973.   458

PR. 20 Geology for planning on the
Sonoma County Coast between the Rus-
sian River and Estero Americano, by
M. E. Huffman, with G. B. Chase,
T. C. Smith, and R. W. Greensfelder.
1973.   459

PR. 21 Geology and geologic hazards
of the Novato area, Marin County,
California, by S. J. Rice, with G. B.
Chase. 1973.   460

PR. 22 Compilation of strong-motion
records recovered from the Santa
Barbara earthquake of 13 August 1978,
by L. D. Porter. 1978.   461

PR. 23 Processed data from the par-
tial strong-motion records of the
Santa Barbara earthquake of 13 August
1978. Preliminary results, by L. D.
Porter, J. T. Ragsdale, and R. D. Mc-
Junkin. 1978.   462

PR. 24 Not yet published.   463

PR. 25 Compilation of strong-motion
records recovered from the Coyote
Lake earthquake of 6 August 1979,

by R. L. Porcella, R. B. Matthiesen, R. D. McJunkin, and J. T. Ragsdale. 1979.   464

Special Publication. 1881-

(SP. 1-32 unavailable for indexing)

SP. 33 Minerals and rocks furnished to California schools by the California Division of Mines and Geology, by S. J. Rice. 1962.   465

SP. 34 Geology of placer deposits, by O. P. Jenkins. 1970.   466

SP. 35 Preliminary report and geologic guide to Franciscan melanges of the Morro Bay, San Simeon area, California, by K. J. Hsu. 1969.   467

SP. 36 Preliminary bibliography, Lake Tahoe Basin, California-Nevada, by R. A. Matthews and C. Schwarz. 1969.   468

SP. 37 Not published?   469

SP. 38 Site characteristics of southern California strong-motion earthquake stations, by C. M. Duke and D. J. Leeds. 1962.   470

SP. 39 Earthquakes: Be prepared! by W. B. Clark and C. J. Hauge. 1973.   471

SP. 40 Legal guide for California prospectors and miners, by C. Gilmore and R. M. Stewart. 1962. Addenda 1970 and 1973, by R. M. Stewart.   472

SP. 41 Basic placer mining (Rev. from B. 135). 1946.   473

SP. 42 Index to maps of special studies zones, by E. W. Hart and others. 1973.   474

SP. 43 Mineral producers in California for 1971, by F. F. Davis. 1974.   475

SP. 44 Potential disposal sites for Class 1 (Toxic) waste in southern

California, by J. L. Burnett and G. C. Taylor. 1973.   476

SP. 45 Meeting the earthquake challenge. Final report to the Legislature by the Joint Committee on Seismic Safety. 1974.   477

SP. 46 Second report of the Governor's earthquake council. 1974. 478

SP. 47 Active fault mapping and evaluation program; ten year program to implement Alquist-Priolo Special Zones Act. 1976.   479

SP. 48 Second report on the strong-motion instrumentation program. n. d.   480

SP. 49 California jade; a collection of reprints. 1976.   481

SP. 50 Colemanite deposits near Kramer Junction, San Bernardino County, California, by J. R. Evans and T. P. Anderson. 1976.   482

SP. 51 State policy for surface mining and reclamation practice, by the California State Mining and Geology Board. 1977. Rev. 1979.   483

SP. 52 Earthquake catalog of California, January 1, 1900-December 31, 1974, by C. R. Real, T. R. Toppozada, and D. L. Parke. 1978.   484

SP. 53 Mines and mineral producers in California during 1978, by J. S. Rapp. 1979.   485

SP. 54 Technical review of the seismic safety of the Auburn Damsite, by C. F. Bacon and J. F. Davis. 1979.   486

SP. 55 Activities of the strong motion instrumentation program of August 3, 1977 to September 15, 1978, by T. M. Wootton. 1979.   487

SP. 56 Geologic evaluation of the General Electric Test Reactor Site, Vallecitos, Alameda County, California, by S. Rice, E. Stephens, and C. Real. 1979.   488

SP. 57 Proposed earthquake safety

programs and activities of the Department of Conservation, fiscal years 1982 through 1986, by J. F. Davis and others. 1980. 489

Special Report. 1950-

SR. 73 Economic geology of the Panamint Butte (15') Quadrangle and the Modoc District, Inyo County, California, by W. E. Hall and H. G. Stephens. 1963. 490

SR. 74 Index to graduate theses on California geology to December 31, 1961, by C. W. Jennings and R. G. Strand. 1963. 491

SR. 75 Not published? 492

SR. 76 Recent volcanism at Amboy Crater, San Bernardino County, California, by R. B. Parker. 1963. 493

SR. 77 Geology and mineral deposits of the Ord Mountain District, San Bernardino County, California, by F. H. Weber. 1963. 494

SR. 78 Age and structural relationships of the Franciscan Formation in the Montara Mountain Quadrangle, San Mateo County, California, by R. L. Darrow. 1963. 495

SR. 79 Xenotime mineralization in the southern Music Valley area, Riverside County, California, by J. R. Evans. 1964. 496

SR. 80 Geology and mineral resources of Mt. Diablo, Contra Costa County, California, by E. H. Pampeyan. 1964. 497

SR. 81 Geology of the Lockwood Valley area, Kern and Ventura Counties, California, by M. F. Carman. 1964. 498

SR. 82 Not published? 499

SR. 83 A geologic reconnaissance in the southeastern Mojave Desert (San Bernardino County), by A. M. Bassett and D. H. Kupfer. 1964. 500

SR. 84 Stratigraphy, structure, and mineral deposits in the Oro Grande Series near Victorville (San Bernardino County), California, by O. E. Bowen and W. E. Ver Planck. 1965. 501

SR. 85 Economic geology of the French Gulch (15') Quadrangle, Shasta and Trinity Counties, California, by J. P. Albers and C. H. Sandberg. 1965. 502

SR. 86 Short contributions to California geology, ed. by J. L. Burnett. 1964. 503

Geology of the central part of the Ramona pegmatite district, San Diego County, California, by D. R. Simpson. 504

Tertiary stratigraphy of the Church Creek area, Monterey County, California, by W. R. Dickinson. 505

Magnetic study of the Island Mountain mine area, Trinity County, California, by R. H. Chapman. 506

SR. 87 Expansible shale resources of the San Jose-Gilroy area (Santa Clara County), California, by J. L. Burnett. 1965. 507

SR. 88 Geology of the Queen of Sheba Lead Mine, Death Valley (Inyo County), California, by P. K. Morton. 1965. 508

SR. 89 Not published? 509

SR. 90 The California Division of Mines and Geology Gravity Base Station Network, by R. H. Chapman. 1966. 510

SR. 91 Short contributions to California geology. 1967. 511

Petrography of six granitic intrusive units in the Yosemite Valley area (Mariposa County), California, by A. R. Smith. 512

The origin of Tuscan Buttes and the volume of the Tuscan Formation in (Tehama County) northern California, by P. A. Lydon. 513

The Plutonic and metamorphic rocks of the Ben Lomond Mountain area, Santa Cruz County, California, by G. W. Leo.　514

Chittenden (Santa Cruz County), California, earthquake of September 14, 1963, by E. E. Brabb.　515

Notes on the types of California species of the foraminiferal genus Orthokarstenia Dietrich, 1935, by J. J. Graham and D. K. Clark.　516

SR. 92　Short contributions to California geology. 1967.　517

Mineralogy of the Kalkar Quarry, Santa Cruz (Santa Cruz County), California, by E. B. Gross, C. W. Chesterman, G. Dunning, and J. F. Cooper.　518

A test of chemical variability and field sampling methods, Lakeview Mountain Tonalite, Lakeview Mountains (Riverside County), southern California batholith, by A. K. Baird, D. B. McIntyre, E. E. Welday, and D. M. Morton.　519

The effects of provenance and basin-edge topography on sedimentation in the basal Castaic Formation (Upper Miocene, marine), Los Angeles County, California, by R. J. Stanton.　520

Sedimentary rocks of Late Precambrian and Cambrian age in the southern Salt Spring Hills, southeastern Death Valley (Inyo County), California, by B. W. Troxel.　521

Reconnaissance geology of the Helena (7½') Quadrangle, Trinity County, California, by D. P. Cox.　522

SR. 93　Early Pliocene sedimentary history of the Los Angeles Basin (Los Angeles and Orange Counties), California, by B. L. Conrey. 1967.　523

SR. 94　Geology of the Desert Hot Springs-Upper Coachella Valley area (Riverside County), California, by R. J. Proctor. 1968.　524

SR. 95　Talc deposits of the southern Death Valley-Kingston Range Region (Inyo and San Bernardino Counties), California, by L. A. Wright. 1968.　525

SR. 96　Geologic reconnaissance of the Slate Range, San Bernardino and Inyo Counties, California, by G. I. Smith, B. Troxel, C. H. Gray, and R. von Huene. 1968.　526

SR. 97　Geologic and engineering aspects of San Francisco Bay Fill, ed. by H. B. Goldman. 1969.　527

Geology of San Francisco Bay, by H. B. Goldman, including Geologic features of earthquakes in the Bay area, by G. B. Oakeshott.　528

Salt, sand, and shells, by H. B. Goldman.　529

Bay mud developments and related structural foundations, by C. H. Lee and M. Praszker.　530

Seismic problems in the use of fills in San Francisco Bay, by H. B. Seed.　531

Seismic risk to buildings and structures on filled lands in San Francisco Bay, by K. V. Steinbrugge.　532

The safety of fills, by Board of Consultants, Bay Conservation and Development Commission.　533

SR. 98　Natural slope stability as related to geology, San Clemente area, Orange and San Diego Counties, California, by R. P. Blanc and G. B. Cleveland. 1968.　534

SR. 99　Geology of the Dry Mountains (15') Quadrangle, Inyo County, California, by B. C. Burchfiel. 1969.　535

SR. 100　Short contributions to California geology. 1969.　536

Clay mobility, Portuguese Bend (Los Angeles County), California, by P. F. Kerr and I. M. Drew.　537

The chemical fingerprinting of acid

volcanic rocks, by R. N. Jack and
I. S. E. Carmichael.   538

Cretaceous and Eocene coccoliths
at San Diego (San Diego County),
California, by D. Bukry and M. P.
Kennedy.   539

Stratigraphy and petrology of the
Lost Burro Formation, Panamint
Range (Inyo County), California,
by D. H. Zenger and E. F. Pearson.   540

Rapid method of sampling diatomaceous earth, by G. B. Cleveland.
          541

SR. 101 Geology of the Elysian
Park-Repetto Hills area, Los An-
geles County, California, by D. L.
Lamar. 1970.   542

SR. 102 Index to geologic maps of
California, 1965-1968, by E. W.
Kiessling. 1972.   543

SR. 103 Trace elements in the
Plumas Copper Belt, Plumas County,
California, by A. R. Smith. 1971.
          544

SR. 104 Upper Cretaceous strati-
graphy on the west side of the San
Joaquin Valley, Stanislaus and San
Joaquin Counties, California, by
C. C. Bishop. 1971.   545

SR. 105 Geology of parts of the
Azusa and Mount Wilson Quadrangles,
San Gabriel Mountains, Los Angeles
County, California, by D. M. Mor-
ton. 1973.   546

SR. 106 Geologic features, Death
Valley, California; revised papers
from a guidebook on the Death
Valley Region, California and
Nevada, published in 1974, ed. by
B. W. Troxel and L. A. Wright.
1976.   547

SR. 107 Potential seismic hazards
in Santa Clara County, California,
by T. H. Rogers and J. W. Williams.
1974.   548

SR. 108 First annual report of the
strong-motion instrumentation pro-

gram, 1972-1973. 1974.   549

SR. 109 Geology of the Dana Point
($7\frac{1}{2}$ ') Quadrangle, Orange County,
California, by W. J. Edgington. 1974.
          550

SR. 110 Geology of the $S\frac{1}{2}$ El Toro
($7\frac{1}{2}$ ') Quadrangle, Orange County,
California, by D. L. Fife. 1974.
          551

SR. 111 Geology and engineering geologic aspects of the $S\frac{1}{2}$ Canada Gobernadora ($7\frac{1}{2}$ ') Quadrangle, Orange
County, California, by P. K. Morton.
1974.   552

SR. 112 Geology and engineering geologic aspects of the San Juan Capis-
trano ($7\frac{1}{2}$ ') Quadrangle, Orange County,
California, by P. K. Morton, W. J.
Edgington, and D. L. Fife. 1974. 553

SR. 113 Geologic hazards in southwestern San Bernardino County, California, by D. L. Fife and others.
1976.   554

SR. 114 A review of the geology and
earthquake history of the Newport-
Inglewood structural zone, southern
California, by A. G. Barrows. 1974.
          555

SR. 115 Index to graduate theses and
dissertations on California geology,
1962 through 1972, by G. C. Taylor.
1974.   556

SR. 116 Geophysical study of the
Clear Lake region, California, by
R. H. Chapman. 1975.   557

SR. 117 Geophysical investigation in
the Ione area, Amador, Sacramento,
and Calaveras Counties, California,
by R. H. Chapman and C. C. Bishop.
1975.   558

SR. 118 San Andreas Fault in southern California, by J. C. Crowell.
1975.   559

SR. 119 Landsliding in marine terrace terrain, California, by G. B.
Cleveland. 1975.   560

SR. 120 Geology for planning in So-

noma County, by M. E. Huffman and C. F. Armstrong. 1980. 561

SR. 121 Sand and gravel resources of the Sacramento area, California, by J. S. Rapp. 1975. 562

SR. 122 Engineering geology of the Geysers Geothermal Resource Area, Lake, Mendocino, and Sonoma Counties, California, by C. F. Bacon and others. 1976. 563

SR. 123 Character and recency of faulting, San Diego metropolitan area, California, by M. P. Kennedy and others. 1975. 564

SR. 124 Oroville, California, earthquake, 1 August 1975, ed. by R. W. Sherburne and C. J. Hauge. 1975. 565

SR. 125 Mines and mineral deposits in Death Valley National Monument, California, by J. R. Evans, G. C. Taylor, and J. S. Rapp. 1976. 566

SR. 126 Geology and engineering geologic aspects of the South Half Tustin Quadrangle, Orange County, California, by R. V. Miller and S. S. Tan. 1976. 567

SR. 127 Geology and engineering geologic aspects of the Laguna Beach Quadrangle, Orange County, California, by S. S. Tan and W. J. Edgington. 1976. 568

SR. 128 K-Feldspar in Upper Mesozoic sandstone units near Atascadero, Santa Lucia Range, San Luis Obispo County, California, by E. W. Hart. 1977. 569

SR. 129 Short contributions to California geology. 1977. 570

SR. 130 Index to geologic maps of California, 1969-1975, by E. W. Kiessling and D. H. Peterson. 1977. 571

SR. 131 Recency and character of faulting along the Elsinore Fault zone in southern Riverside County,

California, by M. P. Kennedy. 1977. 572

SR. 132 Not yet published. 573

SR. 133 Clay mineralogy and slope stability, by G. A. Borchardt. 1977. 574

SR. 134 Erosion along Dry Creek, Sonoma County, California, by G. B. Cleveland and F. R. Kelley. 1977. 575

SR. 135 Seismicity of California, 1900-1931, by T. R. Toppozada, D. L. Plarke, and C. T. Higgins. 1978. 576

SR. 136 Landsliding and mudflows at Wrightwood, San Bernardino County, California, by D. M. Morton, M. P. Kennedy, R. H. Campbell, A. G. Barrows, J. E. Kahle, and R. F. Yerkes. 1979. 577

SR. 137 San Gregorio-Hosgri Fault Zone, California, by E. A. Silver and W. R. Normark. 1978. 578

SR. 138 Not yet published. 579

SR. 139 Aggregates in the greater Los Angeles area, California, by J. R. Evans, T. P. Anderson, M. W. Manson, R. L. Maud, W. B. Clark, and D. L. Fife. 1979. 580

SR. 140 Not yet published. 581

SR. 141 Not yet published. 582

SR. 142 Geology and slope stability in selected parts of the Geysers Geothermal area; a guide to geologic features indicative of stable and unstable terraine in areas underlain by Franciscan and related rocks, by T. L. Bedrossian. 1980. 583

SR. 143 Mineral land classification of the greater Los Angeles area, by T. P. Anderson, R. C. Lloyd, W. B. Clark, R. V. Miller, R. Corbaley, S. Kohler, and M. M. Bushnell. 1979. 584

SR. 144 Processed data from the

strong-motion records of the Santa
Barbara earthquake of 13 August
1978. Final results, in 3 volumes,
by L. D. Porter, J. T. Ragsdale,
and R. D. McJunkin. 1979.   585

Bulletin. 1910-

B. 32 Prairie, Peak and Plateau-- a guide to the geology of Colorado, by J. Chronic and H. Chronic. 1972. 586

B. 33 Bibliography of hydrogeologic reports in Colorado, by R. H. Pearl. 1971. 587

B. 34A Bibliography, coal resources in Colorado, by R. D. Holt. 1972. 588

B. 35 Proceedings of a symposium on geothermal energy and Colorado, by R. H. Pearl, ed. 1974. 589

B. 36 Geologic control of supply and quality of water in the mountainous part of Jefferson County, Colorado, by W. E. Hofstra and D. C. Hall. 1975. 590

B. 37 Bibliography and index of Colorado geology, 1875 to 1975, by the American Geological Institute. 1976. 591

B. 38 Guidelines and methods for detailed snow avalanche hazard investigations in Colorado, by A. I. Mears. 1976. 592

B. 39 An appraisal of Colorado's geothermal resources, by J. K. Barrett and R. H. Pearl. 1978. 593

B. 40 Radioactive mineral occurrences of Colorado and bibliography, by J. L. Nelson-Moore, D. B. Collins, and A. L. Hornbaker. 1978. 594

B. 41 Bibliography and index of publications related to coal in Colorado, 1972-77, by H. B. Fender,

D. C. Jones, and D. K. Murray. 1978. 595

B. 42 Water resources of Boulder County, Colorado, by D. C. Hall, D. Hillier, D. Cain, and E. L. Boyd. 1980. 596

Environmental Geology Report. 1972-

EGR. 1 Geologic aspects, soils and related foundation problems, Denver Metropolitan area, Colorado, by J. L. Hamilton and W. G. Owens. 1972. 597

EGR. 2 Engineering geology report for Planning District 1, State of Colorado, by F. M. Fox & Associates. 1974. 598

EGR. 3 Engineering geology report for Planning Districts 7 and 13, State of Colorado, by F. M. Fox & Associates. 1974. 599

EGR. 4 Engineering geology report for Planning District 9, State of Colorado, by F. M. Fox & Associates. 1974. 600

EGR. 5 Not published? 601

EGR. 6 Environmental and engineering geology of the Windsor Study Area, Larimer and Weld Counties, Colorado, by D. C. Shelton and W. P. Rogers. 1980? 602

EGR. 7 Potentially swelling soil and rock in the Front Range Urban Corridor, Colorado, by S. S. Hart. 1974. 603

EGR. 8 Roaring Fork and Crystal Valleys--an environmental and engineering geology study: Eagle, Garfield, Gunnison, and Pitkin Counties, Col-

31

orado, by F. M. Fox & Associates. 1974. 604

EGR. 9 Ground subsidence and land-use considerations over coal mines in the Boulder-Weld Coal Field, Colorado, by Amuedo and Ivey. 1975. 605

EGR. 10 Geologic hazards, geomorphic features, and land-use implications in the area of the 1976 Big Thompson Flood, Larimer County, Colorado, by J. M. Soule, W. P. Rogers, and D. C. Shelton. 1976. 606

EGR. 11 Promises and problems of a "new" uranium mining method: in situ solution mining, by R. M. Kirkham. 1979. 607

EGR. 12 Energy resources of the Denver and Cheyenne Basins, Colorado--resource characteristics, developmental potential, and environmental problems, by R. M. Kirkham and L. R. Ladwig. 1980. 608

Information Series. 1976-

IS. 1 Coal mines and coal fields of Colorado, by D. C. Jones. 1976. 609

IS. 2 Coal mines of Colorado, statistical data, by D. C. Jones and D. K. Murray. 1976. 610

IS. 3 Oil and gas fields of Colorado, statistical data, by D. C. Jones and D. K. Murray. 1976. 611

IS. 4 Map showing thermal springs, wells and heat-flow contours in Colorado, by J. K. Barrett, R. H. Pearl, and A. J. Pennington. 1976. 612

IS. 5 Geologic hazards in the Crested Butte-Gunnison area, Gunnison County, Colorado, by J. M. Soule. 1976. 613

IS. 6 Hydrogeological data of thermal springs and wells in Colorado, by J. K. Barrett and R. H. Pearl. 1976. 614

IS. 7 Colorado coal analyses, 1975 (analyses of 64 samples collected in 1975), by D. L. Boreck, D. C. Jones, D. K. Murray, J. E. Schultz, and D. C. Suek. 1977. 615

IS. 8 Debris-flow hazard analysis and mitigation--an example from Glenwood Springs, Colorado, by A. I. Mears. 1977. 616

IS. 9 Geothermal energy development in Colorado: processes, promises & problems, by B. A. Coe. 1978. 617

IS. 10 Not yet published. 618

IS. 11 Not yet published. 619

IS. 12 Hydrogeologic and stratigraphic data pertinent to uranium mining, Cheyenne Basin, Colorado, by R. M. Kirkham, W. J. O'Leary, and J. W. Warner. 1980. 620

IS. 13 Chemical analyses of water wells in selected strippable coal and lignite areas, Denver Basin, Colorado, by R. M. Kirkham and W. J. O'Leary. 1980. 621

IS. 14 Hazardous wastes in Colorado: a preliminary evaluation of generation and geologic criteria for disposal, by J. L. Hynes and C. J. Sutton. 1980. 622

IS. 15 Regulation of geothermal energy development in Colorado, by B. A. Coe and N. A. Forman. 1980. 623

Resource Series. 1977-

RS. 1 Geology of Rocky Mountain coal--a symposium, 1976. ed. by D. K. Murray. 1977. 624

RS. 2 Mineral resources survey of Mesa County--a model study, by S. D. Schwochow. 1978. 625

RS. 3 Colorado coal directory and source book, 1978, by L. C. Dawson and D. K. Murray. 1978. 626

RS. 4 Proceedings of the second symposium on the geology of Rocky Mountain coal--1977, ed. by H. E. Hodgson. 1978. 627

RS. 5 Coal resources of the Denver and Cheyenne Basins, Colorado, by R. M. Kirkham and L. R. Ladwig. 1979. 628

RS. 6 Colorado's hydrothermal resource base--an assessment, by R. H. Pearl. 1979. 629

RS. 7 Evaluation of coking coals in Colorado, by S. M. Goolsby, N. S. Reade, and D. K. Murray. 1979. 630

RS. 8 Proceedings of the fifteenth forum on geology of industrial minerals; theme: industrial minerals in Colorado and the Rocky Mountain Region; Golden, Colorado, June 13-15, 1979, ed. by S. D. Schwochow. 1980. 631

RS. 9 Conservation of methane from mined /minable coal beds, Colorado, by D. L. Boreck and M. T. Strever. 1980. 632

RS. 10 Proceedings of the fourth symposium on the geology of Rocky Mountain coal, 1980, ed. by L. M. Carter. 1980. 633

RS. 11 Rare-earth pegmatites of the South Platte District, Colorado, by W. B. Simmons and E. W. Heinrich. 1980. 634

Special Publication. 1970-

SP. 1 The governor's first conference on environmental geology; proceedings, 1969. 1970. 635

SP. 2 Geothermal resources of Colorado, by R. H. Pearl. 1972. 636

SP. 3 1972 summary of coal resources in Colorado, by A. L. Hornbaker and R. D. Holt. 1973. 637

SP. 4 Geology of ground water resources in Colorado--an introduction, by R. H. Pearl. 1974. 638

SP. 5A Sand, gravel, and quarry aggregate resources, Colorado Front Range counties, by S. D. Schwochow, R. R. Shroba, and P. C. Wicklein. 1974. 639

SP. 5B Atlas of sand, gravel, and quarry aggregate resources, Colorado Front Range counties, by S. D. Schwochow, R. R. Shroba, and P. C. Wicklein. 1975. 640

SP. 6 Guidelines and criteria for identification and land-use controls of geologic hazard and mineral resource areas, by W. P. Rogers and others. 1974. 641

SP. 7 Colorado avalanche area studies and guidelines for avalanche-hazard planning, by A. I. Mears. 1979. 642

SP. 8 Proceedings, governor's third conference on environmental geology--geologic factors in land-use planning, House Bill 1041, ed. by D. C. Shelton. 1977. 643

SP. 9 Summary of coal resources of Colorado, by A. L. Hornbaker, R. D. Holt, and D. K. Murray. 1976 (Supercedes SP. 13). 644

SP. 10 Hydrogeologic and geothermal investigation of Pagosa Hot Springs, Colorado, by M. J. Galloway. 1980. 645

SP. 11 Home construction on shrinking and swelling soils, by W. G. Holtz and S. S. Hart. 1978. 646

SP. 12 Nature's building codes--geology and construction in Colorado, by D. C. Shelton and D. Prouty. 1979. 647

SP. 13 1979 Summary of coal resources in Colorado, by D. K. Murray. 1980. 648

CONNECTICUT

Bulletin. 1904-

B. 93 Guide to the insects of Connecticut. Part VI: The Diptera or true flies of Connecticut. Eighth fascicle: Scatopsidae and Hyperoscelidae, by E. F. Cook, Blepharoceridae and Deuterophleblidae, by C. P. Alexander, and Dixidae, by W. R. Nowell. 1963.   649

B. 94 Marine sedimentary environments in the vicinity of the Norwalk Islands, Connecticut, by W. C. Ellis. 1962.   650

B. 95 Twenty-ninth and thirtieth biennial reports of the Commissioners of the State Geological and Natural History Survey, 1959-1962. 1963.   651

B. 96 Fossils of the Connecticut Valley, by E. H. Colbert. Rev. ed. 1970.   652

B. 97 Guide to the insects of Connecticut. Part VI: The Diptera or the true flies of Connecticut. Ninth fascicle: Simullidae and Thaumaleidae, by A. Stone. 1964.   653

B. 98 Thirty-first biennial report of the Commissioners of the State Geological and Natural History Survey, 1963-1965. 1965.   654

B. 99 The climate of Connecticut, by J. J. Brumbach. 1965.   655

B. 100 Thirty-second biennial report of the Commissioners of the State Geological and Natural History Survey, 1965-1967. 1967.
656

B. 101 Freshwater fishes of Connecticut, by W. R. Whitworth, P. L.

Berrien, and W. T. Keller. 1968.
657

B. 102 Thirty-third biennial report of the Commissioners of the State Geological and Natural History Survey, 1967-1969. 1969.   658

B. 103 Connecticut's venomous snakes, by R. C. Petersen. 1970.   659

B. 104 Thirty-fourth biennial report of the Commissioners of the State Geological and Natural History Survey, 1969-1971. 1971.   660

B. 105 Saltwater fishes of Connecticut, by K. S. Thomson, W. H. Weed, A. G. Taruski, and D. E. Simanek. 2nd ed. 1978.   661

B. 106 Diatoms of the streams of eastern Connecticut, by E. W. Hansmann. 1973.   662

B. 107 Guide to the insects of Connecticut. Part VII: The Plecoptera or Stoneflies of Connecticut, by S. W. Hitchcock. 1974.   663

B. 108 An annotated checklist of Connecticut seaweeds, by C. W. Schneider, M. M. Suyemoto, and C. Yarish. 1979.   664

Guidebook. 1965-

GB. 1 Postglacial stratigraphy and morphology of coastal Connecticut, by A. L. Bloom and C. W. Ellis. 1965.   665

GB. 2 Guidebook for fieldtrips in Connecticut, New England Intercollegiate Geological Conference, ed. by P. M. Orville. 1968.   666

GB. 3 Stratigraphy and structure of

the Triassic strata of the Gaillard Graben, south-central Connecticut, by J. E. Sanders. 1970.   667

GB. 4 Guide to the Mesozoic red-beds of central Connecticut, by J. F. Hubert, A. A. Reed, W. L. Dowdall, and J. M. Gilchrist. 1978.   668

Quadrangle Report. 1951-

QR. 13 The bedrock geology of the Deep River Quadrangle, by L. Lundgren. 1963.   669

QR. 14 The surficial geology of the Branford Quadrangle, by R. F. Flint. 1964.   670

QR. 15 The bedrock geology of the Essex Quadrangle, by L. Lundgren. 1964.   671

QR. 16 The bedrock geology of the Collinsville Quadrangle, by R. S. Stanley. 1964.   672

QR. 17 The bedrock geology of the West Torrington Quadrangle, by R. M. Gates and N. I. Christensen. 1965.   673

QR. 18 The surficial geology of the New Haven and Woodmont Quadrangles, by R. F. Flint. 1965.   674

QR. 19 The bedrock geology of the Hamburg Quadrangle, by L. Lundgren. 1966.   675

QR. 20 The surficial geology of the Hartford South Quadrangle, by R. E. Deane. 1967.   676

QR. 21 The bedrock geology of the Old Lyme Quadrangle, by L. Lundgren. 1967.   677

QR. 22 The bedrock geology of the Waterbury Quadrangle, by R. M. Gates and C. W. Martin. 1967.   678

QR. 23 The surficial geology of the Ansonia and Milford Quadrangles, by R. F. Flint. 1968.   679

QR. 24 The bedrock geology of the Long Hill and Bridgeport Quadrangles, by W. P. Crowley. 1968.   680

QR. 25 The bedrock geology of the Torrington Quadrangle, by C. W. Martin. 1970.   681

QR. 26 The surficial geology of the Spring Hill Quadrangle, by P. H. Rahn. 1971.   682

QR. 27 The bedrock geology of the Moodus and Colchester Quadrangles, by L. Lundgren, L. Ashmead, and G. L. Synder. 1971.   683

QR. 28 The surficial geology of the Guilford and Clinton Quadrangles, by R. F. Flint. 1971.   684

QR. 29 The bedrock geology of the Clinton Quadrangle, by L. Lundgren and R. F. Thurrell. 1973.   685

QR. 30 The bedrock geology of the Southbury Quadrangle, by R. B. Scott. 1974.   686

QR. 31 The surficial geology of the Essex and Old Lyme Quadrangles, by R. F. Flint. 1975.   687

QR. 32 The bedrock geology of the South Canaan Quadrangle, by R. M. Gates. 1975.   688

QR. 33 The bedrock geology of the Newtown Quadrangle, by R. S. Stanley. 1976.   689

QR. 34 The bedrock geology of the Norwalk North and Norwalk South Quadrangles, by R. L. Kroll. 1977.   690

QR. 35 The surficial geology of the Naugatuck Quadrangle, by R. F. Flint. 1978.   691

QR. 36 The surficial geology of the Haddam Quadrangle, by R. F. Flint. 1978.   692

QR. 37 The bedrock geology of the Haddam Quadrangle, by L. Lundgren. 1979.   693

QR. 38 The bedrock geology of the

Sharon Quadrangle, by R. M. Gates.
1979. 694

Report of Investigation. 1961-

RI. 1 Geological and economic aspects of beach erosion along the Connecticut coast, by J. E. Sanders and C. W. Ellis. 1961. 695

RI. 2 Foxon Fault and Gaillard Graben in the Triassic of southern Connecticut, by J. E. Sanders, C. V. Guidotti, and P. Wilde. 1963. 696

RI. 3 Stratigraphy and petrology of the Hampden Basalt in central Connecticut, by R. W. Chapman. 1965. 697

RI. 4 Stratigraphy and structure of the western part of the New Haven Quadrangle, Connecticut, by H. R. Burger. 1967. 698

RI. 5 Contributions to geochronology in Connecticut, by R. L. Armstrong, J. M. Barton, J. R. Besancon, D. G. Brookins, S. W. Carmalt, and W. P. Crowley. 1970. 699

RI. 6 Rare and endangered species of Connecticut and their habitats, by J. J. Dowhan and R. J. Craig. 1976. 700

RI. 7 An exploited population of muskrats with unusual biomass, productivity, and body size, by H. R. Smith and P. A. Jordan. 1976. 701

RI. 8 Preliminary checklist of the vascular flora of Connecticut (Growing without cultivation), by J. J. Dowhan. 1979. 702

RI. 9 Lichens and air pollution: a study in Connecticut, by K. J. Metzler. 1980. 703

DELAWARE

Bulletin. 1953-

B. 10 Salinity of the Delaware Estuary, by B. Cohen and L. T. McCarthy. 1963.   704

B. 11 Ground-water resources of southern New Castle County, Delaware, by D. R. Rima, O. J. Coskery, and P. W. Anderson. 1964.   705

B. 12 Columbia (Pleistocene) sediments of Delaware, by R. R. Jordan. 1964.   706

B. 13 Geology, hydrology, and geophysics of Columbia sediments in the Middletown-Odessa area, Delaware, by N. Spoljaric and K. D. Woodruff. 1970.   707

B. 14 Hydrology of the Columbia (Pleistocene) deposits of Delaware: an appraisal of a regional water-table aquifer, by R. H. Johnston. 1973.   708

B. 15 Digital model of the unconfined aquifer in central and southeastern Delaware, by R. H. Johnston. 1977.   709

Report of Investigations. 1957-

RI. 7 An invertebrate macrofauna from the Upper Cretaceous of Delaware, by H. G. Richards and E. Shapiro. 1963.   710

RI. 8 Evaluation of the water resources of Delaware, by W. W. Baker, R. D. Varrin, J. J. Groot, and R. R. Jordan. 1966.   711

RI. 9 Ground-water levels in Delaware, January 1962-June 1966, by K. D. Woodruff. 1967.   712

RI. 10 Pleistocene channels of New Castle County, Delaware, by N. Spoljaric. 1967.   713

RI. 11 An evaluation of the resistivity and seismic refraction techniques in the search of Pleistocene channels in Delaware, by W. E. Bonini. 1967.   714

RI. 12 Quantitative lithofacies analysis of Potomac Formation, Delaware, by N. Spoljaric. 1967.   715

RI. 13 The occurrence of saline ground water in Delaware aquifers, by K. D. Woodruff. 1969.   716

RI. 14 Delaware clay resources, by T. E. Pickett. 1970.   717

RI. 15 General ground-water quality in fresh-water aquifers of Delaware, by K. D. Woodruff. 1970.   718

RI. 16 Application of geophysics to highway design in the Piedmont of Delaware, by K. D. Woodruff. 1971.   719

RI. 17 Ground-water geology of the Delaware Atlantic seashore, by J. C. Miller. 1971.   720

RI. 18 Geology and ground water, University of Delaware, Newark, Delaware, by K. D. Woodruff, J. C. Miller, R. R. Jordan, N. Spoljaric, and T. E. Pickett. 1972.   721

RI. 19 Geology of the Fall Zone in Delaware, by N. Spoljaric. 1972.   722

RI. 20 Nitrate contamination of the water-table aquifer in Delaware, by J. C. Miller. 1972.   723

RI. 21 Guide to common Cretaceous

fossils of Delaware, by T. E. Pickett. 1972.   724

RI. 22 Hydrogeology of selected sites in the Greater Newark area, Delaware, by J. H. Talley. 1974.   725

RI. 23 Cretaceous and Tertiary section, deep test, well, Greenwood, Delaware, by J. H. Talley. 1975.   726

RI. 24 Relation of ground water to surface water in four small basins of the Delaware coastal plain, by R. H. Johnston. 1976.   727

RI. 25 Selected logging data and examples of geophysical logs for the coastal plain of Delaware, by K. D. Woodruff. 1976.   728

RI. 26 Hydraulic characteristics of the Piney Point Aquifer and overlying confining bed near Dover, Delaware, by P. P. Leahy. 1976.   729

RI. 27 Ground-water investigations in the Delaware Piedmont for the City of Newark, 1976, by W. F. Hahn. 1977.   730

RI. 28 Well and aquifer tests, Laird Tract Well Field, Newark, Delaware, by J. H. Talley and W. F. Hahn. 1978.   731

RI. 29 A digital model of the Piney Point Aquifer in Kent County, Delaware, by P. P. Leahy. 1979.   732

RI. 30 Ground-water levels in Delaware, July 1966-December 1977, by J. H. Talley. 1978.   733

RI. 31 Hydrocarbon resource potential of the Baltimore Canyon Trough, by R. N. Benson. 1979.   734

RI. 32 Removal of metals from laboratory solutions and landfill leachate by Greensand filters, by N. Spoljaric and W. A. Crawford. 1979.   735

RI. 33 Exploring, drilling, and producing petroleum offshore, by N. Spoljaric. 1979.   736

RI. 34 Long-term chemical-quality changes in selected Delaware streams, by S. W. McKenzie. 1979.   737

Bulletin. 1908-

B. 44 Osteology and paleontology of the Passerine birds of Reddick, Florida, by J. H. Hamon. 1964. 738

B. 45 The regional lithostratigraphic analysis of Paleocene and Eocene rocks of Florida, by C. S. Chen. 1965. 739

B. 46 Geology of Escambia and Santa Rosa Counties, western Florida Panhandle, by O. T. Marsh. 1966. 740

B. 47 Geology and ground-water resources of Leon County, Florida, by C. W. Hendry and C. R. Sproul. 1966. 741

B. 48 Geology of Jefferson County, Florida, by J. W. Yon. 1966. 742

B. 49 Geology of Dixie and Gilchrist Counties, Florida, by H. S. Puri, J. W. Yon, and W. R. Oglesby. 1967. 743

B. 50 Mineral resource study of Holmes, Walton, and Washington Counties, by J. W. Yon and C. W. Hendry. 1970. 744

B. 51 Geomorphology of the Florida Peninsula, by W. A. White. 1970. 745

B. 52 Ancient sea level stands in Florida, by E. C. Pirkle, W. H. Yoho, and C. W. Hendry. 1970. 746

B. 53 Corals from the Chippola and Jackson Bluff formations of Florida, by N. E. Weisbord. 1972. 747

B. 54 Suwannee limestone in Hernando and Pasco Coun

nando and Pasco Counties, Florida, by J. W. Yon and C. W. Hendry. 1972. Part 1. 748

Petrography of the Suwannee limestone, by A. E. Randazzo. 1972. Part 2. 749

B. 55 Igneous and metamorphic basement rocks of Florida, by C. Milton. 1972. 750

B. 56 New and little-known corals from the Tampa formation of Florida, by N. E. Weisbord. 1973. 751

B. 57 Geology of Bay County, Florida, by W. Schmidt and M. W. Clark. 1980. 752

Information Circular. 1948-

IC. 37 Ground-water records of Flagler, Putnam, and St. Johns Counties, Florida, by B. J. Bermes, G. W. Leve, and G. R. Tarver. 1963. 753

IC. 38 Records of wells and water resources data in Polk County, Florida, by H. G. Stewart. 1963. 754

IC. 39 Surface-water resources of St. Johns, Flagler, and Putnam Counties, by W. E. Kenner and J. W. Crooks. 1963. 755

IC. 40 Maps showing depths of selected lakes in Florida, by W. E. Kenner. 1964. 756

IC. 41 Interim report on the water resources of Orange County, Florida, by W. F. Lichtler, W. Anderson, and B. F. Joyner. 1964. 757

IC. 42 Summary of Florida petroleum production and exploration in 1962, by C. Babcock. 1964. 758

IC. 43 Water-resources data for Alachua, Bradford, Clay, and Union Counties, Florida, by W. E. Clark, R. H. Musgrove, C. G. Menke, and J. W. Cagle. 1964. 759

IC. 44 Water-resources records of Hillsborough County, Florida, by C. G. Menke, E. W. Meridith, and W. S. Wetterhall. 1964. 760

IC. 45 Summary of Florida petroleum production and exploration in 1963, by C. Babcock. 1965. 761

IC. 46 Ceramic clay investigations in Alachua, Clay, and Putnam Counties, Florida, by R. C. Hickman and H. P. Hamlin. 1965. 762

IC. 47 Control of lake levels in Orange County, Florida, by W. Anderson, W. F. Lichtler, and B. F. Joyner. 1965. 763

IC. 48 Water levels in artesian and non-artesian aquifers in Florida, 1961-62, by H. G. Healy. 1966. 764

IC. 49 Florida petroleum exploration, production, and prospects, 1964, by C. Babcock. 1966. 765

IC. 50 Water resource records of Escambia and Santa Rosa Counties, Florida, by R. H. Musgrove, J. T. Barraclough, and R. G. Grantham. 1966. 766

IC. 51 Groundwater in the Immokalee area, Collier County, Florida, by H. J. McCoy. 1967. 767

IC. 52 Water levels in artesian and non-artesian aquifers in Florida, 1963-64, by H. G. Healy. 1968. 768

IC. 53 Groundwater resource data of Charlotte, DeSoto, and Hardee Counties, Florida, by M. I. Kaufman and N. P. Dion. 1968. 769

IC. 54 Oil and gas activities in Florida--1965, by C. Babcock. 1968. 770

IC. 55 Oil and gas activities in Florida, 1966, by C. Babcock. 1968. 771

IC. 56 Test well exploration in the Myakka River Basin area, Florida, by H. Sutcliffe and B. F. Joyner. 1968. 772

IC. 57 Water resource records of the Econfina Creek Basin area, Florida, by R. H. Musgrove, J. B. Foster, and L. G. Toler. 1968. 773

IC. 58 Production and utilization of water in the metropolitan area of Jacksonville, Florida, by G. W. Leve and D. A. Goolsby. 1969. 774

IC. 59 Seepage tests in L-D1 Borrow Canal at Lake Okeechobee, Florida, by F. W. Meyer and J. E. Hull. 1969. 775

IC. 60 Geology of the Upper Cretaceous clastic section, northern Peninsular Florida, by C. Babcock. 1969. 776

IC. 61 Water levels in artesian and nonartesian aquifers of Florida, 1965-66, by H. G. Healy. 1969. 777

IC. 62 A test of flushing procedures to control salt-water intrusion at the W. P. Franklin Dam near Ft. Myers, Florida. 778

The magnitude and extent of saltwater contamination in the Caloosahatchee River between La Belle and Olga, Florida, by D. H. Boggess. 1970. 779

IC. 63 Oil and gas activities in Florida, by C. Babcock. 1967. 780

IC. 64 Report on geophysical and television explorations in City of Jacksonville water wells, by G. W. Leve. 1970. 781

IC. 65 Oil and gas activities in Florida, by C. Babcock. 1970. 782

IC. 66 Directory of mineral producers in Florida, by E. L. Maxwell. 1968. 783

IC. 67 Selected water resources of

Okaloosa County, Florida, by J. B. Foster and C. A. Pascale. 1971.
784

IC. 68 Water levels in artesian and non-artesian aquifers of Florida, 1967-68, by H. G. Healy. 1970.
785

IC. 69 Selected flow characteristics of Florida streams and canals, by R. C. Heath and E. T. Wimberly. 1971.
786

IC. 70 The beneficial use of zones of high transmissivities in the Florida subsurface for water storage and waste disposal, by R. O. Vernon. 1970.
787

IC. 71 Oil and gas activities in Florida, 1969, by C. Babcock. 1971.
788

IC. 72 Land use conflicts and phosphate mining in Florida, by J. W. Sweeney. 1972.
789

IC. 73 Water levels in artesian and nonartesian aquifers of Florida, 1969-70, by H. G. Healy. 1972.
790

IC. 74 Construction of waste-injection monitor wells near Pensacola, Florida, by J. B. Foster and D. A. Goolsby. 1972.
791

IC. 75 Saline-water intrusion from deep artesian sources in the McGregor Isles area of Lee County, Florida, by C. R. Sproul, D. H. Boggess, and H. J. Woodard. 1972.
792

IC. 76 Hydrologic aspects of freshening Upper Old Tampa Bay, Florida, by J. A. Mann. 1972.
793

IC. 77 Ground water in the Hallandale area, Florida, by H. W. Bearden. 1972.
794

IC. 78 Records of hydrologic data, Walton County, Florida, by C. A. Pascale, C. E. Essig, and R. R. Herring. 1972.
795

IC. 79 Flood of September 20-23, 1969, in the Gadsden County area,

Florida, by W. C. Bridges and D. R. Davis. 1972.
796

IC. 80 Oil and gas activities in Florida, 1970, by C. Babcock. 1972.
797

IC. 81 Public water supplies of selected municipalities in Florida, 1970, by H. G. Healy. 1972.
798

IC. 82 Flow and chemical characteristics of the St. Johns River at Jacksonville, Florida, by W. Anderson and D. A. Goolsby. 1973.
799

IC. 83 Estimated water use in Florida, 1970, by R. W. Pride. 1973.
800

IC. 84 The mineral industry of Florida, 1971, by W. F. Stowasser. 1973.
801

IC. 85 Water levels in artesian and nonartesian aquifers of Florida, 1971-72, by H. G. Healy. 1974.
802

IC. 86 Hydrogeologic characteristics of the surficial aquifer in northwest Hillsboro County, Florida, by W. C. Sinclair. 1974.
803

IC. 87 List of publications. Rev. 1979.
804

IC. 88 The mineral industry of Florida, 1972, by W. F. Stowasser and W. R. Oglesby. 1974.
805

IC. 89 The mineral industry of Florida, 1973, by W. F. Stowasser and C. W. Hendry. 1976.
806

IC. 90 The mineral industry of Florida, 1974, by J. W. Sweeney and C. W. Hendry. 1977.
807

IC. 91 The mineral industry of Florida, 1975, by J. W. Sweeney and C. W. Hendry. 1979.
808

IC. 92 The mineral industry of Florida, 1976, by J. W. Sweeney and C. W. Hendry. 1979.
809

Leaflet. 195?-

L. 1 Your water resources. Rev. 1953.
810

L. 2 Water for thirsty industry--
it's your problem. n. d.   811

L. 3 The Pensacola area's water.
1965.   812

L. 4 Chronological bibliography of
principal published ground water
reports in Florida, prepared by the
U. S. Geological Survey or the Florida Geological Survey, 1908-1963.
1965.   813

L. 5 Water control versus sea-
water intrusion in Broward County,
Florida. 1965.   814

L. 6 Jacksonville's water. 1965.
                                  815

L. 7 Salt intrusion can be controlled. 1966.   816

L. 8 Water in Orange County,
Florida. 1968.   817

L. 9 Large springs of Florida's
sun coast. n. d.   818

Report of Investigations. 1933-

RI. 32 Geology and ground-water
resources of Flagler, Putnam,
and St. Johns Counties, Florida,
by B. J. Bermes, G. W. Leve, and
G. R. Tarver. 1963.   819

RI. 33 Hydrology of Brooklyn Lake
near Keystone Heights, Florida,
by W. E. Clark, R. H. Musgrove,
C. G. Menke, and J. W. Cagle.
1963.   820

RI. 34 Hydrologic reconnaissance
of Pasco and southern Hernando
Counties, Florida, by W. S. Wetterhall. 1964.   821

RI. 35 Water resources of Alachua,
Bradford, Clay, and Union Counties,
Florida, by W. E. Clark, R. H. Mus-
grove, C. G. Menke, and J. W.
Cagle. 1964.   822

RI. 36 Hydrology of the Biscayne
aquifer in the Pompano Beach area,
Broward County, Florida, by G. R.
Tarver. 1964.   823

RI. 37 Geology and ground-water resources of Glades and Hendry Counties, Florida, by H. Klein, M. C.
Schroeder, and W. F. Lichtler. 1964.
                                  824

RI. 38 Possibility of salt-water leakage from proposed intracoastal waterway near Venice, Florida, well field,
by W. E. Clark. 1964.   825

RI. 39 Reconnaissance of springs and
sinks in west-central Florida, by W. S.
Wetterhall. 1965.   826

RI. 40 Water resources of Escambia
and Santa Rosa Counties, Florida, by
R. H. Musgrove, J. T. Barraclough,
and R. G. Grantham. 1965.   827

RI. 41 Water resources of the Econfina Creek Basin, by R. H. Musgrove,
J. B. Foster, and L. G. Toler. 1965.
                                  828

RI. 42 Hydrology of the Green Swamp
area in central Florida, by R. W.
Pride, F. W. Meyer, and R. N. Cherry.
1966.   829

RI. 43 Ground water in Duval and
Nassau Counties, Florida, by G. W.
Leve. 1966.   830

RI. 44 Ground-water study of Polk
County, by H. G. Stewart. 1966.
                                  831

RI. 45 Salt-water study of the Miami
River and its tributaries, Dade County,
Florida, by S. D. Leach and R. G.
Grantham. 1966.   832

RI. 46 Fluoride in water in the Alafia
and Peace River Basins, Florida,
by L. G. Toler. 1967.   833

RI. 47 Hydrologic effects of Area B
flood control plan on urbanization of
Dade County, Florida, by F. A. Kohout and J. H. Hartwell. 1967.   834

RI. 48 Analysis of the water level
fluctuations of Lake Jackson near
Tallahassee, Florida, by G. H. Hughes.
1967.   835

RI. 49 Hydrologic effects of ground-
water pumpage in the Peace and Ala-

fia River Basins, Florida, by M. I. Koufman. 1967. 836

RI. 50 Water resources of Orange County, Florida, by W. F. Lichtler, W. Anderson, and B. F. Joyner. 1968. 837

RI. 51 Chemical quality of waters of Broward County, Florida, by R. G. Grantham and C. B. Sherwood. 1968. 838

RI. 52 Reconnaissance of the ground-water resources of Baker County, Florida, by G. W. Leve. 1968. 839

RI. 53 Low streamflow in the Myakka River Basin area in Florida, by H. N. Flippo and B. F. Joyner. 1968. 840

RI. 54 Water resources of northeast Florida, by L. J. Snell and W. Anderson. 1970. 841

RI. 55 Ground water resources of the Lower Hillsboro Canal area, southeastern Florida, by H. J. McCoy and J. Hardee. 1970. 842

RI. 56 General hydrology of the Middle Gulf area, Florida, by R. N. Cherry, J. W. Stewart, and J. A. Mann. 1970. 843

RI. 57 Evaluation of quantity and quality of the water resources of Volusia County, Florida, by D. D. Knochenmus and M. E. Beard. 1971. 844

RI. 58 Seepage beneath Hoover Dike, southern shore of Lake Okeechobee, Florida, by F. W. Meyer. 1971. 845

RI. 59 The shallow-aquifer system in Duval County, Florida, by R. W. Fairchild. 1972. 846

RI. 60 Hydrologic effects of water control and management of southeastern Florida, by S. D. Leach, H. Klein, and E. R. Hampton. 1972. 847

RI. 61 Appraisal of water resources

in the east central Florida region, by W. F. Lichtler. 1972. 848

RI. 62 Water available in canals and shallow sediments in St. Lucie County, Florida, by H. W. Bearden. 1972. 849

RI. 63 Hydrology of Western Collier County, Florida, by J. McCoy. 1972. 850

RI. 64 Hydrologic conditions in the Lakeland Ridge area of Polk County, Florida, by A. F. Robertson. 1973. 851

RI. 65 Water resources of Broward County, Florida, by C. B. Sherwood. 1973. 852

RI. 66 Salinity studies in East Glades agricultural area, southeastern Dade County, Florida, by J. E. Hull and F. W. Meyer. 1973. 853

RI. 67 Appraisal of the water resources of eastern Palm Beach County, Florida, by L. F. Land, H. G. Rodis, and J. J. Schneider. 1973. 854

RI. 68 Hydrogeologic aspect of a proposed sanitary landfill near Old Tampa Bay, Florida, by R. N. Cherry and D. P. Brown. 1973. 855

RI. 69 The shallow fresh-water system of Sanibel Island, Lee County, Florida, with emphasis on the sources and effects of saline water, by D. H. Boggess. 1974. 856

RI. 70 Indicators of organic contamination in Plantation Canal, Broward County, Florida, 1971-72, by T. N. Russo. 1974. 857

RI. 71 Chemical and biological conditions of Lake Okeechobee, Florida, 1969-72, by B. F. Joyner. 1974. 858

RI. 72 Hydrologic concepts of artificially recharging the Floridan aquifer in eastern Orange County, Florida-- a feasibility study, by D. D. Knochenmus. 1974. 859

RI. 73 Water balance of Lake Kerr--

44 / FLORIDA

a deductive study of a landlocked
lake in north-central Florida, by
G. H. Hughes. 1974.   860

RI. 74 Hydrologic consequences of
using ground-water to maintain lake
levels affected by water wells near
Tampa, Florida, by J. W. Stewart
and G. H. Hughes. 1974.   861

RI. 75 Evaluation of hydraulic characteristics of a deep artesian aquifer from natural water-level fluctuations, Miami, Florida, by F. W.
Meyer. 1974.   862

RI. 76 Water resources of Walton
County, Florida, by C. A. Pascale.
1974.   863

RI. 77 Ground-water resources of
the Hollywood area, Florida, by
H. W. Bearden. 1974.   864

RI. 78 Appraisal of the water resources of Charlotte County, Florida, by H. Sutcliffe. 1975. 865

RI. 79 Summary of hydrologic conditions and effects of Walt Disney
World development in the Reedy
Creek Improvement District, 1966-
73, by A. L. Putnam. 1975.   866

RI. 80 Water resources of Indian
River County, Florida, by L. J.
Crain, G. H. Hughes, and L. J.
Snell. 1975.   867

RI. 81 Hydrology of three sinkhole
basins in southwestern Seminole
County, Florida, by W. Anderson
and G. H. Hughes. 1975.   868

RI. 82 Hydrologic effects of the
Tampa Bypass Canal System, by
L. H. Motz. 1975.   869

RI. 83 Ground water resources of
DeSoto and Hardee Counties, Florida, by W. E. Wilson. 1977. 870

RI. 84 The Highland heavy-mineral
sand deposit on Trail Ridge in northern peninsular Florida, by E. C.
Pirkle, W. A. Pirkle, and W. H.
Yoho. 1977.   871

RI. 85 The geology of the western

part of Alachua County, Florida, by
K. E. Williams, D. Nichol, and A. F.
Randazzo. 1977.   872

RI. 86 Regional structure and stratigraphy of the limestone outcrop belt
in the Florida Panhandle, by W.
Schmidt and C. Coe. 1978.   873

RI. 87 Not yet published.   874

RI. 88 The limestone, dolomite, and
coquina resources of Florida, by W.
Schmidt and others. 1979.   875

RI. 89 Not yet published.   876

RI. 90 The sand and gravel resources
of Florida, by T. M. Scott and others.
1980.   877

Special Publication. 1956-

SP. 11 Index to water resources data
collection stations in Florida, 1961,
by U. S. Geological Survey and Florida
Geological Survey. 1969.   878

SP. 12 Vertebrate fossil localities in
Florida, by S. J. Olsen. 1965. 879

SP. 13 The water mapping, monitoring, and research program in Florida,
by C. S. Conover, K. A. MacKichan,
and R. W. Pride. 1965.   880

SP. 14 Adventures in geology at Jackson Bluff, by J. W. Yon. 1965. 881

SP. 15 The Dollar Bay formation of
Lower Cretaceous Florida, by G. O.
Winston. 1971.   882

SP. 16 Environmental geology and hydrology, Tallahassee area, Florida,
by the Florida Bureau of Geology.
1972.   883

SP. 17 Proceedings of seventh forum
on geology of industrial minerals, ed.
by H. S. Puri. 1972.   884

SP. 18 Availability of potential utilization of byproduct gypsum in Florida
phosphate operations, by J. W. Sweeney
and B. J. Timmons. 1973.   885

SP. 19 Environmental geology and

hydrology, Tampa area, Florida, by A. P. Wright. 1974. 886

SP. 20 Geologic framework of the high transmissivity zones in south Florida, by H. S. Puri and G. O. Winston. 1974. 887

SP. 21 The geothermal nature of the Floridan Plateau, ed. by D. L. Smith and G. M. Griffin. 1977. 888

SP. 22 Florida: the new uranium producer, by J. W. Sweeney and S. R. Windham. 1979. 889

SP. 23 Guidelines for authors with comments for editorial reviewers, by E. Lane. 1980. 890

SP. 24 Catalogue of invertebrate fossil types at the Bureau of Geology, by C. Shaak. 1980. 891

Bulletin. 1894-

B. 73 Effect of a severe drought, 1954, on stream flow in Georgia, by M. T. Thomson and R. F. Carter. 1963. 892

B. 74 Logs of oil tests in the coastal plain of Georgia, by P. L. Applin and E. R. Applin. 1964. 893

B. 75 The Murphy Syncline in the Tate Quadrangle, Georgia, by W. M. Fairley. 1965. 894

B. 76 Subsurface "basement" rocks of Georgia, by C. Milton and V. J. Hurst. 1965. 895

B. 77 The geology of the Brevard lineament near Atlanta, Georgia, by M. W. Higgins. 1966. 896

B. 78 Specifications in ground waters related to geologic formations in the Broad Quadrangle, Georgia, by C. A. Salotti and J. A. Fouts. 1967. 897

B. 79 Annotated bibliography of Georgia geology through 1959, by H. R. Cramer, A. T. Allen, and J. G. Lester. 1967. 898

B. 80 Precambrian-Paleozoic Appalachian problems. 1969. 899

B. 81 Stratigraphy, paleontology, and economic geology of portions of Perry and Cochran Quadrangles, Georgia, by S. M. Pickering. 1970. 900

B. 82 Stratigraphy and economic geology of the eastern Chatham County phosphate deposit, by J. W. Furlow. 1969. 901

B. 83 The geology of Rabun and Habersham Counties, Georgia, by R. D. Hatcher. 1971. 902

B. 84 Annotated bibliography of Georgia geology, 1960-1964, by H. R. Cramer. 2nd ed. 1972. 903

B. 85 Ultramafic and related rocks in the vicinity of Lake Chatuge, Towns County, Georgia, and Clay County, North Carolina, by M. E. Hartley. 1972. 904

B. 86 The Georgia gravity base net, by R. E. Ziegler and L. M. Dorman. 1976. 905

B. 87 Symposium on the petroleum geology of the Georgia coastal plain, ed. by L. P. Stafford. 1974. 906

B. 88 Trace fossils of the Upper Cretaceous-Lower Tertiary (formerly Tuscaloosa Formation) and basal Jackson Group, east-central Georgia, by C. H. Schroder. 907

B. 89 Abstracts of theses on Georgia geology through 1974, comp. and ed. by F. Moye. 1976. 908

B. 90 Annotated bibliography of Georgia geology, 1965-1970, by H. R. Cramer. 1976. 909

B. 91 Availability of water supplies in northwest Georgia, by C. W. Cressler, M. A. Franklin, and W. G. Hester. 1976. 910

B. 92 Minerals of Georgia: their properties and occurrences, by R. B. Cook. 1978. 911

B. 93 Short contributions to the geology of Georgia, ed. by P. A. Platt. 1978. 912

Circular. 19--

C. 1 List of publications of the Georgia Geologic Survey, comp. and ed. by E. Morrow. 17th ed. 1980.   913

C. 2 Directory of Georgia mineral producers, comp. by W. C. Rozov. 18th ed. 1980.   914

C. 3 The mineral industry of Georgia, by J. D. Cooper and S. M. Pickering. 1977.   915

Educational Series. 1976-

ES. 1 Georgia: a view from space; an interpretation of Landsat-1 imagery, by W. Z. Clark, A. C. Zisa, and R. C. Jones. 1976.   916

Geologic Guide. 1977-

GG. 1 Geologic guide to Sweetwater Creek State Park, by C. E. Abrams and K. I. McConnell. 1977.   917

GG. 2 Geologic guide to Panola Mountain State Park--rock outcrop trail, by R. L. Atkins and M. M. Griffin. 1977.   918

GG. 3 Geologic guide to Panola Mountain State Park--watershed trail, by R. L. Atkins and M. M. Griffin. 1977.   919

GG. 4 Geologic guide to Stone Mountain Park, by R. L. Atkins and L. G. Joyce. 1980.   920

GG. 5 An introduction to caves and caving in Georgia, by B. F. Beck. 1980.   921

Geologic Report. 1971-

GR. 1 Ground water designs and patterns, by J. W. Stewart. 1971.   922

GR. 2 Oil seeps in Georgia: a reinvestigation, by R. H. Sams. 1971.   923

Hydrologic Atlas. 1972-

HA. 1 Water from the principal artesian aquifer in coastal Georgia, by R. E. Krause and D. O. Gregg. 1972.   924

HA. 2 The geohydrology of Ben Hill, Irwin, Tift, Turner, and Worth Counties, Georgia, by T. Watson. 1976.   925

Hydrologic Report. 1974-

HR. 1 Hydrology and chloride contamination of the principal artesian aquifer in Glynn County, Georgia, by R. L. Wait and D. O. Gregg. 1974.   926

HR. 2 Use of water in Georgia, 1970, with projections to 1990, by R. F. Carter and A. M. F. Johnson. 1974.   927

Information Circular. 1933-

IC. 22 Surface-water resources of the Yellow River Basin in Gwinnett County, Georgia, by R. S. Carter and W. B. Gannon. 1962.   928

IC. 23 Interim report on test drilling and water sampling in the Brunswick area, Glynn County, Georgia, by R. L. Wait. 1962.   929

IC. 24 Geology and ground-water resources of Mitchell County, Georgia, by V. Owen. 1963.   930

IC. 25 Subsurface geology of the Georgia coastal plain, by S. M. Herrick and R. C. Vorhis. 1963.   931

IC. 26 Geology and ground-water resources of Dade County, Georgia, by M. G. Croft. 1964.   932

IC. 27 Geology and ground-water resources of the Paleozoic rock area, Chattooga County, Georgia, by C. W. Cressler. 1964.   933

IC. 28 Geology and ground-water resources of Catoosa County, Georgia, by C. W. Cressler. 1963.   934

IC. 29 Geology and ground-water resources of Walker County, Georgia, by C. W. Cressler. 1964.
935

IC. 30 Geology and ground-water resources of crystalline rocks, Dawson County, Georgia, by C. W. Sever. 1964.
936

IC. 31 A subsurface study of Pleistocene deposits in coastal Georgia, by S. M. Herrick. 1965.
937

IC. 32 Ground-water resources of Bainbridge, Georgia, by C. W. Sever. 1965.
938

IC. 33 Ground-water resources and geology of Rockdale County, Georgia, by M. J. McCollum. 1966.
939

IC. 34 Reconnaissance of the ground water and geology of Thomas County, Georgia, by C. W. Sever. 1966.
940

IC. 35 Geology and mineral resources of the Bethesda Church area, Greene County, Georgia, by J. H. Medlin and V. J. Hurst. 1967.
941

IC. 36 Hydraulics of aquifers at Alapaha, Coolidge, Fitzgerald, Montezuma, and Thomasville, Georgia, by C. W. Sever. 1969.
942

IC. 37 A geochemical and geophysical survey of the Gladesville Norite, by R. H. Carpenter and T. C. Hughes. 1970.
943

IC. 38 Petroleum exploration in Georgia, comp. by W. E. Marsalis. 1970.
944

IC. 39 Geology and ground-water resources of Floyd and Polk Counties, by C. W. Cressler. 1970.
945

IC. 40 Palynology of core samples of Paleozoic sediments from beneath the coastal plain of Early County, Georgia, by R. E. McLaughlin. 1970.
946

IC. 41 Chattahoochee anticline, Apalachicola embayment, Gulf Trough and related structural features, southwestern Georgia: fact or fiction, by S. H. Patterson and S. M. Herrick. 1971.
947

IC. 42 A gravity survey of the south-central Georgia Piedmont, by R. H. Carpenter and P. Prather. 1971.
948/9

IC. 43 Copper, lead, and zinc concentration in stream sediment, Metasville Quadrangle, Wilkes and Lincoln Counties, Georgia, by R. H. Carpenter. 1971.
950

IC. 44 Index to geologic mapping in Georgia, by D. E. Lawton and M. G. Pierce. 1972.
951

IC. 45 Effects of ground-water pumping in parts of Liberty and McIntosh Counties, Georgia, 1966-70, by R. E. Krause. 1972.
952

IC. 46 Upper Eocene and Oligocene pectinidae of Georgia and their stratigraphic significance, by L. N. Glawe. 1974.
953

IC. 47 Geology and ground-water resources of Gordon, Whitfield, and Murray Counties, Georgia, by C. W. Cressler. 1974.
954

IC. 48 Quality and availability of ground water in Georgia, by J. L. Sonderegger, L. D. Pollard, and C. W. Cressler. 1978.
955

IC. 49 Twelfth forum on the geology of industrial minerals (1976), comp. by P. A. Platt. 1978.
956

IC. 50 Geohydrology of Bartow, Cherokee, and Forsyth Counties, by C. W. Cressler, H. E. Blanchard, and W. G. Hester. 1979.
957

IC. 51 Petroleum exploration wells in Georgia, by D. E. Swanson and A. Gernazian. 1979.
958

IC. 52 Origin and correlation of the Pumpkinvine Creek Formation: a new unit in the Piedmont of northern Georgia, by K. I. McConnell. 1980. 959

Miscellaneous Publication. 1978-

MP. 1 Natural environment of
Georgia, by C. H. Wharton. 1978.
960

MP. 2 How to mine and prospect
for placer gold, by J. M. West.
1971. 961

MP. 3 Index to core holes--Georgia
coastal plain, 1965-1972, by the
Georgia Department of Natural Resources. 1973. 962

Project Report. 1966-

PR. 1 Introduction, by J. E. Husted,
A. S. Furcron, and F. Bellinger.
1966. 963

PR. 2 Heavy minerals, sand and
clay in Echols County. 1966. 964

PR. 3 Phosphate and sand in Brooks,
Camden, Clinch, Effingham, Lanier,
Lowndes, Thomas, and Ware Coun-
ties. 1966. 965

PR. 4 Phosphorite exploration in
portions of Lowndes, Echols,
Clinch, and Charlton Counties,
Georgia, by N. K. Olson. 1966.
966

PR. 5 Phosphate in Atkinson, Ben
Hill, Berrien, Coffee, Cook, and
Irwin Counties. 1967. 967

PR. 6 Sands and heavy mineral,
and phosphate-sand in Effingham
County. 1967. 968

PR. 7 Phosphate deposits of south-
central Georgia and north-central
peninsular Florida, by C. W. Sever,
J. B. Cathcart, and S. H. Patterson.
1967. 969

PR. 8 Heavy mineral bearing sand
of the coastal region of Georgia,
by J. W. Smith, S. M. Pickering,
and J. R. Landrum. 1967. 970

PR. 9 Phosphate in Chatham,
Echols, Lanier, Lowndes, Jeff Davis, Montgomery, Liberty, Long,
and Wayne Counties. 1967. 971

PR. 10 Phosphate in the Savannah
area. 1968. 972

PR. 11 Phosphorite. 1969. 973

Special Publication. 1963-

SP. 1 Emergency water supplies for
the Atlanta area in a national emergency, by J. W. Stewart and S. M. Herrick. 1963. 974

SP. 2 Illustrated fossils of the Georgia
coastal plain, by H. G. Richards.
1969. 975

Bulletin. 1935-

B. 1 Geology and ground-water resources of the Island of Oahu, Hawaii, by H. T. Stearns and K. N. Vaksvik. 1935. 976

B. 2 Geologic map and guide of the Island of Oahu, Hawaii, by H. T. Stearns. 1939. 977

B. 3 Annotated bibliography and index of geology and water supply of the Island of Oahu, Hawaii, by N. D. Stearns. 1935. 978

B. 4 Records of the drilled wells on the Island of Oahu, Hawaii, by H. T. Stearns and K. N. Vaksvik. 1938. 979

B. 5 Supplement to the geology and ground-water resources of the Island of Oahu, Hawaii, by H. T. Stearns, J. H. Swartz, and G. A. Macdonald. 1940. 980

B. 6 Geology and ground-water resources of the Islands of Lanai and Kahoolawe, Hawaii, by H. T. Stearns, J. H. Swartz, and G. A. Macdonald. 1940. 981

B. 7 Geology and ground-water resources of the Island of Maui, Hawaii, by H. T. Stearns and G. A. Macdonald. 1942. 982

B. 8 Geology of the Hawaiian Islands, by H. T. Stearns. 1946. 983

B. 9 Geology and ground-water resources of the Island of Hawaii, by H. T. Stearns and G. A. Macdonald. 1946. 984

B. 10 Bibliography of the geology and ground-water resources of the Island of Hawaii, annotated and indexed, by G. A. Macdonald. 1947. 985

B. 11 Geology and ground-water resources of the Island of Molokai, Hawaii, by H. T. Stearns and G. A. Macdonald. 1947. 986

B. 12 Geology and ground-water resources of the Island of Niihau, Hawaii, by H. T. Stearns and G. A. Macdonald. 1947. 987

B. 13 Geology and ground-water resources of the Island of Kauai, Hawaii, by G. A. Macdonald, D. A. Davis, and D. C. Cox. 1960. 988

B. 14 Water resources in Hawaii. 1959. 989

B. 15 Flood control and flood water conservation in Hawaii. 2 vols. 1963. 990

B. 16 Water resources development, Molokai. 1966. 991

Circular. 1957-

C. 1 Rights in ground water, by the Legislative Reference Bureau. 1957. 992

C. 2 Ground water legislation, by W. A. Hutchins. 1958. 993

C. 3 Summary of drilling logs and pumping test for Kawaihae Exploratory Well No. 1, South Kohala, Hawaii. 1961. 994

C. 4 Summary of drilling logs and pumping test for Exploratory Well No. 1, Waikapu, Maui, Hawaii. 1961. 995

C. 5 Summary of drilling logs and

pumping test for Exploratory Well No. 2, Hanapepe, Kauai, Hawaii. 1961. 996

C. 6 Summary of drilling logs and pumping test for Kawaihae Exploratory Well No. 2, South Kohala, Hawaii. 1961. 997

C. 7 Summary of drilling logs and pumping test for Exploratory Well No. 4 (USGS Well No. 73), Maka Ridge, Hanalei, Kauai. 1961. 998

C. 8 Summary of drilling logs and pumping test for Exploratory Well No. 3 (USGS Well No. 11), Lihue, Kauai, Hawaii. 1962. 999

C. 9 Summary of drilling logs and pumping test for Lahaina Exploratory Well (USGS Well No. 291), Lahaina, Maui, Hawaii. 1962. 1000

C. 10 Preliminary summary of findings in water resources studies of windward Oahu, Hawaii, by K. J. Takasaki, G. Yamanaga, and R. R. Lubke. 1962. 1001

C. 11 Summary of drilling logs and pumping test for Kamalo Exploratory Well (USGS Well No. 31), Kamalo, Molokai, Hawaii. 1962. 1002

C. 12 Summary of drilling logs and pumping test for Malama-Ki Well 9-9, Malama-Ki, Puna, Hawaii. 1962. 1003

C. 13 Interim report on Hamakua-Kohala Water study. 1963. 1004

C. 14 Preliminary report on the water resources of Kohala Mountain and Mauna Kea, Hawaii, by D. A. Davis and G. Yamanaga. 1963. 1005

C. 15 Investigations of ferruginous bauxite and plastic clay deposits on Kauai and reconnaissance of ferruginous bauxite deposits on Maui, Hawaii, by S. H. Patterson. 1963. 1006

C. 16 Preliminary report on the ground water resources of the Waianae area, by C. P. Zones. 1963. 1007

C. 17 Preliminary report on legal problems to be studied in connection with the Hamakua-Kohala water study, Island of Hawaii. 1963. 1008

C. 18 A brief description of the Upper Hamakua Ditch, Waimea, Island of Hawaii. 1963. 1009

C. 19 Reconnaissance geologic report on proposed Penstock Alinement and Afterbay Reservoir Site, Kokee Project, Hawaii. 1963. 1010

C. 20 Aspects of ground-water storage and depletion along the Molokai Irrigation Tunnel, Molokai, Hawaii, by G. T. Hirashima. 1963. 1011

C. 21 Influence of water-development tunnels on streamflow-groundwater relations in Haiku-Kahaluu area, Oahu, Hawaii, by G. T. Hirashima. 1963. 1012

C. 22 Summary of drilling logs and pumping test for Kamiloloa Well 19, Kamiloloa, Molokai. 1963. 1013

C. 23 Summary of drilling logs and pumping test for Keei Well 12-8, Keei, South Kona, Hawaii. 1963. 1014

C. 24 Summary of drilling logs and pumping test for Kawaihae Well 16, Kawaihae, South Kohala, Hawaii. 1963. 1015

C. 25 Summary of drilling logs and pumping test for Kalaheo Well 24, Kalaheo, Kauai, Hawaii. 1963. 1016

C. 26 Summary of drilling logs and pumping test for Pulama Well 9-10, Pulama, Puna, Hawaii. 1964. 1017

C. 27 Water resources of the Kau District, Hawaii, by D. A. Davis and G. Yamanaga. 1966. 1018

C. 28 Summary of drilling logs and pumping test for Makena Well 68, Makena, Maui, Hawaii. 1964. 1019

C. 29 Summary of drilling logs for Pohakuloa Test Hole 20, Pohakuloa, Hamakua, Hawaii. 1965. 1020

C. 30 A domestic water plan for Ninole, North Hilo District, Hawaii. 1971. 1021

C. 31 Post flood report, storm of January 27-28, 1971, Islands of Oahu, Maui and Hawaii. 1971. 1022

C. 32 Flood prevention by soil conservation service. 1965. 1023

C. 33 Reconnaissance study of sediment transported by streams on the Island of Oahu, Hawaii. 1971. 1024

C. 34 Summary of drilling logs and pumping test for Maui High School Well 35, Hamakuapoko, Maui, Hawaii. 1965. 1025

C. 35 Summary of drilling logs and pumping test for Alaeloa Well 318, Alaeloa, Maui, Hawaii. 1965. 1026

C. 36 Summary of drilling logs for Waimea Test Hole T-19, Waimea, South Kohala, Hawaii. 1965. 1027

C. 37 Summary of drilling logs and pumping test for Lahaina Well 292, Lahaina, Maui, Hawaii. 1965. 1028

C. 38 Application for grant under water resources planning act of 1965. 1966. 1029

C. 39 Preliminary report on the water resources of the Kahuku area, Oahu, Hawaii, by K. J. Takasaki, S. Valenciano, and A. M. Ho. 1966. 1030

C. 40 Drainage criteria meeting. 1966. 1031

C. 41 Summary of drilling log and pumping test for Haena Well 66, Haena, Hawaii. 1967. 1032

C. 42 Summary of drilling log and pumping test for Hanapepe Well 26-1, Hanapepe, Kauai, Hawaii. 1967. 1033

C. 43 Summary of drilling log and pumping test for Maalaea Well 272, Maalaea, Maui, Hawaii. 1967. 1034

C. 44 Summary of drilling log and pumping test for Waimea Well 26, Waimea, Kauai, Hawaii. 1967. 1035

C. 45 Preliminary report on the water resources of the Hilo-Puna area, Hawaii, by D. A. Davis and G. Yamanaga. 1968. 1036

C. 46 Preliminary report on the water resources of the Kona area, Hawaii, by D. A. Davis and G. Yamanaga. 1968. 1037

C. 47 Post flood report, storm of 17 and 18 December 1967, Islands of Kauai and Oahu. 1968. 1038

C. 48 Summary of drilling log and pumping test of Kalaoa Well 12-11, near Keahole, North Kona, Hawaii. 1968. 1039

C. 49 Post flood report, storm of 5 January 1968, Island of Oahu. 1968. 1040

C. 50 Post flood report, storm of 15-16 April 1968, Island of Maui. 1968. 1041

C. 51 Preliminary report on the water resources of the Lahaina District, Maui, Hawaii, by G. Yamanaga and C. J. Huxel. 1969. 1042

C. 52 Post flood report, storm of 3-4 October 1968, Island of Hawaii. 1969. 1043

C. 53 Interim report on Kehena Ditch Water Study, Island of Hawaii. 1969. 1044

C. 54 Preliminary report on geohydrologic exploration for deep well disposal of effluent, Waimanalo Sewage Treatment Plant, Waimanalo, Oahu, by D. Lum. 1969. 1045

C. 55 Post flood report, storms of

November 28 to December 1, 1968, and January 5, 1969, Island of Kauai, Hawaii. 1969.   1046

C. 56 Post flood report, storm of 1 February 1969, Island of Oahu, Hawaii. 1969.   1047

C. 57 Post flood report, storm of 1-2 and 4 December 1969, Hawaiian Islands. 1970.   1048

C. 58 Preliminary report on the water resources of southeast Maui, by K. J. Takasaki. 1971.   1049

C. 59 Hydrologic data networks, Island of Molokai: rainfall stations, stream gage stations, and ground water. 1970.   1050

C. 60 Preliminary report on the water resources of northeast Maui, by K. J. Takasaki and G. Yamanaga. 1970.   1051

C. 61 Preliminary report on the water resources of the Wailuku area, Maui, by G. Yamanaga and C. J. Huxel. 1970.   1052

C. 62 Preliminary report on the water resources of central Maui, Hawaii, by K. J. Takasaki. 1972.   1053

C. 63 Summary of drilling log and pumping test for Kiholo Well 4953-01, North Kona, Hawaii, by D. Lum. 1973.   1054

C. 64 Summary of drilling log and pumping test for Piihonua Well 4306-01, Hilo, Hawaii, by D. Lum. 1973.   1055

C. 65 Summary of drilling log and pumping test for Kepaniwai Well 5332-05 and observation Hole 5332-04, Maui, by D. Lum. 1974.   1056

C. 66 Summary of drilling log and pumping test for Waikapu Well 5130-02, Waikapu, west Maui, Hawaii, by D. Lum. 1974.   1057

C. 67 Geologic description of quar-ries on Oahu, Hawaii, by H. T. Stearns. 1974.   1058

C. 68 Post flood report, storm of April 19, 1974, Islands of Kauai, Oahu, and Maui. 1974.   1059

C. 69 Hydrologic data, summaries of 1971 and 1972 rainfall data, compiled from State Network Stations. 1976.   1060

Report. 1954-

R. 1 Methods for development and delivery of water for irrigation of Hawaiian Homes Commission lands at Hoolehua, Island of Molokai, by H. A. R. Austin and H. T. Stearns. 1954.   1061

R. 2 General reference guide for irrigation project planning in Hawaii. 1957.   1062

R. 3 Kokee Irrigation Project progress report, Island of Kauai. 1957. 1063

R. 4 Kula Irrigation Project feasibility report, Island of Maui. 1957.   1064

R. 5 Waianae Irrigation Project feasibility report, Island of Oahu. 1957.   1065

R. 6 Growth of population in the Kona District, Island of Hawaii, by the Stanford Research Institute. 1958.   1066

R. 7 Report on estimated costs of operating the Waimanalo Irrigation System to serve the existing farm lots subdivision and the University of Hawaii Experimental Farm, by L. H. Herschler. 1958.   1067

R. 8 Kona rainfall, by W. J. Taliaferro. 1958.   1068

R. 9 Kona rainfall (supplement), by W. J. Taliaferro. 1959.   1069

R. 10 Kona Irrigation Project feasibility study, Island of Hawaii. 1959.   1070

R. 11 Progress report on Hawaii Water Authority's plans for the Kona Water System. 1959.   1071

R. 12 Rainfall of the Hawaiian Islands, by W. J. Taliaferro. 1959.   1072

R. 13 Rainfall, tanks, catchment, and family use of water, by C. K. Wentworth. 1959.   1073

R. 14 A report on a water supply for the proposed Ahuamoa Camp, Island of Hawaii. 1960.   1074

R. 15 An inventory of basic water resources data: Molokai. 1961.   1075

R. 16 Molokai Project, loan application report. 1961.   1076

R. 17 Pan evaporation data, State of Hawaii. 1961.   1077

R. 18 A domestic water plan for Kaunakakai-Pukoo, Island of Molokai. 1962.   1078

R. 19 Relationship of the State Water Program and the U. S. Geological Survey. 1962.   1079

R. 20 Report on Lihue Water System, Island of Kauai. 1962. 1080

R. 21 Improvements to county water system, Lahaina District, Maui. 1963.   1081

R. 22 Kokee Water Project, Island of Kauai, Hawaii: a report on the feasibility of water development. 1964.   1082

R. 22A Kokee Water Project, Island of Kauai, Hawaii: geology supplement, by M. H. Logan and D. Lum. 1966.   1083

R. 23 Ocean outfall report, Waimanalo core development, by R. M. Towill Corp. , 1964.   1084

R. 24 Report on oceanographic study for Kapaa Ocean sewer outfall, Kapaa, Kauai, by Sunn, Low,

Tom & Hara, Inc. , 1964.   1085

R. 25 A water development plan for south Kohala-Hamakua, Island of Hawaii. 1965.   1086

R. 26 Floods of December 1964-February 1965 in Hawaii, by S. Hoffard. 1965.   1087

R. 27 Flow characteristics of selected streams in Hawaii, by G. T. Hirashima. 1965.   1088

R. 28 Effects of water withdrawals by tunnels, Waihee Valley, Oahu, Hawaii, by G. T. Hirashima. 1965.   1089

R. 29 Water supply investigations for Laupahoehoe Water System, south Hamakua, Island of Hawaii. 1966.   1090

R. 30 Flood of March 24, 1967, Kihei and Olowalu areas, Island of Maui, by S. H. Hoffard. 1968.   1091

R. 31 Geologic-hydrologic investigations for deep-well disposal, Waimanalo, Oahu, Hawaii, by D. Lum. n. d.   1092

R. 32 Waikolu and Pelekunu Valleys Water Resources Feasibility Study, Island of Molokai, by Parsons-Brinckerhoff-Hirota Associates. 1969.   1093

R. 33 A water source development plan for the Lahaina District, Island of Maui, by Belt, Collins & Associates. 1969.   1094

R. 34 An inventory of basic water resources data, Island of Hawaii. 1970.   1095

R. 35 Kohakohau Dam Engineering Feasibility, south Kohala Water Project, Hawaii, by Parsons Brinckerhoff-Hirota Associates, 1970.   1096

R. 36 Flood frequencies for selected streams in Hawaii. 1970.   1097

R. 37 Flood hazard information, Island of Hawaii. 1970.   1098

R. 38 Water for Kihei-Makena, Island

of Maui. 1970.   1099

R. 39 Flood hazard information,
Island of Maui. 1971.   1100

R. 40 A general plan for domestic
water, Island of Kauai. 1972.
                              1101

R. 41 Soil survey interpretations,
Kauai. 1972.   1102

R. 42 Climatologic stations in Hawaii. 1973.   1103

R. 43 Soil survey interpretations,
Molokai. 1972.   1104

R. 44 Soil survey interpretations,
Lanai. 1972.   1105

R. 45 Soil survey interpretations,
Maui. 1972.   1106

R. 46 Not yet published.   1107

R. 47 Water resources summary,
Island of Hawaii, by D. A. Davis
and G. Yamanaga. 1973.   1108

R. 48 Chemical quality of ground
water, by L. A. Swain. 1973.
                              1109

R. 49 Flood hazard information,
Island of Kauai. 1973.   1110

R. 50 Water desalting in Hawaii,
by J. M. Duncan and B. J. Garrick.
1974.   1111

R. 51 Pan evaporation in Hawaii,
1894-1970. 1973.   1112

R. 52 Environmental impact state-
ment, Kohakohau Dam Project,
South Kohala Water Project, Island
of Hawaii, by Parsons-Brinckerhoff-
Hirota Associates, 1975.   1113

R. 53 Availability of ground water
for irrigation on the Kekaha-Mana
coastal plain, Island of Kauai, by
R. J. Burt. 1976.   1114

R. 54 Kahakuloa water study. 1977.
                              1115

R. 55 Specific problem analysis

summary report, 1975 National Assessment of Water and Related Land Resources, Hawaii Region--20 December
1977.   1116

R. 56 Innovative approaches to storm
water designs to protect our beaches
and coastal waters from sedimentation, by Walter Lum Associates, Inc.
1977.   1117

R. 57 Solar radiation in Hawaii,
1932-1975, by K. T. S. How. 1978.
                              1118

R. 58 Water use in Hawaii, 1975.
1978.   1119

R. 59 Waialeale hydropower study, by
Belt, Collins & Associates, 1978.
                              1120

Bulletin. 1920-

B. 22 Gold camps and silver cities, by M. W. Wells. 1963. 1121

B. 23 Distribution and economic potential of Idaho carbonate rocks, by C. N. Savage. 1969. 1122

County Report. 1956-1967.

CR. 6 Geology and mineral resources of Bonner County, by C. N. Savage. 1967. 1123

Earth Science Series. 1967-1969.

ESS. 1 Idaho earth science: geology, fossils, climate, water, and soils, by S. H. Ross and C. N. Savage. 1967. 1124

ESS. 2 Introduction to Idaho caves and caving, by S. H. Ross. 1969. 1125

Information Circular. 1957-

IC. 17 Mean-dip maps--indicators of deformation intensity: a contribution to geometrics, by W. R. Greenwood. 1967. 1126

IC. 18 Reconnaissance geology of the Selway-Bitterroot Wilderness Area, by W. R. Greenwood and D. A. Morrison. 1967. 1127

IC. 19 Abstracts of publications of the Idaho Bureau of Mines and Geology, by S. H. Ross and C. N. Savage. 2nd ed. 1976. 1128

IC. 20 Lexicon of Idaho geologic names, by C. N. Savage. 1968. 1129

IC. 21 Computer analysis of zircon morphology data: a contribution to geometrics, by M. R. Hedberg and W. R. Greenwood. 1970. 1130

IC. 22 Delineation of mineral belts of northern and central Idaho, by W. R. Greenwood. 1972. 1131

IC. 23 Guide for the location of water wells in Latah County, Idaho, by D. R. Ralston. 1972. 1132

IC. 24 A dictionary of stream names of the St. Joe National Forest, by J. F. Hoffman. 1973. 1133

IC. 25 Evaluation of phosphate resources in southeastern Idaho, by J. D. Powell. 1974. 1134

IC. 26 Reconnaissance geochemistry of the Bighorn Crags, by C. R. Knowles. 1975. 1135

IC. 27 Geochemical processing at the Idaho Bureau of Mines and Geology, by J. Galbraith. 1975. 1136

IC. 28 Guidebook for the Later Tertiary stratigraphy and paleobotany of the Weiser area, Idaho, by C. J. Smiley, S. M. I. Shah, and R. W. Jones. 1975. 1137

Guidebook for the geology and scenery of the Snake River on the Idaho-Oregon border from Brownlee Dam to Hells Canyon Dam, comp. by P. J. B. Miller. 1975. 1138

IC. 29 A brief geological survey of the east Thunder Mountain Mining District, Valley County, Idaho, by S. S. Shannon and S. J. Reynolds. 1975. 1139

IC. 30 Distribution of precipitation in Little Long Valley and Dry Valley,

Caribou County, Idaho, by D. R. Ralston and E. W. Trihey. 1975. 1140

IC. 31 Not yet published. 1141

IC. 32 Graduate theses on the geology of Idaho, 1900-1977, by M. P. Gaston. 1979. 1142

IC. 33 Guidebook and road log to the St. Maries River (Clarkia) fossil area of northern Idaho, by C. J. Smiley and W. C. Rember. 1979. 1143

IC. 34 Geologic section and road log across the Idaho batholith, by R. R. Reid, E. Bittner, W. R. Greenwood, S. Ludington, K. Lund, W. E. Motzer, and M. Toth. 1979. 1144

Mineral Resources Report. 1945-1970.

MRR. 9 The Oneida perlite deposit, by W. W. Staley. 1962. 1145

MRR. 10 Economic geology of carbonate rocks adjacent to the Snake River south of Lewiston, Idaho, by C. N. Savage. 1965. 1146

MRR. 11 Mineralogy of the Lemhi Pass thorium and rare-earth deposits, by S. R. Austin, D. L. Hetland, and B. J. Sharp. 1970. 1147

Pamphlet. 1921-

P. 128 Geology of the Clearwater embayment, by J. G. Bond. 1963. 1148

P. 129 Reconnaissance geology of the Sawtooth Range, by R. R. Reid. 1963. 1149

P. 130 Geology along U. S. Highway 93 in Idaho, by C. P. Ross. 1963. 1150

P. 131 Mining history of south-central Idaho, by C. P. Ross. 1963. 1151

P. 132 Gravity anomalies in Idaho, by W. E. Bonini. 1963. 1152

P. 133 The Coeur d'Alene Mining District in 1963. 1963. 1153

P. 134 Geologic history of Pend Oreille Lake region in north Idaho, by C. N. Savage. 1965. 1154

P. 135 Volcanic construction materials in Idaho, by R. R. Asher. 1965. 1155

P. 136 Geology of the Oxbow on Snake River near Homestead, Oregon, by H. T. Stearns and A. L. Anderson. 1966. 1156

P. 137 Interpretation of short term water fluctuations in the Moscow Basin, Latah County, Idaho, by D. Sokol. 1966. 1157

P. 138 Geology and mineral resources of a portion of the Silver City region, Owyhee County, Idaho, by R. R. Asher. 1968. 1158

P. 139 Idaho's minerals industry--a flow-of-product analysis, by M. L. Newell and R. D. Peterson. 1968. 1159

P. 140 Stratigraphy and distribution of basalt, Benewah County, Idaho, by D. T. Bishop. 1969. 1160

P. 141 Ground-water flow systems and the origin of evaporite deposits, by R. E. Williams. 1968. 1161

P. 142 Bedrock geology of the Pioneer Mountains, Blaine and Custer Counties, central Idaho, by J. H. Dover. 1969. 1162

P. 143 Feasibility of re-use of treated wastewater for irrigation, fertilization, and ground-water re-charge in Idaho, by R. E. Williams, D. D. Eler, and A. T. Wallace. 1969. 1163

P. 144 Applicability of mathematical models of ground-water flow systems to hydrogeochemical exploration, by R. E. Williams. 1970. 1164

P. 145 Hydrogeological aspects of the

selection of refuse disposal sites in Idaho, by R. E. Williams and A. T. Wallace. 1970. 1165

P. 146 Geology and geochemical exploration of the Vienna District, Blaine and Camas Counties, Idaho, by S. S. Shannon. 1971. 1166

P. 147 Evaluation of minerals and mineral potential of the Salmon River drainage basin in Idaho, by C. N. Savage. 1970. 1167

P. 148 The effects of drain wells on the ground-water quality of the Snake River Plain, by D. E. Abegglen, A. T. Wallace, and R. E. Williams. 1970. 1168

P. 149 Effect of industrial and domestic effluents on the water quality of the Coeur d'Alene River Basin, by L. L. Mink, R. E. Williams, and A. T. Wallace. 1971. 1169

P. 150 Geothermal potential of Idaho, by S. H. Ross. 1971. 1170

P. 151 Cenozoic geology of the Reynolds Creek Experimental Watershed, Owyhee County, Idaho, by D. H. McIntyre. 1972. 1171

P. 152 Community perception in the Coeur d'Alene Mining District, by L. E. Ellsworth. 1972. 1172

P. 153 Moscow Basin ground water studies, by R. W. Jones and S. H. Ross. 1972. 1173

P. 154 Reconnaissance geology of the Selway-Bitterroot Wilderness Area, by W. R. Greenwood and D. A. Morrison. 1973. 1174

P. 155 Trends in the phosphate industry of Idaho and the western phosphate field, by R. L. Day. 1973. 1175

P. 156 Geology and mineral resources of the Lakeview Mining District, Idaho, by P. Kun. 1974. 1176

P. 157 Air photography and satellite image interpretation for Linears mapping and geologic evaluation, by W. B. Hall and T. H. Walsh. 1974. 1177

P. 158 Analysis of the impact of legal constraints on ground-water resource development in Idaho, by D. R. Ralston, D. L. Grant, H. L. Schatz, and D. Goldman. 1974. 1178

P. 159 A preliminary evaluation of ground water in Upper Dry Valley and Little Long Valley, Caribou County, Idaho, by K. A. Sylvester. 1975. 1179

P. 160 Geologic field guide to the Quaternary volcanics of the south-central Snake River Plain, Idaho, by R. Greeley and J. S. King. 1975. 1180

P. 161 Geology and ore deposits of the Silver City-DeLamar-Flint region, Owyhee County, Idaho, by A. J. Pansze. 1975. 1181

P. 162 Reconnaissance geology and geochemistry of the Silver City-South Mountain region, Owyee County, Idaho, by E. H. Bennett and J. H. Galbraith. 1975. 1182

P. 163 Reconnaissance geology and geochemistry of the Mt. Pend Oreille Quadrangle and surrounding areas, by E. H. Bennett, R. S. Kopp, and J. H. Galbraith. 1975. 1183

P. 164 Settling ponds as a mining wastewater treatment facility, by R. E. Williams and L. L. Mink. 1975. 1184

P. 165 Sources and causes of acid mine drainage, by B. D. Trexler, D. R. Ralston, D. R. Reece, and R. E. Williams. 1975. 1185

P. 166 Reconnaissance geology and geochemistry of the South Mountain-Juniper Mountain region, Owyhee County, Idaho, by E. H. Bennett. 1976. 1186

P. 167 Reconnaissance geology and geochemistry of the Blackbird Mountain-Panther Creek region, Lemhi

County, Idaho, by E. H. Bennett.
1977.   1187

Special Report.   1964-1973.

SR. 1 Mineral and water resources
of Idaho, by the U. S. Geological
Survey, the Idaho Bureau of Mines
and Geology, Idaho Department of
Highways, and the Idaho Department of Reclamation.   1964. 1188

SR. 2 Belt symposium, 1973.
Vol. 1.   1189

SR. 3 Belt symposium, 1973.
Vol. 2.   1190

Bulletin. 1906-

B. 89 Champlainian (Middle Ordovician) Series in Illinois, by J. S. Templeton and H. B. Willman. 1963. 1191

B. 90 Spores in strata of Late Pennsylvanian cyclothems in the Illinois Basin, by R. A. Peppers. 1964. 1192

B. 91 Handbook on limestone and dolomite for Illinois quarry operators, by J. E. Lamar. 1967. 1193

B. 92 Bibliography and index of Illinois geology through 1965, by H. B. Willman, J. A. Simon, B. M. Lynch, and V. A. Langenheim. 1968. 1194

B. 93 Correlation and palynology of coals in the Carbondale and Spoon Formations (Pennsylvanian) of the northeastern part of the Illinois Basin, by R. A. Peppers. 1970. 1195

B. 94 Pleistocene stratigraphy of Illinois, by H. B. Willman and J. C. Frye. 1970. 1196

B. 95 Handbook of Illinois stratigraphy, by H. B. Willman, E. Atherton, T. C. Buschbach, C. Collinson, J. C. Frye, M. E. Hopkins, J. A. Lineback, and J. A. Simon. 1975. 1197

Circular. 1932-

C. 342 Geology of the Illinois fluorspar district. Part 1: Saline mines, Cave in Rock, Dekoven, and Repton Quadrangles, by J. W. Baxter, P. E.

Potter, and F. L. Doyle. 1963. 1198

C. 343 Collection and preparation of conodonts through mass production techniques, by C. Collinson. 1963. 1199

C. 344 Earthquakes and crustal movement is related to water load in the Mississippi Valley region, by L. D. McGinnis. 1963. 1200

C. 345 Impact resistance of Illinois limestones and dolomites, by R. D. Harvey. 1963. 1201

C. 346 Limestone resources of the Lower Kaskaskia Valley, by J. C. Bradbury. 1963. 1202

C. 347 Mineralogy of glacial tills and their weathering profiles in Illinois. Part I: Glacial tills, by H. B. Willman, H. D. Glass, and J. C. Frye. 1963. 1203

C. 348 Strippable coal reserves of Illinois. Part 5A: Fulton, Henry, Knox, Peoria, Stark, Tazewell, and parts of Bureau, Marshall, Mercer, and Warren Counties, by W. H. Smith and D. J. Berggren. 1963. 1204

C. 349 Wapella East Oil Pool, DeWitt County, Illinois--a Silurian reef, by R. H. Howard. 1963. 1205

C. 350 Studies on the ultrafine structure of some Illinois coals, by J. S. Machin, F. Staplin, and D. L. Deadmore. 1963. 1206

C. 351 Structural framework of southernmost Illinois, by C. A. Ross. 1963. 1207

C. 352 Buff-burning clay resources

of southwestern and southern Illinois, by W. E. Parham and W. A. White. 1963. 1208

C. 353 Buff-burning clay resources of western Illinois, by W. A. White. 1963. 1209

C. 354 Relationship of gravity anomalies to a drift-filled bedrock valley system in northern Illinois, by L. D. McGinnis, J. P. Kempton, and P. C. Heigold. 1963. 1210

C. 355 Conodonts from the St. Louis Formation (Valmeyeran Series) of Illinois, Indiana, and Missouri, by C. B. Rexroad and C. Collinson. 1963. 1211

C. 356 Subsurface stratigraphy of the Pleistocene deposits of central northern Illinois, by J. P. Kempton. 1963. 1212

C. 357 Mineral production in Illinois in 1962, by W. L. Busch. 1963. 1213

C. 358 Some economic aspects of the Illinois oil industry, by W. H. Voskuil and H. E. Risser. 1963. 1214

C. 359 Sand and gravel resources of northeastern Illinois, by G. E. Ekblaw and J. E. Lamar. 1964. 1215

C. 360 Geology of the Paducah and Smithland Quadrangles in Illinois, by C. A. Ross. 1964. 1216

C. 361 Conodonts from the Devonian Lingle and Alto Formations of southern Illinois, by R. W. Orr. 1964. 1217

C. 362 Carper sand oil production in St. James, Wilberton, and St. Paul Pools, Fayette County, Illinois, by D. L. Stevenson. 1964. 1218

C. 363 Aeromagnetic study of the Hardin County area, Illinois, by L. D. McGinnis and J. C. Bradbury. 1964. 1219

C. 364 Cretaceous deposits and the

Illinoian glacial boundary in western Illinois, by J. C. Frye, H. B. Willman, and H. D. Glass. 1964. 1220

C. 365 Composition of the ash of Illinois coals, by O. W. Rees. 1964. 1221

C. 366 Predicting coke stability from petrographic analysis of Illinois coals, by J. A. Harrison, H. W. Jackman, and J. A. Simon. 1964. 1222

C. 367 Sand and gravel resources of De Kalb County, by R. C. Anderson. 1964. 1223

C. 368 Deep oil possibilities of the Illinois Basin, by A. H. Bell, E. Atherton, T. C. Buschbach, and D. H. Swann. 1964. 1224

C. 369 Geologic significance of the gravity field in the De Witt-McLean County area, Illinois, by P. C. Heigold, L. D. McGinnis, and R. H. Howard. 1964. 1225

C. 370 Mississippian limestone resources in Fulton, McDonough, and Schuyler Counties, Illinois, by R. D. Harvey. 1964. 1226

C. 371 Illinois clay resources for lightweight ceramic block, by W. A. White and N. R. O'Brien. 1964. 1227

C. 372 Chlorine in Illinois coal, by H. J. Gluskoter and O. W. Rees. 1964. 1228

C. 373 Mineral production in Illinois in 1963, by W. L. Busch. 1964. 1229

C. 374 Strippable coal reserves of Illinois. Part 4: Adams, Brown, Calhoun, Hancock, McDonough, Pike, Schuyler, and the southern parts of Henderson and Warren Counties, by D. L. Reinertsen. 1964. 1230

C. 375 Coke crushing characteristics, by H. W. Jackman and R. J. Helfinstine. 1964. 1231

C. 376 Electrical earth resistivity surveying in Illinois, by M. B. Buhle

and J. E. Brueckmann. 1964.
1232

C. 377 Niagaran reefs and oil accumulation in the De Witt-McLean
County area, Illinois, by R. H. Howard. 1964. 1233

C. 378 Stratigraphy and petrography
of Illinoian and Kansan drift in central Illinois, by W. H. Johnson.
1964. 1234

C. 379 Dolomite resources of
Boone and De Kalb Counties, by
J. C. Bradbury. 1965. 1235

C. 380 Subsurface geology and coal
resources of the Pennsylvanian system in Clark and Edgar Counties,
Illinois, by K. E. Clegg. 1965.
1236

C. 381 Sand and gravel resources
of Peoria County, by R. C. Anderson and R. E. Hunter. 1965.
1237

C. 382 The Precambrian basement
of Illinois, by J. C. Bradbury and
E. Atherton. 1965. 1238

C. 383 The Sangamon Arch, by
L. L. Whiting and D. L. Stevenson.
1965. 1239

C. 384 Pre-Pennsylvanian Evansville Valley and Caseyville (Pennsylvanian) sedimentation in the Illinois
Basin, by P. E. Potter and G. A.
Desborough. 1965. 1240

C. 385 Areal geology of the Illinois
fluorspar district. Part 2: Karbers
Ridge and Rosiclare Quadrangles,
by J. W. Baxter and G. A. Desborough. 1965. 1241

C. 386 The Borden siltstone (Mississippian) delta in southwestern
Illinois, by D. H. Swann, J. A. Lineback, and E. Frund. 1965. 1242

C. 387 Ashmore gas area, Coles
County, Illinois, W. F. Meents.
1965. 1243

C. 388 Conodonts from the Keokuk,
Warsaw, and Salem Formations

(Mississippian) of Illinois, by C. B.
Rexroad and C. Collinson. 1965.
1244

C. 389 Grand Tower limestone (Devonian) of southern Illinois, by W. F.
Meents and D. H. Swann. 1965.
1245

C. 390 Limestone resources of Madison County, Illinois, by J. W. Baxter.
1965. 1246

C. 391 Feldspar in Illinois sands: a
further study, by R. E. Hunter. 1965.
1247

C. 392 Mineral production in Illinois
in 1964, by W. L. Busch. 1965. 1248

C. 393 The origin of saline formation
waters. II: Isotopic fractionation by
shale micropore systems, by D. L.
Graf, I. Friedman, and W. F. Meents.
1965. 1249

C. 394 Deep stratigraphic test well
near Rock Island, Illinois, by T. C.
Buschbach. 1965. 1250

C. 395 Geology of Freeport Quadrangle, Illinois, by F. L. Doyle.
1965. 1251

C. 396 Sulfur retention in bituminous
coal ash, by O. W. Rees, N. F. Shimp,
C. W. Beeler, J. K. Kuhn, and R. J.
Helfinstine. 1966. 1252

C. 397 The origin of saline formation
waters. III: Calcium chloride waters,
by D. L. Graf, W. F. Meents, I. Friedman, and N. F. Shimp. 1966. 1253

C. 398 Gravity base station network
in Illinois, by L. D. McGinnis. 1966.
1254

C. 399 Sand and gravel resources of
Tazewell County, Illinois, by R. E.
Hunter. 1966. 1255

C. 400 Mineralogy of glacial tills and
their weathering profiles in Illinois.
Part II: Weathering profiles, by H. B.
Willman, H. D. Glass, and J. C. Frye.
1966. 1256

C. 401 Deep-water sediments adjacent

to the Borden siltstone (Mississippian) delta in southern Illinois, by J. A. Lineback. 1966. 1257

C. 402 Heavy minerals in sands along the Wabash River, by R. E. Hunter 1966. 1258

C. 403 Ironton and Galesville (Cambrian) sandstone in Illinois and adjacent areas, by G. H. Emrich. 1966. 1259

C. 404 Electron microscope study of microtexture and grain surfaces in limestones, by R. D. Harvey. 1966. 1260

C. 405 Long-term dimensional changes in Illinois bricks and other clay products, by J. S. Hosking, W. A. White, and W. E. Parham. 1966. 1261

C. 406 Bedrock aquifers of northeastern Illinois, by G. M. Hughes, P. Kraatz, and R. A. Landon. 1966. 1262

C. 407 Mineral production in Illinois in 1965 and summary of Illinois mineral production by commodities, 1941-1965, by W. L. Busch. 1966. 1263

C. 408 One-dimensional disorder in carbonates, by D. L. Graf, C. R. Blyth, and R. S. Stemmler. 1967. 1264

C. 409 Hydrogeology of glacial deposits of the Mahomet Bedrock Valley in east-central Illinois, by D. A. Stephenson. 1967. 1265

C. 410 Production and consumption of mineral fuels in Illinois, 1933-1964, by R. F. Severson. 1967. 1266

C. 411 Description of Late Pennsylvanian strata from deep diamond drill cores in the southern part of the Illinois Basin, by W. H. Smith and G. E. Smith. 1967. 1267

C. 412 A survey of the coking properties of Illinois coals, by

H. W. Jackman and R. J. Helfinstine. 1967. 1268

C. 413 Areal geology of the Illinois fluorspar district. Part 3: Herod and Shetlerville Quadrangles, by J. W. Baxter, G. A. Desborough, and C. W. Shaw. 1967. 1269

C. 414 Sand and gravel resources along the Rock River in Illinois, by R. C. Anderson. 1967. 1270

C. 415 Thermal expansion of certain Illinois limestones and dolomites, by R. D. Harvey. 1967. 1271

C. 416 Glacial drift in Illinois: thickness and character, by K. Piskin and R. E. Bergstrom. 1967. 1272

C. 417 Sand and gravel resources of Boone County, Illinois, by R. E. Hunter and J. P. Kempton. 1967. 1273

C. 418 Mineral production in Illinois in 1966, by W. L. Busch. 1968. 1274

C. 419 Strippable coal reserves of Illinois. Part 6: La Salle, Livingston, Grundy, Kankakee, Will, Putnam, and parts of Bureau and Marshall Counties, by W. H. Smith. 1968. 1275

C. 420 Fluorspar in Illinois, by J. C. Bradbury, G. C. Finger, and R. L. Major. 1968. 1276

C. 421 Geochemical trends in Chesterian (Upper Mississippian) Waltersburg crudes of the Illinois Basin, by R. F. Mast, N. F. Shimp, and P. A. Witherspoon. 1968. 1277

C. 422 Geology related to land use in the Hennepin region, by M. R. McComas. 1968. 1278

C. 423 Drying and preheating coals before coking. Part I: Individual coals, by H. W. Jackman and R. J. Helfinstine. 1968. 1279

C. 424 Geology and oil production in the Tuscola area, Illinois, by H. M.

Bristol and R. Prescott. 1968.
1280

C. 425 Turbidites and other sandstone bodies in the Borden Siltstone (Mississippian) in Illinois, by J. A. Lineback. 1968. 1281

C. 426 Feasibility of subsurface disposal of industrial wastes in Illinois, by R. E. Bergstrom. 1968. 1282

C. 427 Mineral zonation of Woodfordian loesses of Illinois, by J. C. Frye, H. D. Glass, and H. B. Willman. 1968. 1283

C. 428 Heavy minerals of the Cretaceous and Tertiary sands of extreme southern Illinois, by R. E. Hunter. 1968. 1284

C. 429 Petrography of Pennsylvanian underclays in Illinois and their application to some mineral industries, by I. E. Odom and W. E. Parham. 1968. 1285

C. 430 Gasification and liquefaction --their potential impact on various aspects of the coal industry, by H. E. Risser. 1968. 1286

C. 431 Harrisburg (No. 5) coal reserves of southeastern Illinois, by M. E. Hopkins. 1968. 1287

C. 432 Sulfur in Illinois coals, by H. J. Gluskoter and J. A. Simon. 1968. 1288

C. 433 Temperature prospecting for shallow glacial and alluvial aquifers in Illinois, by K. Cartwright. 1968. 1289

C. 434 Drying and preheating coals before coking. Part 2: Coal blends, by H. W. Jackman and R. J. Helfinstine. 1968. 1290

C. 435 Mineral production in Illinois in 1967, by W. L. Busch. 1968. 1291

C. 436 Oil production from the Ste. Genevieve limestone in the Exchange area, Marion County,

Illinois, by D. L. Stevenson. 1969.
1292

C. 437 Glacial tills of northwestern Illinois, by J. C. Frye, H. D. Glass, J. P. Kempton, and H. B. Willman. 1969. 1293

C. 438 Geology for planning in McHenry County, by J. E. Hackett and M. R. McComas. 1969. 1294

C. 439 Strippable coal reserves of Illinois. Part 5B: Mercer, Rock Island, Warren, and parts of Henderson and Henry Counties, by T. K. Searight and W. H. Smith. 1969. 1295

C. 440 High-level glacial outwash in the driftless area of northwestern Illinois, by H. B. Willman and J. C. Frye. 1969. 1296

C. 441 The Middle Devonian strata of southern Illinois, by W. G. North. 1969. 1297

C. 442 Glacial geology of the Vandalia, Illinois, region, by A. M. Jacobs and J. A. Lineback. 1969. 1298

C. 443 Radiographic exposure guides for mud, sandstone, limestone, and shale, by G. S. Fraser and A. T. James. 1969. 1299

C. 444 A study of the surface areas of particulate microcrystalline silica and silica sand, by R. N. Leamnson, J. Thomas, and H. P. Ehrlinger. 1969. 1300

C. 445 Palynology and petrography of a Middle Devonian coal in Illinois, by R. A. Peppers and H. H. Damberger. 1969. 1301

C. 446 Sand and gravel resources of Macon County, Illinois, by N. C. Hester and R. C. Anderson. 1969. 1302

C. 447 Mineral production in Illinois in 1968, by W. L. Busch. 1969. 1303

C. 448 Limestone and dolomite resources of Jersey County, Illinois, by J. W. Baxter. 1970. 1304

C. 449 Heat drying coals at moderate

temperatures before coking, by
H. W. Jackman and R. J. Helfinstine. 1970. 1305

C. 450 A gravity survey of extreme
southeastern Illinois, by P. C. Heigold. 1970. 1306

C. 451 ILLIMAP--a computer-based
mapping system for Illinois, by
D. H. Swann, P. B. Dumontelle,
R. F. Mast, and L. H. Van Dyke.
1970. 1307

C. 452 Sand and gravel resources
of Sangamon County, Illinois, by
N. C. Hester. 1970. 1308

C. 453 Preheating coal blends as
a means of increasing coke strength,
by H. W. Jackman and R. J. Helfinstine. 1970. 1309

C. 454 Trend-surface analysis of
the structure of the Ste. Genevieve
limestone in the Effingham, Illinois,
area, by D. L. Stevenson. 1970.
1310

C. 455 Mineral production in Illinois in 1969, by W. L. Busch.
1971. 1311

C. 456 Stratigraphy of the glacial
deposits at the National Accelerator
Laboratory site, Batavia, Illinois,
by R. A. Landon and J. P. Kempton.
1971. 1312

C. 457 Old glacial drift near Danville, Illinois, by W. H. Johnson.
1971. 1313

C. 458 Paleogeologic map of the
sub-Pennsylvanian Chesterian (Upper Mississippian) surface in the
Illinois Basin, by H. M. Bristol
and R. H. Howard. 1971. 1314

C. 459 Glacial drift of the Shelbyville moraine at Shelbyville, Illinois, by W. H. Johnson, H. D. Glass,
D. L. Gross, and S. R. Moran.
1971. 1315

C. 460 Summary of the geology of
the Chicago area, by H. B. Willman. 1971. 1316

C. 461 Illinoian and Kansan molluscan
faunas of Illinois, by A. B. Leonard,
J. C. Frye, and W. H. Johnson. 1971.
1317

C. 462 Sulfur reduction of Illinois
coals--washability studies. Part 1,
by R. J. Helfinstine, N. F. Shimp,
J. A. Simon, and M. E. Hopkins. 1971.
1318

C. 463 Some fern-like fructifications
and their spores from the Mazon
Creek compression flora of Illinois
(Pennsylvanian), by H. W. Pfefferkorn,
R. A. Peppers, and T. L. Phillips.
1971. 1319

C. 464 Sand and gravel resources of
Mason County, Illinois, by T. C. Labotka and N. C. Hester. 1971. 1320

C. 465 Geology for planning in St.
Clair County, Illinois, by A. M. Jacobs, comp. 1971. 1321

C. 466 Deltaic sedimentation in glacial Lake Douglas, by G. S. Fraser
and J. C. Steinmetz. 1971. 1322

C. 467 Farmdalian lake deposits and
faunas in northern Illinois, by H. B.
Willman, A. B. Leonard, and J. C.
Frye. 1971. 1323

C. 468 Properties of chert related
to its reactivity in an alkaline environment, by G. S. Fraser, R. D. Harvey,
and P. C. Heigold. 1972. 1324

C. 469 Mineral production in Illinois
in 1970, by W. L. Busch. 1972. 1325

C. 470 Hydrodynamics in deep aquifers of the Illinois Basin, by D. C.
Bond. 1972. 1326

C. 471 Geology and paleontology of
Late Pleistocene Lake Saline, south-
eastern Illinois, by J. C. Frye, A. B.
Leonard, H. B. Willman, and H. D.
Glass. 1972. 1327

C. 472 Limestone resources of Scott
County, Illinois, by J. W. Baxter.
1972. 1328

C. 473 Subsurface geology and coal

resources of the Pennsylvanian system in De Witt, McLean, and Platt Counties, Illinois, by K. E. Glegg. 1972. 1329

C. 474 Lateral gradation of the Salem and St. Louis limestones (Middle Mississippian) in Illinois, by J. A. Lineback. 1972. 1330

C. 475 Velocity of sound in sediments cored from southern Lake Michigan, by M. L. Silver and J. A. Lineback. 1972. 1331

C. 476 Occurrence and distribution of minerals in Illinois coals, by C. P. Rao and H. J. Gluskoter. 1973. 1332

C. 477 Mineral production in Illinois in 1971 and summary of Illinois mineral production by commodities, 1941-1970, by W. L. Busch. 1973. 1333

C. 478 Geology along the Illinois waterway--a basis for environmental planning, by H. B. Willman. 1973. 1334

C. 479 Rock stratigraphy of the Silurian system in northeastern and northwestern Illinois, by H. B. Willman. 1973. 1335

C. 480 Development of paleobotany in the Illinois Basin, by T. L. Phillips, H. W. Pfefferkorn, and R. A. Peppers. 1973. 1336

C. 481 Geology for planning in Lake County, Illinois, by J. I. Larsen. 1973. 1337

C. 482 The effect of buried Niagaran reefs on overlying strata in southwestern Illinois, by D. L. Stevenson. 1973. 1338

C. 483 Vanadium in Devonian, Silurian, and Ordovician crude oils of Illinois, by R. F. Mast, R. R. Ruch, and W. F. Meents. 1973. 1339

C. 484 Sulfur reduction of Illinois coals--washability studies. Part 2, by R. J. Helfinstine, N. F. Shimp,

M. E. Hopkins, and J. S. Simon. 1974. 1340

C. 485 Earliest Wisconsinan sediments and soils, by J. C. Frye, L. R. Follmer, H. G. Glass, J. M. Masters, and H. B. Willman. 1974. 1341

C. 486 The Late Woodfordian Jules soil and associated molluscan faunas, by J. C. Frye, A. B. Leonard, H. B. Willman, H. D. Glass, and L. R. Follmer. 1974. 1342

C. 487 Chronology and molluscan paleontology of two Post-Woodfordian bogs in northeastern Illinois, by A. B. Leonard. 1974. 1343

C. 488 A seismic refraction survey of the Meredosia Channel area of northwestern Illinois, by L. D. McGinnis and P. C. Heigold. 1974. 1344

C. 489 Reserves of the Herrin (No. 6) coal in the Fairfield Basin in southeastern Illinois, by G. J. Allgaier and M. E. Hopkins. 1975. 1345

C. 490 Glacial drift in Illinois: thickness and character, by K. Piskin and R. E. Bergstrom. 1975. 1346

C. 491 Plum River fault zone of northwestern Illinois, by D. R. Kolata and T. C. Buschbach. 1976. 1347

C. 492 Pennsylvanian tree fern compressions Caulopteris Megaphyton and Artisophyton gen. nov. in Illinois, by H. W. Pfefferkorn. 1976. 1348

C. 493 Internal surface area, moisture content, and porosity of Illinois coals: variations with coal rank, by J. Thomas and H. H. Damberger.. 1976. 1349

C. 494 The gravity field and tectonics of Illinois, by L. D. McGinnis, P. C. Heigold, C. P. Ervin, and M. Heidari. 1976. 1350

C. 495 An aeromagnetic survey of southwestern Illinois, by P. C. Heigold. 1976. 1351

C. 496 Holocene palynology and sed-

imentation of southern Lake Michigan, by J. E. King, J. A. Lineback, and D. L. Gross. 1976. 1352

C. 497 Geology for planning in the Springfield-Decatur region, Illinois, by R. E. Bergstrom, K. Piskin, and L. R. Follmer. 1977. 1353

C. 498 Mineral matter in the Springfield-Harrisbur (No. 5) Coal Member in the Illinois Basin, by C. R. Ward. 1977. 1354

C. 499 Trace elements in coal: occurrence and distribution, by H. J. Gluskoter, R. R. Ruch, W. G. Miller, R. A. Cahill, G. B. Dreher, and J. K. Kuhn. 1977. 1355

C. 500 An Early Pennsylvanian flora with Megalopteris and Noeggerathiales from west central Illinois, by R. L. Leary and H. W. Pfefferkorn. 1977. 1356

C. 501 Aromatic fluorine chemistry at the Illinois State Geological Survey--research notes, 1934-1976, by R. H. Shiley, D. R. Dickerson, and G. C. Finger. 1978. 1357

C. 502 The Platteville and Galena Groups in northern Illinois, by H. B. Willman and D. R. Kolata. 1978. 1358

C. 503 Sand and gravel and peat resources in northeastern Illinois, by J. M. Masters. 1978. 1359

C. 504 Reserves and resources of surface-minable coal in Illinois, by C. G. Treworgy, L. E. Bengal, and A. G. Dingwell. 1978. 1360

C. 505 The Sandwich Fault Zone of northern Illinois, by D. R. Kolata, T. C. Buschbach, and J. D. Treworgy. 1978. 1361

C. 506 Glacial geology of north-central and western Champaign County, Illinois, by J. T. Wickham. 1979. 1362

C. 507 A seismic refraction survey of the Lower Illinois Valley bottom-lands, by P. C. Heigold and R. W. Ringler. 1979. 1363

C. 508 Biological screening of some fluoroaromatic compounds, by I. R. H. Shiley, D. R. Dickerson, J. R. Willaw, and C. Grunewald. 1979. 1364

C. 509 The Wabash Valley fault system in southeastern Illinois, by H. M. Bristol and J. D. Treworgy. 1979. 1365

Educational Series. 1927-

ES. 1 The story of the geologic making of southern Illinois, by S. Weller. 1927. 1366

ES. 2 The Rook River country of northern Illinois, by D. Rolfe. 1929- 1367

ES. 3 Typical rocks and minerals in Illinois, by G. E. Ekblaw and D. L. Carroll. 1931. 1368

ES. 4 Guide for beginning fossil hunters, by C. W. Collinson. 1956. Rev. 1959. 1369

ES. 5 Guide to rocks and minerals of Illinois. 1959. 1370

ES. 6 Field book--Pennsylvanian plant fossils of Illinois, by C. Collinson and R. Skartvedt. 1960. 1371

ES. 7 Guide to the geologic map of Illinois. 1961. 1372

ES. 8 Industrial minerals and metals of Illinois, by J. E. Lamar. 1965. 1373

ES. 9 Inside Illinois--mineral resources. 1965. 1374

ES. 10 History of Illinois mineral industries, by H. E. Risser and R. L. Major. 1968. 1375

ES. 11 Fossil peat from the Illinois Basin--a guide to the study of coal balls of Pennsylvanian age, by T. L. Phillips, M. J. Avcin, and D. J. Berggren. 1976. 1376

Environmental Geology Notes.
1965-

EGN. 1 Controlled drilling program in northeastern Illinois, by J. E. Hackett and G. M. Hughes. 1965.
1377

EGN. 2 Data from controlled drilling program in Du Page County, Illinois, by J. I. Larsen and C. R. Lund. 1965.
1378

EGN. 3 Activities in environmental geology in northeastern Illinois, by J. I. Larsen and J. E. Hackett. 1965.
1379

EGN. 4 Geological and geophysical investigations for a ground-water supply at Macomb, Illinois, by K. Cartwright and D. A. Stephenson. 1965.
1380

EGN. 5 Problems in providing minerals for an expanding population, by H. E. Risser. 1965.
1381

EGN. 6 Data from controlled drilling program in Kane, Kendall, and De Kalb Counties, Illinois, by C. R. Lund. 1965.
1382

EGN. 7 Data from controlled drilling program in McHenry County, Illinois, by C. R. Lund. 1965.
1383

EGN. 8 An application of geologic information to land use in the Chicago Metropolitan region, by J. E. Hackett. 1966.
1384

EGN. 9 Data from controlled drilling program in Lake County and the northern part of Cook County, Illinois, by C. R. Lund. 1966. 1385

EGN. 10 Data from controlled drilling program in Will and southern Cook Counties, Illinois, comp. by C. R. Lund. 1966.
1386

EGN. 11 Ground-water supplies along the Interstate Highway system in Illinois, by K. Cartwright. 1966.
1387

EGN. 12 Effects of a soap, a detergent and a water softener on the plasticity of earth materials, by W. A. White and S. M. Bremser. 1966.
1388

EGN. 13 Geologic factors in dam and reservoir planning, by W. C. Smith. 1966.
1389

EGN. 14 Geologic studies as an aid to ground-water management, by R. A. Landon. 1967.
1390

EGN. 15 Hydrogeology at Shelbyville, Illinois--a basis for water resources planning, by K. Cartwright and P. Kraatz. 1967.
1391

EGN. 16 Urban expansion--an opportunity and a challenge to industrial mineral producers, by H. E. Risser and R. L. Major. 1967.
1392

EGN. 17 Selection of refuse disposal sites in northeastern Illinois, by G. M. Hughes. 1967.
1393

EGN. 18 Geological information for managing the environment, by J. C. Frye. 1967.
1394

EGN. 19 Geology and engineering characteristics of some surface materials in McHenry County, Illinois, by W. C. Smith. 1968.
1395

EGN. 20 Disposal of wastes: scientific and administrative considerations, by R. E. Bergstrom. 1968.
1396

EGN. 21 Mineralogy and petrography of carbonate rocks related to control of sulfur dioxide in flue gases--a preliminary report, by R. D. Harvey. 1968.
1397

EGN. 22 Geologic factors in community development at Naperville, Illinois, by J. E. Hackett. 1968.
1398

EGN. 23 Effects of waste effluents on the plasticity of earth materials, by W. A. White and M. K. Kyriazis. 1968.
1399

EGN. 24 Notes on the earthquake of

November 9, 1968, in southern Illinois, by P. C. Heigold. 1968.
1400

EGN. 25 Preliminary geological evaluation of dam and reservoir sites in McHenry County, Illinois, by W. C. Smith. 1969. 1401

EGN. 26 Hydrogeologic data from four landfills in northeastern Illinois, by G. M. Hughes, R. A. Landon, and R. N. Farvolden. 1969.
1402

EGN. 27 Evaluating sanitary landfill sites in Illinois, by K. Cartwright and F. B. Sherman. 1969.
1403

EGN. 28 Radiocarbon dating at the Illinois State Geological Survey, by S. M. Kim, R. R. Ruch, and J. P. Kempton. 1969. 1404

EGN. 29 Coordinated mapping of geology and soils for land-use planning, by M. R. McComas, K. C. Hinkley, and J. P. Kempton. 1969.
1405

EGN. 30 Studies of Lake Michigan bottom sediments. No. 1: Preliminary stratigraphy of unconsolidated sediments from the southwestern part of Lake Michigan, by D. L. Gross, J. A. Lineback, W. A. White, N. J. Ayer, C. Collinson, and H. V. Leland. 1970. 1406

EGN. 31 Geologic investigation of the site for an environmental pollution study, by P. B. DuMontelle. 1970. 1407

EGN. 32 Studies of Lake Michigan bottom sediments. No. 2: Distribution of major, minor, and trace constituents in unconsolidated sediments from southern Lake Michigan, by N. F. Shimp, H. V. Leland, and W. A. White. 1970. 1408

EGN. 33 Geology for planning in De Kalb County, Illinois, comp. by D. L. Gross. 1970. 1409

EGN. 34 Sulfur reduction of Illinois coals--washability tests, by R. J.

Helfinstine, J. A. Simon, N. F. Shimp, and M. E. Hopkins. 1970. 1410

EGN. 35 Studies of Lake Michigan bottom sediments. No. 3: Stratigraphy of unconsolidated sediments in the southern part of Lake Michigan, by J. A. Lineback, N. J. Ayer, and D. L. Gross. 1970. 1411

EGN. 36 Geology for planning at Crescent City, Illinois, comp. by R. E. Bergstrom. 1970. 1412

EGN. 37 Studies of Lake Michigan bottom sediments. No. 4: Distribution of arsenic in unconsolidated sediments from southern Lake Michigan, by R. R. Ruch, E. J. Kennedy, and N. F. Shimp. 1970. 1413

EGN. 38 Petrographic and mineralogical characteristics of carbonate rocks related to sorption of sulfur oxides in flue gases, by R. D. Harvey. 1970.
1414

EGN. 39 Studies of Lake Michigan bottom sediments. No. 5: Phosphorus content in unconsolidated sediments from southern Lake Michigan, by J. A. Schleicher and J. K. Kuhn. 1970.
1415

EGN. 40 Power and the environment --a potential crisis in energy supply, by H. E. Risser. 1970. 1416

EGN. 41 Studies of Lake Michigan bottom sediments. No. 6: Trace element and organic carbon accumulation in the most recent sediments of southern Lake Michigan, by N. F. Shimp, J. A. Schleicher, R. R. Ruch, D. B. Heck, and H. V. Leland. 1971. 1417

EGN. 42 A geologist views the environment, by J. C. Frye. 1971.
1418

EGN. 43 Mercury content of Illinois coals, by R. R. Ruch, H. J. Gluskoter, and E. J. Kennedy. 1971. 1419

EGN. 44 Studies of Lake Michigan bottom sediments. No. 7: Distribution of mercury in unconsolidated sediments from southern Lake Michigan, by E. J. Kennedy, R. R. Ruch, and

N. F. Shimp. 1971. 1420

EGN. 45 Summary of findings on solid waste disposal sites in northeastern Illinois, by G. M. Hughes, R. A. Landon, and R. N. Farvolden. 1971. 1421

EGN. 46 Land-use problems in Illinois, by R. E. Bergstrom. 1971. 1422

EGN. 47 Studies of Lake Michigan bottom sediments. No. 8: High-resolution seismic profiles and gravity cores of sediments in southern Lake Michigan, by J. A. Lineback, D. L. Gross, R. P. Meyer, and W. L. Unger. 1971. 1423

EGN. 48 Landslides along the Illinois River Valley south and west of La Salle and Peru, Illinois, by P. B. DuMontelle, N. C. Hester, and R. E. Cole. 1971. 1424

EGN. 49 Environmental quality control and minerals, by H. E. Risser. 1971. 1425

EGN. 50 Petrographic properties of carbonate rocks related to their sorption of sulfur dioxide, by R. D. Harvey and J. C. Steinmetz. 1971. 1426

EGN. 51 Hydrogeologic considerations in the siting and design of landfills, by G. M. Hughes. 1972. 1427

EGN. 52 Preliminary geologic investigations of rock tunnel sites for flood and pollution control in the Greater Chicago Area, by T. C. Buschbach and G. E. Heim. 1972. 1428

EGN. 53 Data from controlled drilling program in Du Page, Kane, and Kendall Counties, Illinois, by P. C. Reed. 1972. 1429

EGN. 54 Studies of Lake Michigan bottom sediments. No. 9: Geologic cross sections derived from seismic profiles and sediment cores from southern Lake Michigan, by J. A. Lineback, D. L. Gross, and R. P.

Meyer. 1972. 1430

EGN. 55 Use of carbonate rocks for control of sulfur dioxide in flue gases. Part I: Petrographic characteristics and physical properties of marls, shells, and their calcines, by R. D. Harvey, R. R. Frost, and J. Thomas. 1972. 1431

EGN. 56 Trace elements in bottom sediments from Upper Peoria Lake, Middle Illinois River--a pilot project, by C. Collinson and N. F. Shimp. 1972. 1432

EGN. 57 Geology, soils, and hydrogeology of Volo Bog and vicinity, Lake County, Illinois, by M. R. McComas, J. P. Kempton, and K. C. Hinkley. 1972. 1433

EGN. 58 Studies of Lake Michigan bottom sediments. No. 10: Depositional patterns, facies, and trace element accumulation in the Waukegan Member of the Late Pleistocene Lake Michigan formation in southern Lake Michigan, by J. A. Lineback and D. L. Gross. 1972. 1434

EGN. 59 Notes on the earthquake of September 15, 1972, in northern Illinois, by P. C. Heigold. 1972. 1435

EGN. 60 Major, minor, and trace elements in sediments of Late Pleistocene Lake Saline compared with those in Lake Michigan sediments, by J. C. Frye and N. F. Shimp. 1973. 1436

EGN. 61 Occurrence and distribution of potentially volatile trace elements in coal: an interim report, by R. R. Ruch, H. J. Gluskoter, and N. F. Shimp. 1973. 1437

EGN. 62 Energy supply problems for the 1970s and beyond, by H. E. Risser. 1973. 1438

EGN. 63 Sedimentology of a beach ridge complex and its significance in land-use planning, by N. C. Hester and G. S. Fraser. 1973. 1439

EGN. 64 The U. S. energy dilemma: the gap between today's requirements and tomorrow's potential, by H. E.

Risser. 1973. 1440

EGN. 65 Survey of Illinois crude oils for trace concentrations of mercury and selenium, by R. F. Mast and R. R. Ruch. 1973.
1441

EGN. 66 Comparison of oxidation and reduction methods in the determination of forms of sulfur in coal, by J. K. Kuhn, L. B. Kohlenberger, and N. F. Shimp. 1973.
1442

EGN. 67 Sediment distribution in a beach ridge complex and its applications to artificial beach replenishment, by G. S. Fraser and N. C. Hester. 1974. 1443

EGN. 68 Lake marls, chalk, and other carbonate rocks with high dissolution rates in SO_2-scrubbing liquors, by R. D. Harvey, R. R. Frost, and J. Thomas. 1974.
1444

EGN. 69 Studies of Lake Michigan bottom sediments. No. 11: Glacial tills under Lake Michigan, by J. A. Lineback, D. L. Gross, and R. P. Meyer. 1974. 1445

EGN. 70 Land resources--its use and analysis, by J. C. Frye. 1974.
1446

EGN. 71 Data from controlled drilling program in Lee and Ogle Counties, Illinois, comp. by P. C. Reed. 1974. 1447

EGN. 72 Occurrence and distribution of potentially volatile trace elements in coal: a final report, by R. R. Ruch, H. J. Gluskoter, and N. F. Shimp. 1974. 1448

EGN. 73 Illinois geology from space, by J. A. Lineback. 1975.
1449

EGN. 74 Studies of Lake Michigan bottom sediments. No. 12: A side-scan sonar investigation of small-scale features on the floor of southern Lake Michigan, by J. M. Berkson, J. A. Lineback, and D. L.

Gross. 1975. 1450

EGN. 75 Data from controlled drilling program in Kane County, Illinois, by P. C. Reed. 1975. 1451

EGN. 76 Bluff erosion, recession rates, and volumetric losses on the Lake Michigan shore in Illinois, by R. C. Berg and C. Collinson. 1976.
1452

EGN. 77 Data from controlled drilling program in Boone and De Kalb Counties, Illinois, by P. C. Reed. 1976.
1453

EGN. 78 Attenuation of pollutants in municipal landfill leachate by clay minerals. Part 1: Column leaching and field verification, by R. A. Griffin, K. Cartwright, N. F. Shimp, J. D. Steele, R. R. Ruch, W. A. White, G. M. Hughes, and R. H. Gilkeson. 1976. 1454

EGN. 79 Attenuation of pollutants in municipal landfill leachate by clay minerals. Part 2: Heavy-metal absorption, by R. A. Griffin, R. R. Frost, A. K. Au, G. D. Robinson, and N. F. Shimp. 1977. 1455

EGN. 80 Supplement to the final report on the hydrogeology of solid waste disposal sites in northeastern Illinois, by G. M. Hughes, J. A. Schleicher, and K. Cartwright. 1976. 1456

EGN. 81 Ground-water contamination: problems and remedial action, by D. E. Lindorff and K. Cartwright. 1977. 1457

EGN. 82 Major, minor, and trace elements of bottom sediments in Lake Du Quoin, Johnston City Lake, and Little Grassy Lake in southern Illinois, by G. B. Dreher, C. B. Muchmore, and D. W. Stover. 1977.
1458

EGN. 83 Geology for planning in De Witt County, Illinois, by C. S. Hunt and J. P. Kempton. 1977. 1459

EGN. 84 Sediments of Lake Michigan, by J. T. Wickham, D. L. Gross, J. A.

Lineback, and R. L. Thomas. 1978.
1460

EGN. 85 Ground-water geology of selected wetlands in Union and Alexander Counties, Illinois, by E. D. McKay, A. Elzeftawy, and K. Cartwright. 1979. 1461

EGN. 86 Attenuation of water-soluble polychlorinated biphenyls by earth materials, by R. A. Griffin, E. S. K. Chian, M. C. Lee, R. R. Clark, A. K. Au, H. Meng, and M. L. Miller. 1979. 1462

Guidebook Series. 1950-

GB. 1 Field conference on Niagaran reefs in the Chicago region, by H. B. Willman, H. A. Lowenstam, and L. E. Workman. 1950. 1463

GB. 2 Central northern Illinois, by J. S. Templeton and H. B. Willman. 1952. 1464

GB. 3 Guide to the structure and Paleozoic stratigraphy along the Lincoln fold in western Illinois, by C. W. Collinson, D. H. Swann, and H. B. Willman. 1954. 1465

GB. 4 The Niagaran reef at Thornton, Illinois, by H. A. Lowenstam, H. B. Willman, and D. H. Swann. 1956. 1466

GB. 5 Loess stratigraphy, Wisconsinan classification and accretiongleys in central western Illinois, by J. C. Frye and H. B. Willman. 1963. 1467

GB. 6 Western Illinois, by C. Collinson. 1964. 1468

GB. 7 Sedimentary structure and morphology of Late Paleozoic sand bodies in southern Illinois, by J. A. Simon and M. E. Hopkins. 1966. 1469

GB. 8 Depositional environments in parts of the Carbondale formation--western and northern Illinois, by W. H. Smith, R. B. Nance, M. E. Hopkins, R. G. Johnson, and C. W.

Shabica. 1970. 1470

GB. 9 Pleistocene stratigraphy of east-central Illinois, by W. H. Johnson, L. R. Follmer, D. L. Gross, and A. M. Jacobs. 1972. 1471

GB. 10 Pennsylvanian conodont assemblages from La Salle County, northern Illinois, by C. Collinson, M. J. Avcin, R. D. Norby, and G. K. Merrill. 1972. 1472

GB. 11 A geologic excursion to fluorspar mines in Hardin and Pope Counties, Illinois. Illinois-Kentucky Mining District and adjacent Upper Mississippi embayment, by J. W. Baxter, C. Bradbury, and N. C. Hester. 1973. 1473

GB. 12 Coastal geology, sedimentology and management of Chicago and the northshore, by C. Collinson, J. A. Lineback, P. B. DuMontelle, and D. C. Brown. 1977. 1474

GB. 13 Wisconsinan, Sangamonian, and Illinoian stratigraphy in central Illinois, by J. A. Lineback, L. R. Follmer, H. B. Willman, E. D. McKay, J. E. King, and N. G. Miller. 1979. 1475

GB. 14 Tri-state geological field conference, by J. D. Treworgy, E. D. McKay, and J. T. Wickham. 1979. 1476

GB. 15 Depositional and structural history of the Pennsylvanian system of the Illinois Basin. Part 1: Road log and descriptions of stops, by J. E. Palmer and R. R. Dutcher, eds. 1979. 1477

GB. 15A Depositional and structural history of the Pennsylvanian system of the Illinois Basin. Part 2: Invited papers, ed. by J. E. Palmer and R. R. Dutcher. 1979. 1478

Report of Investigation. 1924-

RI. 216 Classification of the Genevievian and Chesterian (Late Mississippian) rocks of Illinois, by D. H. Swann. 1963. 1479

RI. 217 Late Paleozoic sandstones of the Illinois Basin, by P. E. Potter. 1963. 1480

RI. 218   Cambrian and Ordovician strata of northeastern Illinois, by T. C. Buschbach.   1964.   1481

RI. 219   Crustal tectonics and Precambrian basement in northeastern Illinois, by L. D. McGinnis.   1966.
1482

RI. 220   Chemistry, uses, and limitations of coal analyses, by O. W. Rees.   1966.   1483

RI. 221   Groundwater geology of the Rock Island, Monmouth, Galesburg, and Kewanee area, Illinois, by J. E. Brueckmann and R. E. Bergstrom.   1968.   1484

Bulletin. 1948-

B. 25 Pleistocene formations in Indiana, by W. J. Wayne. 1963. 1485

B. 26 Bryozoans from the Glen Dean Limestone (Middle Chester) of southern Indiana and Kentucky, by T. G. Perry and A. S. Horowitz. 1963. 1486

B. 27 High-calcium limestone and dolomite in Indiana, by D. J. McGregor. 1963. 1487

B. 28 Geology of Marion County, Indiana, by W. Harrison. 1963. 1488

B. 29 The Stanley Cemetery flora (Early Pennsylvanian) of Greene County, Indiana, by J. M. Wood. 1963. 1489

B. 30 Conodont zones in the Rockford Limestone and the lower part of the New Providence Shale (Mississippian) in Indiana, by C. B. Rexroad and A. J. Scott. 1964. 1490

B. 31 Clays and shales of Indiana, by J. L. Harrison and H. H. Murray. 1964. 1491

B. 32 The Silurian formations of northern Indiana, by A. P. Pinsak and R. H. Shaver. 1964. 1492

B. 33 Trepostomatous bryozoan fauna of the upper part of the White-water Formation (Cincinnatian) of eastern Indiana and western Ohio, by J. Utgaard and T. G. Perry. 1964. 1493

B. 34 Crinoids from the Glen Dean Limestone (Middle Chester) of south-

ern Indiana and Kentucky, by A. S. Horowitz. 1965. 1494

B. 35 Conodonts from the Menard Formation (Chester Series) of the Illinois Basic, by C. B. Rexroad and R. S. Nicoll. 1965. 1495

B. 36 Stratigraphy and conodont paleontology of the Brassfield (Silurian) in the Cincinnati Arch area, by C. B. Rexroad. 1967. 1496

B. 37 Crushed stone resources of the Devonian and Silurian carbonate rocks of Indiana, by R. R. French. 1967. 1497

B. 38 Arenaceous foraminiferida and zonation of the Silurian rocks of northern Indiana, by M. C. Mound. 1968. 1498

B. 39 Geology and mineral resources of Washington County, Indiana, by J. A. Sunderman. 1968. 1499

B. 40 Stratigraphy and conodont paleontology of the Salamonie Dolomite and Lee Creek Member of the Brassfield limestone (Silurian) in southeastern Indiana and adjacent Kentucky, by R. S. Nicoll and C. B. Rexroad, 1968. 1500

B. 41 Conodonts from the Jacobs Chapel Bed (Mississippian) of the New Albany shale in southern Indiana, by C. B. Rexroad. 1969. 1501

B. 42A Gypsum resources of Indiana, by R. R. French and L. F. Rooney. 1969. 1502

B. 42B High-calcium limestone and high-magnesium dolomite resources of Indiana, by L. F. Rooney. 1970. 1503

B. 42C Dimension limestone resources

of Indiana, by L. F. Rooney. 1970.
1504

B. 42D Sand and gravel resources of Indiana, by D. D. Carr and W. M. Webb. 1970.　1505

B. 42E Native Indiana iron ores and 19th century ironworks, by W. J. Wayne. 1970.　1506

B. 42F Specialty sand resources of Indiana, by D. D. Carr. 1971.
1507

B. 42G Marl resources of Indiana, by W. J. Wayne. 1971.　1508

B. 42H Crushed stone aggregate resources of Indiana, by D. D. Carr, R. R. French, and C. H. Ault. 1971.　1509

B. 42I Coal resources of Indiana, by C. E. Wier. 1973.　1510

B. 42J The lime industry of Indiana, by C. H. Ault, L. F. Rooney, and M. V. Palmer. 1974.　1511

B. 42K Cement raw material resources of Indiana, by L. F. Rooney and D. D. Carr, with R. R. French and J. A. Sunderman. 1975. 1512

B. 42L Clay and shale resources of Indiana, by G. S. Austin. 1975.
1513

B. 42M Dimension sandstone resources of Indiana, by D. D. Carr and C. B. Hatfield. 1975.　1514

B. 42N Petroleum industry in Indiana, by G. L. Carpenter, T. A. Dawson, and S. J. Keller. 1975.
1515

B. 42O Peat resources of Indiana, by A. F. Schneider and M. C. Moore, 1978.　1516

B. 43 Compendium of rock-unit stratigraphy in Indiana, by R. H. Shaver and others. 1970.　1517

B. 44 Stratigraphy of the New Albany shale in Indiana, by J. A. Lineback. 1970.　1518

B. 45 Conodonts from Middle Devonian strata of the Michigan Basin, by R. W. Orr. 1971.　1519

B. 46 Applied geology of industrial limestone and dolomite, by L. F. Rooney and D. D. Carr. 1971. 1520

B. 47 Subsurface stratigraphy of the West Baden Group in Indiana, by D. M. Sullivan. 1972.　1521

B. 48 Geometry and origin of Oolite bodies in the Ste. Genevieve limestone (Mississippian) in the Illinois Basin, by D. D. Carr. 1973.　1522

B. 49 Characteristics of Late Wisconsinan tills in eastern Indiana, by A. M. Gooding. 1973.　1523

B. 50 Silurian and Devonian rocks in Indiana southwest of the Cincinnati Arch, by L. E. Becker. 1974. 1524

B. 51 Stratigraphy and conodont paleontology of the Sanders Group (Mississippian) in Indiana and adjacent Kentucky, by R. S. Nicoll and C. B. Rexroad. 1975.　1525

B. 52 Crushed stone resources of the Blue River Group (Mississippian) of Indiana, by D. D. Carr, R. K. Leininger, and M. V. Golde. 1978. 1526

B. 53 Stratigraphy of the Detroit River Formation (Middle Devonian) of northern Indiana, by E. J. Doheny, J. B. Droste, and R. H. Shaver. 1975.
1527

B. 54 Geology of the Catlin-Mansfield area, Parke and Putnam Counties, Indiana, by H. C. Hutchison. 1976.
1528

B. 55 Reduction of sulfur in Indiana coal by washability techniques, by C. E. Wier and H. C. Hutchison. 1977.
1529

B. 56 Polypora M'Coy from the Devonian of southeastern Indiana, by J. F. Stratton and A. S. Horowitz. 1977.　1530

B. 57 Pre-Knox (Cambrian) stratigraphy in Indiana, by L. E. Becker,

A. J. Hreha, and T. A. Dawson.
1978.   1531

B. 58 Stratigraphy and conodont
paleontology of the Cataract Formation and the Salamonie Dolomite
(Silurian) in northeastern Indiana,
by C. B. Rexroad. 1980.   1532

Circular.   1952-1970

C. 9 Pages from the geologic past
of Marion County, by W. Harrison.
1963.   1533

C. 10 Geology of the falls of the
Ohio River, by R. L. Powell.
1970.   1534

Field Conference Guidebook.   1947-

FCG. 11 Geomorphology and ground-
water hydrology of the Mitchell
Plain and Crawford Upland in southern Indiana, by H. H. Gray and R.
L. Powell. 1965.   1535

FCG. 12 Excursions in Indiana geology, by A. M. Burger, C. B. Rexroad, A. F. Schneider, and R. H.
Shaver. 1966.   1536

FCG. 13 A field guide to the Mt.
Carmel Fault of southern Indiana,
by R. H. Shaver and G. S. Austin.
1972.   1537

Mineral Economics Series.   1955-

MES. 9 Oil development and production in Indiana during 1962, by
G. L. Carpenter and S. J. Keller.
1964.   1538

MES. 10 Oil development and production in Indiana during 1963, by
G. L. Carpenter. 1964.   1539

MES. 11 Oil development and production in Indiana during 1964, by
G. L. Carpenter. 1965.   1540

MES. 12 Oil development and production in Indiana during 1965, by
G. L. Carpenter. 1966.   1541

MES. 13 Oil development and production in Indiana during 1966, by G. L.
Carpenter. 1968.   1542

MES. 14 Oil development and production in Indiana during 1967, by G. L.
Carpenter and S. J. Keller. 1968.
                                1543

MES. 15 Oil development and production in Indiana during 1968, by G. L.
Carpenter and S. J. Keller. 1969.
                                1544

MES. 16 Oil development and production in Indiana during 1969, by G. L.
Carpenter and S. J. Keller. 1970.
                                1545

MES. 17 Oil development and production in Indiana during 1970, by G. L.
Carpenter and S. J. Keller. 1972.
                                1546

MES. 18 Oil development and production in Indiana during 1971, by G. L.
Carpenter and S. J. Keller. 1973.
                                1547

MES. 19 Oil development and production in Indiana during 1972, by G. L.
Carpenter and S. J. Keller. 1974.
                                1548

MES. 20 Oil development and production in Indiana during 1973, by G. L.
Carpenter and S. J. Keller. 1974.
                                1549

MES. 21 Oil development and production in Indiana during 1974, by G. L.
Carpenter and S. J. Keller. 1975.
                                1550

MES. 22 Oil development and production in Indiana during 1975, by G. L.
Carpenter and S. J. Keller. 1976.
                                1551

MES. 23 Oil development and production in Indiana during 1976, by G. L.
Carpenter and S. J. Keller. 1977.
                                1552

MES. 24 Oil development and production in Indiana during 1977, by G. L.
Carpenter and S. J. Keller. 1978.
                                1553

MES. 25 Oil development and production in Indiana during 1978, by G. L. Carpenter and S. J. Keller. 1979. 1554

MES. 26 Oil development and production in Indiana during 1979, by G. L. Carpenter, B. D. Keith, and S. J. Keller. 1980. 1555

Occasional Paper. 1974-

OP. 1 Glacial lake sediments in Salt Creek Valley near Bedford, Indiana, by H. H. Gray. 1974. 1556

OP. 2 Haney limestone (Stephensport Group, Chesterian, Mississippian) in Indiana, by H. H. Gray. 1974. 1557

OP. 3 The Muscatatuck Group (New Middle Devonian name) in Indiana, by R. H. Shaver. 1974. 1558

OP. 4 Age and origin of stone quarried near Fort Wayne in the mid-1800's, by M. C. Moore and C. B. Rexroad. 1974. 1559

OP. 5 Age of the Detroit River Formation in Indiana, by J. B. Droste and R. W. Orr. 1974. 1560

OP. 6 Glossary of building stone and masonry terms, by J. B. Patton. 1974. 1561

OP. 7 Ctenoconularia Delphiensis, a new species of the conulata from the New Albany shale (Upper Devonian) at Delphi, Indiana, by D. G. Maroney and R. W. Orr. 1974. 1562

OP. 8 Distribution and significance of some ice-disintegration features in west-central Indiana, by N. K. Bleuer. 1974. 1563

OP. 9 Sedimentation in Lake Lemon, Monroe County, Indiana, by E. J. Hartke and J. R. Hill. 1974. 1564

OP. 10 FORTRAN program for the upward and downward continuation and derivatives of potential fields, by A. J. Rudman and R. F. Blakely. 1975. 1565

OP. 11 The Stone Creek section, a historical key to the glacial stratigraphy of west-central Indiana, by N. K. Bleuer. 1975. 1566

OP. 12 Some environmental geologic factors as aids to planning in Hendricks County, Indiana, by J. R. Hill and G. S. Austin. 1975. 1567

OP. 13 FORTRAN program for generation of synthetic seismograms, by A. J. Rudman and R. F. Blakely. 1976. 1568

OP. 14 FORTRAN program for correlation of stratigraphic time series, by A. J. Rudman and R. F. Blakely. 1976. 1569

OP. 15 The Limberlost dolomite of Indiana, a key to the great Silurian facies in the southern Great Lakes area, by J. B. Droste and R. H. Shaver. 1976. 1570

OP. 16 The seismicity of Indiana described by return periods of earthquake intensities, by R. F. Blakely and M. M. Varma. 1976. 1571

OP. 17 Environmental geologic maps for land use evaluations in Morgan County, Indiana, by E. J. Hartke and J. R. Hill. 1976. 1572

OP. 18 Environmental geologic maps for land use evaluations in Johnson County, Indiana, by J. R. Hill. 1976. 1573

OP. 19 Silurian reefs in southwestern Indiana and their relation to petroleum accumulation, by L. E. Becker and S. J. Keller. 1976. 1574

OP. 20 Pyrite in the Coxville sandstone member, Linton formation, and its effect on acid mine conditions near Latta, Greene County, Indiana, by V. P. Wiram. 1976. 1575

OP. 21 Table of key lines in x-ray powder diffraction patterns of minerals

in clays and associated rocks, by
P. Y. Chen. 1977.   1576

OP. 22 FORTRAN program for
generation of earth tide gravity
values, by A. J. Rudman, R. Ziegler, and R. F. Blakely. 1977.
                                   1577

OP. 23 FORTRAN program for reduction of gravimeter observations
to Bouguer anomaly, by B. D. Kwon,
A. J. Rudman, and R. F. Blakely.
1977.   1578

OP. 24 Late Silurian and Early
Devonian sedimentologic history of
southwestern Indiana, by L. E.
Becker and J. B. Droste. 1978.
                                   1579

OP. 25 Buffalo Wallow group--
Upper Chesterian (Mississippian)
of southern Indiana, by H. H. Gray.
1978.   1580

OP. 26 FORTRAN program for correlation of stratigraphic time series.
Part 2: Power spectral analysis,
by B. D. Kwon, R. F. Blakely, and
A. J. Rudman. 1978.   1581

OP. 27 Geologic story of the Lower
Wabash Valley with emphasis on the
New Harmony area, by R. H. Shaver.
1979.   1582

OP. 28 The Flatwoods region of
Owen and Monroe Counties, Indiana,
by C. A. Malott. 1979.   1583

OP. 29 Application of finite-element
analysis to terrestrial heat flow,
by T. C. Lee, A. J. Rudman, and
A. Sjoreen. 1980.   1584

OP. 30 Stratigraphy and oil fields
in the Salem limestone and associated rocks in Indiana, by S. J.
Keller and L. E. Becker. 1980.
                                   1585

OP. 31 Post-Knox unconformity--
significance at Unionport Gas-Storage
Project and relationship to petroleum exploration in Indiana, by S. J.
Keller and T. F. Abdulkareem.
1980.   1586

Report of Progress. 1946-

RP. 27 Lightweight aggregate potential
of the New Albany shale in northwestern Indiana, by L. F. Rooney and J. A.
Sunderman. 1964.   1587

RP. 28 The Crawfordsville and Knightstown moraines in Indiana, by W. J.
Wayne. 1965.   1588

RP. 29 The Sanders Group and subjacent
Muldraugh Formation (Mississippian)
in Indiana, by N. M. Smith, 1965.
                                   1589

RP. 30 Glacial lake deposits in southern Indiana--engineering problems and
land use, by H. H. Gray. 1971. 1590

RP. 31 Some Pennsylvanian Kirkbyacean ostracods of Indiana and midcontinent series terminology, by R. H.
Shaver and S. G. Smith. 1974. 1591

Special Report. 1963-

SR. 1 Underground storage of natural
gas in Indiana, by T. A. Dawson and
G. L. Carpenter. 1963.   1592

SR. 2 Geology of the Upper Patoka
drainage basin, by H. H. Gray. 1963.
                                   1593

SR. 3 Geology of the Upper East Fork
drainage basin, Indiana, by A. F. Schneider and H. H. Gray. 1966.   1594

SR. 4 Sand and gravel resources in
eastern Johnson County and western
Shelby County, Indiana, by D. D. Carr.
1966.   1595

SR. 5 Geologic consideration in planning solid-waste disposal sites in Indiana, by N. K. Bleuer. 1970.   1596

SR. 6 Coal strip-mined land in Indiana, by R. L. Powell. 1972.   1597

SR. 7 Lithostratigraphy of the Maquoketa Group (Ordovician) in Indiana,
by H. H. Gray. 1972.   1598

SR. 8 The Indiana Dunes--legacy of
sand, by J. R. Hill. 1974.   1599

SR. 9 Pyrite in the Springfield Coal Member (V), Petersburg Formation, Sullivan County, Indiana, by I. U. Khawaja. 1975. 1600

SR. 10 Urban geology of Madison County, Indiana, by W. J. Wayne. 1975. 1601

SR. 11 Environmental geology of Lake and Porter Counties, Indiana --an aid to planning, by E. J. Hartke, J. R. Hill, and M. Reshkin. 1975. 1602

SR. 12 Environmental geology of the Evansville area, southwestern Indiana, by W. T. Straw, H. H. Gray, and R. L. Powell. 1977. 1603

SR. 13 Environmental geology of Allen County, Indiana, by N. K. Bleuer and M. C. Moore. 1978. 1604

SR. 14 Geology as a contribution to land use planning in LaPorte County, Indiana, by J. R. Hill, D. D. Carr, E. J. Hartke, and C. B. Rexroad. 1979. 1605

SR. 15 The search for a Silurian reef model: Great Lakes area, by R. H. Shaver and others. 1978. 1606

SR. 16 Conodonts from the Louisville limestone and the Wabash Formation (Silurian) in Clark County, Indiana, and Jefferson County, Kentucky, by C. B. Rexroad, A. V. Noland, and C. A. Pollock. 1978. 1607

SR. 17 The Plummer Field, Greene County, Indiana, by J. A. Noel. 1979. 1608

SR. 18 Middle Devonian chitinozoa of Indiana, by R. P. Wright. 1980. 1609

SR. 19 Geology for environmental planning in Marion County, Indiana, by E. J. Hartke and others. 1980. 1610

State Park Guide. 1974-

SPG. 1 Geologic story of Pokagon State Park, legacy of Indiana's ice age, by N. K. Bleuer. 1974. 1611

SPG. 2 Geologic story of Clifty Falls State Park, by C. B. Rexroad. 1975. 1612

SPG. 3 Geologic story of McCormicks Creek State Park, by H. H. Gray. 1977. 1613

SPG. 4 Geologic story of Shades State Park, by K. H. P. Bridges. 1977. 1614

SPG. 5 Geologic story of Turkey Run State Park, by K. H. P. Bridges, 1977. 1615

SPG. 6 Geologic story of Versailles State Park, by C. B. Rexroad. 1977. 1616

SPG. 7 Geologic story of Spring Mill State Park, by C. B. Rexroad and L. M. Gray. 1979. 1617

SPG. 8 Geologic story of Indiana Dunes State Park, by J. R. Hill. 1979. 1618

Miscellaneous Publication. 1918-

MP. 1 A description of the region about Camp Dodge, by J. H. Lees. 1918. 1619

MP. 2 Sketch of the geology of Iowa, by G. F. Kay and J. H. Lees. 1920. 1620

MP. 3 The Pleistocene geology of Iowa. Part 1: The Pre-Illinoian Pleistocene geology of Iowa, by G. F. Kay and E. T. Apfel. 1621

Part 2: The Pre-Illinoian and Post-Illinoian Pleistocene geology of Iowa, by G. F. Kay and J. B. Graham. 1622

Part 3: Bibliography of the Pleistocene geology of Iowa. 1623

MP. 4 Highway construction materials from the consolidated rocks of southwestern Iowa, by H. G. Hershey, C. N. Brown, O. J. Van Eck, and R. C. Northup. 1960. 1624

MP. 5 Bibliography of the geology of Iowa, 1960-1964, by P. J. Horick, J. C. Prior, and E. E. Hinman. 1967. 1625

MP. 6 Summary of oil and gas tests in Iowa. Rev. annually. 1626

Public Information Circular. 1971-

PIC. 1 Optimal well-plugging procedures, by O. J. Van Eck. 1971. 1627

PIC. 2 Resume of oil exploration and potential in Iowa, by M. C. Parker. 1971. 1628

PIC. 3 Proceedings seminar in applied remote sensing, May 8-12, 1972. 1629

PIC. 4 Hydrogeologic considerations in solid waste storage in Iowa, by S. J. Tuthill, D. L. Gordon, and F. H. Dorheim. 1972. 1630

PIC. 5 Proceedings, eighth forum on geology of industrial minerals, April 12-14, 1972. 1973. 1631

PIC. 6 Development and testing of operational flood mapping techniques, by B. E. Hoyer. 1979. 1632

PIC. 7 Summary of multispectral flood inundation mapping in Iowa, by B. E. Hoyer, G. R. Hallberg, and J. V. Taranik. 1974. 1633

PIC. 8 Guide to aerial imagery of Iowa, by R. R. Anderson, B. E. Hoyer, and J. V. Taranik. 1974. 1634

PIC. 9 Not published? 1635

PIC. 10 Underground mining in Iowa, exclusive of coals, by F. H. Dorheim. 1979. 1636

PIC. 11 Plugging procedures for domestic wells, by O. J. Van Eck. 1978. 1637

Report of Investigations. 1964-

RI. 1 Stratigraphy of the Osage Series in southeastern Iowa, by S. E. Harris and M. C. Parker. 1964. 1638

RI. 2 Glacial drift in the "Driftless Area" of northeast Iowa, by A. C. Trowbridge. 1966. 1639

RI. 3 Preliminary report on basement complex rocks of Iowa, by W. H. Yoho. 1967. 1640

RI. 4 LaPorte City chert--a Devonian subsurface formation in Iowa, by M. C. Parker. 1967.
1641

RI. 5 A new Upper Devonian cystoid attached to a discontinuity surface, by D. L. Koch and H. L. Strimple. 1968.. 1642

RI. 6 A biothermal facies in the Silurian of eastern Iowa, by E. E. Hinman. 1968. 1643

RI. 7 Iowan drift problem, northeastern Iowa, by R. V. Ruhe, W. P. Dietz, T. E. Fenton, and G. F. Hall. 1968. 1644

RI. 8 Iowa gravity base station network, by D. H. Hase, R. B. Campbell, and O. J. Van Eck. 1969.
1645

RI. 9 Yellow Spring Group of the Upper Devonian in Iowa, by F. H. Dorheim, D. L. Koch, and M. C. Parker. 1969. 1646

RI. 10 Stratigraphy of the Upper Devonian shell rock formation of north-central Iowa, by D. L. Koch. 1970. 1647

RI. 11 Gravity survey of the Randalia magnetic anomaly in Fayette County, Iowa, by J. L. Gilmore. 1976. 1648

RI. 12 The Plum River Fault Zone in east-central Iowa, by B. J. Bunker and G. A. Ludvigson. 1979.
1649

Special Report. 1979-

SR. 1 Changes in the channel area of the Missouri River in Iowa, 1879-1976, by G. R. Hallberg, J. M. Harbaugh, and P. M. Witinok. 1979. 1650

Technical Information Series. 1976-

TIS. 1 A thermal model for the surface temperature of materials on the earth's surface, by L. K. Kuiper.

1976. 1651

TIS. 2 Summary of ADP drill hole information. Part I: Northeast Iowa, by D. J. Gockel. 1976. 1652

TIS. 3 A late-glacial pollen sequence from northeast Iowa; Summer Bog revisited, by G. R. Hallberg and K. L. Van Zant. 1976. 1653

TIS. 4 Land-use in Iowa: 1976; an explanation of the map, by R. R. Anderson. 1976. 1654

TIS. 5 Not published? 1655

TIS. 6 Fluvial sediment data for Iowa: suspended-sediment concentrations, loads and sizes; bed-material sizes; and reservoir siltation, by J. R. Schuetz and W. J. Matthes. 1977.
1656

TIS. 7 Highway soil engineering data for major soils in Iowa, by G. A. Miller, J. D. Highland, and G. R. Hallberg. 1978. 1657

TIS. 8 Standard procedures for evaluation of Quaternary materials in Iowa, ed. by G. R. Hallberg. 1978.
1658

TIS. 9 Not yet published. 1659

TIS. 10 Not yet published. 1660

TIS. 11 Status of hydrogeologic studies in northwest Iowa, by B. J. Bunker and G. A. Ludvigson. 1979. 1661

Technical Paper. 1929-

TP. 4 Coal resources of Iowa, by E. R. Landis and O. J. Van Eck. 1965. 1662

TP. 5 Chemical analyses of selected Iowa coals, including trace element data, by M. J. Avcin and J. R. Hatch. 1979. 1663

TP. 6 An introduction to the stratigraphic palynology of the Cherokee Group (Pennsylvanian) coals of Iowa, by R. L. Ravn. 1979. 1664

TP. 7 Stratigraphic ranges of mio-

spores in coals of the Des Moines Series of southern Iowa, by R. L. Ravn. 1979. 1665

WSB. 13 Geohydrology of the Silurian-Devonian carbonate units in the Cedar Rapids area, by K. D. Wahl and B. J. Bunker. 1979. 1677

Water Atlas. 1965-

WA. 1 The water story in central Iowa, by F. R. Twenter and R. W. Coble. 1965.

WA. 2 Availability of ground water in Decatur County, Iowa, by J. W. Cagle and W. L. Steinhilber. 1967. 1667

WA. 3 Availability of ground water in Wayne County, Iowa, by J. W. Cagle. 1969. 1668

WA. 4 Water resources of southeast Iowa, by R. W. Coble. 1971. 1669

WA. 5 Water resources of south-central Iowa, by J. W. Cagle and A. J. Heinitz. 1978. 1670

WA. 6 Water resources of east-central Iowa, by K. D. Wahl and others. 1978. 1671

Water Supply Bulletin. 1942-

WSB. 8 Surface water resources of Iowa from October 1, 1955, to September 30, 1960, by R. E. Myers. 1963. 1672

WSB. 9 Geology and ground-water resources of Cerro Gordo County, Iowa, by H. G. Hershey, K. D. Wahl, and W. L. Steinhilber. 1970. 1673 1673

WSB. 10 Geology and ground-water resources of Linn County, Iowa, by R. E. Hansen. 1970. 1674

WSB. 11 Geohydrology of Muscatine Island, Muscatine County, Iowa, by R. E. Hansen and W. L. Steinhilber. 1977. 1675

WSB. 12 Alluvial ground-water resources of the Floyd River Basin, by K. D. Wahl and others. 1979. 1676

Bulletin. 1913-

B. 161 Geology and ground-water resources of Wallace County, Kansas, by W. G. Hodson. 1963.
1678

B. 162 The geologic history of Kansas, by D. F. Merriam. 1963.
1679

B. 163 Geology of Franklin County, Kansas, by S. M. Ball, M. M. Ball, and D. J. Laughlin. 1963. 1680

B. 164 Paleoecology and biostratigraphy of the Red Eagle Cyclothem (Lower Permian) in Kansas, by A. W. McCrone. 1963. 1681

B. 165 1963 reports of studies.
1682

Part 1: Preliminary report of conodonts of the Meramecian Stage (Upper Mississippian) from the subsurface of western Kansas, by T. L. Thompson and E. D. Coebel. 1683

Part 2: Sources of error in thermoluminescence studies, by J. M. McNellis. 1684

Part 3: Test-hole exploration for light-firing clay in Cloud and Ellsworth Counties, Kansas, by N. Plummer, C. S. Edmonds, and M. P. Bauleke. 1685

Part 4: Kansas basement rocks committee report for 1962 and additional Precambrian wells, by V. B. Cole, D. F. Merriam, and W. W. Hambleton. 1686

Part 5: Preliminary report on the beneficiation of some Kansas clays and shales, by W. E. Hill, W. B. Hladik, and W. N. Waugh. 1687

Part 6: New serial micropeel technique, by S. Honjo. 1688

Part 7: Rates of solution of limestone using the chelating properties of Versene (EDTA) compounds, by W. E. Hill and E. D. Goebel. 1689

B. 166 Oil and gas developments in Kansas during 1962, by E. D. Goebel, P. L. Hilpman, M. O. Oros, and D. L. Beene. 1963. 1690

B. 167 Ground-water levels in observation wells in Kansas, 1962, by M. E. Broeker and J. D. Winslow. 1963. 1691

B. 168 Geohydrology of Grant and Stanton Counties, Kansas, by S. W. Fader, E. D. Gutentag, D. H. Lobmeyer, and W. R. Meyer. 1964.
1692

B. 169 Symposium on cyclic sedimentation, ed. by D. F. Merriam. 1964. 2 vols. 1693

B. 170 1964 reports of studies. 1694

Part 1: Archaeolithophyllum, an abundant calcareous alga in limestones of the Lansing Group (Pennsylvanian), southeastern Kansas, by J. L. Wray.
1695

Part 2: Precambrian-Paleozoic contact in two wells in northwestern Kansas, by R. W. Scott and M. N. McElroy. 1696

Part 3: Mathematical conversion of section, township, and range notation to Cartesian coordinates, by D. I. Good. 1697

Part 4: Activities of the Kansas Basement Rocks Committee in 1963 and additional Precambrian wells, by V. B.

Cole, D. F. Merriam, and W. W. Hambleton. 1698

Part 5: Paleoecology of the Council Grove Group (Lower Permian) in Kansas, based on microfossil assemblages, by N. G. Lane. 1699

Part 6: Upper Pennsylvanian calcareous rocks cored in two wells in Rawlins and Stafford Counties, Kansas, by J. W. Harbaugh and W. Davie. 1700

Part 7: The amenability of Kansas clays to alumina extraction by hydrochloric acid treatment, by W. N. Waugh, W. E. Hill, O. K. Galle, and W. B. Hladik. 1701

B. 171 A computer method of four-variable analysis illustrated by a study of oil-gravity variations in southeastern Kansas, by J. W. Harbaugh. 1964. 1702

B. 172 Oil and gas developments in Kansas during 1963, by P. L. Hilpman, M. O. Oros, D. L. Beene, and E. D. Goebel. 1964. 1703

B. 173 Ground-water levels in observation wells in Kansas, 1963, by M. E. Broeker and J. D. Winslow. 1964. 1704

B. 174 Geology and ground-water resources in Trego County, Kansas, by W. G. Hodson. 1965. 1705

B. 175 1965 reports of studies. 1706

Part 1: Internal structures of "homogeneous" sandstones, by W. K. Hamblin. 1707

Part 2: Ice-push deformation in northeastern Kansas, by L. F. Dellwig and A. D. Baldwin. 1708

Part 3: Solubility of twenty minerals in selected versene (EDTA) solutions, by W. E. Hill and D. R. Evans. 1709

Part 4: The Pleasanton Group (Upper Pennsylvanian) in Kansas, by J. M. Jewett, P. A. Emery, and

D. A. Hatcher. 1710

B. 176 Geohydrology of Sedgwick County, Kansas, by C. W. Lane and D. E. Miller. 1965. 1711

B. 177 Ground-water levels in observation wells in Kansas, 1964, by M. E. Broeker and J. D. Winslow. 1965. 1712

B. 178 Stratigraphy of the Graneros Shale (Upper Cretaceous) in central Kansas, by D. E. Hattin. 1965. 1713

B. 179 Oil and gas developments in Kansas during 1964, by D. L. Beene and M. O. Oros. 1965. 1714

B. 180 1966 reports of studies. 1715

Part 1: Compositional variance in the Plattsmouth limestone member (Pennsylvanian) in Kansas, by O. K. Galle and W. N. Waugh. 1716

Part 2: Analyses of high-calcium chert-free beds in the Keokuk limestone, Cherokee County, Kansas, by T. C. Waugh. 1717

B. 181 Geology and ground-water resources of Miami County, Kansas, by D. E. Miller. 1966. 1718

B. 182 Bibliography and index of ground water in Kansas, by R. S. Roberts and W. G. Hodson. 1966. 1719

B. 183 Geology and ground-water resources of Neosho County, Kansas, by W. L. Jungmann. 1966. 1720

B. 184 Ground-water levels in observation wells in Kansas, 1965, by M. E. Broeker and J. D. Winslow. 1966. 1721

B. 185 Oil and gas developments in Kansas during 1965, by D. L. Beene and M. O. Oros. 1967. 1722

B. 186 Geology and ground-water resources of Brown County, Kansas, by C. K. Bayne and W. H. Schoewe. 1967. 1723

B. 187 1967 reports of studies. 1724

Part 1: Short papers on research in 1966, ed. by D. E. Zeller. 1725

Part 2: Progress report on the ground-water hydrology of the Equus-beds area, Kansas--1966, by G. J. Stramel. 1726

Part 3: Stratigraphy and depositional environment of the Elgin sandstone (Pennsylvanian) in southcentral Kansas, by S. L. Brown. 1727

Part 4: Recovery of phosphate from the Cabaniss and Pleasanton shales of Kansas, by K. E. Rose and R. G. Hardy. 1728

Part 5: Sedimentary characteristics of Pleistocene deposits, Neosho River Valley, southeastern Kansas, by S. Jamhindikar. 1729

B. 188 Ground water in the Republican River area, Cloud, Jewell, and Republic Counties, Kansas, by S. W. Fader. 1968. 1730

B. 189 The stratigraphic succession in Kansas, ed. by D. E. Zeller. 1968. 1731

B. 190 Oil and gas developments in Kansas during 1966, by M. O. Oros and D. L. Beene. 1968. 1732

B. 191 1968 reports of studies. 1733

Part 1: Not published? 1734

Part 2: A study of the joint patterns in gently dipping sedimentary rocks of south-central Kansas, by J. R. Ward. 1735

B. 192 Conodonts and stratigraphy of the Meramecian Stage (Upper Mississippian) in Kansas, by T. L. Thompson and E. D. Goebel. 1968. 1736

B. 193 Geology and ground-water resources of Linn County, Kansas, by W. J. Seevers. 1969. 1737

B. 194 1969 reports of studies. 1738

Part 1: Short papers on research in 1968, ed. by D. E. Zeller. 1739

Part 2: Marine paleocurrent directions in limestones of the Kansas City Group (Upper Pennsylvanian) in eastern Kansas, by W. K. Hamblin. 1740

B. 195 Geology and ground-water resources of Allen County, Kansas, by D. E. Miller. 1969. 1741

B. 196 Geology and ground-water resources of Decatur County, Kansas, by W. G. Hodson. 1969. 1742

B. 197 Depositional facies of Toronto limestone member (Oread limestone, Pennsylvanian), subsurface marker unit in Kansas, by A. R. Troell. 1969. 1743

B. 198 Algal-bank complex in Wyandotte limestone (Late Pennsylvanian) in eastern Kansas, by D. J. Crowley. 1969. 1744

B. 199 1970 reports of studies. 1745

Part 1: Short papers on research in 1969, ed. by D. E. Zeller. 1746

Part 2: Rb-Sr geochronologic investigation of basic and ultrabasic xenoliths from the Stockdale kimberlite, Riley County, Kansas, by D. G. Brookins and M. J. Woods. 1747

Part 3: High pressure mineral reactions in a pyroxenite granulite nodule from the Stockdale kimberlite, Riley County, Kansas, by D. G. Brookins and M. J. Woods. 1748

Part 4: Factors governing emplacement of Riley County, Kansas, kimberlites, by D. G. Brookins. 1749

Part 5: Inventory of industrial, metallic, and solid-fuel minerals in Kansas, by R. G. Hardy. 1750

B. 200 The kimberlites of Riley County, Kansas, by D. G. Brookins. 1970. 1751

B. 201 Geology and ground-water resources of Ellsworth County, central Kansas, by C. K. Bayne, P. C. Franks, and W. Ives. 1971. 1752

B. 202 1971 reports of studies. 1753

Part 1: Short papers on research in 1970, ed. by J. A. Crossfield. 1754

Part 2: Techniques for determining coccolith abundance in shaly chalk of Greenhorn limestone (Upper Cretaceous) of Kansas, by D. E. Hattin and D. A. Darko. 1971. 1755

Part 3: Patrognathus and Siphonodella (conodanta) from the Kinderhookian (Lower Mississippian) of western Kansas by southwestern Nebraska, by G. Klapper. 1971. 1756

Part 4: Geohydrology of Jefferson County, northeastern Kansas, by J. D. Winslow. 1972. 1757

B. 203 Geology and ground-water resources of Johnson County, northeastern Kansas, by H. G. O'Connor. 1971. 1758

B. 204 1972 reports of studies. 1759

Part 1: Short papers on research in 1971, ed. by J. A. Kellogg. 1972. 1760

Part 2: Gravity and magnetic survey of an abandoned lead and zinc mine in Linn County, Kansas, by H. Yarger and S. Z. Jarjur. 1972. 1761

Part 3: Modeling discharge and conservation water quality in the lower Kansas River Basin, by W. J. O'Brien, P. B. MacRoberts, E. C. Pogge, and R. L. Smith. 1972. 1762

Part 4: Structural geology of the Manhattan, Kansas, area, by J. R. Chelikowsky. 1972. 1763

B. 205 Geology and ground-water resources of Pratt County: south-central Kansas, by D. W. Layton and D. W. Berry. 1973. 1764

B. 206 1973-74 reports of studies. 1765

Part 1: Carbonate facies of the Swope limestone (Upper Pennsylvanian), southeast Kansas, by J. H. Mossler. 1973. 1766

Part 2: Ground-water in the Kansas River Valley: Junction City to Kansas City, Kansas, by S. W. Fader. 1974. 1767

Part 3: Geology and hydrology of Rice County, central Kansas, by C. K. Bayne and J. R. Ward. 1974. 1768

B. 207 Geology and ground-water resources of Rush County, central Kansas, by J. M. McNellis. 1974. 1769

B. 208 Using soils of Kansas for waste disposal, by G. W. Olson. 1974. 1770

B. 209 Stratigraphy and depositional environment of Greenhorn limestone (Upper Cretaceous) of Kansas, by D. E. Hattin. 1975. 1771

B. 210 Stratigraphic and depositional framework of the Stanton formation in southeastern Kansas, by P. H. Heckel. 1975. 1772

B. 211 1976 reports of studies. 1773

Part 1: Pleistocene drainage reversal in the upper Tuttle Creek Reservoir area of Kansas, by J. R. Chelikowsky. 1976. 1774

Part 2: Geology and structure of Cheyenne Bottoms, Barton County, Kansas, by C. K. Bayne. 1977. 1775

Part 3: Age and structure of subsurface beds in Cherokee County, Kansas --implications from Endothyrid foraminifera and conodonts, by D. E. Nodine-Zeller and T. L. Thompson. 1977. 1776

Part 4: Short papers on research in 1976-77, ed. by G. A. Waldron. 1978. 1777

B. 212 Criteria for making and interpreting a soil profile description: a compilation of the official USDA procedure and nomenclature for describing soils, by G. W. Olson. 1976. 1778

B. 213 Bibliography and index to Kansas geology, by the American Geological Institute. 1977. 1779

B. 214 Land subsidence in central Kansas related to sale dissolution, by R. F. Walters. 1978. 1780

B. 215 Deposition of evaporites and red beds of the Nippewalla Group, Permian, western Kansas, by K. Holdoway. 1978. 1781

B. 216 Stratigraphy of the Lower Permian Wreford megacyclothem in southernmost Kansas and northern Oklahoma, by A. B. Lutz-Garihan and R. J. Cuffey. 1979. 1782

Chemical Quality Series. 1975-

CQS. 1 Changes in chemical quality of water, Cedar Bluff Irrigation District area, west-central Kansas, by R. B. Leonard. 1975. 1783

CQS. 2 Chemical quality of irrigation waters in west-central Kansas, by L. R. Hathaway, L. M. Magnuson, B. L. Carr, O. K. Galle, and T. C. Waugh. 1975. 1784

CQS. 3 Saline water in the Little Arkansas River Basin area, south-central Kansas, by R. B. Leonard and M. K. Kleinschmidt. 1976. 1785

CQS. 4 Chemical quality of irrigation waters in Hamilton, Kearny, Finney, and northern Gray Counties, by L. R. Hathaway, B. L. Carr, O. K. Galle, L. M. Magnuson, T. C. Waugh, and H. P. Dickey. 1977. 1786

CQS. 5 Ground water from Lower

Cretaceous rocks in Kansas, by K. M. Keene and C. K. Bayne. 1978. 1787

CQS. 6 Chemical quality of irrigation waters in southwestern Kansas, by L. R. Hathaway, B. L. Carr, M. A. Flanagan, O. K. Galle, T. C. Waugh, H. P. Dickey, and L. M. Magnuson. 1978. 1788

CQS. 7 Chemical quality of irrigation waters in Ford County and the Great Bend Prairie of Kansas, by L. R. Hathaway, O. K. Galle, T. C. Waugh, and H. P. Dickey. 1978. 1789

Educational Series. 1975-

ES. 1 Ancient life found in Kansas rocks (common fossils of Kansas), by R. B. Williams. 1975. 1790

ES. 2 Kansas rocks and minerals, by L. L. Tolsted and A. Swineford. 1957. 1791

ES. 3 Kansas clays for the ceramic hobbyist, by D. A. Grisafe and M. Bauleke. 1977. 1792

ES. 4 Kansas Geological Survey publications: a complete list and cross-index, by R. A. Hardy, B. Welter, and C. K. Bayne, T. McClain, D. Schlobaum, and G. Waldron. 1980. 1793

ES. 5 Kansas landscapes: a geologic diary, by F. W. Wilson. 1978. 1794

Energy Resources Series. 1973-

ERS. 1 Secondary recovery and pressure maintenance operations in Kansas 1972, by R. L. Dilts. 1973. 1795

ERS. 2 1972 oil and gas production in Kansas, by D. L. Beene. 1974. 1796

ERS. 3 Enhanced oil-recovery operations in Kansas, 1973, by M. O. Oros and D. K. Saile. 1974. 1797

ERS. 4 1973 oil and gas production in Kansas, by D. L. Beene. 1975. 1798

ERS. 5 Enhanced oil-recovery operations in Kansas, 1974, by M. O. Oros and D. K. Saile. 1975. 1799

ERS. 6 1974 oil and gas production in Kansas, by D. L. Beene. 1976. 1800

ERS. 7 Enhanced oil recovery operations in Kansas, 1975, by D. K. Saile and M. O. Oros. 1976. 1801

ERS. 8 1975 oil and gas production in Kansas, by D. L. Beene. 1977. 1802

ERS. 9 Enhanced recovery operations in Kansas, 1976, by M. O. Oros, D. K. Saile, R. Sherman, and C. Woods. 1977. 1803

ERS. 10 1976 oil and gas production in Kansas, by D. L. Beene. 1977. 1804

ERS. 11 Enhanced oil recovery operations in Kansas, 1977, by M. O. Oros, D. K. Saile, and R. Sherman. 1978. 1805

ERS. 12 1977 oil and gas production in Kansas, by D. L. Beene. 1980. 1806

ERS. 13 Oil and gas in eastern Kansas: a 25-year update, by M. O. Oros. 1807

Environmental Geology Series. 1977-

EGS. 1 Contemporary Kansas maps: selected products for map users, by T. McClain. 1977. 1808

EGS. 2 List of earthquake intensities for Kansas, 1867-1967, by S. M. DuBois and F. W. Wilson. 1978. 1809

Geology Series. 1975-

GS. 1 Description of the surficial rocks in Cherokee County, southeastern Kansas, by W. J. Seevers. 1975. 1810

GS. 2 Potential uranium host rocks and structures in the central Great Plains, by E. J. Zeller, G. Dreschhoff, E. Angino, K. Holdoway, W. Hakes, G. Jayaprakash, and K. Crisler. 1976. 1811

Ground-Water Releases. 197?-

GWR. 1 Not published? 1812

GWR. 2 Not published? 1813

GWR. 3 Ground-water levels in observation wells in Kansas, 1966-70, by M. E. Broeker and J. M. McNellis. 1973. 1814

GWR. 4 Hydrogeologic data from Greeley, Wichita, Scott, and Lane Counties, Kansas, by L. E. Stullken, E. C. Weakly, E. D. Gutentag, and S. E. Slagle. 1974. 1815

GWR. 5 Hydrogeologic data from the Great Bend Prairie, south-central Kansas, by L. E. Stullken and S. W. Fader. 1976. 1816

GWR. 6 Ground-water levels in observation wells in Kansas, 1971-75, by M. E. Broeker, H. J. McIntyre, and J. M. McNellis. 1980. 1817

Ground Water Series. 1974-

GWS. 1 Geology and ground-water resources of Montgomery County, southeastern Kansas, by H. G. O'Connor. 1974. 1818

GWS. 2 Geohydrology of Nemaha County, northeastern Kansas, by J. R. Ward. 1974. 1819

Guidebook Series. 1976-

GBS. 1 Guidebook: 24th annual meeting, Midwestern Friends by the Pleistocene (Meade County, Kansas), by C. K. Bayne, J. Boellstorff, and B. Miller. 1976. 1820

GBS. 2 Guidebook: Upper Pennsylvanian cyclothemic limestone facies in eastern Kansas, by P. H. Heckel.

1978. 1821

GBS. 3 Guidebook: Upper Cretaceous stratigraphy and depositional environments of western Kansas, by D. E. Hattin and C. T. Siemers. 1978. 1822

Irrigation Series. 1976-

IS. 1 Ground-water resources of Lane and Scott Counties, western Kansas, by E. D. Gutentag and L. E. Stullken. 1976. 1823

IS. 2 Ground-water resources of Greeley and Wichita Counties, western Kansas, by S. E. Slagle and E. C. Weakly. 1976. 1824

IS. 3 Water-resources reconnaissance of Ness County, west-central Kansas, by E. D. Jenkins and M. E. Pabst. 1977. 1825

IS. 4 Geohydrology of the Great Bend Prairie, south-central Kansas, by S. W. Fader and L. E. Stullken. 1978. 1826

IS. 5 Water in the Dakota Formation, Hodgeman and northern Ford Counties, southwestern Kansas, by D. H. Lobmeyer and E. C. Weakly. 1979. 1827

Mineral Resources Series. 1973-

MRS. 1 Kansas mineral industry report, 1972, by R. G. Hardy. 1973. 1828

MRS. 2 Kansas mineral industry report, 1973, with directory of Kansas mineral producers, by D. A. Grisafe. 1974. 1829

MRS. 3 Kansas mineral industries report, 1974, by P. Berendsen. 1975. 1830

MRS. 4 Kansas building limestones, by D. A. Grisafe. 1977. 1831

MRS. 5 An evaluation of the strippable coal reserves in Kansas, by L. L. Brady, D. B. Adams, and N. D.

Livingston. 1977. 1832

MRS. 6 Kansas mineral industry report, 1975-1977, by C. Zarley and D. Collins. 1978. 1833

MRS. 7 A preliminary report on the environmental effects of strip mining and reclamation in southeast Kansas, by C. Zarley, L. Brady, R. G. Hardy, D. Grisafe, H. Dickey, K. Keene, and C. Bass. 1980. 1834

MRS. 8 Directory of Kansas mineral producers, 1978, by H. Wolfe, L. Brady, and G. Romero. 1978. 1835

Series on Spatial Analysis. 1977-

SSA. 1 SURFACE II Graphics System (revision 1), by R. J. Sampson. 1977. 1836

SSA. 2 Optimum mapping techniques using regionalized variable theory, by R. A. Olea. 1975. 1837

SSA. 3 Measuring spatial dependence with semivariograms, by R. A. Olea. 1977. 1838

Special Distribution Publication. 1962-

SDP. 1 The Kansas mineral industry, 1962, by G. Muilenburg, R. G. Hardy, and A. Hornbaker. 1962. 1839

SDP. 2 Economic development for Kansas; a sector report on its mineral and water resources. 1962. 1840

SDP. 3 BALGOL program for trend-surface mapping using an IBM 7090 computer, by J. W. Harbaugh. 1963. 1841

SDP. 4 FORTRAN II program for coefficient of association (MATCH-COEFF) using an IBM 1620 computer, by R. L. Kaesler, F. W. Preston, and D. Good. 1963. 1842

SDP. 5 Activities of State Geological Survey of Kansas, biennium ending June 30, 1963, by G. Muilenburg. 1963. 1843

SDP. 6 Secondary recovery operations in Kansas during 1962, by the Kansas Recovery Committee. 1964.
1844

SDP. 7 The Kansas mineral industry. 1963.
1845

SDP. 8 Annotated bibliography of the Kansas Precambrian, by D. F. Merriam. 1964.
1846

SDP. 9 BALGOL programs for calculation of distance coefficients and correlation coefficients using an IBM 7090 computer, by J. W. Harbaugh. 1964.
1847

SDP. 10 Water-level changes in Grant and Stanton Counties, 1939-1964, by J. W. Winslow, C. E. Nuzman, and S. W. Fader. 1964.
1848

SDP. 11 Trend-surface analysis of regional and residual components of geologic structure in Kansas, by D. F. Merriam and J. W. Harbaugh. 1964.
1849

SDP. 12 FORTRAN and FAP programs for calculating and plotting time-trend curves using an IBM 7090 and 7094/1401 computer system, by W. T. Fox. 1964.
1850

SDP. 13 FORTRAN program for factor and vector analysis of geologic data using an IBM 7090 or 7094/1401 computer system, by V. Manson and J. Imbrie. 1964.
1851

SDP. 14 FORTRAN II trend-surface program for the IBM 1620, by D. I. Good. 1964.
1852

SDP. 15 Application of factor analysis to petrologic variations of Americus limestone (Lower Permian), Kansas and Oklahoma, by J. W. Harbaugh and F. D. Emirmen. 1964.
1853

SDP. 16 Secondary recovery operations in Kansas, 1963, by the Kansas Secondary Recovery and Pressure Maintenance Committee. 1964.
1854

SDP. 17 The Kansas mineral industry, 1964, by A. L. Hornbaker and R. G. Hardy. 1964.
1855

SDP. 18 Water-level changes in Grant and Stanton Counties, Kansas, 1939-1965, by C. E. Nuzman and W. R. Meyer. 1965.
1856

SDP. 19 Kansas industrial mineral development--a feasibility study for the production of filter aids from Kansas volcanic ash, by R. G. Hardy, M. P. Bauleke, A. L. Hornbaker, W. R. Hess, and W. B. Hladik. 1965.
1857

SDP. 20 Secondary recovery and pressure maintenance operations in Kansas, 1964, by the Kansas Secondary Recovery and Pressure Maintenance Committee. 1965.
1858

SDP. 21 Oil and gas fields and production in Kansas--1964, by M. O. Oros and D. L. Beene. 1965.
1859

SDP. 22 Logs of wells and test holes in Sedgwick County, Kansas, by C. W. Lane and D. E. Miller. 1965.
1860

SDP. 23 ALGOL program for cross-association of nonnumeric sequences using a medium-size computer, by M. J. Sakin, P. H. A. Sneach, and D. F. Merriam. 1965.
1861

SDP. 24 BALGOL programs and geologic application for single and double Fourier series using IBM 7090/7094 computers, by F. W. Preston and J. W. Harbaugh. 1965.
1862

SDP. 25 Final report of the Kansas Geological Society Basement Rock Committee and list of Kansas wells drilled into Precambrian rocks, by V. B. Cole, D. F. Merriam, and W. W. Hambleton. 1965.
1863

SDP. 26 FORTRAN II trend-surface program with unrestricted input for the IBM 1620 computer, by R. J. Sampson and J. C. Davis. 1966.
1864

SDP. 27 Application of factor analysis to a facies study of the Leavenworth limestone (Pennsylvanian-Virgilian) of Kansas and environs, by D. F.

Toomey. 1966. 1865

SDP. 28 FORTRAN II program for standard-size analysis of unconsolidated sediments using an IBM 1620 computer, by J. W. Pierce and D. I. Good. 1966. 1866

SDP. 29 Electronic simulation of ground-water hydrology in the Kansas River Valley near Topeka, Kansas, by J. D. Winslow and C. E. Nuzman. 1966. 1867

SDP. 30 Secondary recovery and pressure maintenance operations in Kansas, 1965, by R. L. Dilts. 1966. 1868

SDP. 31 New dimensions for mineral resources studies, by W. W. Hambleton. 1966. 1869

SDP. 32 Secondary recovery and pressure maintenance operations in Kansas, 1966, by R. L. Dilts. 1967. 1870

SDP. 33 Notes on the shape of the truncated cone of depression in the vicinity of an infinite well field, by S. W. Fader. 1967. 1871

SDP. 34 A symposium on industrial mineral exploration and development: proceedings, third forum on geology of industrial minerals, April 5-7, 1967, ed. by E. E. Angino and R. G. Hardy. 1967. 1872

SDP. 35 Kansas mineral industry, 1967, with directory of Kansas mineral producers, by A. L. Hornbaker and R. G. Hardy. 1968. 1873

SDP. 36 Secondary recovery and pressure maintenance operations in Kansas, 1967, by R. L. Dilts. 1968. 1874

SDP. 37 Water-level changes in Grant and Stanton Counties, Kansas, 1938-1968, by J. D. Winslow, H. E. McGovern, and H. L. Mackey. 1968. 1875

SDP. 38 The unit regional-value concept and its application to Kansas, by J. C. Griffiths. 1969. 1876

SDP. 39 FORTRAN IV program for synthesis and plotting of water-quality data, by L. H. Ropes, C. O. Morgan, and J. M. McNellis. 1969. 1877

SDP. 40 Kansas mineral industry, 1968, by A. L. Hornbaker and R. G. Hardy. 1969. 1878

SDP. 41 Oil and gas production in Kansas--1967, by D. L. Beene. 1969. 1879

SDP. 42 FORTRAN IV programs, KANS for the conversion of general land office locations to latitude and longitude coordinates, by C. O. Morgan and J. M. McNellis. 1969. 1880

SDP. 43 Stiff diagrams of water-quality data programmed for the digital computer, by C. O. Morgan and J. M. McNellis. 1969. 1881

SDP. 44 The world petroleum industry and its impact on mid-continent oil and gas economics, by C. A. Heller. 1969. 1882

SDP. 45 Modified piper diagrams by the digital computer, by J. M. McNellis and C. O. Morgan. 1969. 1883

SDP. 46 Secondary recovery and pressure maintenance operations in Kansas, 1968, by R. L. Dilts. 1969. 1884

SDP. 47 Application of computer techniques to seepage-salinity surveys in Kansas, by R. B. Leonard and C. O. Morgan. 1970. 1885

SDP. 48 Brief descriptions of and examples of output from computer programs developed for use with water data in Kansas, by B. H. Lowell, C. O. Morgan, and J. M. McNellis. 1970. 1886

SDP. 49 Secondary recovery and pressure maintenance operations in Kansas, 1969, by R. L. Dilts. 1970. 1887

SDP. 50 Oil and gas production in Kansas during 1968, by D. L. Beene.

1970. 1888

SDP. 51 Kansas mineral industry report, 1969, with directory of Kansas mineral producers, by A. L. Hornbaker, R. F. Walters, J. Berberick, and J. Willoughby. 1970. 1889

SDP. 52 Towns and minerals in southeastern Kansas, a study in regional industrialization, 1890-1930, by J. G. Clark. 1970. 1890

SDP. 53 Pleistocene stratigraphy of Missouri River Valley along the Kansas-Missouri border, by C. K. Bayne, H. G. O'Connor, S. N. Davis, and W. B. Howe. 1971. 1891

SDP. 54 Oil and gas production in Kansas during 1969, by D. L. Beene. 1971. 1892

SDP. 55 Kansas mineral industry report, 1970, by L. A. Flueckinger, M. O. Oros, and E. E. Angino. 1970. 1893

SDP. 56 Proceedings of the 17th annual conference of the Mid-Continent Research and Development Council: Implementation for Change. 1971. 1894

SDP. 57 Secondary recovery and pressure maintenance operations in Kansas, by R. L. Dilts. 1970. 1895

SDP. 58 Mined-land redevelopment workshop, Pittsburg, Kansas, May 26 & 27, 1971. 1971. 1896

SDP. 59 Oil and gas production in Kansas during 1970, by D. L. Beene. 1971. 1897

SDP. 60 Supplemental areas for storage of radioactive wastes in Kansas, by C. K. Bayne. 1972. 1898

SDP. 61 Kansas mineral industry report. 1971, by L. L. Brady and others. 1899

SDP. 62 Secondary recovery and pressure maintenance operations

in Kansas, 1971, by R. L. Dilts. 1972. 1900

SDP. 63 A model for assessing mineral resource advantage, by F. Miller, J. Berberick, R. G. Hardy, and M. P. Bauleke. 1972. 1901

SDP. 64 Oil and gas production in Kansas during 1971, by D. L. Beene. 1973. 1902

SDP. 65 1972 mined-land workshop proceedings. 1972. 1903

Subsurface Geology Series. 1974-

SGS. 1 List of wells drilled into Precambrian rocks, by V. B. Cole and W. J. Ebanks. 1974. 1904

SGS. 2 Subsurface Ordovician-Cambrian rocks in Kansas with maps showing thickness of potentially oil-bearing rocks, by V. B. Cole. 1975. 1905

KENTUCKY

Series IX: 1948-1958

Reprint. 1950-1958

R. 1 Oil and gas developments in Kentucky in 1949, by E. B. Wood. 1950.   1906

R. 2 Regional aspects of Silurian and Devonian subsurface stratigraphy in Kentucky, by L. B. Freeman. 1951.   1907

R. 3 Oil and gas developments in Kentucky in 1950, by E. B. Wood and M. A. McCarville. 1952. 1908

R. 4 Devonian shale gas production in central Appalachian area, by R. N. Thomas. 1951.   1909

R. 5 Summary of secondary recovery operations in Kentucky to 1951, by D. J. Jones and others. 1952.   1910

R. 6 Oil and gas developments in Kentucky in 1951, by E. B. Wood and M. A. McCarville. 1952. 1911

R. 7 Relationship of natural gas occurrence and production in eastern Kentucky to joints and fractures in Devonian bituminous shale, by C. D. Hunter and D. M. Young. 1953.   1912

R. 8 Oil and gas developments in Kentucky in 1952, by F. H. Walker. 1953.   1913

R. 9 Oil and gas developments in Kentucky in 1953, by F. H. Walker, C. D. Hunter, and J. B. Cathey. 1954.   1914

R. 10 Oil and gas developments in Kentucky in 1954, by F. H. Walker,

E. O. Ray, and J. B. Cathey. 1955. 1915

R. 11 Development of natural gas fields of eastern Kentucky, by C. D. Hunter. 1955.   1916

Exploration extensive in eastern Kentucky, by F. H. Walker. 1955. 1917

R. 12 Oil and gas developments in Kentucky in 1955, by F. H. Walker, E. O. Ray, and J. B. Cathey. 1956. 1918

R. 13 The mineral industry of Kentucky (1953), by R. H. Mote and A. Kaufman. 1956.   1919

R. 14 Oil and gas developments in Kentucky in 1956, by F. H. Walker, D. J. Jones, and E. O. Ray. 1957. 1920

R. 15 The mineral industry of Kentucky (1954), by A. H. Reed and A. C. McFarlan. 1956.   1921

R. 16 The mineral industry of Kentucky (1955), by A. H. Reed and A. C. McFarlan. 1956.   1922

R. 17 Summary of secondary recovery operations in Kentucky in 1957, by P. McGrain and others. 1958. 1923

R. 18 Chester cross-bedding and sandstone trends in Illinois Basin, by P. E. Potter and others. 1958.   1924

R. 19 Oil and gas developments in Kentucky in 1957, by E. Nosow and E. O. Ray. 1958.   1925

Series X. 1959-1978

Bulletin. 1963-1978

B. 1 Oil and gas geology of Muhlen-

93

94 / KENTUCKY

berg County, Kentucky, by W. D.
Rose. 1963.   1926

B. 2 The Silurian formations of
east-central Kentucky and adjacent
Ohio, by C. B. Rexroad, E. R.
Branson, M. O. Smith, C. Summerson, and A. J. Boucot. 1965.
                              1927

B. 3 Fuller's earth resources of
the Jackson Purchase region, Kentucky, by P. McGrain. 1965.
                              1928

B. 4 Limestone resources in the
Appalachian region of Kentucky, by
P. McGrain and G. R. Dever.
1967.   1929

B. 5 High-calcium and low-mag-
nesium limestone resources in the
region of the Lower Cumberland
Tennessee and Ohio Valleys, western Kentucky, by G. R. Dever and
P. McGrain. 1969.   1930

County Report. 1966-1978

CR. 1 Economic geology of Allen
County, Kentucky, by E. R. Branson. 1966.   1931

CR. 2 Economic geology of Calloway County, Kentucky, by P. Mc-
Grain. 1968.   1932

CR. 3 Economic geology of Simpson County, Kentucky, by E. R.
Branson. 1969.   1933

CR. 4 Economic geology of Han-
cock County, Kentucky, by P. Mc-
Grain, H. R. Schwalb, and G. E.
Smith. 1970.   1934

CR. 5 Economic geology of Marshall County, Kentucky, by P. Mc-
Grain. 1970.   1935

CR. 6 Economic geology of Warren
County, Kentucky, by P. McGrain
and D. G. Sutton. 1973.   1936

CR. 7 Economic geology of Mc-
Cracken County, Kentucky, by P.
McGrain. 1978.   1937

Information Circular. 1963-1977

IC. 11 Coal reserves in portions of
Butler, Edmonson, Grayson, Muhlenberg, Ohio, and Warren Counties,
Kentucky, by A. T. Mullins and others.
1963.   1938

IC. 12 A deep fresh water aquifer in
New Cypress Pool, Muhlenberg
County, Kentucky, corroborated by
geophysical logs, by E. N. Wilson and
J. A. Van Couvering. 1965.   1939

IC. 13 Gypsum and anhydrite in the
St. Louis limestone in northwestern
Kentucky, by P. McGrain and W. L.
Helton. 1964.   1940

IC. 14 A deposit of high-calcium
limestone near Barkley Lake, Kentucky,
by P. McGrain. 1964.   1941

IC. 15 Pumping test of an Eocene
aquifer near Mayfield, Kentucky, by
J. H. Morgan. 1967.   1942

IC. 16 Engineering geology of the
Calvert City Quadrangle, Livingston
and Marshall Counties, Kentucky, by
W. I. Finch. 1968.   1943

IC. 17 Catalog of Devonian and deeper
wells in western Kentucky, by H. R.
Schwalb. 1969.   1944

IC. 18 Effects of pumping from the
Ohio River Valley alluvium between
Carrollton and Ghent, Kentucky, by
D. V. Whitesides and P. D. Ryder.
1969.   1945

IC. 19 Catalog of well samples and
cores on file at Kentucky Geological
Survey, by H. R. Schwalb and J. G.
Smith. 1970.   1946

IC. 20 Public and industrial water
supplies of Kentucky, 1968-69, by
D. S. Mull, R. V. Cushman, and T. W.
Lambert. 1971.   1947

IC. 21 Yields and specific capacities
of bedrock wells in Kentucky, by D. V.
Whitesides. 1971.   1948

IC. 22 High-carbonate rock in the
High Bridge Group (Middle Ordovician),

Boone County, Kentucky, by G. R. Dever. 1974. 1949

IC. 23 Coal production in Kentucky, 1790-1975, by J. C. Currens and G. E. Smith. 1977. 1950

Report of Investigations. 1963-1978

RI. 5 Water resources of eastern Kentucky--progress report, by G. A. Kirkpatrick, W. E. Price, and R. A. Madison. 1963. 1951

RI. 6 Pennsylvanian cross sections in western Kentucky--coals of the Lower Carbondale Formation. Part 1, by G. E. Smith. 1967. 1952

RI. 7 Industrial sand in Pike County, Kentucky, by R. P. Hollenbeck, J. S. Browning, and T. L. McVay. 1967. 1953

RI. 8 High-purity limestones at Somerset, Kentucky, by P. Mc-Grain and G. R. Dever. 1967. 1954

RI. 9 Water resources of the Middlesboro area, Kentucky, by D. S. Mull and R. J. Pickering. 1968. 1955

RI. 10 Paleozoic geology of the Jackson Purchase region, Kentucky, with reference to petroleum possibilities, by H. R. Schwalb. 1969. 1956

RI. 11 Bethel sandstone (Mississippian) of western Kentucky and south-central Indiana, a submarine-channel fill, by Sedimentation Seminar (Indiana University and University of Cincinnati), 1969. 1957

RI. 12 Miscellaneous analyses of Kentucky clays and shales for 1960-1970, by P. McGrain and T. A. Kendall. 1972. 1958

RI. 13 Sedimentology of the Mississippian Knifley sandstone and Cane Valley limestone of south-central Kentucky, by Sedimentation Seminar (University of Cincinnati

and Indiana University), 1972. 1959

RI. 14 Preliminary report of the oil and gas possibilities between Pine and Cumberland Mountains, southeastern Kentucky, by A. J. Froelich. 1973. 1960

RI. 15 Hydrology and geology of deep sandstone aquifers of Pennsylvanian age in part of the western coal field region, Kentucky, by R. W. Davis, R. O. Plebuch, and H. M. Whitman. 1974. 1961

RI. 16 Oil and gas in Butler County, Kentucky, by H. R. Schwalb. 1975. 1962

RI. 17 Water in a limestone terrane in the Bowling Green area, Warren County, Kentucky, by T. W. Lambert. 1976. 1963

RI. 18 Tidal-flat carbonate environments in the High Bridge Group (Middle Ordovician) of central Kentucky, by E. R. Cressman and M. C. Noger. 1976. 1964

RI. 19 Tar sands (rock asphalt) of Kentucky--a review, by P. McGrain. 1976. 1965

RI. 20 A geologic profile of Sloans Valley, Pulaski County, Kentucky, by C. A. Malott and P. McGrain. 1977. 1966

RI. 21 Sedimentology of the Kyrock sandstone (Pennsylvanian) in the Brownsville paleovalley, Edmonson and Hart Counties, Kentucky, by Sedimentation Seminar. 1978. 1967

Reprint. 1959-1978

R. 1 The mineral industry of Kentucky (1957), by A. H. Reed, P. Mc-Grain, and M. E. Rivers. 1959. 1968

R. 2 Oil and gas developments in Kentucky in 1958, by E. Nosow. 1959. 1969

R. 3 Services of state geological surveys to the structural clay products

industry, by P. McGrain. 1959.
1970

R. 4 The mineral industry of Kentucky (1958), by A. H. Reed, P. McGrain, and M. E. Rivers. 1960.
1971

R. 5 Geology of the clay deposits in the Olive Hill District, Kentucky, by S. H. Patterson and J. W. Hosterman. 1960. 1972

R. 6 Oil and gas developments in Kentucky in 1959, by E. Nosow. 1960. 1973

R. 7 The mineral industry of Kentucky (1959), by A. H. Reed, P. McGrain, and M. E. Rivers. 1961.
1974

R. 8 Oil and gas developments in Kentucky in 1960, by E. Nosow. 1961. 1975

R. 9 The mineral industry of Kentucky (1960), by A. H. Reed, P. McGrain, and M. E. Rivers. 1962.
1976

R. 10 Oil and gas developments in Kentucky in 1961, by E. Nosow. 1962. 1977

R. 11 The mineral industry of Kentucky (1961), by H. L. Riley, P. McGrain, and M. E. Rivers. 1962.
1978

R. 12 Oil and gas developments in Kentucky in 1962, by E. Nosow. 1963. 1979

R. 13 The mineral industry of Kentucky (1962), by H. L. Riley, P. McGrain, and M. E. Rivers. 1963.
1980

R. 14 Oil and gas developments in Kentucky in 1963, by E. Nosow. 1964. 1981

R. 15 Mineral paragenesis and zoning in the central Kentucky mineral district, by J. L. Jolly and A. V. Heyl. 1964. 1982

R. 16 The mineral industry of

Kentucky (1963), by H. L. Riley and P. McGrain. 1964. 1983

R. 17 Oil and gas developments in east-central states in 1964, by G. L. Carpenter and others. 1965. 1984

R. 18 The mineral industry of Kentucky (1964), by H. L. Riley and P. McGrain. 1965. 1985

R. 19 Oil and gas developments in east-central states in 1965, by H. C. Milhous and others. 1966. 1986

R. 20 Geology of cement raw materials in Kentucky, by P. McGrain and G. R. Dever. 1966. 1987

R. 21 Some sources of ceramic materials in Kentucky, by P. McGrain. 1966. 1988

R. 22 The mineral industry of Kentucky (1965), by H. L. Riley and P. McGrain. 1967. 1989

R. 23 Oil and gas developments in east-central states in 1966, by E. Nosow and others. 1967. 1990

R. 24 The mineral industry of Kentucky (1966), by H. L. Riley and P. McGrain. 1968. 1991

R. 25 Oil and gas developments in east-central states in 1967, by J. Van Den Berg and others. 1968. 1992

R. 26 Natural gas in Illinois Basin, by D. C. Bond and others. 1968.
1993

R. 27 The mineral industry of Kentucky (1967), by H. L. Riley and P. McGrain. 1968. 1994

R. 28 Origin of the Jeptha Knob structure, Kentucky, by C. R. Seeger. 1969. 1995

R. 29 The American Upper Ordovician standard X. Upper Maysville and Richmond conodonts from the Cincinnati region of Ohio, Indiana, and Kentucky, by J. J. Kohut and W. C. Sweet. 1969. 1996

R. 30 Oil and gas developments in

east-central states in 1968, by J. Van Den Berg and others. 1969.
1997

R. 31 Extension of Lost River chert across parts of Kentucky, by P. McGrain. 1969.
1998

R. 32 The mineral industry of Kentucky, 1968, by H. L. Riley and P. McGrain. 1970.
1999

R. 33 Oil and gas developments in east-central states in 1969, by J. Van Den Berg and others. 1970.
2000

R. 34 The mineral industry of Kentucky, 1969, by S. A. Friedman and P. McGrain. 1971.
2001

R. 35 Oil and gas developments in east-central states in 1970, by G. L. Carpenter and others. 1971.
2002

R. 36 Greensburg consolidated oil pool, Green and Taylor Counties, Kentucky, by H. R. Schwalb and E. N. Wilson. 1972.
2003

R. 37 The mineral industry of Kentucky, 1970, by H. L. Riley and P. McGrain. 1972.
2004

R. 38 Oil and gas developments in east-central states in 1971, by E. Nosow and others. 1972.
2005

R. 39 Compositional variations in high-calcium limestone deposits in western Kentucky, by G. R. Dever and P. McGrain. 1973.
2006

R. 40 The mineral industry of Kentucky, 1971, by H. L. Riley and P. McGrain. 1973.
2007

R. 41 Oil and gas developments in east-central states in 1972, by A. T. Statler and others. 1973.
2008

R. 42 Oil and gas developments in east-central states in 1973, by J. Van Den Berg and others. 1974.
2009

R. 43 The mineral industry of Kentucky, 1972, by H. L. Riley and P. McGrain. 1974.
2010

R. 44 Oil and gas developments in east-central states in 1974, by G. L. Carpenter and others. 1975.
2011

R. 45 The mineral industry of Kentucky, 1973, by L. H. Frey and P. McGrain. 1976.
2012

R. 46 Oil and gas developments in east-central states in 1975, by E. Nosow and others. 1976.
2013

R. 47 The mineral industry of Kentucky, 1974, by W. T. Boyd and P. McGrain. 1977.
2014

R. 48 Oil and gas developments in east-central states in 1976, by A. T. Statler, J. Van Den Berg, G. L. Carpenter, and E. Nosow. 1978.
2015

R. 49 Industrial limestone resources along the Ohio River Valley of Kentucky, by G. R. Dever, P. McGrain, and G. W. Ellsworth. 1978.
2016

Special Publication. 1964-1978

SP. 8 Proceedings of the technical sessions, Kentucky Oil and Gas Association Annual meetings, 1962 and 1963. 1964.
2017

Technical Session, 1962:

A. Recent developments in oil and gas exploration in southwestern Butler County, Kentucky, by B. D. Nuttall.
2018

B. Does the oil and gas industry need drilling contractors? by H. Eakle.
2019

C. Gas development, production, and estimated ultimate recovery of Devonian shale in eastern Kentucky, by C. D. Hunter.
2020

D. Recent developments in well completions, by A. A. Baker.
2021

E. New developments in Appalachian

logging services, by J. B. Gehr.
2022

F. The determination of the bonding quality of cement to casing, by J. E. Gwinn. 2023

G. Oil production in Kentucky for 1961. 2024

Technical Session, 1963:

A. The Bon Harbor Gas Pool, Daviess County, Kentucky--a Waltersburg anomaly, by R. L. Norris. 2025

B. Oil exploration in Hopkins County, Kentucky, relative to the Hanson pool, by G. J. Hennessy. 2026

C. Thermal recovery by in situ combustion, by J. D. Alexander. 2027

D. The application of fracturing in the Oil Springs Pool, by J. V. George. 2028

E. Geology of the St. Peter sandstone in Clark and Estill Counties, Kentucky, by W. H. McGuire and P. Howell. 2029

F. Completion techniques and results of recent Berea sand development in Pike County, Kentucky, by G. R. Thomas and J. B. Lauffer. 2030

G. Oil production in Kentucky for 1962. 2031

SP. 9 Kentucky's rocks and minerals, by W. L. Helton. 1964. 2032

SP. 10 Proceedings of the technical sessions, Kentucky Oil and Gas Association Annual Meeting, June 4-5, 1964. 1965. 2033

A. Secondary recovery by inert gas injection in Spring Grove Pool, Union County, Kentucky, by W. M. Duchscherer. 2034

B. Application of well logging to

underground gas storage in western Kentucky, by J. A. Brown. 2035

C. Highlights of western Kentucky oil and gas exploration in 1963 and early 1964, by W. D. Rose. 2036

D. Improved secondary recovery by control of water mobility, by W. F. Seifert. 2037

E. Relationship of oil and gas production to major gravity anomalies in western Kentucky, by W. H. McGuire. 2038

F. Some aspects of waterflooding in the Big Sinking Field, Lee County, Kentucky, by W. Stack. 2039

G. Porosity zones in the Knox of northeastern Kentucky, by W. H. McGuire. 2040

H. Oil and gas conservation activities, by F. H. Walker. 2041

I. The new tax law--problems and opportunities, by P. S. Townsend. 2042

J. Resume of drilling activity in eastern Kentucky in 1963, by E. B. Jenkins. 2043

K. Oil production in Kentucky for 1963. 2044

SP. 11 Geology of the Cumberland Falls State Park area, by P. McGrain. 1955. Reprinted with minor corrections, 1966. 2045

SP. 12 Geology of the Carter and Cascade Caves area, by P. McGrain. 1954. Reprinted with minor corrections, 1966. 2046

SP. 13 The geologic story of Bernheim Forest, by P. McGrain. 1967. 2047

SP. 14 Proceedings of the technical sessions, Kentucky Oil and Gas Association annual meeting, June 3-4, 1965. 1967. 2048

A. Limited-entry well completion technique, by R. B. Sublett. 2049

B. The application of new geologic maps to the economic growth of Kentucky, by P. McGrain. 2050

C. Western Kentucky's Bethel channel--the largest continuous reservoir in the Illinois Basin, by D.W. Reynolds and J.K. Vincent. 2051

D. The geological procedures used in the evaluation of Camp Breckinridge, by J. Avila. 2052

E. Cambrian play in southwestern Ontario, by J.A. Pounder. 2053

F. Geology of the L.S. Bales well, Lee County, Virginia--a Cambrian and Ordovician test, by L.D. Harris. 2054

G. Trempealeau remnants defined by dipmeter surveys, by R.W. Bashe. 2055

H. Using geologic maps in oil and gas exploration, by E.N. Wilson. 2056

I. Oil production in Kentucky for 1965. 2057

SP. 15 Proceedings of the technical sessions, Kentucky Oil and Gas Association Annual Meetings, 1966 and 1967. 1968. 2058

Technical Sessions, 1966:

A. Future prospective areas of western Kentucky, by F.L. Stanonis. 2059

B. Petroleum potential of the Rough Creek tectonic element in Kentucky, by E.E. Rehn. 2060

C. Petroleum--where it came from, where it is found, how it got there, by D.C. Bond. 2061

D. Operational aspects of the Fry in situ project, by W.W. Smith. 2062

E. Shale stimulation--a new approach, by E.O. Ray. 2063

F. Oil and gas conservation activities, by F.H. Walker. 2064

G. Subsurface geology of the Sulphur Lick-Tompkinsville area, Monroe County, Kentucky, by P. Perkins. 2065

H. Some aspects of drilling activities in eastern Kentucky in 1965-1966, by E.N. Wilson. 2066

I. Exploration objectives in the Cambro-Ordovician of Kentucky, by W.H. McGuire. 2067

Technical Sessions, 1967:

A. A progress report on increasing permeability by explosives and heat, by W.E. Eckard. 2068

B. Pennsylvanian structure of western Kentucky, by A.T. Mullins. 2069

C. Applications and the future of the computer in the petroleum industry, by C.S. Riley. 2070

D. Oil-field technology applied to subsurface disposal of industrial fluid wastes, by F.L. Stanonis. 2071

E. Geophysical exploration in eastern Kentucky, by W.H. McGuire. 2072

F. Oil-field application of detergents, by T. Halloran. 2073

G. Low-mobility cement slurries reduce gas cutting and fall-back, by B.B. Bradford, G.A. Sullivan, and L.B. Spangle. 2074

H. Newburg gas development, Kanawha and Jackson Counties, West Virginia, by G. Haney. 2075

I. Oil production in Kentucky for 1964-66. 2076

SP. 16 Water in Kentucky, by R.A. Krieger, R.V. Cushman, and N.O. Thomas. 1969. 2077

SP. 17 Proceedings of the technical sessions, Kentucky Oil and Gas Association Annual Meeting, June 6-7,

1968. 1969. 2078

A. Operations research and economic analysis application to the study of exploration decisions, by R. A. Franzoni and J. E. Green. 2079

B. Deep (Cambro-Ordovician) exploration in western Kentucky, by H. R. Schwalb. 2080

C. Using the drill-stem test as an exploratory tool, by S. J. Bateman. 2081

D. Geology and drilling history of the Newburg sand in West Virginia, by D. G. Patchen. 2082

E. Some aspects of recent production research and possible effects on petroleum recovery, by J. Pasini. 2083

F. Resume of current activity of the oil industry northeast of the Mississippi River--geographic, geologic--and magnitude of recent years' discoveries, by C. D. Fenstermaker. 2084

G. Cartographic activities of the geologic mapping program in Kentucky, by J. T. Hopkins. 2085

H. Oil production in Kentucky for 1967 and 1968. 2086

SP. 18 Proceedings of the technical sessions, Kentucky Oil and Gas Association Annual Meeting, June 5-6, 1969. 1969. 2087

A. Geologic history of the Cambrian system in the Appalachian Basin, by E. J. Webb. 2088

B. Drilling conditions and problems in eastern Kentucky, by G. B. Putman. 2089

C. Some important factors to consider when planning a fracturing treatment, by A. R. Jennings, H. N. Black, and W. T. Malone. 2090

D. Methods and considerations in appraising a coal property, by

S. E. Fish. 2091

E. Some petroleum prospects of the Cincinnati Arch province, by J. B. Patton and T. A. Dawson. 2092

F. Application of electronic computers to reservoir engineering, by H. D. Griffith. 2093

G. Bethel sandstone (Mississippian) of western Kentucky and south-central Indiana, a submarine channel fill (abstracted version), by J. F. Friberg and others. 2094

H. Oil production in Kentucky for 1968. 2095

SP. 19 Bibliography of coal in Kentucky, 1970. 2096

SP. 20 Water in the economy of the Jackson Purchase region of Kentucky, by R. W. Davis, T. W. Lambert, and A. J. Hansen. 1971. 2097

SP. 21 Proceedings of the technical sessions, Kentucky Oil and Gas Association Annual Meetings, 1970 and 1971. 1972. 2098

Technical Sessions, 1970:

A. "Brown shale" problems in eastern Kentucky, by J. L. Wilson. 2099

B. Delivery performance of fractured shale wells, by E. O. Ray. 2100

C. Geology and economics of Knox dolomite oil production in Gradyville east field, Adair County, Kentucky, by J. H. Perkins. 2101

D. Bottom-hole percussion tools--where and how to use them, by S. C. Berube and R. N. Young. 2102

E. Sixty years of exploration in Logan County, south-central Kentucky, by H. Schwalb. 2103

Technical Sessions, 1971:

A. Cambro-Ordovician structural

and stratigraphic relationships of
a portion of the Rome trough, by
J. D. Silberman. 2104

B. Notes on "Corniferous" production in eastern Kentucky, by P. M.
Miles. 2105

C. Stratigraphic relationships of
certain Mississippian-age pools in
southeastern Kentucky and northeastern Tennessee, by E. J. Webb.
2106

D. External corrosion of well
casings in salt sands of the Lee
Formation, by H. L. Baldridge.
2107

E. Inorganic geochemical prospecting for oil and gas accumulations,
by A. C. Johnson. 2108

F. Independents should automate,
by N. I. Lieberman. 2109

G. Petroleum exploration opportunities in Butler County, Kentucky,
by H. Schwalb. 2110

H. Frac pad acidizing in carbonate
reservoirs, by D. R. Wieland.
2111

I. Oil production in Kentucky for
1969-1970. 2112

SP. 22 A symposium on the geology
of fluorspar (proceedings of the
ninth forum on geology of industrial
minerals). 1974. 2113

A. Fluorine resources--an over-
view, by G. Montgomery. 2114

B. The environments of deposition
of fluorspar, by R. M. Grogan and
others. 2115

C. Geology of the Derbyshire
fluorspar deposits, United Kingdom,
by J. E. Mason. 2116

D. Geology of Mexican fluorspar
deposits, by G. W. Pickard. 2117

E. Geology of fluorspar deposits of
the western United States, by R. G.
Worl. 2118

F. Some fluorite-barite deposits in
the Mississippi Valley in relation to
major structures and zonation, by A.
V. Heyl. 2119

G. Illinois-Kentucky fluorspar district, by R. D. Trace. 2120

H. Structure of the fault systems in
the Illinois-Kentucky fluorspar district, by J. W. Hook. 2121

I. Geology and history of Pennwalt
Corporation's Dyers Hill Mine, Liv-
ingston County, Kentucky, by J. S.
Tibbs. 2122

J. The Eagle-Babb-Barnes fluorspar
project, Crittenden County, Kentucky,
by F. B. Moodie and P. McGrain.
2123

SP. 23 Bibliography of industrial and
metallic minerals in Kentucky through
August 1973, by P. McGrain and J. C.
Currens. 1975. 2124

SP. 24 Scenic geology of Pine Mountain in Kentucky, by P. McGrain.
1975. 2125

SP. 25 Topography of Kentucky, by
P. McGrain and J. C. Currens. 1978.
2126

Thesis Series. 1966-1975

TS. 1 Geology and successful farm
ponds in the inner blue grass region
of Kentucky, by M. O. Smith. 1966.
2127

TS. 2 Silurian-Devonian stratigraphy
of Pulaski County, Kentucky, by W. L.
Helton. 1968. 2128

TS. 3 Relation of fracture traces,
joints, and ground-water occurrence
in the area of the Bryantsville Quadrangle, central Kentucky, by G. T.
Hine. 1970. 2129

TS. 4 Lithofacies and biofacies of the
Haney limestone (Mississippian), Illi-
nois, Indiana, and Kentucky, by J. W.
Vincent. 1975. 2130

Series XI. 1979-

County Report. 1979-

CR. 1 Economic geology of Lincoln County, Kentucky, by P. McGrain. 1979. 2131

Information Circular. 1979-

IC. 1 A Pennsylvanian channel in Henderson and Webster Counties, Kentucky, by J. G. Beard and A. D. Williamson. 1979. 2132

Reprint. 1978-

R. 1 The mineral industry of Kentucky, by W. T. Boyd and P. McGrain. 1978. 2133

R. 2 Oil and gas developments in east-central states in 1977, by J. Van Den Berg and others. 1978. 2134

R. 3 Recognition of Lapies-type features in the Kentucky karst-- an example of applied geomorphology, by P. McGrain. 1979. 2135

R. 4 Oil and gas developments in east-central states in 1978, by G. L. Carpenter and others. 1979. 2136

R. 5 The Mississippian and Pennsylvanian (Carboniferous) systems in Kentucky, by C. L. Rice, E. G. Sable, G. R. Dever, and T. M. Kehn. 1980. 2137

Special Publication. 1979-

SP. 1 Bibliography of karst geology in Kentucky, by J. C. Currens and P. McGrain. 1979. 2138

Anthropological Studies. 1935-1969

AS. 1 Ceramic decoration sequence at an old Indian village site near Sicily Island, Louisiana, by J. A. Ford. 1935. 2139

AS. 2 Analysis of Indian village site collections from Louisiana and Mississippi, by J. A. Ford. 1936. 2140

AS. 3 Crooks site, a Marksville Period burial mound in La Salle Parish, Louisiana, by J. A. Ford and G. R. Willey. 1940. 2141

AS. 4 A bibliography relative to Indians of the State of Louisiana, by R. W. Neuman and L. A. Simmons. 1969. 2142

Clay Resources Bulletin. 1967-

CRB. 1 Test data and evaluation of miscellaneous clays, by L. H. Dixon. 1967. 2143

CRB. 2 Occurrence, test data, and evaluation of clay for making lightweight aggregate, by L. H. Dixon and M. E. Tyrrell. 1969. 2144

CRB. 3 Occurrence, test data, and evaluation of clay for making structural clay products, by L. H. Dixon and M. E. Tyrrell. 1972. 2145

Folio Series. 1960-

F. 1 Paleontology of the L. L. &E. et al. Well Unit 1-L No. 1, by E. A. Butler. 1960. 2146

F. 2 St. John's bentonite report, Claiborne Parish, Louisiana, by

C. O. Durham, F. L. O'Bryan, T. K. Smith, and W. S. White. 1962. 2147

F. 3 Basic Mesozoic study in Louisiana, the northern coastal region, and the Gulf Basin Province, by E. G. Anderson. 1979. 2148

Geological Bulletin. 1931-

GB. 39 Ostracoda and correlation of the Upper and Middle Frio from Louisiana to Florida, by E. A. Butler. 1963. 2149

GB. 40 Type saline bayou ostracoda of Louisiana, by R. C. Howe. 1963. 2150

GB. 41 Iron ore of central north Louisiana, by C. O. Durham. 1964. 2151

GB. 42 Cenozoic cyclic deposition in the subsurface of central Louisiana, by L. H. Dixon. 1965. 2152

GB. 43 Computer-aided subsurface structural analysis of the Miocene formations of the Bayou Carlin-Lake Sand areas, south Louisiana, by M. B. Kumar. Pt. I. 1977. 2153

Production mechanisms of the geopressured gas reservoirs of Lake Sand Field, south Louisiana, by M. B. Kumar. Pt. II. 1977. 2154

Mineral Resources Bulletin. 1969-

MRB. 1 Some silica sands of southeast Louisiana, by H. L. Roland. 1969. 2155

MRB. 2 Lignite--evaluation of near-surface deposits in northwest Louisiana, by H. L. Roland, G. M. Jenkins,

and D. E. Pope. 1976.   2156

MRB. 3 Gamma ray surveys of
Louisiana, progress report, 1971,
by L. H. Dixon. 1978.   2157

Paleontological Studies.   1959-

PS. 1 Miocene-Upper Oligocene
foraminifera of Louisiana, by A.
Butler. Pt. 1. 1959.   2158

Resources Information Series.   1980-

RIS. 1 A list of Louisiana oil and
gas fields and salt domes (including the offshore areas) showing
code numbers and official observations, by C. P. Stanfield. 1980.
     2159

RIS. 2 Bibliography and index of
theses and dissertations on the
geology of Louisiana, by M. J.
Nault. 1980.   2160

Water Resources Bulletin.   1960-

WRB. 1 Ground water in Louisiana,
by J. R. Rollo. 1960.   2161

WRB. 2 Ground water conditions
in the Baton Rouge area, 1954-59,
with special reference to increased
pumpage, by C. O. Morgan. 1962.
     2162

WRB. 3 Water resources of Sabine
Parish, Louisiana, by L. V. Page,
R. Newcome, and G. D. Graeff.
1963.   2163

WRB. 4 Water resources of Nat-
chitoches Parish, Louisiana, by
R. Newcome, L. V. Page, and R.
Sloss. 1963.   2164

WRB. 5 Water resources of Bossier and Caddo Parishes, Louisiana, by L. V. Page and H. G. May.
1964.   2165

WRB. 6 Water resources of Vernon
Parish, Louisiana, by J. E. Rogers
and A. J. Calandro. 1965.   2166

WRB. 7 Ground water in the Geismar-
Gonzales area, Ascension Parish,
Louisiana, by R. A. Long. 1965.
     2167

WRB. 8 Water resources of Rapides
Parish, Louisiana, by R. Newcome
and R. Sloss. 1966.   2168

WRB. 9 Ground-water resources of
the Greater New Orleans area, Louisiana, by J. R. Rollo. 1966.   2169

WRB. 10 Effects of ground-water
withdrawals on water levels and salt-
water encroachment in southwestern
Louisiana, by A. H. Harder, C. Kilburn, H. M. Whitman, and S. M. Rogers. 1967.   2170

WRB. 11 Water resources of Pointe
Coupee Parish, Louisiana, by M. D.
Winner, M. J. Forbes, and W. L.
Broussard. 1968.   2171

WRB. 12 Water resources of the Lake
Pontchartrain area, Louisiana, by G.
T. Cardwell, M. J. Forbes, and M. W.
Gaydos. 1967.   2172

WRB. 13 Salt-water encroachment in
aquifers of the Baton Rouge area,
Louisiana, by J. R. Rollo. 1969. 2173

WRB. 14 Water resources of Ouachita
Parish, Louisiana, by J. E. Rogers,
A. J. Calandro, and M. W. Gaydos.
1972.   2174

WRB. 15 Ground-water resources of
Avoyelles Parish, Louisiana, by J. R.
Marie. 1971.   2175

WRB. 16 Ground-water in the Plaque-
mine-White Castle area, Iberville Parish, Louisiana, by C. D. Whitman.
1972.   2176

WRB. 17 Water resources of Union
Parish, Louisiana, by J. L. Snider,
A. J. Calandro, and W. J. Shampine.
1972.   2177

WRB. 18 Ground-water resources of
the Norco area, Louisiana, by R. L.
Hosman. 1972.   2178

RWB. 19 Ground-water resources of

Morehouse Parish, Louisiana, by
T. H. Sanford. 1972.   2179

WRB. 20 Geohydrology of the Evangeline and Jasper aquifers of southwestern Louisiana, by M. S. Whitfield. 1975.   2180

Water Resource Pamphlet. 1954-

WRP. 10 Water levels and water-level contour maps for southwestern Louisiana, 1959, and Spring 1960, with a discussion of groundwater withdrawals, by A. H. Harder. 1961.   2181

WRP. 11 Water levels in southwestern Louisiana, April 1960 to April 1961, with a discussion of water-level trends from 1950 to 1960. 1962.   2182

WRP. 12 Ground-water conditions in southwestern Louisiana, 1961, and 1962, with a discussion of the Chicot aquifer in the coastal area. 1963.   2183

WRP. 13 Gas and brackish water in fresh-water aquifers, Lake Charles area. 1963.   2184

Pumpage of water in Louisiana, 1960, by J. L. Snider and M. J. Forbes. 1961.   2185

Sources of emergency water supply in the Alexandria area, Louisiana, by R. Newcome. 1961.   2186

Emergency ground-water supplies in Calcasieu Parish, Louisiana, by G. W. Swindel and A. L. Hodges. n. d.   2187

Emergency ground-water supplies in the Monroe area, Louisiana, by J. L. Snider. 1962.   2188

WRP. 14 Methane in the fresh-water aquifers of southwestern Louisiana and theoretical explosion hazards, by A. H. Harder, H. M. Whitman, and S. M. Rogers. 1965.   2189

WRP. 15 Feasibility of a scavenger-

well system as a solution to the problem of vertical salt-water encroachment, by R. A. Long. 1965.   2190

WRP. 16 Estimating water quality from electrical logs in southwestern Louisiana, by H. M. Whitman. 1965.   2191

WRP. 17 Salt-water encroachment, Baton Rouge area, Louisiana, by R. R. Meyer and J. R. Rollo. 1965.   2192

WRP. 18 Progress report on the availability of fresh water, Lake Pontchartrain area, Louisiana, by G. T. Cardwell, M. J. Forbes, and M. W. Gaydos. 1966.   2193

WRP. 19 Calculation of water quality from electrical logs, theory and practice, by A. N. Turcan. 1966.   2194

WRP. 20 Pumpage of water in Louisiana, by P. P. Bieber and M. J. Forbes. 1966.   2195

WRP. 21 Water resources of the Lettsworth-Innis-Batchelor area, Pointe Coupee Parish, by A. H. Harder, V. B. Sauer, and W. L. Broussard. 1968.   2196

WRP. 22 Water-level trends in southeastern Louisiana, by D. C. Dial. 1968.   2197

WRP. 23 Water resources of northwestern St. Landry Parish and vicinity, Louisiana, by R. L. Hosman, W. L. Broussard, and A. J. Calandro. 1970.   2198

WRP. 24 Water resources of the Slagle-Simpson-Flatwoods area, Louisiana, by C. D. Whitman, A. J. Calandro, and W. L. Broussard. 1970. 2199

WRP. 25 Water resources of the Belmont-Marthaville-Robeline area, Louisiana, by A. J. Calandro, W. L. Broussard, and R. L. Hosman. 1970.   2200

WRP. 26 Pumpage of water in Louisiana, by D. C. Dial. 1970.   2201

WRP. 27 Ground-water pumpage

and related effects, southwestern Louisiana, 1970, with a section
on surface-water withdrawals, by
A. L. Zack. 1971. 2202

Bulletin. 1944-

B. 8 Never published? 2203

B. 9 Never published? 2204

B. 10 Glacial lake and glacial
marine clays of the Farmington
area, Maine, origin and possible
use as lightweight aggregate, by
D. W. Caldwell. 1959. 2205

B. 11 The geology of Sebago Lake
State Park, by A. L. Bloom. 1959.
2206

B. 12 The geology of Baxter State
Park and Mt. Katahdin, by D. W.
Caldwell. 2nd ed. 1972. 2207

B. 13 Never published? 2208

B. 14 The geology of southern
York County, Maine, by A. M. Hussey. 1962. 2209

B. 15 Reconnaissance bedrock
geology on the Presque Isle Quadrangle, Maine, by A. J. Boucot,
M. T. Field, R. Fletcher, W. H.
Forbes, R. S. Naylor, and L. Pavlides. 1974. 2210

B. 16 Geology of the Bryant Pond
Quadrangle, Maine, by C. V. Guidotti. 1965. 2211

B. 17 The geology of Mount Blue
State Park, by K. A. Pankiwskj.
1965. 2212

B. 18 Contributions to the geology
of Maine: papers by A. J. Boucot,
F. M. Beck, R. G. Doyle, B. Hall,
and R. Gilman. 1966. 2213

B. 19 Reconnaissance and economic
geology of the northwestern Knox
County marble belt, by E. S.

Cheney. 1967. 2214

B. 20 Stratigraphy, structural geology,
and metamorphism of the Waterville-
Vassalboro area, Maine, by P. H.
Osberg. 1968. 2215

B. 21 Geology of the Moose River and
Roach River synclinoria, northwestern
Maine, by A. J. Boucot and E. W.
Heath. 1969. 2216

B. 22 Stratigraphy of the southern end
of the Munsongun anticlinorium, Maine,
by B. A. Hall. 1970. 2217

B. 23 Shorter contributions to Maine
geology: papers by H. N. Andrews,
and A. E. Kasper, D. C. Roy, W. H.
Forbes, K. A. Pankiwskj, G. M. Boone,
A. J. Boucot, P. D. Fullagar, M. L.
Bottino, R. A. Gilman, and A. M.
Hussey. 1970. 2218

B. 24 Metamorphic stratigraphy, petrology, and structural geology of the
Little Bigelow Mountain map area,
western Maine, by G. M. Boone.
1973. 2219

B. 25 History of sedimentation in
Montsweag Bay, by D. Schnitker.
1972. 2220

Physical Resources Series. 1974-

PRS. 1 Physical resources of Knox
County, Maine, by W. B. Caswell.
1974. 2221

State Park Geologic Series. 1959-

SPGS. 3 The geology of Mount Blue
State Park, by K. A. Pankiwskj. 1965.
2222

Archeological Studies. 1973-

AS. 1 An archeological sequence in the middle Chesapeake region, by H. T. Wright. 1973.　　2223

Basic-Data Report. 1966-

BDR. 1 Records of wells and springs in Baltimore County, Maryland, by C. P. Laughlin. 1966.
　　2224

BDR. 2 Records of wells and springs, chemical analyses, and selected well logs in Charles County, Maryland, by T. H. Slaughter and C. P. Laughlin. 1966.
　　2225

BDR. 3 Hydrogeologic data from the Janes Island State Park test well (1,514 feet), Somerset County, Maryland, by H. J. Hansen. 1967.
　　2226

BDR. 4 Southern Maryland--records of selected wells, water levels, and chemical analyses of water, by J. M. Weigle and W. E. Webb. 1970.
　　2227

U. S. G. S. hydrologic atlas HA-365: water resources of southern Maryland, by J. M. Weigle, W. E. Webb, and R. A. Gardner. 1970.　　2228

BDR. 5 Deep wells of Maryland, by J. Edwards. 1970.　　2229

BDR. 6 Worcester County groundwater information: well records, pumpage, chemical quality data, and selected well logs, by R. C. Lucas. 1972.　　2230

BDR. 7 Harford County groundwater information: well records,

chemical quality data, and pumpage, by L. J. Nutter and M. J. Smigaj. 1975.　　2231

BDR. 8 Anne Arundel County groundwater information: selected well records, chemical quality data, pumpage, appropriation data, and selected well logs, by R. C. Lucas. 1976.
　　2232

Bulletin. 1944-

B. 27 Spores and pollen of the Potomac Group of Maryland, by G. J. Brenner. 1963.　　2233

B. 28 Copper, zinc, lead, iron, cobalt, and barite deposits in the Piedmont Upland of Maryland, by A. V. Heyl and N. C. Pearre. 1965. 2234

B. 29 Ground water in Prince George's County, by F. K. Mack. 1966. 2235

B. 30 Availability of ground water in Charles County, by T. H. Slaughter, E. G. Otton, and C. P. Laughlin. 1968.　　2236

B. 31 Geohydrology of channel-fill deposits near Salisbury, Maryland, by F. K. Mack, W. O. Thomas, and J. M. Weigle. 1972.　　2237

B. 32 Ground-water resources in Harford County, by L. J. Nutter. 1977.　　2238

Educational Series. 1964-

ES. 1 Proposed shore erosion program, by T. H. Slaughter. 1964.
　　2239

ES. 2 Water in Maryland: a review of the Free State's liquid assets, by

P. N. Walker. 1970.   2240

ES. 3   Caves of Maryland, by D. Slifer and R. Franz. 1971. 2241

ES. 4   Collecting fossils in Maryland, by J. D. Glaser. 1979. 2242

Geologic Guidebook. 1968-

GG. 1   Coastal Plain geology of southern Maryland, by J. D. Glaser. 1968.   2243

GG. 2   New interpretations of the eastern Piedmont geology of Maryland, by W. P. Crowley, M. W. Higgins, T. Bastian, and S. Olsen. 1971.   2244

GG. 3   Environmental history of Maryland Miocene, by R. E. Gernant, T. G. Gibson, and F. C. Whitmore. 1971.   2245

GG. 4   The Piedmont crystalline rocks at Bear Island, Potomac River, Maryland, by G. W. Fisher. 1975.   2246

GG. 5   Selected examples of carbonate sedimentation, Lower Paleozoic of Maryland, by J. Reinhardt and L. A. Hardie. 1976.   2247

Information Circular. 1964-

IC. 1   The mineral industry of Maryland in 1963, by N. A. Eilertsen. 1964.   2248

IC. 2   The mineral industry of Maryland in 1964, by N. A. Eilertsen. 1966.   2249

IC. 3   Directory of mineral producers in Maryland, by J. Edwards. 1966.   2250

IC. 4   The electric log: geophysics' contribution to ground-water prospecting and evaluation, by H. J. Hansen. 1967.   2251

IC. 5   The mineral industry of Maryland in 1965, by M. E. Hinkle. 1967.   2252

IC. 6   The mineral industry of Maryland in 1966, by E. L. Hemingway. 1967.   2253

IC. 7   The mineral industry of Maryland in 1967, by C. D. Edgerton. 1969.   2254

IC. 8   The mineral industry of Maryland in 1968, by C. D. Edgerton. 1970.   2255

IC. 9   Traveltime and concentration attenuation of a soluble dye in the Monocacy River, Maryland, by K. R. Taylor. 1970.   2256

IC. 10   The mineral industry of Maryland in 1969, by C. D. Edgerton. 1971.   2257

IC. 11   Directory of mineral producers in Maryland, by J. Edwards. 1971. Addendum, 1976.   2258

IC. 12   Traveltime and concentration attenuation of a soluble dye in Antietam and Conococheague Creeks, Maryland, by K. R. Taylor and W. B. Solley. 1971.   2259

IC. 13   Seismic evidence for high angle reverse faulting in the coastal plain of Prince George's and Charles Counties, Maryland, by F. H. Jacobeen. 1972.   2260

IC. 14   The mineral industry of Maryland in 1970, by D. C. Wininger. 1972.   2261

IC. 15   The mineral industry of Maryland in 1971, by C. L. Klingman. 1973.   2262

IC. 16   Well yields in the bedrock aquifers of Maryland, by L. J. Nutter. 1974.   2263

IC. 17   The mineral industry of Maryland in 1972, by C. L. Klingman. 1974.   2264

IC. 18   The mineral industry of Maryland in 1973, by C. L. Klingman. 1975.   2265

IC. 19   Chemical and physical erosion in the south mountain anticlinorium,

Maryland, by A. E. Godfrey. 1975.
2266

IC. 20 A digital simulation model of the Aquia Aquifer in southern Maryland, by G. W. Kapple and H. J. Hansen. 1976. 2267

IC. 21 1975 annual report of the Maryland Geological Survey. 1976.
2268

IC. 22 Coal reserves in Maryland --potential for future development, by K. N. Weaver, J. M. Coffroth, and J. Edwards. 1976. 2269

IC. 23 Reconnaissance survey of the Severn River sediments, by R. T. Kerhin. 1977. 2270

IC. 24 The mineral industry of Maryland in 1974, by J. A. Sutton. 1977. 2271

IC. 25 The mineral industry of Maryland in 1975, by J. A. Sutton. 1978. 2272

IC. 26 1976 annual report of the Maryland Geological Survey. 1978.
2273

IC. 27 1977 annual report of the Maryland Geological Survey. 1978.
2274

IC. 28 1978 annual report of the Maryland Geological Survey. 1979.
2275

IC. 29 Equilibrium, cyclicity, and problems of scale--Maryland's Piedmont landscape, by E. T. Cleaves and J. E. Costa. 1979.
2276

IC. 30 Directory of mineral producers in Maryland, 1979, by K. R. Kuff. 1980. 2277

IC. 31 The mineral industry of Maryland in 1976, by J. A. Sutton and M. N. Dimmitt. 1976. 2278

Report of Investigations. 1965-

RI. 1 Chemical quality of water and

trace elements in the Patuxent River Basin, by S. G. Heidel and W. W. Frenier. 1965. 2279

RI. 2 Pleistocene stratigraphy of the Salisbury area, Maryland, and its relationship to the lower eastern shore: a subsurface approach, by H. J. Hansen. 1966. 2280

RI. 3 Water resources of the Salisbury area, Maryland, by D. H. Boggess and S. G. Heidel. 1968. 2281

RI. 4 Expandable clay in St. Mary's formation of southern Maryland, by M. M. Knechtel, H. P. Hamlin, and J. W. Hosterman. 1968. 2282

RI. 5 Chemical quality reconnaissance of water of Maryland streams, by J. D. Thomas. 1966. 2283

RI. 6 Revision of stratigraphic nomenclature--Glenarm series of the Appalachian Piedmont, by D. L. Southwick and G. W. Fisher. 1968. 2284

RI. 7 Geophysical log cross-section network of the Cretaceous sediments of southern Maryland, by H. J. Hansen. 1968. 2285

RI. 8 Piedmont and coastal plain geology along the Susquehanna aqueduct: Baltimore to Aberdeen, Maryland, by E. T. Cleaves. 1968. 2286

RI. 9 Chemical and physical character of municipal water supplies in Maryland, by J. D. Thomas and S. G. Heidel. 1969. 2287

RI. 10 Ground-water occurrence in the Maryland Piedmont, by L. J. Nutter and E. G. Otton. 1969. 2288

RI. 11 Petrology and origin of Potomac and Magothy (Cretaceous) sediments, Middle Atlantic coastal plain, by J. D. Glaser. 1969. 2289

RI. 12 Paleoecology of the Choptank Formation (Miocene) of Maryland and Virginia, by R. E. Gernant. 1970.
2290

RI. 13 Extent of brackish water in the tidal rivers of Maryland, by W. E.

Webb and S. G. Heidel. 1970. 2291

RI. 14 Geologic and hydrologic factors bearing on subsurface storage of liquid wastes in Maryland, by E. G. Otton. 1970. 2292

RI. 15 Geology and mineral resources of southern Maryland, by J. D. Glaser. 1971. 2293

RI. 16 Flow characteristics of Maryland streams, by P. N. Walker. 1971. 2294

RI. 17 Water resources of Dorchester and Talbot Counties, Maryland, with special emphasis on the ground-water potential of the Cambridge and Easton area, by F. K. Mack, W. E. Webb, and R. A. Gardner. 1971. 2295

RI. 18 Solid-waste disposal in the geohydrologic environment of Maryland, by E. G. Otton. 1972. 2296

RI. 19 Hydrogeology of the carbonate rocks, Frederick and Hagerstown Valleys, Maryland, by L. J. Nutter. 1973. 2297

RI. 20 Hydrogeology of the formation and neutralization of acid water draining from underground coal mines of western Maryland, by E. F. Hollyday and S. W. McKenzie. 1973. 2298

RI. 21 Sedimentary facies of the Aquia formation in the subsurface of the Maryland coastal plain, by H. J. Hansen. 1974. 2299

RI. 22 An evaluation of the Magothy Aquifer in the Annapolis area, Maryland, by F. K. Mack. 1974. 2300

RI. 23 Stratigraphy, sedimentology, and Cambro-Ordovician paleogeography of the Frederick Valley, Maryland, by J. Reinhardt. 1974. 2301

RI. 24 Availability of fresh ground water in northeastern Worcester County, Maryland, with special emphasis on the Ocean City area, by J. M. Weigle. 1974. 2302

RI. 25 Petrologic and chemical investigation of chemical weathering in mafic rocks, eastern Piedmont of Maryland, by E. T. Cleaves. 1974. 2303

RI. 26 Hydrogeology of the Triassic rocks of Maryland, by L. J. Nutter. 1975. 2304

RI. 27 The geology of the crystalline rocks near Baltimore and its bearing on the evolution of the eastern Maryland Piedmont, by W. P. Crowley. 1976. 2305

RI. 28 Digital simulation and prediction of water levels in the Magothy Aquifer in southern Maryland, by F. K. Mack and R. J. Mandle. 1977. 2306

RI. 29 Upper Cretaceous (Senonian) and Paleocene (Danian) pinchouts on the south flank of the Salisbury embayment, Maryland, and their relationship to antecedent basement structures, by H. J. Hansen. 1978. 2307

RI. 30 New data bearing on the structural significance of the Upper Chesapeake Bay magnetic anomaly, by J. Edwards and H. J. Hansen. 1979. 2308

RI. 31 Simulated changes in water level in the Piney Point Aquifer in Maryland, by J. F. Williams. 1979. 2309

RI. 32 Geologic interpretations of aeromagnetic maps of the crystalline rocks in the Appalachians, northern Virginia to New Jersey, by G. W. Fisher, M. W. Higgins, and I. Zietz. 1980. 2310

RI. 33 A quasi three-dimensional finite-difference ground-water flow model with a field application, by G. Achmad and J. M. Weigle. 1979. 2311

MASSACHUSETTS

No activity.

Bulletin. 1964-

B. 1 Our rock riches; a selected collection of reprinted articles on Michigan's mineral resources by various authors. 1964.   2312

B. 2 Rocks and minerals of Michigan, by O. F. Poindexter, H. M. Martin, and S. G. Bergquist. Rev. ed. 1965.   2313

B. 3 Michigan's Au Sable River, today and tomorrow, by G. E. Hendrickson. 1966.   2314

B. 4 The glacial lakes around Michigan, by R. W. Kelley and W. R. Farrand. 1967.   2315

B. 5 Meteorites of Michigan, by V. D. Chamberlain. 1968.   2316

B. 6 Mineralogy of Michigan, by E. W. Heinrich. 1976.   2317

Circular. 1963-

C. 1 Michigan geological sourcebook, by R. W. Kelley and E. A. Kirkby. 1963.   2318

C. 2 Geological contributions to Michigan conservation; classified and annotated list, 1935-1964, by E. A. Kirkby. Rev. ed. 1964.   2319

C. 3 Mineral resources, depletion or deterioration? by W. Been. 1966.   2320

C. 4 Outline of procedures for determining gas-oil ratios and subsurface pressures, by J. S. Lorenz and T. L. Culver. 1966.   2321

C. 5 Sources of geological infor-

mation in Michigan, by R. W. Kelley and E. A. Kirkby. 1967.   2322

C. 6 Role of geology in state government, by G. E. Eddy. 1966.   2323

C. 7 Index to Michigan geologic theses, by E. A. Kirkby. 1967.   2324

C. 8 Geologic map index of Michigan, 1843-1962, by E. A. Kirkby. 1970.   2325

C. 9 Not published?   2326

C. 10 Mineral well act and general rules governing mineral well operations promulgated under Act No. 315, Public Acts of 1969. 1972.   2327

C. 11 Michigan's industrial sand resources, by J. D. Lewis. 1975.   2328

C. 12 One thousand million tons, by R. C. Reed. 1975.   2329

C. 13 Mines reclamation act and rules. 1978.   2330

C. 14 Seismic disturbances in Michigan, by D. M. Bricker. 1977.   2331

C. 15 Michigan's oil and gas regulations. 1980.   2332

C. 16 Annotated list of the publications of the Michigan Geological Survey, 1838-1977, indexed by author, mineral, and county, by W. W. Currie. 1978.   2333

C. 17 Blowout prevention; equipment, use, and testing, by D. T. Bertalan. 1979.   2334

Pamphlet. 196?-

P. 1 Biographical sketch of Douglass

Houghton, Michigan's first state geologist, 1837-1845, by H. Wallin. 1966. 2335

P. 2 Not published? 2336

P. 3 Guide to Michigan fossils, by R. W. Kelley. 1962. 2337

P. 4 Not published? 2338

P. 5 Geologic sketch of Michigan sand dunes, by R. W. Kelley. 1971. 2339

P. 6 Collecting rocks, minerals and fossils in Michigan, by S. E. Wilson. 1976. 2340

P. 7 Michigan's sand dunes, by S. E. Wilson. 1980. 2341

Report of Investigation. 1963-

RI. 1 Regional gravity and magnetic anomaly maps of the southern peninsula of Michigan, by W. J. Hinze. 1963. 2342

RI. 2 Michigan's Silurian oil and gas pools, by G. D. Ells. 1967. 2343

RI. 3 Geology for land and groundwater development in Wayne County, Michigan, by A. J. Mozola. 1969. 2344

RI. 4 Geologic and magnetic data for northern Iron River area, Michifan, by H. L. James. 1967. 2345

RI. 5 Geology and magnetic data for central Iron River area, Michigan, by C. E. Dutton. 1969. 2346

RI. 6 Geology and magnetic data for southeastern Iron River area, Michigan, by H. L. James and K. L. Weir. 1969. 2347

RI. 7 Geology and magnetic data between Iron River and Crystal Falls, Michigan, by H. L. James, F. J. Pettijohn, and L. D. Clark. 1970. 2348

RI. 8 Geology and magnetic data

for northern Crystal Falls area, Michigan, by F. J. Pettijohn. 1970. 2349

RI. 9 Geology and magnetic data for southern Crystal Falls area, Michigan, by F. J. Pettijohn. 1972. 2350

RI. 10 Geology and magnetic data for Alpha-B Rule River and Panola Plains areas, Michigan, by F. J. Pettijohn and others. 1969. 2351

RI. 11 Geology and magnetic data for northeastern Crystal Falls area, Michigan, by K. L. Weir. 1971. 2352

RI. 12 Geologic interpretation of aeromagnetic data in western upper peninsula of Michigan, by W. M. Meshref and W. J. Hinze. 1970. 2353

RI. 13 Geology for environmental planning in Monroe County, Michigan, by A. J. Mozola. 1970. 2354

RI. 14 Gravity and aeromagnetic anomaly maps of the southern peninsula of Michigan, by W. J. Hinze and others. 1971. 2355

RI. 15 Subsurface geology of Barry County, Michigan, by R. T. Lilienthal. 1975. 2356

RI. 16 A ground investigation of an aeromagnetic anomaly, north central Dickinson County, Michigan, by D. W. Snider. 1977. 2357

RI. 17 Establishing vegetation on alkaline iron and copper tailings, by S. G. Shetron and others. 1977. 2358

RI. 18 Drill core investigation of the Fiborn limestone member in Schoolcraft, Mackinac, and Chippewa Counties, Michigan, by A. Johnson and H. Sorensen. 1978. 2359

RI. 19 Stratigraphic cross-sections of the Michigan Basin, by R. T. Lilienthal. 1978. 2360

RI. 20 An economic study of coastal sand dune mining in Michigan, by Ayres, Lewis, Norris, & May, Inc.,

and M. J. Chapman. 1978.   2361

RI. 21 Economic geology of the
sand and sandstone resources of
Michigan, by E. W. Heinrich.   1979.
2362

RI. 22 Stratigraphic cross sections
extending from Devonian Antrim
shale to Mississippian Sunbury
shale in the Michigan Basin, by
G. D. Ells.   1979.   2363

RI. 23 Dune type inventory and
barrier dune classification study of
Michigan's Lake Michigan shore,
by W. R. Buckler.   1979.   2364

RI. 24 The Yellow Dog Peridotite
and a possible buried igneous complex of Lower Keweenawan Age in
the northern peninsula of Michigan,
by J. S. Klasner and others.   1979.
2365

RI. 25 Potassium salts (potash) of
the Salina A-1 evaporite in the
Michigan Basin, by R. C. Elowski.
1981.   2366

Water Information Series.   1971-

WIS. 1 Upper Rifle River Basin,
northeastern lower Michigan, by
R. L. Knutilla and others.   1971.
2367

WIS. 2 Flowing wells in Michigan,
1974, by W. B. Allen.   1977. 2368

WIS. 3 Not yet published.   2369

WIS. 4 Hydrology and recreation
on the cold-water rivers of Michigan's upper peninsula, by G. E.
Hendrickson, R. L. Knutilla, and
C. J. Doonan.   1973.   2370

WIS. 5 Compilation of miscellaneous stream flow measurements
for Michigan streams through
September 1970, by R. L. Knutilla.
1974.   2371

Water Investigation.   1963-

WI. 1 Reconnaissance of the ground-

water resources of Alger County,
Michigan, by K. E. Vanlier.   1963.
2372

WI. 2 Ground-water in Menominee
County, by K. E. Vanlier.   1963.
2373

WI. 3 Water resources of Van Buren
County, Michigan, by P. R. Giroux
and others.   1964.   2374

WI. 4 Ground-water resources of the
Battle Creek area, Michigan, by K. E.
Vanlier.   1966.   2375

WI. 5 Ground-water resources of
Dickinson County, Michigan, by G. E.
Hendrickson and C. J. Doonan.   1966.
2376

WI. 6 Water resources of Branch
County, Michigan, by P. R. Giroux
and others.   1966.   2377

WI. 7 Ground water in Iron County,
Michigan, by C. J. Doonan and G. E.
Hendrickson.   1967.   2378

WI. 8 Ground-water in Gogebic County,
Michigan, by C. J. Doonan and G. E.
Hendrickson.   1968.   2379

WI. 9 Ground water in Ontonagon
County, Michigan, by C. J. Doonan
and G. E. Hendrickson.   1969.   2380

WI. 10 Ground water and geology of
Keweenaw Peninsula, Michigan, by
C. J. Doonan.   1970.   2381

WI. 11 Ground water and geology of
Baraga County, Michigan, by C. J.
Doonan and J. R. Byerlay.   1973.
2382

Bulletin. 1889-

B. 44 Geology of the Duluth gabbro complex near Duluth, Minnesota, by R. B. Taylor. 1964.
2383

B. 45 Progressive contact metamorphism of the Biwabik ironformation, Mesabi Range, Minnesota, by B. M. French. 1968.
2384

Educational Series. 196?-

ES. 1 Guide to fossil collecting in Minnesota, by R. K. Hogberg, R. E. Sloan, and S. P. Tufford. Rev. ed. 1967.
2385

ES. 2 Guide to mineral collecting in Minnesota, by G. R. Rapp and D. T. Wallace. Rev. ed. 1979.
2386

ES. 3 Geologic sketch of the Tower-Soudan State Park, by P. K. Sims and G. B. Morey. 1966.
2387

ES. 4 Guide to the caves of Minnesota, by R. K. Hogberg and T. N. Bayer. 1967.
2388

ES. 5 Environmental geology of the Twin Cities metropolitan area, by R. K. Hogberg. 1971.
2389

Field Trip Guidebook. 1968-

FTG. 1 A geological field trip in the Rochester, Minnesota, area, by G. S. Austin. 1968.
2390

FTG. 2 Field trip guidebook for Lower Precambrian volcanic-sedimentary rocks of the Vermilion District, Minnesota, prepared for the annual meeting of the Geological Society of America and Associated Societies, Minneapolis. 1972.
2391

FTG. 3 Field trip guidebook for Precambrian rocks of the North Shore Volcanic Group, northeastern Minnesota, prepared for the annual meeting of the Geological Society of America and Associated Societies, Minneapolis. 1972.
2392

FTG. 4 Field trip guidebook for Paleozoic and Mesozoic rocks of southeastern Minnesota, prepared for the annual meeting of the Geological Society of America and Associated Societies, Minneapolis. 1972.
2393

FTG. 5 Field trip guidebook for Precambrian migmatitic terrane of the Minnesota River Valley, prepared for the annual meeting of the Geological Society of America and Associated Societies, Minneapolis. 1972.
2394

FTG. 6 Field trip guidebook for Precambrian geology of northwestern Cook County, Minnesota, prepared for the annual meeting of the Geological Society of America and Associated Societies, Minneapolis. 1972.
2395

FTG. 7 Field trip guidebook for geomorphology and Quaternary stratigraphy of western Minnesota and eastern South Dakota, prepared for the annual meeting of the Geological Society of America and Associated Societies, Minneapolis. 1972.
2396

FTG. 8 Field trip guidebook for hydrogeology of the Twin Cities artesian basin, prepared for the annual meeting of the Geological Society of America and Associated Societies, Minneapolis. 1972.
2397

Information Circular. 1962-

IC. 1 Directory of Minnesota mineral producers, 1962, by R. K. Hogberg. 1964.   2398

IC. 2 Chemical analyses of igneous rocks, by A. P. Ruotsala and S. P. Tufford. 1965.   2399

IC. 3 Not published.   2400

IC. 4 Directory of Minnesota mineral producers, 1964, comp. by R. K. Hogberg. 1966.   2401

IC. 5 Directory of Minnesota industrial mineral producers, 1967, comp. by R. K. Hogberg. 1969.   2402

IC. 6 Paleozoic lithostratigraphic nomenclature for southeastern Minnesota, by G. S. Austin. 1969.   2403

IC. 7 Summary of field work, 1969, ed. by P. K. Sims and I. Westfall. 1969.   2404

IC. 8 Summary of field work, 1970, ed. by P. K. Sims and I. Westfall. 1970.   2405

IC. 9 Instructions for using the Minnesota system for storage and retrieval of geologic log data, by J. H. Mossler, T. C. Winter, and S. P. Tufford. 1971.   2406

IC. 10 Copper and nickel resources in the Duluth complex, northeastern Minnesota, by B. Bonnichsen. 1974.   2407

IC. 11 The basis for a continental drilling program in Minnesota, by G. B. Morey. 1976.   2408

IC. 12 Selected bibliography of Cuyuna range geology, mining, and metallurgy, by R. J. Beltrame. 1977.   2409

IC. 13 Directory of information on Cuyuna range geology and mining, by R. J. Beltrame. 1977.   2410

IC. 14 Index to geophysical investigations in Minnesota, by R. J. Beltrame. 1978.   2411

Report of Investigations. 1963-

RI. 1 Geologic interpretation of magnetic map of McLeod County, Minnesota, by P. K. Sims and G. S. Austin. 1963.   2412

RI. 2 Geology of clay deposits, Red Wing area, Goodhue and Wabasha Counties, Minnesota, by G. S. Austin. 1963.   2413

RI. 3 Kaolin clay resources of the Minnesota River Valley in Brown, Redwood, and Renville Counties; a preliminary report, by W. E. Parham and R. K. Hogberg. 1964.   2414

RI. 4 Interpretation of Lake Washington magnetic anomaly, Meeker County, Minnesota, by P. K. Sims, G. S. Austin, and R. J. Ikola. 1965.   2415

RI. 5 The Cretaceous system in Minnesota, by R. E. Sloan. 1964.   2416

RI. 6 Groundwater contribution to streamflow and its relation to basin characteristics in Minnesota, by E. A. Ackroyd, W. C. Walton, and D. L. Hills. 1967.   2417

RI. 7 Stratigraphy and petrology of the type Fond du Lac Formation, Duluth, Minnesota, by G. B. Morey. 1967.   2418

RI. 8 K-Ar ages for hornblende from granites and gneisses and for basaltic intrusives in Minnesota, by G. N. Hanson. 1968.   2419

RI. 9 The Duluth complex in the Gabbro Lake quadrangle, Minnesota, by W. C. Phinney. 1969.   2420

RI. 10 Clay mineralogy, fabric, and industrial uses of the shale of the Decorah Formation, southeastern Minnesota, by W. E. Parham and G. S. Austin. 1969.   2421

RI. 11 Seismic studies over the midcontinent gravity high in Minnesota and northwestern Wisconsin, by H. M.

Mooney, P. R. Farnham, S. H. Johnson, G. Volz, and C. Craddock. 1970. 2422

RI. 12 Deep stratigraphic test well near Hollandale, Minnesota, by G. S. Austin. 1970. 2423

RI. 13 Sedimentology of the Middle Precambrian Thomson Formation, east-central Minnesota, by G. B. Morey and R. W. Ojakangas. 1970. 2424

RI. 14 Stratigraphy of the Lower Precambrian rocks in the Vermilion District, northeastern Minnesota, by G. B. Morey, J. C. Green, R. W. Ojakangas, and P. K. Sims. 1970. 2425

RI. 15 Quaternary geologic map index of Minnesota, by J. E. Goebel. 1976. 2426

RI. 16 Revised Keweenawan subsurface stratigraphy, southeastern Minnesota, by G. B. Morey. 1977. 2427

RI. 17 Geology, sulfide mineralization, and geochemistry of the Birchdale-Indus area, Koochiching County, northwestern Minnesota, by R. W. Ojakangas, D. G. Meineke, and W. H. Listerud. 1977. 2428

RI. 18 Cedar Valley Formation (Devonian) of Minnesota and northern Iowa, by J. H. Mossler. 1978. 2429

RI. 19 Results of subsurface investigations in northwestern Minnesota, by J. H. Mossler. 1978. 2430

RI. 20 Geology of the Deer Lake Complex, Itasca County, Minnesota, by J. L. Berkley, G. R. Himmelberg, and E. M. Ripley. 1978. 2431

RI. 20A Cumulus mineralogy and petrology of the Deer Lake Complex, Itasca County, Minnesota, by J. L. Berkley and G. R. Himmelberg. 1978. 2432

RI. 20B Sulfide minerals in the layered sills of the Deer Lake

Complex, by E. M. Ripley. 1978. 2433

RI. 21 Lower and Middle Precambrian stratigraphic nomenclature for east-central Minnesota, by G. B. Morey. 1978. 2434

RI. 22 Regional approach to estimating the ground-water resources of Minnesota, by R. Kanivetsky. 1979. 2435

RI. 23 Earthquake history of Minnesota, by H. M. Mooney. 1979. 2436

Special Publication. 1964-1970

SP. 1 History of the Minnesota Geological Survey, by G. M. Schwartz and P. K. Sims. 1964. 2437

SP. 2 Geology and origin of iron ore deposits of the Zenith Mine, Vermilion District, Minnesota, by J. F. Machamer. 1968. 2438

SP. 3 Ostracoda of the Dubuque and Maquoketa Formations of Minnesota and northern Iowa, by J. H. Burr and F. M. Swain. 1965. 2439

SP. 4 The Middle and Upper Ordovician conodont faunas of Minnesota, by G. F. Webers. 1966. 2440

SP. 5 Geology of Precambrian rocks, Granite Falls-Montevideo area, southwestern Minnesota, by G. R. Himmelberg. 1968. 2441

SP. 6 The cryptosome Bryozoa from the Middle Ordovician Decorah shale, Minnesota, by O. L. Karklins. 1969. 2442

SP. 7 The geology of the Middle Precambrian Rove Formation, in northeastern Minnesota, by G. B. Morey. 1969. 2443

SP. 8 The geology of the Isaac Lake Quadrangle, St. Louis County, Minnesota, by W. L. Griffin and G. B. Morey. 1969. 2444

SP. 9 Will not be published. 2445

SP. 10 Clay mineralogy and geology

of Minnesota's kaolin clays, by
W. E. Parham. 1970.   2446

SP. 11 Glacial and vegetational
history of northeastern Minnesota,
by H. E. Wright, W. A. Watts, S.
Jelgersma, J. C. B. Waddington,
J. Ogawa, and T. C. Winter. 1969.
                             2447

SP. 12 Will not be published. 2448

SP. 13 Lower Precambrian rocks
of the Gabbro Lake Quadrangle,
northeastern Minnesota, by J. C.
Green. 1970.   2449

Bulletin. 1907-

B. 95 Jasper County resources,
by D. A. DeVries and others. 1963.
2450

Oil and gas production, by H. M.
Morse.   2451

Subsurface stratigraphy, by W. H.
Moore.   2452

Ground-water geology and subsurface structure, by M. K. Kern.
2453

(B. 96 in base index)

B. 97 Mississippi geologic research
papers--1962. 1963.   2454

Economic potential of alumina-rich
clays and bauxite in Mississippi,
by M. K. Kern.   2455

Stratigraphic implications from studies of the Mesozoic of central and
southern Mississippi, by W. H.
Moore.   2456

Land snails from the loess of Mississippi, by L. Hubricht.   2457

Pleistocene land snails of southern
Mississippi and adjacent Louisiana,
by L. Hubricht.   2458

Problem of desiccation sinking at
Clarksdale, by T. W. Lusk.   2459

Geologic history and oil and gas
possibilities of Mississippi, by
E. H. Rainwater.   2460

B. 98 Geologic study along Highway 25 from Starkville to Carthage,
by T. W. Lusk. 1963.   2461

B. 99 Attala County mineral re-

sources, by W. S. Parks and others.
1963.   2462

Attala County ceramic tests, by T. E.
McCutcheon.   2463

Attala County subsurface geology, by
W. H. Moore.   2464

Water resources of Attala County,
Mississippi, by B. E. Wasson. 2465

B. 100 The Mississippi Geological
Survey, a centennial. 1963. 2466

Report of the geological reconnaissance
of parts of the counties of Yazoo, Issaquena, Washington, Holmes, Bolivar,
Tallahatchie, Coahoma, Mississippi,
during the months of October and November 1870, by E. A. Smith. 2467

History of the Mississippi Geological
Survey, by E. N. Lowe.   2468

Memorial to Ephraim N. Lowe (1864-
1933), by W. C. Morse.   2469

Address at recognition dinner honor-
ing Dr. W. C. Morse and Prof. F. E.
Vestal, upon their retirements, by
F. F. Mellen.   2470

Memorial to William Clifford Morse,
by F. E. Vestal.   2471

Employment in the mining industry in
Mississippi, by W. T. Hankins. 2472

The present course of the Mississippi
Geological Survey, by F. F. Mellen
and W. S. Parks.   2473

The Survey's enabling act.   2474

List of State geologists.   2475

Index to Bulletins 1-99, by W. S.
Parks.   2476

B. 101 An investigation of Mississippi iron ores, by M. K. Kern. 1963. 2477

B. 102 Mississippi geologic research papers--1963. 1964. 2478

Regional stratigraphy of the Midway and Wilcox in Mississippi, by E. H. Rainwater. 2479

Late Pleistocene and recent history of Mississippi sound between Beauvoir and Ship Island, by E. H. Rainwater. 2480

Geology of the northeast quarter of the West Point, Mississippi, quadrangle, and related bentonites, by T. F. Torries. 2481

B. 103 Survey of lightweight aggregate materials of Mississippi. 1964. 2482

Geology, by W. S. Parks. 2483

Economics, by C. A. McLeod. 2484

Tests, by A. G. Wehr. 2485

B. 104 Mississippi geologic research papers--1964. 1964. 2486

Type localities sampling program, by W. H. Moore, W. S. Parks, and M. K. Kern. 2487

Hilgard as a geologist, by H. V. Howe. 2488

Plant microfossils from the Eocene Cockfield Formation, Hinds County, Mississippi, by D. W. Engelhardt. 2489

Current projects of the Mississippi Geological Survey, by F. F. Mellen. 2490

Well logging by Mississippi Geological Survey, by A. R. Bicker and F. F. Mellen. 2491

B. 105 Hinds County mineral resources, by W. H. Moore and others. 1965. 2492

Foreword, by F. F. Mellen. 2493

Hinds County structural geology, by A. R. Bicker. 2494

Hinds County water resources, by A. R. Bicker, W. S. Parks, and W. H. Moore. 2495

Hinds County clay tests, by T. E. McCutcheon. 2496

Hinds County mineral industries, by W. S. Parks. 2497

B. 106 Sediments and microfauna off the coasts of Mississippi and adjacent states, by C. F. Upshaw, W. B. Creath, and F. L. Brooks. 1966. 2498

B. 107 Claiborne County geology and mineral resources, by A. R. Bicker and others. 1966. 2499

Subsurface stratigraphy of Claiborne County, by T. H. Dinkins. 2500

Claiborne County structural geology, by C. H. Williams. 2501

Claiborne County clay tests, by T. E. McCutcheon. 2502

B. 108 George County geology and mineral resources, by C. H. Williams and others. 1967. 2503

George County clay tests, by T. E. McCutcheon. 2504

Subsurface stratigraphy of George County, by T. H. Dinkins. 2505

B. 109 Jurassic stratigraphy of Mississippi. 1968. 2506

Jurassic stratigraphy of central and southern Mississippi, by T. H. Dinkins. 2507

A study of the Jurassic sediments in portions of Mississippi and Alabama, by M. L. Oxley and others. 2508

B. 110 Copiah County geology and mineral resources, by A. R. Bicker and others. 1969. 2509

Water resources of Copiah County, by T. N. Shows. 2510

Subsurface stratigraphy of Copiah County, by T. H. Dinkins.　2511

Copiah County clay tests, by T. E. McCutcheon.　2512

B. 111　Loess investigations in Mississippi, by J. O. Snowden and others.　1968.　2513

Pyrophysical (ceramic) and plastic properties of Mississippi loess, by T. E. McCutcheon.　2514

Pulmonate gastropods in the loess, by L. J. Hubricht.　2515

Forests of west central Mississippi as affected by loess, by C. D. Caplenor and others.　2516

B. 112　Economic minerals of Mississippi, by A. R. Bicker.　1970.　2517

B. 113　Water resources of Mississippi, by T. N. Shows.　1970. 2518

B. 114　The Jackson Eocene ostracoda of Mississippi, by W. J. Huff. 1970.　2519

B. 115　Rankin County geology and mineral resources, by W. T. Baughman and others.　1971.　2520

Rankin County clay tests, by T. E. McCutcheon.　2521

Rankin County mineral industries, by A. R. Bicker.　2522

Subsurface stratigraphy of Rankin County, by T. H. Dinkins.　2523

Water resources of Rankin County, by T. N. Shows.　2524

B. 116　Smith County geology and mineral resources, by E. E. Luper and others.　1973.　2525

Smith County clay tests, by R. Angurarohita.　2526

Water resources of Smith County, by W. T. Baughman.　2527

B. 117　Wayne County geology and

mineral resources, by J. H. May and others.　1974.　2528

Water resources of Wayne County, by W. T. Baughman and J. E. McCarty. 2529

Wayne County clay tests, by W. B. Hall and R. C. Glenn.　2530

Test results from Bureau of Mines and chemical analyses supervised by Dr. James P. Minyard.　2531

B. 118　Mississippi geologic names, by S. C. Childress.　1973.　2532

B. 119　Tinsley Field, 1939-1974; a commemorative bulletin, by W. H. Moore.　1974.　2533

B. 120　The mollusca of the Moodys Branch Formation, Mississippi, by D. T. Dockery.　1977.　2534

B. 121　Not yet published.　2535

B. 122　The invertebrate macropaleontology of Clarke County, Mississippi, by D. T. Dockery.　1980.　2536

Environmental Geology Series.　1973-

EGS. 1　Environmental geology of the Pocahontas, Clinton, Raymond, and Brownsville Quadrangle, Hinds County, Mississippi, by J. W. Green and M. Bograd.　1973.　2537

EGS. 2　Environmental geology of the Jackson, Jackson SE, Madison, and Ridgeland Quadrangles, Hinds County, Mississippi, by J. W. Green and S. C. Childress.　1975.　2538

EGS. 3　Not yet published.　2539

EGS. 4　Geology and man in Adams County, Mississippi, by S. C. Childress, M. Bograd, and J. C. Marble. 1976.　2540

Information Series.　1971?-

IS. 71-1　Electrical logs of water wells and test holes on file at the Mississippi Geological Survey, by

M. E. McGregor, T. N. Shows, and
W. T. Baughman. 1971.   2541

IS. 72-1 Mineral producers directory,
by A. R. Bicker. 1972.   2542

IS. 72-2 Fossil and mineral collecting localities of Mississippi, by
L. P. Pitts and M. Bograd. 1972.
2543

IS. 72-3 An investigation of water
supply problems at Alcorn College,
Lorman, Mississippi, by W. T.
Baughman. 1972.   2544

IS. 72-4 Salt water disposal wells
in Mississippi, by A. R. Bicker.
1972.   2545

IS. 73-1 Acker Lake landslide,
Monroe County, Mississippi, by
T. J. Laswell, D. M. Keady, and
E. E. Russell. 1973.   2546

IS. 74-1 An investigation of the Tertiary lignites of Mississippi, by
D. R. Williamson. 1976.   2547

IS. 77-1 Agricultural lime in central Mississippi, by A. R. Bicker
and J. H. May. 1977.   2548

IS. 79-1 Electrical logs of water
wells and test holes on file at the
Bureau of Geology and Energy Resources, by W. D. Easom, J. P.
Bradshaw, and R. J. Smith. 1979.
2549

Educational Series. 1962-

(ES. 1 in base index)

ES. 2 Environmental geology in towne and country, by W. C. Hayes and J. D. Vineyard. 1969.   2549a

ES. 3 An introduction to Missouri's geologic environment, by L. N. Stout and D. Hoffman. 1973.
2549b

ES. 4 Geologic wonders and curiosities of Missouri, by T. R. Beveridge. 1978.   2549c

ES. 5 Water in Missouri, by B. Harris. 1979.   2550

ES. 6 Diaspore--a depleted nonrenewable mineral resource of Missouri, by W. D. Keller. 1979.
2551

Engineering Geology Series. 1968-

EGS. 1 Engineering geology of the Maxville Quadrangle, Jefferson and St. Louis Counties, Missouri, by E. E. Lutzen. 1968.   2552

EGS. 2 Engineering geology of the Creve Coeur Quadrangle, St. Louis County, Missouri, by J. D. Rockaway and E. F. Lutzen. 1970.
2553

EGS. 3 Engineering geology criteria applicable to sewage treatment locations in Missouri, by J. W. Whitfield and others. 1971.   2554

EGS. 4 Engineering geology of St. Louis County, Missouri, by E. E. Lutzen and J. D. Rockaway. 1971.   2555

EGS. 5 Groundwater contamination and sinkhole collapse induced by leaky impoundments in soluble rock terrain, by T. J. Aley, J. H. Williams, and J. W. Massello. 1972.   2556

Information Circular. 1946-

IC. 17 Northwest Missouri oil and gas logs (1945-1963), comp. by J. Wells and E. McCracken. 1964.   2557

IC. 18 Bibliography of the geology of Missouri, 1966, by J. D. Vineyard. 1967.   2558

IC. 19 Bibliography of the geology of Missouri, 1967, by J. D. Vineyard. 1968.   2559

IC. 20 Missouri directory of mineral producers and processors, by J. A. Martin. 1969.   2560

IC. 21 Bibliography of the geology of Missouri, 1968, by J. D. Vineyard. 1969.   2561

IC. 22 Index to Missouri areal geologic maps, 1890-1969, by L. N. Stout. 1969.   2562

IC. 23 Bibliography of the geology of Missouri, 1969, by J. D. Vineyard. 1970.   2563

IC. 24 Bibliography of the geology of Missouri, 1970, by J. D. Vineyard. 1971.   2564

IC. 25 Bibliography of the geology of Missouri, 1971, by J. D. Vineyard. 1972.   2565

IC. 26 Minerals in the economy of Missouri, by U. S. Bureau of Mines. 1978.   2566

Miscellaneous Publication. 1918-

MP. 1 Oil and gas possibilities in the Belton area, by M. E. Wilson. 1918. 2567

MP. 2 Chemical analyses of river and spring waters, by H. W. Mundt and W. D. Turner. 1926. 2568

MP. 3 Paleontology of Late Cambrian and Early Ordovician formations in Missouri, by E. O. Ulrich, A. F. Foerste, and J. Bridge. n. d. 2569

MP. 4 Brown iron ore locations of Howell County, Missouri, investigated in 1953. Rev. 1954. 2570

MP. 5 Brown iron ore locations of Oregon County, Missouri, investigated in 1953. Rev. 1954. 2571

MP. 6 An introduction to the geologic history of Missouri, by T. R. Beveridge. 1955. Rev. 1978. 2572

MP. 7 Prospecting in Missouri, by T. R. Beveridge and others. 1956. 2573

MP. 8 The truth about ground water, by G. A. Muilenburg. 1956. 2574

MP. 9 Barite mining and production in Missouri, by G. A. Muilenburg. 1957. 2575

MP. 10 A brief discussion of some of the determinations made in a chemical water analysis, by W. B. Russell. 1957. 2576

MP. 11 Guidebook, field trip, 42nd annual meeting, the American Association of Petroleum Geologists, by R. D. Knight and J. W. Koenig. 1957. 2577

MP. 12 List of brown iron deposits, Wayne County, Missouri, comp. by W. C. Hayes. 1957. 2578

MP. 13 Catalogue of the caves of Missouri, comp. by J. D. Vineyard,

D. Stevens, B. Wills, and J. W. Koenig. 1957. 2579

MP. 14 Water wells in the glacial drift of north Missouri, by W. B. Russell. 1957. 2580

MP. 15 Bibliography of the Missouri Precambrian, comp. by B. Wills and N. Bertram. 1959. 2581

MP. 16 Chemical analyses, Precambrian rocks of Missouri, comp. by W. C. Hayes. 1959. 2582

MP. 17 Geology and exploration of Missouri iron deposits, by W. C. Hayes. 1959. Rev. 1961. 2583

MP. 18 Guidebook to the geology of the Rolla area, emphasizing solution phenomena, by T. R. Beveridge and W. C. Hayes. 1960. 2584

MP. 19 Kaolin deposits near Glen Allen, Bollinger County, Missouri, by A. C. Tennissen. 1960. 2585

MP. 20 Report writing manual for the Missouri Geological Survey and Water Resources, by J. W. Koenig. 1960. 2586

MP. 21 Composite stratigraphic column for Missouri. 1961. 2587

MP. 22 Mineral conservation in Missouri, by T. R. Beveridge. 1963. 2588

MP. 23 Directory of Missouri mineral producers and processors, by W. C. Hayes. 1964. 2589

MP. 24 Pleistocene stratigraphy of Missouri River Valley along the Kansas-Missouri border, by State Geological Survey of Kansas and Missouri Geological Survey. 1971. 2590

MP. 25 Oil for today and tomorrow, by Interstate Oil Compact Commission, 1971. 2591

MP. 26 Physical constraints to urban development in nineteen selected areas in Missouri, by W. E. Collins. 1974. 2592

MP. 27 Annotated bibliography of Missouri Precambrian, 1959-1973, by E. B. Kisvarsanyi. 1973. 2593

MP. 28 Interim report: sources of test data base information for a natural resources information system, comp, by D. L. Rath. 1975. 2594

MP. 29 The resources of St. Charles County, Missouri--land, water, and minerals, by Missouri Geological Survey. 1975. 2595

MP. 30 Works in print on water, by Missouri Division of Geology and Land Survey. 1979. 2596

MP. 31 Geographic location referencing and display considerations for proposed electronic data process ing of Missouri natural resources information, by D. Hoffman. 1976. 2597

MP. 32 Rules and regulations of Missouri Oil and Gas Council, by Missouri Oil and Gas Council. 1980. 2598

MP. 33 Inventory of strippable tar sands in southwestern Missouri, by J. S. Wells. 1977. 2599

MP. 34 Works in print on minerals, by Missouri Division of Geology and Land Survey. 1977. 2600

MP. 35 Remote sensing applications to Missouri environmental resources information system, by Missouri Division of Geology and Land Survey. 1978. 2601

MP. 36 Guide to mineral resources along I-44, Rolla to St. Louis, Missouri, by A. W. Herbrank. 1978. 2602

MP. 37 The role of coal in Missouri. 1979. 2603

Report (2nd Series). 1903-

R. 40. 1 Supplement 1 The Mississippian system, by T. L. Thompson

and K. H. Anderson. 1976. 2604

R. 41 Geomorphic history of the Ozarks of Missouri, by J. H. Bretz. 1965. 2605

R. 42 Bibliography of the geology of Missouri, 1955-1965, by J. D. Vineyard, J. W. Koenig, and B. L. Happel. 1967. 2606

R. 43 Mineral and water resources of Missouri, by the staffs of the Missouri Geological Survey and U. S. Geological Survey. 1967. 2607

Report of Investigations. 1945-

RI. 28 Desmoinesian fusulinids of Missouri, by D. G. Bebout. 1963. 2608

RI. 29 Mineral commodities of Putnam County, by R. J. Gentile. 1965. 2609

RI. 30 Guidebook, 1965 annual meeting, Geological Society of America, crypto-explosive structures in Missouri, by F. G. Snyder, J. H. Williams, and others. 1965. 2610

RI. 31 Guidebook, field trip, 1965 annual meeting, Geological Society of America, Geology of the Kansas City Group at Kansas City, by E. J. Parizek and R. J. Gentile. 1965. 2611

RI. 32 A new Pennsylvanian dibunophyllid coral from Missouri, by G. H. Fraunfelter. 1965. 2612

RI. 33 Midcontinent Pennsylvanian edestidae, by J. W. Koenig. 1965. 2613

RI. 34 Guidebook to Middle Ordovician and Mississippian strata, St. Louis and St. Charles Counties, Missouri, 1966 annual meeting, American Association of Petroleum Geologists, by J. A. Martin and J. S. Wells. 1966. 2614

RI. 35 The Grundel mastodon, by M. G. Mehl, J. A. Martin, J. H. Williams, and R. A. Marshall. 1966. 2615

RI. 36 Clay mineralogy and ceramic properties of Lower Cabaniss underclays in western Missouri, by A. C. Tennissen. 1967. 2616

RI. 37 Guidebook to the geology between Springfield and Branson, Missouri, emphasizing stratigraphy and cavern development, by J. D. Vineyard and L. D. Fellows. 1967. 2617

RI. 38 The Elsey Formation and its relationship to the Grand Falls chert, by C. E. Robertson. 1967. 2618

RI. 39 Conodont zonation of Lower Osagean rocks (Lower Mississippian) of southwestern Missouri, by T. L. Thompson. 1967. 2619

RI. 40 Mineral commodities of Macon and Randolph Counties, by R. J. Gentile. 1967. 2620

RI. 41 Planar stromatolite and burrowed carbonate mud facies in Cambrian strata of the St. Francois Mountain area, by W. B. Howe. 1968. 2621

RI. 42 The Ferrelview Formation (Pleistocene) of Missouri, by W. B. Howe and G. E. Heim. 1968. 2622

RI. 43 Lapilli tuffs and associated pyroclastic sediments in Upper Cambrian strata along Dent Branch, Washington County, Missouri, by R. E. Wagner and E. B. Kisvarsanyi. 1969. 2623

RI. 44 Exposed Precambrian rocks in southeast Missouri, by C. Tolman and F. Robertson. 1969 (Contribution to Precambrian Geology No. 1) 2624

RI. 45 Stratigraphy and conodont biostratigraphy of Kinderhookian and Osagean (Lower Mississippian) rocks of southwestern Missouri and adjacent areas, by T. L. Thompson and L. D. Fellows. 1969. 2625

RI. 46 Ash-flow tuffs of Precambrian age in southeast Missouri, by

R. E. Anderson, 1970. (Contribution to Precambrian Geology No. 2) 2626

RI. 47 Mafic intrusive rocks of Precambrian age in southeast Missouri, by D. H. Amos and G. A. Desborough. 1970. (Contribution to Precambrian geology No. 3) 2627

RI. 48 Evaluation of Missouri's coal resources, by C. E. Robertson. 1971. 2628

RI. 49 Structural features of Missouri, by M. H. McCracken. 1971. 2629

RI. 50 Conodont biostratigraphy of Chesterian strata in southwestern Missouri, by T. L. Thompson. 1972. 2630

RI. 51 Petrochemistry of a Precambrian igneous province, St. Francois Mountains, Missouri, by E. B. Kisvarsanyi. 1972. (Contribution to Precambrian geology No. 4) 2631

RI. 52 Correlation of Cambrian strata of the Ozark and Upper Mississippi Valley regions, by W. B. Howe, V. E. Kurtz, and K. H. Anderson. 1972. 2632

RI. 53 Barite ore potential of four tailings ponds, Washington County Barite District, Missouri, by H. M. Wharton. 1972. 2633

RI. 54 Mineable coal reserves of Missouri, by C. E. Robertson. 1973. 2634

RI. 55 Traverse in Late Cambrian strata from the St. Francois Mountains, Missouri, to Delaware County, Oklahoma, by K. H. Anderson, J. L. Thacker, V. E. Kurtz, and P. E. Gerdemann. 1975. 2635

RI. 56 Data on Precambrian in drillholes of Missouri including rock type and surface configuration, by E. B. Kisvarsanyi. 1975. (Contribution to Precambrian geology No. 5) 2636

RI. 57 Studies in stratigarphy, by Missouri Geological Survey. 1975. 2637

RI. 58 Guidebook to the geology and ore deposits of selected mines in the Viburnum trend, Missouri, by H. M. Wharton and others, ed. by J. D. Vineyard, B. Harris, and L. D. Stout. 1975.   2638

RI. 59 The geology of Bates County, Missouri, by R. J. Gentile.   1976.
                                          2639

RI. 60 Chemical analyses of selected Missouri coal and some statistical implications, by W. K. Wedge, D. M. S. Bhatia, and A. W. Rueff. 1976.   2640

RI. 61 Studies in Precambrian geology with a guide to selected parts of the St. Francois Mountains, Missouri, ed. by E. B. Kisvarsanyi. 1976.   2641

RI. 62 Guidebook to the geology along Interstate 55 in Missouri, by J. L. Thacker and I. R. Satterfield. 1977.   2642

RI. 63 Chemical composition of Missouri coals, by W. K. Wedge and J. R. Hatch. 1980.   2643

Special Publication.   1969-

SP. 1 Missouri minerals--resources, production, and forecasts, by H. M. Wharton and others. 1969. 2644

Water Resources Report.   1956-

WRR. 1 Water possibilities from the glacial drift of Grundy County, by D. L. Fuller and W. B. Russell. 1956.   2645

WRR. 2 Water possibilities from the glacial drift of Mercer County, by J. R. McMillen and W. B. Russell. 1956.   2646

WRR. 3 Water possibilities from the glacial drift of Harrison County, by D. L. Fuller, J. R. McMillen, and W. B. Russell. 1956. 2647

WRR. 4 Water possibilities from the

glacial drift of Putnam County, by D. L. Fuller and J. R. McMillen. 1956.   2648

WRR. 5 Water possibilities from the glacial drift of Worth County, by D. L. Fuller and others. 1956.   2649

WRR. 6 Water possibilities from the glacial drift of Livingston County, by D. L. Fuller and others. 1956. 2650

WRR. 7 Water possibilities from the glacial drift of Gentry County, by D. L. Fuller and others. 1956. 2651

WRR. 8 Water possibilities from the glacial drift of DeKalb County, by D. L. Fuller and others. 1957. 2652

WRR. 9 Water possibilities from the glacial drift of Daviess County, by D. L. Fuller and others. 1957. 2653

WRR. 10 Water possibilities from the glacial drift of Sullivan County, by D. L. Fuller and others. 1957. 2654

WRR. 11 Water possibilities from the glacial drift of Linn County, by D. L. Fuller and others. 1957.   2655

WRR. 12 Water possibilities from the glacial drift of Chariton County, by D. L. Fuller and others. 1957. 2656

WRR. 13 Water possibilities from the glacial drift of Carroll County, by D. L. Fuller and others. 1957. 2657

WRR. 14 Water possibilities from the glacial drift of Buchanan County, by D. L. Fuller and others. 1957. 2658

WRR. 15 Water possibilities from the glacial drift of Andrew County, by D. L. Fuller and others. 1957. 2659

WRR. 16 Preliminary report on the groundwater resources of Nodaway County, Missouri, by G. E. Heim, J. A. Martin, and W. B. Howe. 1959.
                                          2660

WRR. 17 Preliminary report, groundwater resources of Holt County, by G. E. Heim, J. A. Martin, and W. B. Howe, 1960.   2661

WRR. 18 Preliminary report, groundwater resources of Atchison County, by G. E. Heim, J. A. Martin, and W. B. Howe. 1960. 2662

WRR. 19 Floods of June 17-18, 1964, in Jefferson, Ste. Genevieve, and St. Francois Counties, Missouri, by M. S. Petersen. 1965. 2663

WRR. 20 Low-flow characteristics of Missouri streams, by J. Skelton. 1966. 2664

WRR. 21 Floods of July 18-23, 1965, in northwestern Missouri, by J. E. Bowie and E. E. Gann. 1967. 2665

WRR. 22 Storage requirements to augment low flows of Missouri streams, by J. Skelton and J. H. Williams. 1968. 2666

WRR. 23 Magnitude and frequency of Missouri floods, by E. H. Sandhaus and J. Skelton. 1968. 2667

WRR. 24 Water resources of the Joplin area, Missouri, by G. L. Feder and others. 1969. 2668

WRR. 25 Base-flow recession characteristics and seasonal low-flow frequency characteristics for Missouri streams, by J. Skelton. 1970. 2669

WRR. 26 Groundwater resources of Saline County, Missouri, by J. C. Miller. 1972. 2670

WRR. 27 Carryover storage requirements for reservoir design in Missouri, by J. Skelton. 1971. 2671

WRR. 28 Flood volume design data for Missouri streams, by J. Skelton. 1973. 2672

WRR. 29 Springs of Missouri, by J. D. Vineyard, G. L. Feder, W. L. Pflieger, and R. G. Lipscomb. 1974. 2673

WRR. 30 Water resources of the St. Louis area, Missouri, by D. E. Miller and others. 1974. 2674

WRR. 31 A guide for the geologic and hydrologic evaluation of small lake sites in Missouri, by T. J. Dean, J. H. Barks, and J. H. Williams. 1976. 2675

WRR. 32 Missouri stream and spring-flow characteristics: low-flow frequency and flow duration, by J. Skelton. 1976. 2676

WRR. 33 Water-quality characteristics of six small lakes in Missouri, by J. H. Barks. 1976. 2677

WRR. 34 Water resources and geology of the Springfield area, Missouri, by L. F. Emmett and others. 1978. 2678

Bulletin. 1919-

B. 33 Directory of known mining enterprises, 1962, by R. D. Geach and J. M. Chelini. 1963.   2679

A list of active coal mines, by T. Morgan. 1963.   2680

The mineral industry of Montana in 1962, by F. B. Fulkerson and R. W. Knostman. 1963.   2681

B. 34 Mines and mineral deposits (except fuels), Sanders County, Montana, by F. A. Crowley. 1963.   2682

B. 35 Ore deposits of the northern part of the Park (Indian Creek) district, Broadwater County, Montana, by E. M. Schell. 1963.   2683

B. 36 Geologic investigations in the Kootenai-Flathead area, northwest Montana, No. 5, western Flathead County, and part of Lincoln County, by W. M. Johns, A. G. Smith, W. C. Barnes, E. H. Gilmour, and W. D. Page. 1963.   2684

B. 37 Basic water data report No. 1, Missoula Valley, Montana, by A. Brietkrietz. 1964.   2685

B. 38 Directory of mining enterprises, 1963, by R. D. Geach. 1964.   2686

A list of active coal mines, by T. Morgan. 1964.   2687

The mineral industry of Montana in 1963, by F. B. Fulkerson, A. J. Kauffman, and R. W. Knostman. 1964.   2688

B. 39 Handbook for small mining enterprises in Montana, by U. M.

Sahinen, F. N. Earll, G. G. Griswold, F. H. Kelly, K. S. Stout, R. I. Smith, W. A. Vine, and D. J. Emblen. 1964.   2689

B. 40 Geology and ore deposits of the Clinton mining district, Missoula County, Montana, by D. E. Hintzman. 1964.   2690

B. 41 Economic geology and geochemical study of Winston mining district, Broadwater County, Montana, by F. N. Earll. 1964.   2691

B. 42 Geologic investigations in the Kootenai-Flathead area, northwest Montana, No. 6, southeastern Flathead County and northern Lake County, by W. M. Johns. 1964.   2692

B. 43 Geology of the Garnet Mountain quadrangle, Gallatin County, Montana, by W. J. McMannis and R. A. Chadwick. 1964.   2693

B. 44 Limestone, dolomite, and travertine in Montana, by J. M. Chelini. 1965.   2694

B. 45 Progress report on clays and shales of Montana, 1962-64, by J. M. Chelini, R. I. Smith, and D. C. Lawson. 1965.   2695

B. 46 Directory of mining enterprises, 1964, by R. D. Geach. 1965.   2696

A list of active coal mines, by T. Morgan. 1965.   2697

The mineral industry in Montana in 1964, by R. W. Knostman and W. N. Hale. 1965.   2698

B. 47 Geology and ground-water resources of the Missoula Basin, Montana, by R. G. McMurtrey, R. L. Konizeski, and A. Brietkrietz. 1965.   2699

B. 48 Geochemical reconnaissance stream-sediment sampling in Flathead and Lincoln Counties, Montana, by U. M. Sahinen, W. M. Johns, and D. C. Lawson. 1965. 2700

B. 49 Directory of mining enterprises, 1965, by R. D. Geach. 1966. 2701

A list of active coal mines, by T. Morgan. 1966. 2702

The mineral industry of Montana in 1965, by R. P. Collins, W. N. Hale, and R. W. Knostman. 1966. 2703

B. 50 Geology and ground-water resources of western and southern parts of Judith Basin, Montana, by E. A. Zimmerman. 1966. 2 vols. 2704

B. 51 The recovery of elemental sulfur from sulfide ores, by F. Habashi. 1966. 2705

B. 52 Geology and ground-water resources of the Cascade-Ulm area, Montana, by R. D. Fox. 1966. 2706

B. 53 Basic water data report No. 3, Kalispell Valley, Montana, by A. Brietkrietz. 1966. 2707

B. 54 Some high-purity quartz deposits in Montana, by J. M. Chelini. 1966. 2708

B. 55 Progress report on clays and shales of Montana, 1964-65, by J. M. Chelini, R. I. Smith, and F. P. Jones. 1966. 2709

B. 56 Stratigraphy and economic geology of the Great Falls-Lewistown coal field, central Montana, by A. J. Silverman and W. L. Harris. 1967. 2710

B. 57 Water levels and artesian pressures in observation wells in Montana, by R. G. McMurtrey and T. E. Reed. 1967. 2711

B. 58 Directory of mining enterprises, 1966, by R. D. Geach.

1967. 2712

A list of active coal mines, by T. Morgan. 1967. 2713

The mineral industry of Montana in 1966, by R. P. Collins, W. N. Hale, and R. W. Knostman. 1967. 2714

B. 59 Kinetics and mechanism of gold and silver dissolution in cyanide solution, by F. Habashi. 1967. 2715

B. 60 Water resources of the Cut Bank area, Glacier and Toole Counties, Montana, by E. A. Zimmerman. 1967. 2716

B. 61 Geochemical investigations in Lincoln and Flathead Counties, Montana, by U. M. Sahinen, W. M. Johns, and D. C. Lawson. 1967. 2717

B. 62 Market study and compendium of data on industrial minerals and rocks in Montana, by J. M. Chelini. 1967. Parts A and B. 2718

B. 63 Mining methods and equipment illustrated, by K. S. Stout. 1967. 2719

B. 64 Geology and ore deposits of the Castle Mountains mining district, Meagher County, Montana, by A. S. Winters, 1968. 2720

B. 65 Water levels and artesian pressures in observation wells in Montana through 1967, by R. G. McMurtrey and T. E. Reed. 1968. 2721

B. 66 Ground-water resources of the northern Powder River Valley, southeastern Montana, by O. J. Taylor. 1968. 2722

B. 67 Directory of mining enterprises, 1967, by R. D. Geach. 1968. 2723

A list of active coal mines, by T. Morgan. 1968. 2724

The mineral industry of Montana in 1967, by R. P. Collins, W. N. Hale, and F. V. Carrillo. 1968. 2725

B. 68 Geology and ground-water resources of the Kalispell Valley, north-

western Montana, by R. L. Konizeski, A. Brietkrietz, and R. G.
McMurtrey. 1968.   2726

B. 69 Strippable coal deposits on
state land, Powder River County,
Montana, by R. E. Matson, G. G.
Dahl, and J. W. Blumer. 1968.
                           2727

B. 70 Progress report on clays
and shales of Montana, 1966-67,
by R. B. Berg, D. C. Lawson, F. P.
Jones, and R. I. Smith. 1968.
                           2728

B. 71 Water levels and artesian
pressures in observation wells in
Montana through 1968, by T. E. Reed
and R. G. McMurtrey. 1969. 2729

B. 72 Directory of mining enterprises, 1968, with a list of active
coal mines, 1968, by K. T. Bondurant
and D. C. Lawson. 1969.   2730

The mineral industry of Montana in
1968, by F. V. Carrillo, W. N.
Hale, and M. A. McComb. 1969.
                           2731

B. 73 Geology and coal resources
of the Foster Creek coal deposits,
eastern Montana, by E. H. Gilmour
and L. A. Williams. 1969. 2732

B. 74 Bentonite in Montana, by
R. B. Berg. 1969.   2733

B. 75 Hydrogeology of the Upper
Silver Bow Creek drainage area,
Montana, by M. K. Botz. 1969.
                           2734

B. 76 Water levels and artesian
pressure in observation wells in
Montana, 1966-69, by T. E. Reed
and R. G. McMurtrey. 1970. 2735

B. 77 Directory of mining enterprises, 1969, with a list of active
coal mines, 1969, by M. Hansen.
1970.   2736

The mineral industry of Montana
in 1969, by F. V. Carrillo, W. N.
Hale, and M. A. McComb. 1970.
                           2737

B. 78 Strippable coal deposits, Mc-
Cone County, Montana, by R. E. Matson. 1970.   2738

B. 79 Geology and mineral deposits
of Lincoln and Flathead Counties,
Montana, by W. M. Johns. 1970.
                           2739

B. 80 Progress report on clays and
shales of Montana, 1968-69, by R. B.
Berg, D. C. Lawson, F. P. Jones,
and R. I. Smith. 1970.   2740

B. 81 Surficial geology and water resources of the Tobacco and upper
Stillwater River Valleys, northwestern
Montana, by D. L. Coffin, A. Brietkrietz, and R. G. McMurtrey. 1971.
                           2741

B. 82 Directory of mining enterprises,
1970, with a list of active coal mines,
1970, by M. Hansen. 1971. 2742

The mineral industry of Montana in
1970, by F. V. Carrillo, W. N. Hale,
and M. A. McComb. 1971.   2743

B. 83 Strippable coal in the Moorhead
coal field, Montana, by R. E. Matson.
1971.   2744

B. 84 Mines and mineral deposits of
the southern Flint Creek Range,
Montana, by F. N. Earll. 1972. 2745

B. 85 Mines and mineral deposits
(except fuels), Beaverhead County,
Montana, by R. D. Geach. 1972.
                           2746

B. 86 Directory of mining enterprises,
1971, with a list of active coal mines,
1971, by D. C. Lawson. 1972. 2747

The mineral industry of Montana in
1971, by the Division of Nonferrous
Metals, U. S. Department of the Interior, Bureau of Mines. 1972. 2748

B. 87 Geology and water resources
of eastern part of Judith Basin, Montana, by R. D. Feltis. 1973. 2749

B. 88 Directory of mining enterprises,
1972, with a list of active coal mines,
1972, by D. C. Lawson. 1973. 2750

The mineral industry of Montana in 1972, by the Division of Non-ferrous Metals, U.S. Department of the Interior, Bureau of Mines. 1973. 2751

B. 89 Progress report on clays and shales of Montana, 1970, by R. B. Berg, D. C. Lawson, F. P. Jones, and R. I. Smith. 1973. 2752

B. 90 Characteristics and uses of Montana fly ash, by F. J. Quilici. 1973. 2753

B. 91 Quality and reserves of strippable coal, selected deposits, southeastern Montana, by R. E. Matson and J. W. Blumer. 1973. 2754

B. 92 Directory of mining enter-prises, 1973, with a list of active coal mines, 1973, by D. C. Lawson. 1974. 2755

The mineral industry of Montana in 1973, by the Division of Nonfer-rous Metals, U. S. Department of the Interior, Bureau of Mines. 1974. 2756

B. 93 Hydrologic effects of strip coal mining in southeastern Montana --emphasis: one year of mining near Decker, by W. A. Van Voast. 1974. 2757

B. 94 Building stone in Montana, by R. B. Berg. 1974. 2758

B. 95 Directory of mining enter-prises, 1974, with a list of active coal mines, 1974, by D. C. Lawson. 1975. 2759

The mineral industry of Montana in 1974, by the Division of Non-ferrous Metals, U. S. Department of the Interior, Bureau of Mines. 1975. 2760

B. 96 Soil and rock mechanics illustrated, by K. S. Stout. 1975. 2761

B. 97 Hydrogeologic aspects of

existing and proposed strip coal mines near Decker, southeastern Montana, by W. A. Van Voast and R. B. Hedges. 1975. 2762

B. 98 Metallic mineral deposits of Powell County, Montana, by H. G. McClernan. 1976. 2763

B. 99 Handbook for small mining en-terprises, by F. N. Earll, K. S. Stout, G. G. Griswold, R. I. Smith, F. H. Kelly, D. J. Emblen, W. A. Vine, and D. H. Dahlem. 1976. 2764

B. 100 Directory of mining enterprises, 1975, with a list of coal mines, 1975, by D. C. Lawson. 1976. 2765

The mineral industry of Montana in 1975, by the Division of Nonferrous Metals, U. S. Department of the In-terior, Bureau of Mines. 1976. 2766

B. 101 Geology and water resources of northern part of Judith Basin, Montana, by R. D. Feltis. 1977. 2767

B. 102 Hydrogeologic conditions and projections related to mining near Colstrip, southeastern Montana, by W. A. Van Voast, R. B. Hedges, and J. J. McDermott. 1977. 2768

B. 103 Directory of mining enter-prises for 1976, with a list of coal mines, 1976, by D. C. Lawson. 1977. 2769

The mineral industry of Montana in 1976, by the Division of Nonferrous Metals, U. S. Department of the In-terior, Bureau of Mines. 1977. 2770

B. 104 Water resources of the Clark Fork Basin upstream from St. Regis, Montana, by A. J. Boettcher and A. W. Gosling. 1977. 2771

B. 105 Caves of Montana, by N. P. Campbell. 1978. 2772

B. 106 Ground-water resources in the Libby area, northwestern Montana, by A. J. Boettcher and K. R. Wilke. 1978. 2773

B. 107 Directory of Mining enterprises for 1977, by D. C. Lawson. 1978. 2774

The mineral industry of Montana in 1977, by the Division of Nonferrous Metals, U. S. Department of the Interior, Bureau of Mines. 1978. 2775

B. 108 Water resources of the central Powder River area of southeastern Montana, by W. R. Miller. 1979. 2776

B. 109 Directory of mining enterprises for 1978, by D. C. Lawson. 1979. 2777

The mineral industry in Montana in 1978, by the U. S. Bureau of Mines, Montana Liaison Officer, G. Krempasky. 1979. 2778

B. 110 Annotated bibliography of the geothermal resources of Montana, by S. A. Rautio and J. L. Sonderegger. 1980. 2779

B. 111 Directory of mining enterprises, by D. C. Lawson. 1979. 2780

B. 112 Current geological and geophysical studies in Montana, by R. B. Berg. 1980. 2781

B. 113 Mining codes of Montana (annotated), by K. Stout. 1980. 2782

Memoir. 1929-

M. 39 Geology of the Garnet-Bearmouth area, western Montana, by M. E. Kauffman. 1964. 2783

Chapter on metallic resources, by F. N. Earll. 1964. 2784

M. 40 Ground-water resources along Cedar Creek anticline in eastern Montana, by O. J. Taylor. 1965. 2785

M. 41 Bedrock geology of the Sheridan District, Madison County, Montana, by H. R. Burger. 1967. 2786

M. 42 Correlation of Pleistocene glaciation in the Bitterroot Range, Montana, with fluctuations of glacial Lake Missoula, by W. M. Weber. 1972. 2787

M. 43 Petrology of the northeastern border zone of the Idaho batholith, Bitterroot Range, Montana, by R. B. Chase. 1973. 2788

M. 44 Reconnaissance geology of southernmost Ravalli County, Montana, by R. B. Berg. 1977. 2789

M. 45 Talc and chlorite deposits in Montana, by R. B. Berg. 1979. 2790

Special Publication. 1963-

SP. 28 Mineral and water resources of Montana. 1963. 2791

SP. 29 Preliminary compilation, index of unpublished geologic studies in Montana, by J. M. Chelini. 1963. 2792

SP. 30 A comparative tax study of Montana's gas producing industry, by D. H. Harnish. 1964. 2793

SP. 31 Stratigraphic correlations for Montana and adjacent states, by S. L. Groff. 1963. 2794

SP. 32 Biennial report of the Montana Bureau of Mines and Geology, a department of Montana School of Mines, to the Legislative Assembly for the reporting period July 1, 1962, to June 30, 1964. 2795

SP. 33 Mineral potential of eastern Montana--a basis for future growth, report compiled by U. S. Geological Survey and U. S. Bureau of Mines at the request of Senator Mike Mansfield, 89th Congress, 1st Session, Senate Doc. 12, 1965. 2796

SP. 34 Index of unpublished geologic studies in Montana, by J. M. Chelini. 1965. 2797

SP. 35 Reconnaissance ground-water and geologic studies of western Meagher County, Montana, by S. L. Goff. 1965. 2798   2798

SP. 36 Proceedings of the first Montana coal resources symposium, by S. L. Groff. 1966. 2799

SP. 37 Index and bibliography of ground-water studies in Montana, by M. K. Botz and E. W. Bond. 1966. 2800

SP. 38 Selected bibliography of stratigraphy in Montana and adjacent areas, by E. H. Gilmour. 1966. 2801

SP. 39 Index map and bibliography of coal studies in Montana, by E. H. Gilmour and G. G. Dahl. 1966. 2802

SP. 40 Biennial report of the Montana Bureau of Mines and Geology, a department of Montana College of Mineral Science and Technology, for the reporting period July 1, 1964, to June 30, 1966. 2803

SP. 41 Thorium deposits of the Lemhi Pass district, Beaverhead County, Montana, by R. D. Geach and R. E. Matson. 1966. 2804

SP. 42 1966 industrial seminar, western phosphate region, by Montana Bureau of Mines and Geology and Montana State Planning Board. 1967. 2805

SP. 43 Montana coal analyses, by E. H. Gilmour and G. G. Dahl. 1967. 2806

SP. 44 High calcium limestone deposit in the Rattler Gulch area, Granite County, Montana, by J. O. Landreth. 1968. 2807

SP. 45 Biennial report of the Montana Bureau of Mines and Geology, a department of Montana College of Mineral Science and Technology, for the reporting period July 1, 1966, to June 30, 1968. 2808

SP. 46 Current geological and geophysical studies in Montana, by R. B. Berg. 1969. 2809

SP. 47 Painted Rock Lake area, southern Ravalli County, Montana,

by H. G. Fisk. 1969. 2810

SP. 48 Conversion of section-township-range to latitude-longitude, a computer technique, by M. K. Botz. 1969. 2811

SP. 49 Development of a statewide system for computer processing of hydrogeological data, by M. K. Botz. 1970. 2812

SP. 50 Current geological and geophysical studies in Montana, by R. B. Berg. 1970. 2813

SP. 51 Bentonite deposits in the Ingomar-Vananda area, Treasure and Rosebud Counties, Montana, by R. B. Berg. 1970. 2814

SP. 52 Uranium in phosphate rock, by F. Habashi. 1970. 2815

SP. 53 Index of graduate theses on Montana geology, by R. B. Berg. 1971. 2816

SP. 54 Catalog of stratigraphic names for Montana, by C. A. Balster. 1971. 2817

SP. 55 Stratigraphic correlations for Montana and adjacent areas, by C. A. Balster. 2nd ed. 1971. 2818

SP. 56 Current geological and geophysical studies in Montana, by R. B. Berg. 1971. 2819

SP. 57 Hydrology of the west fork drainage of the Gallatin River, southwestern Montana, prior to commercial recreational development, by W. A. Van Voast. 1972. 2820

SP. 58 Current geological and geophysical studies in Montana, by R. B. Berg. 1972. 2821

SP. 59 Current geological and geophysical studies in Montana, by R. B. Berg. 1973. 2822

SP. 60 Structure contour map, Upper Cretaceous, southeastern Montana, by C. A. Balster. 1973. 2823

SP. 61 Geology, stratigraphy, and

biostratigraphy of the north end of Cedar Creek anticline, Dawson County, Montana, by G. A. Bishop. 1973.   2824

SP. 62 Bouguer gravity map of Montana, by W. E. Bonini and R. B. Smith. 1974.   2825

SP. 63 Current geological and geophysical studies in Montana, by R. B. Berg. 1974.   2826

SP. 64 Geochemical exploration of the Stemple Pass area, Lewis and Clark County, Montana, by H. G. McClernan. 1974.   2827

SP. 65 Geothermal map, upper part of Madison Group, Montana, by C. A. Balster. 1974.   2828

SP. 66 Market prospects for Montana coal, by Cameron Engineers, Inc., with a chapter on water, by the U. S. Bureau of Reclamation. 1975.   2829

SP. 67 A study of the influence of seismic shotholes on ground water and aquifers in eastern Montana, by E. W. Bond. 1975.   2830

SP. 68 Current geological and geophysical studies in Montana, by R. B. Berg. 1975.   2831

SP. 69 Geology of the East Pryor Mountain quadrangle, Carbon County, Montana, by D. L. Blackstone. 1975.   2832

SP. 70 Preliminary bibliography and index of the metallic mineral resources of Montana through 1969, by H. G. McClernan. 1975. 2833

SP. 71 Index map and bibliography of coal studies in Montana, by J. M. Pinchock. 1975.   2834

SP. 72 Current geological and geophysical studies in Montana, by R. B. Berg. 1976.   2835

SP. 73 Guidebook--The Tobacco Root Geological Society, 1976 field conference. 1976.   2836

SP. 74 Eleventh industrial minerals forum, 1976.   2837

SP. 75 Principal information on Montana mines, by D. H. Krohn and M. M. Weist. 1977.   2838

SP. 76 Current geological and geophysical studies in Montana, by R. B. Berg. 1977.   2839

SP. 77 Index of graduate theses on Montana geology, by S. J. Czehura and R. B. Berg. 1977.   2840

SP. 78 Current geological and geophysical studies in Montana, by R. B. Berg. 1978.   2841

SP. 79 Geothermal data-base study: Mine-water temperatures, by D. C. Lawson and J. L. Sonderegger. 1978.   2842

SP. 80 Ground water of the Fort Union coal region, eastern Montana, by the Montana Bureau of Mines and Geology and the U. S. Geological Survey. 1978.   2843

SP. 81 Current geological and geophysical studies in Montana, by R. B. Berg. 1979.   2844

SP. 82 Guidebook of the Drummon-Elkhorn areas, west-central Montana, comp. by M. R. Miller. 1980. 2845

NEBRASKA

Bulletin. 1927-

B. 20 Wells drilled for oil and gas in Nebraska during 1960 and 1961, by R. R. Burchett and M. P. Carlson. 1963.　　　　2846

B. 21 Lithostratigraphy and correlation of the Mississippian system in Nebraska, by M. P. Carlson. 1963.　　　　2847

B. 22 Wells drilled for oil and gas in Nebraska during 1962 and 1963, by R. R. Burchett. 1964.　　　2848

B. 23 Revision of the classification of the Pleistocene deposits of Nebraska, by E. C. Reed and V. H. Dreeszen. 1965.　　　2849

B. 24 Wells drilled for oil and gas in Nebraska during 1964 and 1965, by R. R. Burchett. 1966.　　2850

B. 25 Precambrian well data in Nebraska including rock type and surface configuration, by M. P. Carlson. 1967.　　　2851

B. 26 Precambrian rocks in the subsurface of Nebraska, by E. G. Lidiak. 1972.　　　2852

Report of Investigations. 1966-

RI. 1 Twelve maps summarizing the geologic framework of southeastern Nebraska, by R. R. Burchett and M. P. Carlson. 1966.　　　2855

RI. 2 Geology of the liquid propane gas storage facility near 63rd and Oak Streets, Omaha, Nebraska, by R. R. Burchett. 1969.　　　2856

RI. 3 Distribution and subdivision of Precambrian and Lower and Middle Paleozoic rocks in the subsurface of Nebraska, by M. P. Carlson. 1970.　　　2857

RI. 4 Structure of the Tarkio Limestone along the Humboldt Fault Zone in southeastern Nebraska, by R. R. Burchett and J. L. Arrigo. 1978.　　　2858

Paper. 1933-

P. 17 Correlation of formations drilled in test holes for Interstate 480 bridge between Omaha, Nebraska, and Council Bluffs, Iowa, by R. R. Burchett. 1965.　　2853

P. 18 New members of the Gering Formation (Miocene) in western Nebraska, including a geological map of wildcat ridge and related outliers, by C. F. Vondra, C. B. Schultz, and T. M. Stout. 1969.　　　2854

137

Bulletin. 1964-

B. 59 Geology and mineral deposits of Humboldt County, Nevada, by R. Willden. 1964. 2859

(B. 60 in base index)

B. 61 Antimony deposits of Nevada, by E. F. Lawrence. 1963. 2860

B. 62 Geology and mineral deposits of Clark County, Nevada, by C. R. Longwell, E. H. Pampeyan, B. Bowyer, and R. J. Roberts. 1965. 2861

B. 63 Geology and mineral deposits of the Bunkerville mining district, Clark County, Nevada, by L. H. Beal. 1965. 2862

B. 64 Geology and mineral resources of Eureka County, Nevada, by R. J. Roberts, K. M. Montgomery, and R. E. Lehner. 1967. 2863

B. 65 Mineral and water resources of Nevada, by the U. S. Geological Survey and the Nevada Bureau of Mines. 1964. 2864

B. 66 Interpretation of leached outcrops, by R. Blanchard. 1968. 2865

B. 67 Geology of the Rowland quadrangle, Elko County, Nevada, by K. Bushnell. 1967. 2866

B. 68 Geology of the Mount Velma quadrangle, Elko County, Nevada, by J. R. Coash. 1967. 2867

B. 69 Relation of some metal mining districts in the western United States to regional tectonic environments and igneous activity, by S. E. Jerome and D. R. Cook. 1967. 2868

B. 70 Geology and mineral deposits of Washoe and Storey Counties, Nevada, by H. F. Bonham. 1969. 2869

Industrial rock and mineral deposits, by K. G. Papke. 1969. 2870

B. 71 Geology of parts of the Wadsworth and Churchill Butte Quadrangles, Nevada, by R. L. Rose. 1969. 2871

B. 72 Correlation of Great Basin stratigraphic units, by R. L. Langenheim and E. R. Larson. 1973. 2872

B. 72A Index to Bulletin. 72. 1976. 2873

B. 73 Geology and mineral deposits of Lincoln County, Nevada, by C. M. Tschanz and E. H. Pampeyan. 1970. 2874

B. 74 Geology and mineral deposits of the Pamlico mining district, Mineral County, Nevada, by N. L. Archbold and R. R. Paul. 1970. 2875

B. 75 Geology and mineral deposits of Lyon, Douglas, and Ormsby Counties, Nevada, by J. G. Moore. 1969. 2876

B. 76 Montmorillonite, bentonite, and fuller's earth deposits in Nevada, by K. G. Papke. 1970. 2877

B. 77 Geology and mineral deposits of southern Nye County, Nevada, by H. R. Cornwall. 1972. 2878

B. 78 Geology and mineral deposits of Esmeralda County, Nevada, by J. P. Albers and J. H. Stewart. 1972. 2879

B. 79 Erionite and other associated zeolites in Nevada, by K. G. Papke. 1972. 2880

B. 80 Geology of the granite complex of the Eldorado, Newberry, and northern Dead Mountains, Clark County, Nevada, by A. Volborth. 1973. 2881

B. 81 Radioactive mineral occurrences in Nevada, by L. J. Garside. 1973. 2882

B. 82 Forecasts for the future -- minerals, by A. Baker, N. L. Archbold, and W. J. Stoll. 1973. 2883

B. 83 Geology and mineral deposits of Churchill County, Nevada, by R. Willden and R. C. Speed. 1974. 2884

B. 84 Talcose minerals in Nevada: talc, chlorite, and pyrophyllite, by K. G. Papke. 1975. 2885

B. 85 Geology and mineral resources of White Pine County, Nevada, by R. L. Hose, M. C. Blake, and R. M. Smith. 1976. 2886

B. 86 Geology of the Majuba Hill area, Pershing County, Nevada, by W. B. Mackenzie and A. A. Bookstrom. 1976. 2887

B. 87 Evaporites and brines in Nevada playas, by K. G. Papke. 1976. 2888

B. 88 Geology and mineral deposits of Lander County, Nevada, by J. H. Stewart, E. H. McKee, and H. K. Stager. 1977. 2889

B. 89 Geology and mineral deposits of Pershing County, Nevada, by M. G. Johnson. 1977. 2890

B. 90 Ordovician and Devonian trace fossils from Nevada, by C. K. Chamberlain. 1977. 2891

B. 91 Inventory of thermal waters of Nevada, by L. J. Garside and J. H. Schilling. 1979. 2892

B. 92 Geology of the Tonopah, Lone Mountain, Klondike, and Northern Mud Lake quadrangles, Nevada, by H. F. Bonham and L. J. Garside. 1979. 2893

B. 93 Fluorspar in Nevada, by K. G. Papke. 1979. 2894

B. 94 Pluvial lakes and estimated pluvial climates of Nevada, by M. D. Mifflin and M. M. Wheat. 1979. 2895

Report. 1962-

R. 3 Investigation of titanium occurrences in Nevada, by L. H. Beal. 1962. 2896

R. 4 An inventory of barite occurrences in Nevada, by R. C. Horton. 1963. 2897

R. 5 Studies in the hydrometallurgy of mercury sulfide ores, by J. N. Butler. 1963. 2898

R. 6 A-B: X-ray spectrographic determination of major oxides in igneous rocks, and oxygen determination by neutron activation, by A. Volborth. 1963. 2899

C: Semiquantitative X-ray spectrographic determination of 38 elements in a Nevada granite and in standard granite G-1, and quantitative X-ray fluorescent trace analysis of 20 elements in igneous rocks, by Fabbi and Volborth. 1970. 2900

R. 7 Outline of Nevada mining history, by F. C. Lincoln and R. C. Horton. 1964. 2901

R. 8 Bibliography of graduate theses on Nevada geology. 1965. 2902

R. 9 Correlation of the middle and late Quaternary successions of the Lake Lahontan, Lake Bonneville, Wasatch Range, southern Great Plains, and eastern midwest areas, by R. B. Morrison and J. C. Frye. 1965. 2903

R. 10 Isotopic age determination of Nevada rocks, by J. H. Schilling. 1965. 2904

R. 11 Stratigraphy of Tertiary volcanic rocks in eastern Nevada, by E. F. Cook. 1965. 2905

R. 12 Selected readings in mineral economics, by W. H. Voskuil. 1966. 2906

R. 13 Papers presented at the AIME Pacific Southwest Mineral Industry Conference, Sparks, Nevada, May 5-7, 1965. 1966. 2907

Part A: Exploration and mine development in Nevada. 2908

Part B: Industrial minerals and mining-metallurgy analysis. 2909

Part C: North American exploration and mine development. 2910

R. 14 Industrial mineral deposits of Mineral County, Nevada, by N. L. Archbold. 1966. 2911

R. 15 Gravity study of Warm Springs Valley, Washoe County, Nevada, by J. I. Gimlett. 1967. 2912

R. 16 Gold Butte vermiculite deposits, Clark County, Nevada, by F. B. Leighton. 1967. 2913

R. 17 Turquoise deposits of Nevada, by F. R. Morrissey. 1968. 2914

R. 18 Oil and gas developments in Nevada, 1953-1967, by J. H. Schilling and L. J. Garside. 1968. 2915

R. 19 Guidebook to the geology of four Tertiary volcanic centers in central Nevada. 1974. 2916

Road log and trip guide, Black Mountain volcanic center, by D. C. Noble and R. L. Christiansen. 2917

Northumberland caldera and Northumberland tuff, by E. H. McKee. 2918

Tonopah mining district and vicinity, by H. F. Bonham and L. J. Garside. 2919

Goldfield mining district, by R. P. Ashley. 2920

Ages of Tertiary volcanic rocks and hydrothermal precious-metal deposits in central and western Nevada, by M. L. Silberman and E. H. McKee. 2921

R. 20 Interbasin ground-water flow in southern Nevada, by R. L. Naff, G. B. Maxey, and R. F. Kaufman. 1974. 2922

R. 21 Geothermal exploration and development in Nevada through 1973, by L. J. Garside. 1974. 2923

R. 22 Guidebook to the Quaternary geology along the western flank of the Truckee Meadows, Washoe County, Nevada, by E. C. Bingler. 1975. 2924

R. 23 Induced polarization and magnetic surveys of altered Tertiary volcanic rocks near Reno, Nevada, by J. W. Erwin. 1975. 2925

R. 24 Bibliography of Nevada mining and geology, 1966-1970, by M. B. Ansari. 1975. 2926

R. 25 Evaluation of geothermal activity in the Truckee Meadows, Washoe County, Nevada, by R. L. Bateman and R. B. Scheibach. 1975. 2927

R. 26 Guidebook: Las Vegas to Death Valley and return. 1975. 2928

Road log, by K. G. Papke and J. H. Schilling. 2929

Death Valley, borate deposits, by J. M. Barker and J. L. Wilson. 2930

Tenneco's colemanite milling operation, by R. A. Walters. 2931

Grantham Talc Mine. 2932

R. 27 Stratigraphy of the Tertiary ash-flow tuffs in the Yerington district, Nevada, by J. M. and B. H. Proffett. 1976. 2933

R. 28 PETCAL: a BASIC language computer program for petrologic calculations, by E. C. Bingler, D. T.

Trexler, W. R. Kemp, and H. F.
Bonham. 1976. 2934

R. 29 Oil and gas developments in
Nevada, 1968-1976, by L. J. Garside, B. S. Weimer, and I. A. Lutsey. 1977. 2935

R. 30 The Nevada Bureau of Mines
and Geology sample library--an index
to the drill core and cuttings in the
collection, by J. H. Schilling. 1977.
2936

R. 31 Bibliography of graduate
theses on Nevada geology to 1976,
by I. A. Lutsey. 1978. 2937

R. 32 Guidebook to mineral deposits of the central Great Basin. 1978.
2938

R. 33 Papers on mineral deposits
of western North America (papers
presented at the Fifth Quadrennial
Symposium of IAGOD). n. d. 2939

R. 34 Sources of information on
selected industrial minerals, by
K. G. Papke. 1980. 2940

Special Publication. 1975-

SP. 1 Rockhound's map of Nevada.
1975. 2941

SP. 2 Nevada's weather and climate. 1975. 2942

SP. 3 Federal regulators versus
Nevada miners. 1977. 2943

NEW HAMPSHIRE

No new publications issued since base index.

Bulletin. 1911-

B. 66 Geodetic control survey and mapping digest for New Jersey, by G. J. Halasi-Kun. 1980. 2944

B. 67 The story of New Jersey's civil boundaries, 1606-1968, by J. Snyder. 1969. 2945

B. 68 Not yet published. 2946

B. 69 Not yet published. 2947

B. 70 Caves of New Jersey, by R. Dalton. 1976. 2948

B. 71 Proceedings of University seminar on pollution and water resources. Vol. 1, 1967-1968, ed. by G. Halasi-Kun and K. Widmer. 2949

The water-data base for water-management decision, by E. L. Hendricks. 2950

Weather modification and water resources, by B. A. Power. 2951

Managing forest lands for water production, by B. Beattie. 2952

Desalination project in Guantanamo Naval Base, Cuba, by C. Benedek. 2953

Water resources planning and hydrologic risk, by D. P. Loucks. 2954

A comprehensive approach to the problems of pollution and water resources, by L. Czirjak. 2955

Analysis of spit-bar development at Sandy Hook, New Jersey, by W. E. Yasso. 2956

Correlation between precipitation, flood and windbreak phenomena of the mountain--a case study from Central Europe, by G. J. Halasi-Kun. 2957

B. 72A Proceedings of University seminar on pollution and water resources. Vol. II, 1968-1969, ed. by G. Halasi-Kun. 1980. 2958/59

The integration of desalination into water and power problems; some unexplored economic problems, by J. Barnes. 2960

Water resources development of southeast Lower-Saxony, F. R. Germany, by U. Maniak. 2961

Water condensation from the air, by L. Worzel. 2962

Geological subsurface: a computation method of peak discharge from watersheds of area less than 300 square km. in East Czechoslovakia, by G. Halasi-Kun. 2963

Ground water problems on Long Island, New York, by Geraghty & Miller, Inc. 2964

Mathematical programming, computers and large scale water resource systems, by A. Escogbue. 2965

Instream aeration of small polluted rivers (Passaic River in New Jersey), by W. Whipple. 2966

Space station milieu and its water resources management, by L. Slote. 2967

Some experiences with systems analysis and the use of mathematical models in river basin planning, by R. Sadove. 2968

Thermal survey of the waters of New York State, by S. Breslauer and W. Wrobel.   2969

B. 72B Proceedings of University seminar on pollution and water resources. Vol. III, 1969-1970.
2970

Water resources research in the U. S., by R. Renne.   2971

Seasonal sediment yield patterns of United States rivers, by L. Wilson.   2972

Statistically based mathematical water quality model for a nonestuarine river system (Upper Passaic Valley in New Jersey), by M. Tirabassi.   2973

Some reflections on an engineering economic study of the industrial growth potential of the Upper Passaic River Basin, by A. Lesser.
2974

The basic principles and practical consequences of a new concept in strength of materials, by G. Redey.
2975

What's happening to Lake Erie? by J. Jones.   2976

Urban air pollution, by L. Slote.
2977

Financing of water supply and sewerage projects in developing countries, by J. Krombach.   2978

Federal pollution control litigation, by W. Kiechel.   2979

Determination of quality of sediment on the public beaches of Long Island, by Y. Purandare.   2980

B. 72C Proceedings of University seminar on pollution and water resources. Vol. IV, 1970-1971, ed. by G. Halasi-Kun.   2981

Water problems of the mining industry in the U. S., by E. Hayes.
2982

Contrasts and convergence in engineering and economic approaches to water development and pollution control, by J. Butler.   2983

Pollution of the Rhine River and environmental protection problems in the Ruhr area, by U. Maniak.   2984

The role of economics in municipal water supply: theory and practice, by J. Warford.   2985

Problems on pollution and water resources in the N. Y. C. metropolitan area, by M. Lang.   2986

Water resources management in the Tisza Valley in Hungary, by T. Dora and M. Merenyi.   2987

Air pollution--twentieth century plague, by L. Buck.   2988

Some applications of systems analysis to water resources planning, by D. Loucks.   2989

The effects of hydrostatic pressure on the growth of Candida Albicans in a simulated marine environment, by P. Madri.   2990

Hydrogeological aspects of pollution and water resources in urbanized and industrialized areas, by G. Halasi-Kun.   2991

B. 72D Proceedings of University seminar on pollution and water resources. Vol. V, 1971-1972.   2992

Low flow hydrology of Australian streams, by T. McMahon.   2993

Computation of extreme flow and ground water capacity with inadequate hydrologic data in New Jersey, by G. J. Halasi-Kun.   2994

Tropical hydrology, by J. Balek.
2995

Asbestos pollution, by S. Hartman.
2996

Fluorescence spectroscopy in the study of air and water pollution, by A. Yaseen.   2997

The world plan of action for the application of science and technology, by B. Chatel. 2998

Application of the production function in water management, by I. Degen. 2999

The karst artesian water system of the southeastern states, by V. T. Springfield and H. E. LeGrand. 3000

Toxic water pollutants, by J. Fabianek and V. Sajenko. 3001

Some problems of the Papaloapan River Basin, by G. Cruickshank. 3002

B. 72E Proceedings of University seminar on pollution and water resources. Vol. VI, 1973-1975, ed. by G. Halasi-Kun. 3003

Hydrological investigation of the unsaturated zone, by G. Kovacs. 3004

Model ecosystem studies: models of what? by G. Claus, K. Bolander, and P. P. Madri. 3005

The utilization of information about concomitance of water resources and demands in water resources decision making, by M. Domokos. 3006

A physical approach to hydrologic problems, by J. Philip. 3007

Factor analysis of water quality data in New Jersey: evaluation of alternative rotations, by R. Hordon. 3008

Pollution crunch in Japan, by H. Ripman. 3009

Simulation modeling of streams for water quality studies, by A. Goodman. 3010

Micropollution in organism, by A. Szebenyi. 3011

Extreme runoffs in regions of volcanic rocks in central Europe and in northeastern U. S. A. , by G. Halasi-Kun. 3012

Hackensack River--determination of Tertiary sewage treatment requirements for waste water discharge, by R. LoPinto, C. Mattson, and J. LoPinto. 3013

B. 73 Geology and ground water resources of Sussex County and the Warren County portion of the Tocks Island impact area, by J. W. Miller. 1974. 3014

B. 74 New Jersey Land Oriented Reference Data System (LORDS), by K. Widmer and others. 1974. 3015

B. 75 Proceedings of University seminar on pollution and water resources. Selected papers on ocean engineering, in six parts. 3016

B. 75A Proceedings of University seminar on pollution and water resources. Vol. VII, 1972-1973, ed. by G. Halasi-Kun and K. Widmer. 3017

Dispersion and depth of disturbance studies on Foreshore Beach sediment, Sandy Hook, New Jersey, by W. E. Yasso. 3018

Data on the hydrology of Great South Bay, Long Island, New York, by G. Claus. 3019

Shellfish and public health, by W. Jamieson. 3020

Noise is pollution, by G. W. Robin. 3021

Effect of water salinity on the incidence of symbionts of the blue crab, Callinectes Sapidus (Rathbun), by V. M. Scrocco. 3022

Coastal morphology of Brigantine Inlet, New Jersey: history and prediction, 1877-1977, by W. F. Rittschoff. 3023

Currents and sediment migration in Brigantine Inlet, New Jersey, by M. A. Lynch-Blosse. 3024

Parameters of marine pollution, by W. Jamieson. 3025

B. 75B Proceedings of University seminar on pollution and water resources. Vol. VIII, 1974-1975.   3026

Understanding the impact of outer continental shelf development: approaches to design of environmental studies, by A. Hirsch.   3027

Thermal plume field measurements in three dimensions, by J. R. Roney.   3028

Recent progress in wave refraction studies and its application in the mid-Atlantic bight, by C. Yung-Yao.   3029

Beach dynamics and sediment mobility on Sandy Hook, New Jersey, by K. F. Nordstrom, J. R. Allen, and N. P. Psuty.   3030

Phytoplankton bioassays for industrial pollutants in the Hackensack Meadowlands, by R. W. LoPinto and C. P. Mattson.   3031

Sedimentary dynamics of a disturbed estuary-entrance sand shoal; the Shrewsbury entrance area of Sandy Hook Bay, New Jersey, by P. David.   3032

Future energy resources including outer continental shelf development, by S. L. Meisel.   3033

Recent developments in the law of the sea--status after Geneva, 1975, by P. P. Remec.   3034

B. 75C Proceedings of University seminar on pollution and water resources. Vol. IX, 1975-1978, ed. by G. Halasi-Kun.   3035

Federal saline water conversion program (1976), by G. F. Mangan.   3036

Salinity management and the development of the Colorado River Basin; a multidisciplinary problem with international implications, by W. S. Butcher.   3037

Plan nacional hidraulico (Mexican

national water plan, 1973), by G. Cruickshank.   3038

The unanswered challenge: planning to meet the total water resource needs of an urbanized state, by R. D. Ricci. 1977.   3039

Methods for estimating of water pollution load from particular land uses associated with storm runoff, by L. Michna. 1978.   3040

Headland--bay beaches along the western shoreline of Cape Cod Bay, Massachusetts, by W. E. Yasso.   3041

Land oriented data reference data system--LORDS, by G. J. Halasi-Kun. 1978.   3042

Segregation and deposition of particle size-classes by hydrodynamic forces, by D. P. Harper. 1977.   3043

Design optimization of a flue gas desulfurization sludge handling system, by R. W. Goodwin. 1977.   3044

Sediment dynamics and textural facies in the Brigantine Inlet area, New Jersey, by R. F. Krauser and N. K. Coch. 1977.   3045

B. 75D Proceedings of University seminar on pollution and water resources. Vol. X, 1975-1978, ed. by G. Halasi-Kun.   3046

Geodetic control network foundation of the Cadastre, by J. E. Stem and L. S. Baker. 1976.   3047

Present and potential land use mapping in Mexico, by F. G. Simo, H. Garduno, and R. Garcia. 1976.   3048

Geodetic survey activities in New Jersey, by G. J. Halasi-Kun.   3049

The national mapping program and status of mapping in New Jersey, by R. B. Southard. 1978.   3050

Surveying the tidal boundary, by J. P. Weidener. 1978.   3051

Tidal datum and marine boundary surveys, by C. Thurlow. 1978.   3052

Geodetic deformation measurement on larger dams, by S. K. Nazalevicz. 1978.   3053

Calibration base lines for electronic distance measuring instruments in New Jersey and their use, by J. F. Dracup, C. J. Fronczek, and G. J. Halasi-Kun. 1978.   3054

Tide gauging for the 200 mile fisheries limit, by J. P. Weidener. 1978.   3055

Settlement control survey report and method of measurement for the Verrazano Narrows Bridge, NYC, by G. D. Benedicty. 1976.   3056

B. 75E   Proceedings of University seminar on pollution and water resources. Vol. XI, 1975-1978, ed. by G. Halasi-Kun.   3057

Future directions of the program of the Office of Water Research and Technology, by G. D. Cobb. 1978.   3058

Water quality and pollution--issues involved in the development of a national water policy, by G. J. Halasi-Kun. 1978.   3059

The applicability of ultraviolet spectrophotometry for water quality analyses: a review, by M. Hermel, M. S. Tanzer, and G. Claus. 1977.   3060

Efficiency of slipforms in reinforced concrete construction of water towers, by G. Redey. 1978.   3061

Precipitation and snowfall over New Jersey, by D. V. Dunlop. 1978.   3062

Regional geomorphology of the inner New Jersey shelf, by T. F. McKinney. 1975.   3063

Simulation of unsteady flow in natural compound channels, by M. Miloradov. 1978.   3064

Mountainous winter precipitation: a stochastic event based approach, by L. Duckstein, M. Fogel, and

D. Davis. 1977.   3065

Ground water monitoring at solid waste disposal sites--two case studies, by P. H. Roux and D. Miller. 1977.   3066

Some effects of noise pollution on bioacoustics in the sea, by G. W. Robin. 1978.   3067

B. 75F   Proceedings of University seminar on pollution and water resources. Vol. XII, 1978-1979, ed. by G. Halasi-Kun. 1980.   3068

Environmental health and technical cooperation among developing countries, by F. A. Butrico. 1978.   3069

Improving water quality monitoring data, by J. D. Buffington, J. F. Ficket, W. Kirchoff, F. D. Leutner, and R. Morse. 1979.   3070

Potential environmental impacts of ocean thermal energy conversion, by R. P. Stringer and H. S. Rahme. 1978.   3071

On some concepts for solving multiobjective programming problems, by F. Szidarovsky. 1978.   3072

Regional water supply planning--ground water estimate based on hydrogeologic survey in New Jersey, by G. J. Halasi-Kun. 1979.   3073

Alternative approaches to ground water management, by A. Zaporozec. 1979.   3074

Economic review of water and natural resources inventory based on land survey (multipurpose land records and systems, an assessment from an economic point of view), by G. Greulich. 1979.   3075

Water-borne vectors of disease in tropical and subtropical areas: and a novel approach to mosquito control using annual fish, by J. Markofsky and J. Matias. 1979.   3076

Lead industrial monitoring and biological testing, by M. Plechaty. 1979.   3077

New Jersey's tidelands mapping
program, by R. Yunghans. 1979.
3078

The mean high water line: insight
to law, science, and technology,
by A. A. Porro. 1979.   3079

Status of tidal surveying and monuments in New Jersey, by G. J.
Halasi-Kun. 1979.   3080

Medium-term prediction of hydroelectric energy inflows, by L. Ta-
vares. 1979.   3081

Geologic Report. 1959-

GR. 5 Not yet published?   3082

GR. 6 The Ogdensburg-Culvers Gap
moraine, Sussex County, New
Jersey, by H. Herpers. 1961.
3083

GR. 7 Geology of ground water resources of Mercer County, New
Jersey, by K. Widmer. 1965.
3084

GR. 8 Geology as a guide to regional estimates of the water resource, by K. Widmer. 1968.
3085

GR. 9 Bouguer gravity anomaly map
of New Jersey, by W. E. Bonini.
1965.   3086

GR. 10 Water resources resume,
by K. Widmer. 1966.   3087

GR. 11 Geology of the Paleozoic
rocks of the Green Pond outlier,
by S. G. Barnett. 1976.   3088

Bulletin. 1915-

B. 76 Molybdenum resources of New Mexico, by J. H. Schilling. 1965. 3089

(B. 77-78 in base index)

B. 79 Paleozoic and Mesozoic strata of southwestern and south-central New Mexico, by F. E. Kottlowski. 1963. 3090

B. 80 Precambrian geology of La Madera quadrangle, Rio Arriba County, New Mexico, by E. C. Bingler. 1965. 3091

B. 81 Summary of the mineral resources of Bernalillo, Sandoval, and Santa Fe Counties, New Mexico, by W. E. Elston. 1967. 3092

B. 82 The geology of Jarilla Mountains, Otero County, New Mexico, by P. G. Schmidt and C. Craddock. 1964. 3093

B. 83 Mineral deposits of western Grant County, New Mexico, by E. Gillerman. 1964. 3094

B. 84 Geology of the Walnut Wells quadrangle, Hidalgo County, New Mexico, by R. A. Zeller and A. M. Alper. 1965. 3095

B. 85 Stratigraphy and petroleum possibilities of Catron County, New Mexico, by R. W. Foster. 1964. 3096

B. 86 Geology and ore deposits of the Sacramento (High Rolls) mining district, Otero County, New Mexico, by S. E. Jerome, D. D. Campbell, J. S. Wright, and H. E. Vitz. 1965. 3097

B. 87 Mineral and water resources of New Mexico, comp. in cooperation with the U. S. Geological Survey, State Engineer of New Mexico, New Mexico Oil Conservation Commission, and the U. S. Bureau of Mines. 1965. 3098

B. 88 Sources for lightweight shale aggregate in New Mexico, by R. W. Foster. 1966. 3099

B. 89 Geology of the Chama quadrangle, by W. R. Muehlberger. 1967. 3100

B. 90 Bibliography of New Mexico geology and mineral technology, 1961-1965, by T. Ray. 1966. 3101

B. 91 Geology and mineral resources of Rio Arriba County, New Mexico, by E. C. Bingler. 1968. 3102

B. 92 Geology of the Cebolla quadrangle, Rio Arriba County, New Mexico, by H. H. Doney. 1968. 3103

B. 93 Bromine in the Salado Formation, Carlsbad potash district, New Mexico, by S. S. Adams. 1969. 3104

B. 94 Geology and ore deposits of Eagle Nest area, New Mexico, by K. F. Clark and C. B. Read. 1972. 3105

B. 95 Geology and mineral deposits of the Gallinas Mountains, Lincoln and Torrance Counties, New Mexico, by R. M. Perhac. 1970. 3106

B. 96 Geology of the Little Hatchet Mountains, Hidalgo and Grant Counties, New Mexico, by R. A. Zeller. 1970. 3107

B. 97 Geology of San Diego Mountain area, Dona Ana County, New Mexico, By W. R. Seager, J. W. Hawley, and

R. E. Clemons. 1971.   3108

B. 98 Geology of the Fort Sumner sheet, New Mexico, by V. C. Kelley. 1972.   3109

B. 99 Bibliography of New Mexico geology and mineral technology, 1966 through 1970, by M. A. Koehn and H. H. Kochn. 1973.   3110

B. 100 Geology of Souse Springs quadrangle, New Mexico, by R. E. Clemons and W. R. Seager. 1973.   3111

B. 101 Geology of Rincon quadrangle, New Mexico, by W. R. Seager and J. W. Hawley. 1973.   3112

B. 102 Geology of Sierra Alta quadrangle, Dona Ana County, New Mexico, by W. R. Seager, R. E. Clemons, and J. W. Hawley. 1975.   3113

B. 103 Annotated bibliography and mapping index of Precambrian of New Mexico, by J. M. Robertson. 1976.   3114

B. 104 Laws and regulations governing mineral rights in New Mexico, by V. H. Verity and R. J. Young. 1973.   3115

B. 105 Annotated bibliography of Grants uranium region, New Mexico, 1950 to 1972, by F. A. Schilling. 1975.   3116

B. 106 Bibliography of New Mexico geology and mineral technology 1971 through 1975, by J. R. Wright and J. A. Russell. 1977.   3117

B. 107 New Mexico's energy resources '75, by E. C. Arnold and others. 1976.   3118

Circular. 1930-

C. 65 Sulfonate flotation of beryl, by M. C. Fuerstenau and R. B. Bhappu. 1963.   3119

C. 66 Studies on hydrochlorite leaching of molybdenite, by R. B.

Bhappu, D. H. Reynolds, and W. S. Stahmann. 1963.   3120

C. 67 Geology and geochemical survey of a molybdenum deposit near Nogal Peak, Lincoln County, New Mexico, by G. B. Griswold and F. Missaghi. 1964.   3121

C. 68 Niobium-bearing Sanostee heavy mineral deposit, San Juan Basin, northwestern New Mexico, by E. C. Bingler. 1963.   3122

C. 69 How to measure rock pressures: new tools and proved techniques aid mine design, by G. B. Griswold. 1963.   3123

C. 70 Recovery of valuable minerals from pegmatitic ores, by R. B. Bhappu and M. C. Fuerstenau. 1964.   3124

C. 71 Economic geology of coal in New Mexico, by F. E. Kottlowski. 1964.   3125

C. 72 Petroleum developments in New Mexico during 1960, by R. A. Bieberman and M. A. Grandjean. 1965.   3126

C. 73 The role of hydrolysis in sulfonate flotation of quartz, by M. C. Fuerstenau, C. C. Martin, and R. B. Bhappu. 1964.   3127

C. 74 Listing of county mining records, by L. A. File. 1964.   3128

C. 75 Reconnaissance geology of the Little Black Peak quadrangle, Lincoln and Socorro Counties, New Mexico, by C. T. Smith. 1964.   3129

C. 76 Barite deposits of New Mexico, by F. E. Williams, P. V. Fillo, and P. A. Bloom. 1964.   3130

C. 77 Directory of mines of New Mexico, by L. A. File. 1965.   3131

C. 78 An instrumental study of New Mexico earthquakes, by A. R. Sanford. 1965.   3132

C. 79 Design of an apparatus for determining isoelectric point of charge, by R. A. Deju and R. B. Bhappu. 1965.   3133

C. 80 A preliminary report on New Mexico's geothermal energy resources, by W. K. Summers. 1965. 3134

C. 81 Hydrometallurgical recovery of molybdenum from the Questa Mine, by R. B. Bhappu, D. H. Reynolds, R. J. Roman, and D. A. Schwab. 1965. 3135

C. 82 Surface properties of silicate minerals, by R. A. Deju and R. B. Bhappu. 1965. 3136

C. 83 Chemical characteristics of New Mexico's thermal waters--a critique, by W. K. Summers. 1965. 3137

C. 84 County, township, and range locations of New Mexico's mining districts, by L. A. File and S. A. Northrop. 1966. 3138

C. 85 Mercury content of stream sediments--a geochemical survey of the Magdalena mining district, New Mexico, by F. Missaghi. 1966. 3139

C. 86 Study of precipitation of copper on iron from acid solutions, by D. W. Mitchell. 1966. 3140

C. 87 Preliminary investigations of the oil shale potential in New Mexico, by R. W. Foster, P. B. Luce, L. G. Culver, and B. B. Maras. 1966. 3141

C. 88 Index to samples from oil and gas well tests in library at Socorro, New Mexico, July 1, 1961, to July 1, 1966, by R. A. Bieberman and S. Whitmore. 1966. 3142

C. 89 A chemical interpretation of surface phenomena in silicate minerals, by R. A. Deju and R. B. Bhappu. 1966. 3143

C. 90 A correlation between surface phenomena and flotation in silicates, by R. A. Deju and R. B. Bhappu. 1967. 3144

C. 91 Gravity survey in central Socorro County, New Mexico, by

A. R. Sanford. 1968. 3145

C. 92 Geochemical anomalies in the Philmont Ranch region, New Mexico, by F. L. Misaqi. 1968. 3146

C. 93 Structure, stratigraphy, and hydrogeology of the northern Roswell artesian basin, Chaves County, New Mexico, by K. C. Havenor. 1968. 3147

C. 94 Geochemical and biogeochemical studies in the Eagle Nest quadrangle, New Mexico, by F. L. Misaqi. 1968. 3148

C. 95 Water law atlas, by T. A. Garrity and E. T. Nitzschke. 1968. 3149

C. 96 Petroleum developments in New Mexico during 1961, by R. A. Bieberman. 1968. 3150

C. 97 A mathematical and experimental model of the electrical properties of silicate minerals, by R. A. Deju and R. B. Bhappu. 1968. 3151

C. 98 Geothermics--New Mexico's untapped resources, by W. K. Summers. 1968. 3152

C. 99 Chemical mining--a study of leaching agents, by P. H. Johnson and R. B. Bhappu. 1969. 3153

C. 100 The Tierra Amarilla coal field, Rio Arriba County, New Mexico, by E. R. Landis and C. H. Dane. 1969. 3154

C. 101 Exploration for mineral resources, comp. and ed. by F. E. Kottlowski and R. W. Foster. 1969. 3155

C. 102 An instrumental study of New Mexico earthquakes, July 1, 1964, through Dec. 31, 1967, by A. R. Sanford and D. J. Cash. 1969. 3156

C. 103 Geology and beryllium mineralization near Apache Warm Springs, Socorro County, New Mexico, by P. D. Hillard. 1969. 3157

C. 104 Border stratigraphy symposium, ed. by F. E. Kottlowski and

D. V. LeMone. 1969. 3158

C. 105 Evaluation of ground mica products from New Mexico pegmatites, by W. E. Horst and R. B. Bhappu. 1969. 3159

C. 106 Selective flotation of molybdenite from chalcopyrite concentrates by potassium permanganate, by S. J. Escalera and R. B. Bhappu. 1970. 3160

C. 107 Theses and dissertations of the New Mexico Institute of Mining and Technology, 1931-1969, comp. by I. Youngblood and H. H. Koehn. 1970. 3161

C. 108 Analysis of rhenium in molybdenites, by L. M. Goebel. 1970. 3162

C. 109 Index to samples from oil and gas well tests in library at Socorro, New Mexico, July 1, 1966, to June 1, 1970, by R. A. Bieberman. 1970. 3163

C. 110 Test data for New Mexico clay materials. Part 1: Central New Mexico (Bernalillo, Los Alamos, Sandoval, and Santa Fe Counties), by W. L. Hawks. 1970. 3164

C. 111 Computerization of the New Mexico State Bureau of Mines Mineralogical Museum, by J. Renault, R. M. Bonem, and R. Riese. 1970. 3165

C. 112 The dissolution of chalcocite in oxygenated sulfuric acid solution, by W. W. Fisher and R. J. Roman. 1970. 3166

C. 113 Major-element variations in the Potrillo, Carrizozo, and McCartys basalt fields, New Mexico, by J. Renault. 1970. 3167

C. 114 Survey of surface mining in New Mexico, by J. H. Schilling, W. W. Baltosser, G. B. Griswold, W. K. Wagner, L. A. File, F. E. Kottlowski, E. C. Beaumont, and D. H. Baker. 1971. 3168

C. 115 Records of wells and springs in the Socorro and Magdalena areas, Socorro County, New Mexico, 1968, by N. J. Clark and W. K. Summers. 1971. 3169

C. 116 Solubility and enthalpy of the barium and strontium sulfate solid solution series, by E. Brower and J. Renault. 1971. 3170

C. 117 Batch-grinding model, by R. J. Roman and G. W. Becker. 1971. 3171

C. 118 Selected papers from 1970 uranium symposium at Socorro, New Mexico, comp. by R. J. Roman and D. H. Baker. 1971. 3172

C. 119 Bibliography of geophysics for New Mexico through 1970, by D. J. Cash. 1971. 3173

C. 120 Seismicity of the Rio Grande rift in New Mexico, by A. R. Sanford, A. J. Budding, J. P. Hoffman, O. S. Alptekin, C. A. Rush, and T. R. Toppozada. 1972. 3174

C. 121 Geochemical background values in iron-bearing rocks of Rio Arriba County, New Mexico, by D. F. McLeroy. 1972. 3175

C. 122 Structure of Glorieta sandstone in northwest Chaves County, New Mexico (includes comprehensive bibliography), by R. L. Borton. 1972. 3176

C. 123 Mineral deposits of Nogal and Bonito mining districts, New Mexico, by T. B. Thompson. 1973. 3177

C. 124 Ground-water characteristics in a recharge area Magdalena Mountains, Socorro County, New Mexico, by W. K. Summers, G. E. Schwab, and L. A. Brandvold. 1972. 3178

C. 125 Structural geology of northern part of Animas Mountains, Hidalgo County, New Mexico, by J. M. Soule. 1972. 3179

C. 126 Instrumental study of New Mexico earthquakes, January 1968 through June 1971, by T. R. Toppozada

and A. R. Sanford. 1972. 3180

C. 127 Determinative tables of 2O_{cu} and 2O_{fe} for minerals of southwestern United States, by C. W. Walker and J. R. Renault. 1972. 3181

C. 128 Geoscience research projects for New Mexico, 1972, by R. W. Foster and J. A. Meyer. 1972. 3182

C. 129 Morrison Formation of southeastern San Juan Basin, New Mexico, by L. A. Woodward and O. L. Schumacher. 1973. 3183

C. 130 Hydrocarbon potential of pre-Pennsylvanian rocks in Roosevelt County, New Mexico, by W. D. Pitt. 1973. 3184

C. 131 Pennsylvanian system of Chloride Flat, Grant County, New Mexico, by D. V. LeMone, W. E. King, and J. E. Cunningham. 1974. 3185

C. 132 Trace base metals, petrography, and alteration, Tres Hermanas stock, Luna County, New Mexico, by P. Doraibabu and P. D. Proctor. 1973. 3186

C. 133 Middle to late Tertiary geology of the Cedar Hills-Selden Hills area, Dona Ana County, New Mexico, by W. R. Seager and R. E. Clemons. 1975. 3187

C. 134 Coal resources of southern Ute and Ute Mountain Ute Indian Reservations, Colorado and New Mexico, by J. W. Shomaker and R. D. Holt. 1973. 3188

C. 135 Some fishes of the Wild Cow Formation (Pennsylvanian), Manzanita Mountains, New Mexico, by J. Zidek. 1975. 3189

C. 136 Geoscience projects for New Mexico, by J. M. Olsen and R. W. Foster. 1973. 3190

C. 137 Computer program for Monte Carlo economic evaluation of a mineral deposit, by R. J. Roman

and G. W. Becker. 1973. 3191

C. 138 Mineral resources and water requirements for New Mexico minerals industries, by E. F. Sorensen, R. B. Stotelmeyer, and D. H. Baker. 1973. 3192

C. 139 Clay minerals in east-central New Mexico, by H. D. Glass, J. C. Frye, and A. B. Leonard. 1973. 3193

C. 140 Energy crisis symposium, Albuquerque, New Mexico, comp. by W. L. Hiss and J. W. Shomaker. 1973. 3194

C. 141 Petrography and petrogenesis of Tertiary camptonites and diorites, Sacramento Mountains, New Mexico, by G. B. Asquith. 1974. 3195

C. 142 Atomic absorption methods for analysis of some elements in ores and concentrates, by L. A. Brandvold. 1974. 3196

C. 143 Seismicity of proposed radioactive waste disposal site in southeastern New Mexico, by A. R. Sanford and T. R. Toppozada. 1974. 3197

C. 144 Caliche and clay mineral zonation of Ogallala Formation, centraleastern New Mexico, by J. C. Frye, H. D. Glass, A. B. Leonard, and D. D. Coleman. 1974. 3198

C. 145 Late Cenozoic mollusks and sediments, southeastern New Mexico, by A. B. Leonard, J. C. Frye, and H. D. Glass. 1975. 3199

C. 146 Structural geology of Big Hatchet Peak quadrangle, Hidalgo County, New Mexico, by R. A. Zeller. 1975. 3200

C. 147 Geology of Dona Ana Mountains, New Mexico, by W. R. Seager, F. E. Kottlowski, and J. W. Hawley. 1976. 3201

C. 148 New Mexico's energy resources '76--annual report of Office of the State Geologist, by E. C. Arnold and others. 1977. 3202

C. 149 Geology of Potrillo Basalt Field, south-central New Mexico, by J. M. Hoffer. 1976. 3203

C. 150 Sedimentology of braided alluvial interval of Dakota sandstone, northeastern New Mexico, by J. L. Gilbert and G. B. Asquith. 1976. 3204

C. 151 Subsurface temperature data in Jemez Mountains, New Mexico, by M. Reiter, C. Weidman, C. L. Edwards, and H. Hartman. 1976. 3205

C. 152 Geology, petroleum source rocks, and thermal metamorphism in KCM No. 1 Forest Federal Well, Hidalgo County, New Mexico, by S. Thompson and others. 1977. 3206

C. 153 Guidebook to Albuquerque Basin of the Rio Grande rift, New Mexico, by V. C. Kelley, L. A. Woodward, A. M. Kudo, and J. F. Callender. 1976. 3207

C. 154 Guidebook to coal geology of northwest New Mexico, by E. C. Beaumont, J. W. Shomaker, W. J. Stone, and others. 1976. 3208

C. 155 Geologic appraisal of deep coals, San Juan Basin, New Mexico, by J. W. Shomaker and M. R. Whyte. 1977. 3209

C. 156 Volcanoes and related basalts of Albuquerque Basin, New Mexico, by V. C. Kelley and A. M. Kudo. 1978. 3210

C. 157 Sedimentology of Mesa Rica sandstone in Tucumcari Basin, New Mexico, by J. E. Gage and G. B. Asquith. 1977. 3211

C. 158 Geology of central Peloncillo Mountains, Hidalgo County, New Mexico, by A. K. Armstrong, M. L. Silberman, V. R. Todd, W. C. Hoggatt, and R. B. Carten. 1978. 3212

C. 159 Geology and mineral deposits of Ochoan rocks in Delaware Basin and adjacent areas, comp.

by G. S. Austin and others. 1978. 3213

C. 160 Late Cenozoic sediments, molluscan faunas, and clay minerals in northeastern New Mexico, by J. C. Frye, A. B. Leonard, and H. D. Glass. 1978. 3214

C. 161 Paleontology of Ogallala Formation, northeastern New Mexico, by A. B. Leonard and J. C. Frye. 1978. 3215

C. 162 Mercury in New Mexico surface waters, by L. A. Brandvold. 1978. 3216

C. 163 Guidebook to Rio Grande rift in New Mexico and Colorado, comp. by J. W. Hawley and others. 1978. 3217

C. 164 Model beach shoreline in Gallup sandstone (Upper Cretaceous) of northwestern New Mexico, by C. V. Campbell. 1979. 3218

C. 165 Genesis, provenance, and petrography of the Glorieta sandstone of eastern New Mexico, by S. Milner. 1978. 3219

C. 166 Estimates of New Mexico's future oil production including reserves of the 50 largest pools, by R. W. Foster, A. L. Gutsahr, and G. H. Warner. 1978. 3220

C. 167 New Mexico's energy resources '77--Office of the State Geologist, by E. C. Arnold and others. 1978. 3221

C. 168 Geology of Jornada del Muerto coal field, Socorro County, New Mexico, by D. E. Tabet. 1980. 3222

C. 169 Geology of Good Sight Mountains, Luna County, New Mexico, by R. E. Clemons. 1980. 3223

C. 170 Normapolles pollen from Aquilapollenites province, western United States, by R. H. Tschudy. 1980. 3224

C. 171 Not yet published. 3225

C. 172 New Mexico's energy resources;

annual report of the Bureau of Geology, by E. C. Arnold and J. M. Hill. 1979. 3226

Groundwater Report. 1948-1970

GR. 7 Geology and ground-water conditions in eastern Valencia County, New Mexico, by F. B. Titus. 1963. 3227

GR. 8 General occurrence and quality of ground water in Union County, New Mexico, by J. B. Cooper and L. V. Davis. 1967. 3228

GR. 9 Ground-water resources and geology of Quay County, New Mexico, by C. F. Berkstresser and W. A. Mourant. 1966. 3229

GR. 10 Reconnaissance of water resources of De Baca County, New Mexico, by W. A. Mourant and J. W. Shomaker. 1970. 3230

Hydrologic Reports. 1971-

HR. 1 Geology and ground-water resources of central and western Dona Ana County, New Mexico, by W. E. King, J. W. Hawley, A. M. Taylor, and R. P. Wilson. 1971. 3231

HR. 2 Water resources and general geology of Grant County, New Mexico, by F. D. Trauger. 1972. 3232

HR. 3 Water resources of Guadalupe County, New Mexico, by G. A. Dinwiddie and A. Clebsch. 1973. 3233

HR. 4 Catalog of thermal waters in New Mexico, by W. K. Summers. 1976. 3234

Memoir. 1956-

M. 11 Geology of part of the southern Sangre de Cristo Mountains, New Mexico, by J. P. Miller, A. Montgomery, and P. K. Sutherland. 1963. 3235

M. 12 The nautiloid order Ellesmeroceratida (Cephalopoda), by R. H. Flower. 1964. 3236

M. 13 Nautiloid shell morphology, by R. H. Flower. 1964. 3237

M. 14 Sedimentology of the Upper Cretaceous rocks of Todilto Park, New Mexico, by M. E. Willard. 1964. 3238

M. 15 Geology and technology of the Grants uranium region, by V. C. Kelley. 1963. 3239

M. 16 Stratigraphy of the Big Hatchet Mountains area, New Mexico, by R. A. Zeller. 1965. 3240

M. 17 Geology of Pennsylvanian and Wolfcampian rocks in southeast New Mexico, by R. F. Meyer. 1966. 3241

M. 18 Geomorphic surfaces and surficial deposits in southern New Mexico, by R. H. Ruhe. 1967. 3242

M. 19 Part 1: The first great expansion of the Actinoceroids, by R. H. Flower. 1968. 3243

Part 2: Some additional Whitrock cephalopods, by R. H. Flower. 1968. 3244

M. 20 Biostratigraphy and carbonate facies of the Mississippian Arroyo Penasco formation, north-central New Mexico, by A. K. Armstrong. 1967. 3245

M. 21 Part I: Botryceras, a remarkable nautiloid from the second value of New Mexico, by R. H. Flower. 1968. 3246

Part II: An endoceroid from the Mohawkian of Quebec, by R. H. Flower. 1968. 3247

Part III: Endoceroids from the Canadian of Alaska, by R. H. Flower. 1968. 3248

Part IV: A Chazyan cephalopod fauna from Alaska, by R. H. Flower. 1968. 3249

M. 22 Part I: Some El Paso guide fossils, by R. H. Flower. 1968.
3250

Part II: Fossils from the Smith Basin limestone of the Fort Ann region, by R. H. Flower. 1968.
3251

Part III: Fossils from the Fort Ann formation, by R. H. Flower. 1968.
3252

Part IV: Merostomes from the Cassinian portion of the El Paso group, by R. H. Flower. 1968.
3253

M. 23 Part I: The Wolfcampian Joyita uplift in central New Mexico, by F. E. Kottlowski and W. J. Stewart. 1970.
3254

Part II: Fusulinids of the Joyita Hills, Socorro County, central New Mexico, by W. J. Stewart. 1970.
3255

M. 24 Geology of the Pecos County, southeastern New Mexico, by V. C. Kelley. 1971.
3256

M. 25 Strippable low-sulfur coal resources of the San Juan Basin in New Mexico and Colorado, comp. and ed. by J. W. Shomaker, E. C. Beaumont, and F. E. Kottlowski. 1971.
3257

M. 26 Fusulinids Millerella and Eostaffella from the Pennsylvanian of New Mexico and Texas, by W. E. King. 1973.
3258

M. 27 Pennsylvanian brachiopods and biostratigraphy in southern Sangre de Cristo Mountains, New Mexico, by P. K. Sutherland and F. H. Harlow. 1973.
3259

M. 28 Part I: New American Wutinoceratidae with review of Actinoceroid occurrences in eastern hemisphere, by R. H. Flower. 1976.
3260

Part II: Some Whiterock and Chazy endoceroids, by R. H. Flower. 1976.
3261

M. 29 Geology of Sandia Mountains and vicinity, New Mexico, by V. C. Kelley and S. A. Northrop. 1975.
3262

M. 30 Pliocene and Pleistocene deposits and molluscan faunas, east-central New Mexico, by A. B. Leonard and J. C. Frye. 1975.
3263

M. 31 Geology of Cerro de Cristo Rey uplift, Chihuahua and New Mexico, by E. M. P. Lovejoy. 1976. 3264

M. 32 Late Canadian (Zones J, K) cephalopod faunas from southwestern United States, by S. C. Hook and R. H. Flower. 1977.
3265

M. 33 Geology of Albuquerque Basin, New Mexico, by V. C. Kelley. 1977.
3266

M. 34 Fluorspar in New Mexico, by W. N. McAnulty. 1978. 3267

M. 35 Geology and geochemistry of Precambrian rocks, central and south-central New Mexico, by K. C. Condie and A. J. Budding. 1980. 3268

M. 36 Not yet published. 3269

M. 37 Collignoniceras woollgari woollgari (Mantell) ammonite fauna from Upper Cretaceous of western interior United States, by W. A. Cobban and S. C. Hook. 1979. 3270

Scenic Trips to Geologic Past. 1968-

ST. 1 Santa Fe, by B. Baldwin and F. E. Kottlowski. 2nd ed. 1968.
3271

ST. 2 Taos-Red River-Eagle Nest, Circle Drive, by J. H. Schilling. 4th ed. 1968.
3272

ST. 3 Roswell-Capitan-Ruidoso-Bottomless Lakes State Park, by J. E. Allen and F. E. Kottlowski. 2nd ed. 1967.
3273

ST. 4 Southern Zuni Mountains, Zuni-Cibola Trail, by R. W. Foster. 3rd ed. 1971.
3274

ST. 5 Silver City-Santa Rita-
Hurley, by J. H. Schilling. 2nd ed.
1967. 3275

ST. 6 Trail guide to the upper Pecos, by P. K. Sutherland and A.
Montgomery. 3rd ed. 1975. 3276

ST. 7 High Plains-Northeastern
New Mexico (Raton-Capulin Mountain-
Clayton), by W. R. Muehlberger, B.
Baldwin, and R. W. Foster. 3rd
ed. 1967. 3277

ST. 8 Mosaic of New Mexico's
scenery, rocks, and history, ed.
by P. W. Christiansen and F. E.
Kottlowski. 2nd ed. 1967. 3278

ST. 9 Albuquerque--its mountains,
valley, water, and volcanoes, by
V. C. Kelley. Rev. 1974. 3279

ST. 10 Southwestern New Mexico--
Las Cruces, Deming, Lordsburg,
Silver City, Columbus, by R. E.
Clemons, P. W. Christiansen, and
H. L. James. 1980. 3280

ST. 11 Cumbres and Toltec Scenic
Railroad, by H. L. James. 1972.
3281

ST. 12 The story of mining in New
Mexico, by P. W. Christiansen.
1974. 3282

NEW YORK

The publications of this state survey are indexed in Clapp's Museum Publications (see page ix) and are therefore not indexed here.

Bulletin. 1893-

B. 77 Anthophyllite asbestos in North Carolina, by S. G. Conrad, W. F. Wilson, E. P. Allen, and T. J. Wright. 1964.   3283

B. 78 Ostracoda from wells in southeastern United States, by F. M. Swain and P. M. Brown. 1964.   3284

B. 79 Description of the Pungo River Formation in Beaufort County, North Carolina, by J. O. Kimrey. 1965.   3285

B. 80 Pyrophyllite deposits in North Carolina, by J. L. Stuckey. 1967.   3286

B. 81 Geology and mineral resources of Orange County, North Carolina, by E. P. Allen and W. F. Wilson. 1968.   3287

B. 82 Bibliography of North Carolina geology, 1910-1960, by W. F. Wilson. 1975.   3288

B. 83 Geological bibliography of North Carolina's coastal plain, coastal zone, and continental shelf, by S. R. Riggs and M. P. O'Connor. 1975.   3289

B. 84 Metallic mineral deposits of the Carolina slate belt, North Carolina, by P. A. Carpenter. 1976.   3290

B. 85 Buried oyster shell resource evaluation of the eastern region of the Albemarle Sound, by J. L. Sampair. 1976.   3291

B. 86 Geology and mineral resources of Wake County, by J. M. Parker. 1979.   3292

Economic Paper. 1897-

EP. 68 Mineral industry of North Carolina from 1960 through 1967, by J. L. Stuckey. 1970.   3293

Educational Series. 1928-

ES. 4 North Carolina geology and mineral resources: a foundation for progress, by W. F. Wilson, P. A. Carpenter, and S. G. Conrad. 1976. Rev. by W. F. Wilson, 1980.   3294

Information Circular. 1940-

IC. 17 Beryl occurrences in North Carolina, by W. F. Wilson. 1962.   3295

IC. 18 The force of gravity at selected localities in North Carolina, by V. I. Mann. 1963.   3296

IC. 19 Titanium deposits in North Carolina, by L. Williams. 1964.   3297

IC. 20 The feldspar resources of North Carolina, by J. L. Bundy and P. A. Carpenter. 1969.   3298

IC. 21 The gold resources of North Carolina, by P. A. Carpenter. 1972.   3299

IC. 22 Exploratory oil wells of North Carolina, 1925-1976, comp. by J. C. Coffey. 1977.   3300

IC. 23 Diabase dikes of the eastern Piedmont of North Carolina, by E. R. Burt, P. A. Carpenter, R. D. McDaniel, and W. F. Wilson. 1978.   3301

IC. 24 Mineral collecting sites in North Carolina, by W. F. Wilson and

B. J. McKenzie. 1978. 3302

Regional Geology Series. 1975-

RGS. 1 Region J geology: a guide
for North Carolina mineral resource
development and land use planning,
by W. F. Wilson and P. A. Carpenter. 1975. 3303

Special Publication. 1965-

SP. 1 Geology of the Chapel Hill
quadrangle, North Carolina, by V. I.
Mann, T. G. Clarke, L. D. Hayes,
and D. S. Kirstein. 1965. 3304

SP. 2 Triassic flora from the Deep
River Basin, North Carolina, by
R. C. Hope and O. F. Patterson.
1969. 3305

SP. 3 Upper Miocene foraminifera
from near Grimesland, Pitt County,
North Carolina, by D. Schnitker.
1970. 3306

SP. 4 Petrography and stratigraphy
of the Carolina slate belt, Union
County, North Carolina, by A. F.
Randazzo. 1972. 3307

SP. 5 Wrench-style deformation
in rocks of Cretaceous and Paleocene age, North Carolina coastal
plain, by P. M. Brown, D. L.
Brown, T. E. Shufflebarger, and
J. L. Sampair. 1977. 3308

SP. 6 A catalogue of seismic
events recorded at seismograph
station CHC, Chapel Hill, North
Carolina, January 1, 1955, through
December 31, 1970, by D. M. Best
and T. D. Cavanaugh. 1977. 3309

SP. 7 Faunal and diagenetic controls of porosity and permeability
in Tertiary aquifer carbonates,
North Carolina, by P. A. Thayer
and D. A. Textoris. 1977. 3310

NORTH DAKOTA

Bulletin. 1920-

B. 39 The Bakken and Englewood formations of North Dakota and northwestern South Dakota, by J. Kume. 1963. 3311

B. 40 Structural and stratigraphic relationships in the Paleozoic rocks of eastern North Dakota, by F. V. Ballard. 1963. 3312

B. 41 Part I: Geology and ground water resources of Stutsman County, North Carolina, geology, by H. A. Winters. 1963. 3313

Part II: Geology and ground water resources of Stutsman County, North Dakota, ground water basic data, by C. J. Huxel and L. R. Petri. 1963. 3314

Part III: Geology and ground water resources of Stutsman County, North Dakota, ground water and its chemical quality, by C. J. Huxel and L. R. Petri. 1965. 3315

B. 42 Part I: Geology and ground water resources of Burleigh County, North Dakota, geology, by J. Kume and D. E. Hansen. 1965. 3316

Part II: Geology and ground water resources of Burleigh County, North Dakota, ground water basic data, by P. G. Randich. 1965. 3317

Part III: Geology and ground water resources of Burleigh County, North Dakota, ground water resources, by P. G. Randich and J. L. Hatchett. 1966. 3318

B. 43 Part I: Geology and ground water resources of Barnes County, North Dakota, geology, by T. E.

Kelly and D. A. Block. 1967. 3319

Part II: Geology and ground water resources of Barnes County, North Dakota, ground water basic data, by T. E. Kelly. 1965. 3320

Part III: Geology and ground water resources of Barnes County, North Dakota, ground water resources, by T. E. Kelly. 1966. 3321

B. 44 Part I: Geology and ground water resources of Eddy and Foster Counties, North Dakota, geology, by J. P. Bluemle. 1965. 3322

Part II: Geology and ground water resources of Eddy and Foster Counties, North Dakota, ground water basic data, by H. Trapp. 1966. 3323

Part III: Geology and ground water resources of Eddy and Foster Counties, North Dakota, ground water resources, by H. Trapp. 1968. 3324

B. 45 Part I: Geology and ground water resources of Divide County, North Dakota, geology, by D. E. Hansen. 1967. 3325

Part II: Geology and ground water resources of Divide County, North Dakota, ground water basic data, by C. A. Armstrong. 1965. 3326

Part III: Geology and ground water resources, by C. A. Armstrong. 1967. 3327

B. 46 Part I: Geology and ground water resources of Richland County, North Dakota, geology, by C. H. Baker. 1967. 3328

Part II: Geology and ground water resources of Richland County, North Dakota, ground water basic data, by

161

C. H. Baker. 1966.   3329

Part III: Geology and ground water
resources of Richland County,
North Dakota, ground water resources, by C. H. Baker and Q. F.
Paulson. 1967.   3330

B. 47 Part I: Geology and ground
water resources of Cass County,
North Dakota, geology, by R. L.
Klausing. 1968.   3331

Part II: Geology and ground water
resources of Cass County, North
Dakota, ground water basic data,
by R. L. Klausing. 1966.   3332

Part III: Geology and ground water
resources of Cass County, North
Dakota, hydrology, by R. L. Klausing. 1968.   3333

B. 48 Part I: Geology and ground
water resources of Williams County,
North Dakota, geology, by T. F.
Freers. 1970.   3334

Part II: Geology and ground water
resources of Williams County,
North Dakota, ground water basic
data, by C. A. Armstrong. 1967.
                3335

Part III: Geology and ground water
resources of Williams County,
North Dakota, hydrology, by C. A.
Armstrong. 1969.   3336

B. 49 Part I: Geology and ground
water resources of Traill County,
North Dakota, geology, by J. P.
Bluemle. 1967.   3337

Part II: Geology and ground water
resources of Traill County, North
Dakota, ground water basic data,
by H. M. Hansen. 1967.   3338

Part III: Geology and ground water
resources of Traill County, North
Dakota, ground water resources,
by H. M. Jensen and R. L. Klausing.
1971.   3339

B. 50 Part I: Not yet published.
                3340

Part II: Geology and ground water

resources of Renville and Ward Counties, North Dakota, ground water
basic data, by W. A. Pettyjohn. 1968.
                3341

Part III: Ground water resources of
Renville and Ward Counties, by W. A.
Pettyjohn and R. D. Hutchinson. 1971.
                3342

B. 51 Part I: Geology and ground
water resources of Wells County,
North Dakota, geology, by J. P. Bluemle and others. 1967.   3343

Part II: Geology and ground water
resources of Wells County, North Dakota, ground water basic data, by F.
Buturla. 1968.   3344

Part III: Geology and ground water
resources of Wells County, North Dakota, ground water resources, by F.
Buturla. 1970.   3345

B. 52 The Spearfish Formation in the
Williston Basin of western North Dakota, by W. G. Dow. 1967.   3346

B. 53 Part I: Geology and ground
water resources of Grand Forks
County, North Dakota, geology, by
D. E. Hansen and J. Kume. 1970.
                3347

Part II: Geology and ground water
resources of Grand Forks County,
North Dakota, ground water basic
data, by T. E. Kelly. 1968.   3348

Part III: Geology and ground water
resources of Grand Forks County,
North Dakota, ground water resources,
by T. E. Kelly and Q. F. Paulson.
1970.   3349

B. 54 Stratigraphy of the Hell Creek
Formation in North Dakota, by C. I.
Frye. 1969.   3350

B. 55 Part I: Geology of Burke
County, North Dakota, by T. F. Freers.
1973.   3351

Part II: Geology and ground water
resources of Burke and Mountrail
Counties, ground water basic data,
by C. A. Armstrong. 1969.
                3352

Part III: Ground water resources of Burke and Mountrail Counties, by C. A. Armstrong. 1971. 3353

Part IV: Geology of Mountrail County, North Dakota, by L. Clayton. 1972. 3354

B. 56 Part I: Geology of Mercer and Oliver Counties, North Dakota, by C. G. Carlson. 1973. 3355

Part II: Ground water basic data, Mercer and Oliver Counties, by M. G. Croft. 1970. 3356

Part III: Ground-water resources of Mercer and Oliver Counties, North Dakota, by M. G. Croft. 1973. 3357

B. 57 Part I: Geology of Nelson and Walsh Counties, by J. P. Bluemle. 1973. 3358

Part II: Ground-water basic data, Nelson and Walsh Counties, North Dakota, by J. S. Downey. 1971. 3359

Part III: Ground-water resources of Nelson and Walsh Counties, North Dakota, by J. S. Downey. 1973. 3360

B. 58 Geology of Rolette County, North Dakota, by D. E. Deal. 1972. 3361

B. 59 Part I: Geology of Benson and Pierce Counties, North Dakota, by C. G. Carlson and T. F. Freers. 1975. 3362

Part II: Ground-water basic data, Benson and Pierce Counties, North Dakota, by P. G. Randich. 1971. 3363

Part III: Ground-water resources of Benson and Pierce Counties, North Dakota, by P. G. Randich. 1977. 3364

B. 60 Part I: Geology of McLean County, North Dakota, by J. P. Bluemle. 1971. 3365

Part II: Ground water basic data,

McLean County, North Dakota, by R. L. Klausing. 1971. 3366

Part III: Ground-water resources of McLean County, North Dakota, by R. L. Klausing. 1974. 3367

B. 61 Stratigraphy and paleoecology of the Fox Hills Formation (Upper Cretaceous) of North Dakota, by R. M. Feldmann. 1972. 3368

B. 62 Part I: Geology of Cavalier and Pembina Counties, North Dakota, by B. M. Arndt. 1975. 3369

Part II: Ground-water basic data of Cavalier and Pembina Counties, North Dakota, by R. D. Hutchinson. 1973. 3370

Part III: Ground-water resources of Cavalier and Pembina Counties, North Dakota, by R. D. Hutchinson. 1977. 3371

B. 63 Mineral and water resources of North Dakota, by E. R. Landis. 1973. 3372

B. 64 Part I: Geology of Griggs and Steele Counties, North Dakota, by J. P. Bluemle. 1975. 3373

Part II: Ground-water basic data for Griggs and Steele Counties, North Dakota, by J. S. Downey, R. D. Hutchinson, and G. L. Sunderland. 1973. 3374

Part III: Ground-water resources of Griggs and Steele Counties, North Dakota, by J. S. Downey and C. A. Armstrong. 1977. 3375

B. 65 Part I: Geology of Adams and Bowman Counties, North Dakota, by C. G. Carlson. 1979. 3376

Part II: Ground-water basic data for Adams and Bowman Counties, North Dakota, by M. G. Croft. 1974. 3377

Part III: Ground-water resources of Adams and Bowman Counties, North Dakota, by M. G. Croft. 1978. 3378

B. 66 Part I: Not yet published. 3379

Part II: Ground-water basic data
for Emmons County, North Dakota,
by C. A. Armstrong. 1975. 3380

Part III: Ground-water resources
of Emmons County, North Dakota,
by C. A. Armstrong. 1978. 3381

B. 67 Part I: Not yet published.
3382

Part II: Ground-water basic data
for Grant and Sioux Counties,
North Dakota, by P. G. Randich.
1975. 3383

Part III: Ground-water resources
of Grant and Sioux Counties, North
Dakota, by P. G. Randich. 1979.
3384

B. 68 Part I: Not yet published.
3385

Part II: Ground-water basic data
for Dunn County, North Dakota,
by R. L. Klausing. 1976. 3386

Part III: Ground-water resources
of Dunn County, North Dakota, by
R. L. Klausing. 1979. 3387

B. 69 Part I: Geology of Ransom
and Sargent Counties, North Dakota,
by J. P. Bluemle. 1979. 3388

Part II: Ground-water basic data
for Ransom and Sargent Counties,
North Dakota, by C. A. Armstrong.
1979. 3389

B. 70 Part I: Geology of Dickey
and Lamoure Counties, North Dakota, by J. P. Bluemle. 1979.
3390

Part II: Ground-water basic data
for Dickey and Lamoure Counties,
North Dakota, by C. A. Armstrong
and S. P. Luttrell. 1978. 3391

B. 71 Part I: Not yet published.
3392

Part II: Ground water basic data
for Ramsey County, North Dakota,
by R. D. Hutchinson. 1977. 3393

B. 72 Part I: Not yet published.
3394

Part II: Ground-water basic data for
Morton County, North Dakota, by D. J.
Ackerman. 1977. 3395

B. 73 Part I: Not yet published.
3396

Part II: Ground-water basic data for
McIntosh County, North Dakota, by
R. L. Klausing. 1979. 3397

Educational Series. 1972-

ES. 1 Geology along North Dakota
Interstate Highway 94, by J. P. Bluemle. 1972. 3398

ES. 2 Guide to the geology of Northeastern North Dakota, by M. E. Bluemle. 1972. Rev. 1975. 3399

ES. 3 Guide to the geology of southeastern North Dakota, by J. P. Bluemle. 1972. Rev. 1975. 3400

ES. 4 Geology along the South Loop
Road, Theodore Roosevelt National
Memorial Park, by J. P. Bluemle and
A. F. Jacob. 1973. 3401

ES. 5 Some environmental aspects of
strip mining in North Dakota, by
M. K. Wali. 1973. 3402

ES. 6 Guide to the geology of south-
central North Dakota, by J. P. Bluemle. 1973. 3403

ES. 7 Guide to the geology of north-
central North Dakota, by J. P. Bluemle. 1974. 3404

ES. 8 Guide to the geology of northwest North Dakota, by J. P. Bluemle.
1975. 3405

ES. 9 Guide to the geology of southwest North Dakota, by J. P. Bluemle.
1975. 3406

ES. 10 The prairie: land and life,
by J. P. Bluemle. 1975. 3407

ES. 11 The face of North Dakota--
the geologic story, by J. P. Bluemle.
1977. 3408

ES. 12 Flooding in the Grand Forks-

East Grand Forks area, by S.S.
Harrison and J.P. Bluemle. 1980.
3409

ES. 13 Petroleum--a primer for
North Dakota, by E.A. Brostuen.
1980. 3410

Miscellaneous Series. 1957-

MS. 1 Guidebook for geologic field
trip in the Valley City area, North
Dakota, by W.M. Laird and M.
Hansen. 1957. 3411

MS. 2 Guidebook for geologic field
trip in the Minot area, North Dakota,
by W.M. Laird. 1957. 3412

MS. 3 Guidebook for geologic field
trip in the Devils Lake area, North
Dakota, by W.M. Laird. 1957.
3413

MS. 4 Guidebook for geologic field
trip in the Bismarck-Mandan area,
North Dakota, by F.D. Holland.
1957. 3414

MS. 5 Guidebook for geologic field
trip in the Dickinson area, North
Dakota, by F.D. Holland. 1957.
3415

MS. 6 Guidebook for geologic field
trip in the Williston area, North
Dakota, by F.D. Holland. 1957.
3416

MS. 7 Guidebook for geologic field
trip in the Jamestown area, North
Dakota, by F.D. Holland. 1957.
3417

MS. 8 Guidebook for geologic field
trip from Fargo to Valley City,
North Dakota, by F.D. Holland.
1957. 3418

MS. 9 Guidebook for geologic field
trip from Grand Forks to Park
River, North Dakota, by F.D. Holland. 1957. 3419

MS. 10 Guidebook, ninth annual
field conference, Midwestern Friends
of the Pleistocene, by W.M. Laird
and others. 1963. 3420

MS. 11 Crabs from the Cannonball
Formation (Paleocene) of North Dakota,
by F.D. Holland and A. Cvancara.
1958. 3421

MS. 12 Late Wisconsin mollusca from
ice-contact deposits in Logan County,
North Dakota, by L. Clayton. 1961.
3422

MS. 13 A molluscan fauna and Late
Pleistocene climate in southeastern
North Dakota, by S.J. Tuthill. 1961.
3423

MS. 14 The status of paleontology in
North Dakota, by F.D. Holland. 1961.
3424

MS. 15 Iron-cemented glacial drift in
Logan County, North Dakota, by J.W.
Bonneville. 1961. 3425

MS. 16 Ground water, a vital North
Dakota resource, by Q.F. Paulson.
1962. 3426

MS. 17 Selected Devonian possibilities
in North Dakota, by S.B. Anderson.
1963. 3427

MS. 18 North Dakota crude oil inventory as of January 1, 1963, by C.B.
Folsom. 1963. 3428

MS. 19 Geology along the portal pipe
line, Lake Agassiz Plain, by T.F.
Freers and C.G. Carlson. 1963.
3429

MS. 20 Molluscan fossils from upper
glacial Lake Agassiz sediment in Red
Lake County, Minnesota, by S.J. Tuthill. 1963. 3430

MS. 21 Late Pleistocene fish from
lake sediments in Sheridan County,
North Dakota, by N. Sherrod. 1963.
3431

MS. 22 North Dakota crude oil inventory as of January 1, 1964, by C.B.
Folsom. 1964. 3432

MS. 23 Lignite in North Dakota, by
M. Hansen. 1964. 3433

MS. 24 Study of the Spoil Banks associated with lignite strip mining in

North Dakota, by C. G. Carlson and W. M. Laird. 1964. 3434

MS. 25 North Dakota crude oil inventory as of January 1, 1965, by C. B. Folsom. 1965. 3435

MS. 26 Potash in North Dakota, by C. G. Carlson and S. B. Anderson. 1965. 3436

MS. 27 North Dakota crude oil inventory as of January 1, 1966, by C. B. Folsom. 1966. 3437

MS. 28 Sedimentary and tectonic history of North Dakota part of Williston Basin, by C. G. Carlson and S. B. Anderson. 1966. 3438

MS. 29 North Dakota crude oil inventory as of January 1, 1967, by C. B. Folsom. 1967. 3439

MS. 30 Glacial geology of the Missouri Coteau and adjacent areas, ed. by L. Clayton and T. F. Freers. 1967. 3440

MS. 31 Popular geology of Eddy and Foster Counties, North Dakota, by J. P. Bluemle. 1967. 3441

MS. 32 The Dahlen Esker of Grand Forks and Walsh Counties, North Dakota, by J. Kume. 1966. 3442

MS. 33 Ice-thrust bedrock in northwest Cavalier County, North Dakota, by J. P. Bluemle. 1966. 3443

MS. 34 Cross section of Paleozoic rocks of western North Dakota, by C. G. Carlson. 1967. 3444

MS. 35 The flood problem in Grand Forks-East Grant Forks, by S. S. Harrison. 1968. 3445

MS. 36 North Dakota crude oil inventory as of January 1, 1968, by C. B. Folsom. 1968. 3446

MS. 37 Meteorites in North Dakota, a guide to their recognition, by F. R. Karner. 1968. 3447

MS. 38 North Dakota crude oil inventory as of January 1, 1969, by

C. B. Folsom. 1969. 3448

MS. 39 Geology of northeastern North Dakota, ed. by D. T. Pederson and J. R. Reid. 1969. 3449

MS. 40 Geologic field trip from Grand Forks, North Dakota, to Kenora, Ontario, by M. E. Bluemle. 1969. 3450

MS. 41 North Dakota crude oil inventory as of January 1, 1970, by C. B. Folsom. 1970. 3451

MS. 42 Guide to the geology of Burleigh County, North Dakota, by J. P. Bluemle and J. Kume. 1970. 3452

MS. 43 A sedimentologic analysis of the Tongue River-Sentinel Butte interval (Paleocene) of the Williston Basin, western North Dakota, by C. F. Royse. 1970. 3453

MS. 44 Subsurface geology and foundation conditions in Grand Forks, North Dakota, by S. R. Moran. 1972. 3454

MS. 45 North Dakota crude oil inventory as of January 1, 1971, by C. B. Folsom. 1971. 3455

MS. 46 North Dakota crude oil inventory as of January 1, 1972, by C. B. Folsom. 1972. 3456

MS. 47 Newburg-South Westhope oil fields, North Dakota, by H. Marafi. 1972. 3457

MS. 48 Environmental geology and North Dakota, by B. M. Arndt. 1972. 3458

MS. 49 Annotated bibliography of the geology of North Dakota, 1806-1959, by M. W. Scott. 1972. 3459

MS. 50 Depositional environments of the lignite-bearing strata in western North Dakota, ed. by F. T. C. Ting. 1972. 3460

MS. 51 North Dakota crude oil inventory as of January 1, 1973, by C. B. Folsom. 1973. 3461

MS. 52 Late Quaternary stratigraphic

nomenclature, Red River Valley, North Dakota, and Minnesota, by K. L. Harris, S. R. Moran, and L. Clayton. 1974. 3462

MS. 53 Catalog of North Dakota radiocarbon dates, by S. R. Moran, L. Clayton, M. W. Scott, and J. A. Brophy. 1973. 3463

MS. 54 Stratigraphy, origin, and climatic implications of Late Quaternary upland silt in North Dakota, by L. Clayton, S. R. Moran, and W. B. Bickley. 1976. 3464

MS. 55 Development of a pre-mining geological framework for landscape design reclamation in North Dakota, by S. R. Moran, G. H. Groenewold, L. Hemish, and C. Anderson. 1976. 3465

MS. 56 Type and reference sections for a new member of the Fox Hills Formation, Upper Cretaceous (Maestrichtian) in the Missouri Valley region, North and South Dakota, by M. C. Klett and J. M. Erickson. 1979. 3466

MS. 57 Oil exploration and development in the North Dakota Williston Basin, by L. C. Gerhard and S. B. Anderson. 1979. 3467

MS. 58 A history of the North Dakota Geological Survey, by C. B. Folsom. 1980. 3468

Report of Investigations. 1949-

RI. 39 Gravity maps of north central North Dakota, by J. B. Hunt. 1962. 3469

RI. 40 Gravity maps of central North Dakota, by J. B. Hunt. 1963. 3470

RI. 41 The Niobrara Formation of eastern North Dakota: its possibilities for use as a cement rock, by C. G. Carlson. 1964. 3471

RI. 42 A look at the petroleum potential of southwestern North Dakota, by S. B. Anderson. 1966. 3472

RI. 43 A look at the Lower and Middle Madison of northwestern North Dakota, by C. G. Carlson and S. B. Anderson. 1966. 3473

RI. 44 Notes on Pleistocene stratigraphy of North Dakota, by L. Clayton. 1966. 3474

RI. 45 The Tongue River-Sentinel Butte contact in western North Dakota, by C. F. Royse. 1967. 3475

RI. 46 Where North Dakota's best Newcastle sand trends are located, by S. B. Anderson. 1967. 3476

RI. 47 The Newcastle formation in the Williston Basin of North Dakota, by M. Reishus. 1968. 3477

RI. 48 Cement rock possibilities in Paleozoic rocks of eastern North Dakota, by S. B. Anderson and H. C. Haraldson. 1969. 3478

RI. 49 Magnetic anomalies in Pembina County, North Dakota, by W. L. Moore and F. R. Karner. 1969. 3479

RI. 50 Geology for planning at Langdon, North Dakota, by B. M. Arndt. 1972. 3480

RI. 51 Late Cenozoic stratigraphy of the Lake Sakakawea bluffs north and west of Riverdale, North Dakota, by J. H. Ulmer and D. K. Sackreiter. 1973. 3481

RI. 52 A mechanical well log study of the Poplar Interval of the Mississippian Madison Formation in North Dakota, by C. W. Cook. 1976. 3482

RI. 53 Criteria for differentiating the Tongue River and Sentinel Butte Formations (Paleocene), North Dakota, by A. F. Jacob. 1975. 3483

RI. 54 Physical data for land-use planning, Cass County, North Dakota, and Clay County, Minnesota--an inventory of mineral, soil, and water resources, by B. M. Arndt and S. R. Moran. 1974. 3484

RI. 55 Geology of the Fox Hills Formation (Late Cretaceous) in the Williston Basin of North Dakota, with refer-

168 / NORTH DAKOTA

ence to uranium potential, by A. M. Cvancara. 1976. 3485

RI. 56 The stratigraphy and environments of deposition of the Cretaceous Hell Creek Formation (reconnaissance) and the Paleocene Ludlow Formation (detailed), southwestern North Dakota, by W. L. Moore. 1976. 3486

RI. 57 Geology of the Cannonball Formation (Paleocene) in the Williston Basin, with reference to uranium potential, by A. M. Cvancara. 1976. 3487

RI. 58 Geology of the upper part of the Fort Union Group (Paleocene), Williston Basin, with reference to uranium, by A. F. Jacob. 1976. 3488

RI. 59 The Slope (Paleocene) and Bullion Creek (Paleocene) formations of North Dakota, by L. Clayton and others. 1977. 3489

RI. 60 Stratigraphy of offshore sediment, Lake Agassiz, North Dakota, by B. M. Arndt. 1977. 3490

RI. 61 Geology, groundwater hydrology, and hydrogeochemistry of a proposed surface mine and lignite gasification plant site near Dunn Center, North Dakota, by S. R. Moran and others. 1978. 3491

RI. 62 Physical data for land-use planning, Divide, McKenzie, and Williams Counties, North Dakota, by E. A. Brostuen. 1977. 3492

RI. 63 Geologic, hydrologic, and geochemical concepts and techniques in overburden characterization for mined-land reclamation, by S. R. Moran, G. H. Groenewold, and J. A. Cherry. 1978. 3493

RI. 64 Geology and geohydrology of the Knife River Basin and adjacent areas of west-central North Dakota, by G. H. Groenewold and others. 1979. 3494

RI. 65 Late paleocene mammals of the Tongue River formation, western North Dakota, by R. C. Holtzman. 1978. 3495

RI. 66 Depositional environments and paragenetic porosity controls, upper Red River Formation, North Dakota, by W. K. Carroll. 1979. 3496

RI. 67 The Carrington shale facies (Mississippian) and its relationship to the Scallion subinterval in central North Dakota, by P. F. Bjorlie. 1979. 3497

RI. 68 Potash salts in the Williston Basin, U. S. A. , by S. B. Anderson and R. P. Swinehart. 1979. 3498

RI. 69 Explanatory text to accompany the geologic map of North Dakota, by L. Clayton, S. R. Moran, and J. P. Bluemle. 1980. 3499

RI. 70 Geologic and hydrogeologic conditions affecting land use in the Bismarck-Mandan area, by G. H. Groenewold. 1980. 3500

Bulletin. 1900-

B. 61 Geology of Stark County, by
R. M. DeLong and G. W. White.
1963. 3501

B. 62 Part 1: Pleistocene mollusca of Ohio, by A. La Rocque.
1966. 3502

Part 2: Pleistocene mollusca of
Ohio, by A. La Rocque. 1967.
3503

Part 3: Pleistocene mollusca of
Ohio, by A. La Rocque. 1968.
3504

Part 4: Pleistocene mollusca of
Ohio, by A. La Rocque. 1970.
3505

B. 63 Pennsylvanian brachiopods
of Ohio, by M. T. Sturgeon and
R. D. Hoare. 1968. 3506

B. 64 Stratigraphy of the Cambrian
and Lower Ordovician rocks in Ohio,
by A. Janssens. 1973. 3507

B. 65 Ohio--an American heartland,
by A. G. Noble and A. J. Korsok.
1975. 3508

B. 66 Geology and mineral re-
sources of Washington County,
Ohio, by H. R. Collins and B. E.
Smith. 1977. 3509

B. 67 Pennsylvanian marine bival-
via and rostroconchia, by R. D.
Hoare, M. T. Sturgeon, and E. A.
Kindt. 1979. 3510

Educational Leaflet. 1952-

EL. 6 Flint, Ohio's official gem
stone. 1967. 3511

EL. 7 Ohio's glaciers, by M. C. Hansen. 1974. 3512

EL. 8 Coal, by H. R. Collins. 1975.
3513

EL. 9 Earthquakes in Ohio, by M. C.
Hansen. 1975. 3514

EL. 10 Geology in land-capability
analysis, by R. D. Stieglitz, M. L.
Couchot, and R. G. Van Horn. 1977.
3515

EL. 11 Guide to the geology along
U. S. Route 23 between Columbus and
Portsmouth, by M. C. Hansen. 1979.
3516

Geological Notes. 1975-

GN. 1 Preliminary report on potential
hydrocarbon reserves underlying the
Ohio portion of Lake Erie, by M. J.
Clifford. 1975. 3517

GN. 2 Sand and gravel resources of
Madison County, Ohio, by M. L. Couchot. 1975. 3518

GN. 3 Potential natural gas resources
in the Devonian shales in Ohio, by A.
Janssens and W. de Witt. 1976. 3519

GN. 4 Coal resources of a portion of
the Pawpaw Creek watershed, Monroe,
Noble, and Washington Counties, by
R. A. Struble, H. R. Collins, and R. M.
DeLong. 1976. 3520

GN. 5 Limestone in the Tymochtee
dolomite(?) (Upper Silurian), Shawnee
Township, Allen County, Ohio, by
D. A. Stith. 1977. 3521

GN. 6 Extent of till sheets and ice
margins in northeastern Ohio, by
G. W. White. 1979. 3522

Guidebook. 1973-

G. 1 Natural and manmade features affecting the Ohio shore of Lake Erie, by C. H. Carter. 1973.
3523

G. 2 Selected field trips in northeastern Ohio, by R. A. Heimlich and R. M. Feldmann. 1974. 3524

G. 3 Pennsylvanian conodont localities in northeastern Ohio, by G. K. Merrill. 1974. 3525

G. 4 Field guide to the geology of the Hocking Hills State Park region, by M. C. Hansen. 1975. 3526

Information Circular. 1938-

IC. 32 Bibliography of Ohio geology, 1951-1960, by P. Smyth. 1963.
3527

IC. 33 Sauk sequence data sheets, by W. L. Calvert. 1964. 3528

IC. 34 Lake Erie bathythermograph recordings, 1952-1966, comp. by C. E. Herdendorf. 1967. 3529

IC. 35 Subsurface information catalog, 1963-1967, comp. by F. B. Safford. 1969. 3530

IC. 36 Bibliography of Ohio geology, 1961-1965, comp. by P. Smyth. 1969. 3531

IC. 37 Bibliography of Ohio geology, 1966-1970, by P. Smyth. 1972.
3532

IC. 38 Areas of shallow bedrock in part of northwestern Ohio, by D. A. Stith. 1973. 3533

IC. 39 The November 1972 storm on Lake Erie, by C. H. Carter. 1973. 3534

IC. 40 Mercury concentrations in sediments of the Lake Erie Basin, Ohio, by D. A. Stith. 1973. 3535

IC. 41 Hydrogeologic and other considerations related to the selection

of sanitary-landfill sites in Ohio, by G. H. Groenewold. 1974. 3536

IC. 42 Catalog of oil and gas wells in "Newburg" (Silurian) of Ohio, by A. Janssens. 1975. 3537

IC. 43 Subsurface liquid-waste injection in Ohio, by M. J. Clifford. 1975.
3538

IC. 44 Coal production in Ohio--1800-1974, comp. by H. R. Collins. 1976.
3539

IC. 45 Place names directory: northeast Ohio, comp. by M. R. Fitak. 1976. 3540

IC. 46 Subsurface information catalog --1968-1974, comp. by B. J. Adams. 1976. 3541

IC. 47 Analyses of Ohio coals, by G. Botoman and D. A. Stith. 1978. 3542

IC. 48 Bibliography of Ohio geology, 1755-1974, comp. by P. Smyth. 1979.
3543

Miscellaneous Report. 1974-

MR. 1 Tenth forum on geology of industrial minerals proceedings. 3544

Reclamation of pits and quarries. 1974. (Pt. 1) 3545

Carbonate rocks in environmental control. 1974. (Pt. 2) 3546

Report of Investigations. 1947-

RI. 46 Geology of the Silurian producing zones in the Moreland oil pool, Wayne County, northeastern Ohio, by H. G. Multer. 1963. 3547

RI. 47 Summary of oil and gas activity in Ohio during 1962, by W. L. Calvert. 1963. 3548

RI. 48 A cross section of sub-Trenton rocks from Wood County, West Virginia, to Fayette County, Illinois, by W. L. Calvert. 1963. 3549

RI. 49 Sub-Trenton rocks of Ohio in cross sections from West Virginia and Pennsylvania to Michigan, by W. L. Calvert. 1963. 3550

RI. 50 Summary of oil and gas activity in Ohio during 1963, by W. L. Calvert. 1964. 3551

RI. 51 Upper Niagaran and Cayugan stratigraphy of northeastern Ohio and adjacent areas, by J. R. Ulteig. 1964. 3552

RI. 52 Sub-Trenton rocks from Fayette County, Ohio, to Brant County, Ontario, by W. L. Calvert. 1964. 3553

RI. 53 Effects of large structures on the Ohio shore of Lake Erie, by R. P. Hartley. 1964. 3554

RI. 54 Bedrock geology of the Garrettsville quadrangle, by J. O. Fuller. 1965. 3555

RI. 55 Geology of the Kensington quadrangle, Ohio, by R. M. DeLong. 1965. 3556

RI. 56 Summary of oil and gas activity in Ohio during 1964, by D. L. Norling, T. A. DeBrosse, and W. J. Buschman. 1965. 3557

RI. 57 Geology of the Malvern quadrangle, Ohio, by R. M. DeLong. 1965. 3558

RI. 58 Synoptic survey of water properties in the western basin of Lake Erie, by R. P. Hartley, C. E. Herdendorf, and M. Keller. 1966. 3559

RI. 59 Glacial map of Licking County, Ohio, by J. L. Forsyth. 1966. 3560

RI. 60 Geology of the Vermilion West and Berlin Heights quadrangles, Erie and Huron Counties, Ohio, by C. E. Herdendorf. 1966. 3561

RI. 61 Bedrock geology of Wayne County, Ohio, by H. G. Multer. 1967. 3562

RI. 62 Glacial geology of Wayne County, Ohio, by G. W. White. 1967. 3563

RI. 63 Bedrock geology of the South Bloomingville quadrangle, Hocking and Vinton Counties, Ohio, by R. M. DeLong. 1967. 3564

RI. 64 The Precambrian surface of Ohio, by G. L. Owens. 1966. 3565

RI. 65 Bedrock geology of the Minerva quadrangle, Stark, Columbiana, and Carroll Counties, Ohio, by R. M. DeLong. 1967. 3566

RI. 66 Glacial geology of the East Liberty quadrangle, Logan and Union Counties, Ohio, by J. L. Forsyth. 1967. 3567

RI. 67 Glacial geology of Clinton County, Ohio, by J. T. Teller. 1967. 3568

RI. 68 Bedrock geology of the Dellroy quadrangle, Carroll County, Ohio, by R. M. DeLong. 1968. 3569

RI. 69 Glacial geology of the West Mansfield quadrangle, Logan and Union Counties, Ohio, by J. F. Forsyth. 1968. 3570

RI. 70 Stratigraphy of Silurian and pre-Olentangy Devonian rocks of the South Birmingham Pool area, Erie and Lorain Counties, Ohio, by A. Janssens. 1968. 3571

RI. 71 Bedrock geology of the Carrollton quadrangle, Carroll County, Ohio, by R. M. DeLong. 1969. 3572

RI. 72 Bedrock geology of the Caldwell North quadrangle, Guernsey and Noble Counties, Ohio, by B. E. Smith. 1969. 3573

RI. 73 Potential use of Ohio limestones and dolomites for architectural aggregate, by D. A. Stith. 1969. 3574

RI. 74 Water masses and their movements in western Lake Erie, by C. E. Herdendorf. 1969. 3575

RI. 75 The subsurface Silurian-Devonian "Big Lime" of Ohio, by G. L. Owens. 1970.　3576

RI. 76 Sand and gravel resources of the Maumee River estuary, Toledo to Perrysburg, Ohio, by C. E. Herdendorf. 1970.　3577

RI. 77 Bedrock geology of the Madeira quadrangle, Hamilton and Clermont Counties, Ohio, by R. H. Osborne. 1970.　3578

RI. 78 Middle Devonian formations in the subsurface of northwestern Ohio, by A. Janssens. 1970.　3579

RI. 79 Lake Erie physical limnology cruise, midsummer 1967, by C. E. Herdendorf. 1970.　3580

RI. 80 Glacial geology of Trumbull County, by G. W. White. 1971.　3581

RI. 81 Deep-core investigation of low-sulfur coal possibilities in southeastern Ohio, by R. A. Struble, H. R. Collins, and D. L. Kohout. 1971.　3582

RI. 82 Physical characteristics of the reef area of western Lake Erie, by C. E. Herdendorf and L. L. Braidech. 1972.　3583

RI. 83 Bedrock geology of the Addyston quadrangle and part of the Burlington Quadrangle, Hamilton County, Ohio, by J. P. Ford. 1972.　3584

RI. 84 Bedrock geology of the Flint Ridge area, Licking and Muskingum Counties, Ohio, by R. M. DeLong. 1972.　3585

RI. 85 The high-temperature phase of calcium sulfate in relation to SO2 removal by the dry-limestone injection method, by G. W. Kalb. 1972.　3586

RI. 86 High-calcium limestone facies of the Devonian Dundee limestone, northwestern Ohio, by D. A. Stith. 1972. Table of additional analyses (sulfur and phosphorus) added, 1975.　3587

RI. 87 Bedrock geology of the Bergholz quadrangle, Carroll and Jefferson Counties, Ohio, by R. M. DeLong. 1972.　3588

RI. 88 Glacial geology of Richland County, Ohio, by S. M. Totten. 1973.　3589

RI. 89 Potential use of beneficiated Ohio clay in the rubber industry, by R. W. Carlton. 1973.　3590

RI. 90 Silurian rock salt of Ohio, by M. J. Clifford. 1973.　3591

RI. 91 Glacial geology of Holmes County, Ohio, by G. W. White. 1973.　3592

RI. 92 Glacial geology of Highland County, Ohio, by T. E. Rosengreen. 1974.　3593

RI. 93 Bedrock geology of the Cincinnati West quadrangle and part of the Covington quadrangle, Hamilton County, Ohio, by J. P. Ford. 1974.　3594

RI. 94 Bedrock geology of the Cincinnati East quadrangle, Hamilton County, Ohio, by R. H. Osborne. 1974.　3595

RI. 95 Bedrock geology of the Serpent Mound cryptoexplosion structure, Adams, Highland, and Pike Counties, Ohio, by S. P. Reidel. 1975.　3596

RI. 96 The ground-water situation in the Circleville area, Pickaway County, south-central Ohio, by S. E. Norris. 1975.　3597

RI. 97 Potential use of Ohio clays in the well-plugging industry, by R. W. Carlton. 1975.　3598

RI. 98 Land areas in Summit County --geologic suitability for solid-waste disposal, by R. G. Van Horn. 1976.　3599

RI. 99 Lake Erie shore erosion, Lake County, Ohio: setting, processes, and recession rates from 1876 to 1973, by

C. H. Carter. 1976.   3600

RI. 100 Silurian rocks in the subsurface of northwestern Ohio, by
A. Janssens. 1977.   3601

RI. 101 Glacial geology of Ashland
County, Ohio, by G. W. White.
1977.   3602

RI. 102 Sediment-load measurements along the United States shore
of Lake Erie, by C. H. Carter.
1977.   3603

RI. 103 Trace elements in Ohio
coals, by N. F. Knapp. 1977.
     3604

RI. 104 The occurrence of sulfide
and associated minerals in Ohio,
by G. Botoman and R. D. Stieglitz.
1978.   3605

RI. 105 Coal resources of the Pittsburgh (No. 8) coal in the Belmont
field, Ohio, by M. L. Couchot.
1978.   3606

RI. 106 Structure on the Pittsburgh
(No. 8) coal in the Belmont field,
Ohio, comp. by R. M. DeLong.
1978.   3607

RI. 107 Lake Erie shore erosion
and flooding, Lucas County, Ohio,
by D. J. Benson. 1978.   3608

RI. 108 An evaluation of "Newberry"
analysis data on the Brassfield
Formation (Silurian), southwestern
Ohio, by D. A. Stith and R. D.
Stieglitz. 1979.   3609

RI. 109 Surficial materials of Summit County, Ohio, by R. G. Van
Horn. 1979.   3610

RI. 110 Hydraulic properties of a
limestone-dolomite aquifer near
Marion, north-central Ohio, by
S. E. Norris. 1979.   3611

RI. 111 Glacial geology of Champaign County, Ohio, by G. W. White
and S. M. Totten. 1979.   3612

RI. 112 Glacial geology of Ashta-

bula County, Ohio, by G. W. White
and S. M. Totten. 1979.   3613

RI. 113 Chemical composition, stratigraphy, and depositional environments
of the Black River Group (Middle Ordovician), southwestern Ohio, by D. A.
Stith. 1979.   3614

OKLAHOMA

Bulletin.　1908-

B. 94　Early Devonian brachiopods
of Oklahoma.　1963.　　　3615

Pt. I:　Articulate brachiopods of the
Frisco Formation (Devonian), by
T. W.　Amsden and W. P. S.　Ven-
tress.　　　3616

Pt. II:　Articulate brachiopods of
the Sallisaw Formation (Devonian),
by T. W.　Amsden.　　　3617

Pt. III:　Supplement to the Haragan
(Devonian) brachiopods, by T. W.
Amsden.　　　3618

B. 95　Basement rocks and structural evolution of southern Oklahoma, by W. E.　Ham, R. E.　Denison, and C. A.　Merritt.　1964.
　　　3619

(B. 96-97 in base index)

B. 98　The Blaine and related formations of northwestern Oklahoma
and southern Kansas, by R. O.　Fay.
1964.　　　3620

B. 99　Geology and oil and gas resources of Craig County, Oklahoma,
by C. C.　Branson and others.
1965.　　　3621

B. 100　Crinoids of the Hunton
Group (Silurian-Devonian) of Oklahoma, by H. L.　Strimple.　1963.
　　　3622

B. 101　Structure and stratigraphy
of the Rich Mountain area, Oklahoma and Arkansas, by D. R.　Seely.
1963.　　　3623

B. 102　Permian salt and associated
evaporites in the Anadarko basin of
the western Oklahoma-Texas Pan-

handle region, by L.　Jordan and
D. L.　Vosburg.　1963.　　　3624

B. 103　Geology of the eastern part
of Winding Stair Range, Le Flore
County, Oklahoma, by O. D.　Hart.
1963.　　　3625

B. 104　Biostratigraphy and rugose
corals of the Lower Pennsylvanian
Wapanucka Formation in Oklahoma,
by C. L.　Rowett and P. K.　Sutherland.
1964.　　　3626

B. 105　Silurian stratigraphy of northeastern Oklahoma, by T. W.　Amsden
and T. L.　Rowland.　1965.　　　3627

B. 106　Geology and mineral resources
of Woods County, Oklahoma, by R. O.
Fay.　1965.　　　3628

B. 107　Petrology of the Hogshooter
Formation (Missourian), Washington
and Nowata Counties, Oklahoma, by
W. R.　Cronoble and C. J.　Mankin.
1965.　　　3629

B. 108　Ostracodes of the Henryhouse
Formation (Silurian) in Oklahoma, by
R. F.　Lundin.　1965.　　　3630

B. 109　Rugose corals of the Henryhouse Formation (Silurian) in Oklahoma, by P. K.　Sutherland.　1965.
　　　3631

B. 110　Late Cambrian and earliest
Ordovician trilobites, Timbered Hills
and Lower Arbuckle Groups, western
Arbuckle Mountains, Murray County,
Oklahoma, by J. H.　Stitt.　1971.
　　　3632

B. 111　Geology and petroleum of Mc-
Intosh County, Oklahoma.　1967.
　　　3633

Pt. I:　Geology and mineral resources

of McIntosh County, by M. C. Oakes and others.　3634

Pt. II: Petroleum geology of Mc-Intosh County, by T. Koontz. 3635

B. 112 Palynology of the Red Branch Member of the Woodbine Formation (Cenomanian), Bryan County, Oklahoma, by R. W. Hedlund. 1966.
　3636

B. 113 Pennsylvanian fusulinids in the Ardmore Basin, Love and Carter Counties, Oklahoma, by D. E. Waddell. 1966.　3637

B. 114 Geology and mineral resources (exclusive of petroleum) of Custer County, Oklahoma. 1979.
　3638

Pt. I: Stratigraphy and general geology, by R. O. Fay.　3639

Pt. II: Economic geology, by R. O. Fay.　3640

Pt. III: Ground water, by D. L. Hart.　3641

B. 115 Trilobites of the Henryhouse Formation (Silurian) in Oklahoma, by K. S. W. Campbell. 1967.
　3642

B. 116 Ostracodes of the Haragan Formation (Devonian) in Oklahoma, by R. F. Lundin. 1968.　3643

B. 117 Articulate brachiopods of the Viola Formation (Ordovician) in the Arbuckle Mountains, Oklahoma, by L. P. Alberstadt. 1973.
　3644

B. 118 Models of sand and sandstone deposits: a methodology for determining sand genesis and trend, by J. W. Shelton. 1973.　3645

B. 119 Late Ordovician and Early Silurian articulate brachiopods from Oklahoma, southwestern Illinois, and eastern Missouri, by T. W. Amsden. 1975.　3646

B. 120 Geology and mineral resources of Choctaw County, Okla-

homa, by G. G. Huffman, P. P. Alfonsi, R. C. Dalton, A. Duarte-Vivas, and E. L. Jeffries. 1975.　3647

B. 121 Hunton Group (Late Ordovician, Silurian, and Early Devonian) in the Anadarko Basin of Oklahoma, by T. W. Amsden. 1976.　3648

B. 122 Geology and mineral resources (exclusive of petroleum) of Muskogee County, Oklahoma, by M. C. Oakes. 1977.　3649

B. 123 Trilobites of the Haragan, Bois d'Arc, and Frisco Formations (Early Devonian), Arbuckle Mountains region, Oklahoma, by K. S. W. Campbell. 1977.　3650

B. 124 Late Cambrian and earliest Ordovician trilobites, Wichita Mountains area, Oklahoma, by J. H. Stitt. 1977.　3651

B. 125 Articulate brachiopods of the Quarry Mountain Formation (Silurian), eastern Oklahoma, by T. W. Amsden. 1978.　3652

B. 126 Geology and mineral resources of Bryan County, Oklahoma, by G. G. Huffman, T. A. Hart, L. J. Olson, J. D. Currier, and R. W. Ganser. 1979.　3653

B. 127 Cranial anatomy of primitive captorhinid reptiles from the Late Pennsylvanian and Early Permian, Oklahoma and Texas, by M. J. Heaton. 1979.　3654

B. 128 Geology and mineral resources of Noble County, Oklahoma, by J. W. Shelton. 1980.　3655

B. 129 Hunton Group (Late Ordovician, Silurian, and Early Devonian) in the Arkoma Basin of Oklahoma, by T. W. Amsden. 1980.　3656

B. 130 Plant microfossils from the Denton Shale Member of the Bokchito Formation (Lower Cretaceous, Albian) in southern Oklahoma, by F. H. Wingate. 1980.　3657

Circular. 1908-

C. 61 Ground-water resources of the Rush Springs sandstone in the Caddo County area, Oklahoma, by H. J. Tanaka and L. V. Davis. 1963. 3658

C. 62 Petroleum geology of Pawnee County, Oklahoma, by P. H. Clare. 3659

C. 63 Geology and petroleum of Love County, Oklahoma. 1965. 3660

Pt. I: Geology of Love County, by E. A. Frederickson and R. H. Redman. 3661

Pt. II: Petroleum geology of Love County, by J. M. Westheimer. 3662

C. 64 Copper in the Flowerpot Shale (Permian) of the Creta area, Jackson County, Oklahoma, by W. E. Ham and K. S. Johnson. 1977. 3663

C. 65 Geology of the western part of Winding Stair Range, Latimer and Le Flore Counties, Oklahoma, by L. D. Fellows. 1964. 3664

C. 66 Chitons from the Kindblade Formation (Lower Ordovician), Arbuckle Mountains, southern Oklahoma, by A. G. Smith and D. F. Toomey. 1964. 3665

C. 67 The ammonoid family Girtyoceratidae in the southern Midcontinent, by J. A. McCaleb, J. H. Quinn, and W. M. Furnish. 1964. 3666

C. 68 Geology of northern Adair County, Oklahoma, by G. G. Huffman, J. M. Langton, and J. H. Hancock. 1966. 3667

C. 69 Geology of the Cenozoic rocks of Ellis County, Oklahoma, by D. B. Kitts. 1965. 3668

C. 70 New Permian vertebrates from the Chickasha Formation in Oklahoma, by E. C. Olson. 1965. 3669

C. 71 Ground-water resources in Cleveland and Oklahoma Counties, Oklahoma, by P. R. Wood and L. C. Burton. 1968. 3670

C. 72 Studies of Pennsylvanian corals in Oklahoma. 1966. 3671

Pt. I: Tabulate corals of the Wapanucka Formation, by C. L. Rowett. 3672

Pt. II: New species of Dibunophyllum from the Dewey Formation, by J. M. Cocke. 3673

C. 73 Not published? 3674

C. 74 Early Permian vertebrates of Oklahoma, by E. C. Olson. 1967. 3675

C. 75 Geology of the eastern part of the Lynn Mountain syncline, Le Flore County, Oklahoma, by G. Briggs. 1973. 3676

C. 76 Shale and carbonate-rock resources of Osage County, Oklahoma, by W. H. Bellis and T. L. Rowland. 1976. 3677

C. 77 Stratiform copper deposits of the Midcontinent region, a symposium held March 8, 1974, at the South-Central Section meeting of the Geological Society of American, ed. by K. S. Johnson and R. L. Croy. 1977. 3678

C. 78 Calceocrinids from the Bromide Formation (Middle Ordovician) of southern Oklahoma, by J. C. Brower. 1977. 3679

C. 79 Proceedings of thirteenth annual forum on the geology of industrial minerals, ed. by K. S. Johnson and J. A. Russell. 1979. 3680

C. 80 Disposal of industrial wastes in Oklahoma. 1980. 3681

Pt. I: Introduction, by K. S. Johnson, K. V. Luza, and J. F. Roberts. 3682

Pt. II: Surface disposal of industrial wastes in Oklahoma, by K. S. Johnson and K. V. Luza. 3683

Pt. III: Subsurface disposal of industrial wastes in Oklahoma, by K. S. Johnson and J. F. Roberts.
3684

Educational Publication. 1972-

EP. 1 Geology and earth resources of Oklahoma--an atlas of maps and cross sections, by K. S. Johnson, C. C. Branson, N. M. Curtis, W. E. Ham, M. V. Marcher, and J. F. Roberts. 1972. 3685

EP. 2 Introduction, guidelines, and geologic history of Oklahoma. Book I of guidebook for geologic field trips in Oklahoma, by K. S. Johnson. 1971. 3686

EP. 3 Northwest Oklahoma. Book II of guidebook for geologic field trips in Oklahoma, by K. S. Johnson. 1972. 3687

Guidebook. 1953-

GB. 11 Guide to Beavers Bend State Park, by W. D. Pitt and others. 1963. 3688

GB. 12 A guide to the state parks and scenic areas in the Oklahoma Ozarks, by G. G. Huffman, T. A. Cathey, and J. E. Humphrey. 1963. 3689

GB. 13 Sample descriptions and correlations for wells on a cross section from Barber County, Kansas, by Caddo County, Oklahoma, by W. L. Adkison and M. G. Sheldon. 1963. 3690

GB. 14 The composite interpretive method of logging drill cuttings. 2nd ed., ed. by J. C. Maher. 1964. 3691

GB. 15 Guide to Alabaster Cavern and Woodward County, Oklahoma, by A. J. Myers, A. M. Gibson, B. P. Glass, and C. R. Patrick. 1969. 3692

GB. 16 Late Paleozoic conodonts

from the Ouachita and Arbuckle Mountains of Oklahoma, by M. K. Elias. 1966. 3693

GB. 17 Regional geology of the Arbuckle Mountains, Oklahoma, by W. E. Ham, J. H. Stitt, and others. 1969. 3694

GB. 18 Upper Chesterian-Morrowan stratigraphy and the Mississippian-Pennsylvanian boundary in northeastern Oklahoma and northwestern Arkansas; guidebook for field trip no. 5, August 5-7, 1977, North American Paleontological Convention., ed. by P. K. Sutherland and W. L. Manger. 1977. 3695

GB. 19 Mississippian-Pennsylvanian shelf-to-basin transition, Ozark and Ouachita regions, Oklahoma and Arkansas; guidebook for field trip no. 11, May 27-June 1, 1979, ninth international congress of carboniferous stratigraphy and geology, ed. by P. K. Sutherland and W. L. Manger. 1979. 3696

Bulletin. 1937-

B. 55 Quicksilver in Oregon, by
H. C. Brooks. 1963. 3697

B. 56 Report. 14th. 1962/64.
3698

B. 57 State of Oregon lunar geological field conference guidebook, by N. V. Peterson, E. A. Groh, and C. J. Newhouse. 1965. 3699

B. 58 Geology of the Suplee-Izee area, Crook, Grant, and Harney Counties, Oregon, by W. R. Dickinson and L. W. Vigrass. 1965.
3700

B. 59 Report of the State Geologist. 15th. 1964/66. 3701

B. 60 Engineering geology of the Tualatin Valley region, by H. G. Schlicker and R. J. Deacon. 1967.
3702

B. 61 Gold and silver in Oregon, by H. C. Brooks and L. Ramp. 1968. 3703

B. 62 Andesite conference guidebook, by H. M. Dole. 1968. 3704

B. 63 Report of the State Geologist. 16th. 1966/68. 3705

B. 64 Mineral and water resources of Oregon, by the U. S. Geological Survey. 1969. 3706

B. 65 Proceedings of the andesite conference, by A. R. McBirney. 1969. 3707

B. 66 The reconnaissance geology and mineral resources of eastern Klamath County and western Lake County, Oregon, by N. V. Peterson

and J. R. McIntyre. 1970. 3708

B. 67 Bibliography of geology and mineral resources of Oregon, Jan. 1, 1956 to Dec. 31, 1960, by M. Roberts. 4th supplement. 1970. 3709

B. 68 Report of the State Geologist. 17th. 1968/70. 3710

B. 69 Geology of the southwestern Oregon coast west of the 124th meridian, by R. H. Dott. 1971. 3711

B. 70 Geologic formations of western Oregon (west of longitude 121° 31'), by J. D. Beaulieu. 1971. 3712

B. 71 Geology of selected lava tubes in the Bend Area, Oregon, by R. Greeley. 1971. 3713

B. 72 Bedrock geology of the Mitchell Quadrangle, Wheeler County, Oregon, by K. F. Oles and H. Enlows. 1971.
3714

B. 73 Geologic formations of eastern Oregon (east of longitude 121° 30'), by J. D. Beaulieu. 1972. 3715

B. 74 Environmental geology of the coastal region of Tillamook and Clatsop Counties, Oregon, by H. G. Schlicker and others. 1972. 3716

B. 75 Geology and mineral resources of Douglas County, Oregon, by L. Ramp. 1972. 3717

B. 76 Report. 18th. 1970/72. 3718

B. 77 Geologic field trips in northern Oregon and southern Washington, by J. D. Beaulieu. 1973. 3719

B. 78 Bibliography of the geology and mineral resources of Oregon, 1961-70. 5th supplement, by M. S.

Roberts, M. L. Steere, and C. S. Brookhyser. 1973. 3720

B. 79 Environmental geology of inland Tillamook and Clatsop Counties, Oregon, by J. D. Beaulieu. 1973. 3721

B. 80 Geology and mineral resources of Coos County, Oregon, by E. M. Baldwin, J. D. Beaulieu, L. Ramp, and others. 1973. 3722

B. 81 Environmental geology of Lincoln County, Oregon, by H. G. Schlicker and others. 1973. 3723

B. 82 Geologic hazards of Bull Run watershed, Multnomah and Clackamas Counties, by J. D. Beaulieu. 1974. 3724

B. 83 Eocene stratigraphy of southwestern Oregon, by E. M. Baldwin. 1974. 3725

B. 84 Environmental geology of western Linn County, Oregon, by J. D. Beaulieu, P. W. Hughes, and R. K. Mathiot. 1974. 3726

B. 85 Environmental geology of coastal Lane County, Oregon, by H. G. Schlicker and others. 1974. 3727

B. 86 Nineteenth biennial report of the Department, 1972-74. 3728

B. 87 Environmental geology of western Coos and Douglas Counties, Oregon, by J. D. Beaulieu and P. W. Hughes. 1975. 3729

B. 88 Geology and mineral resources of the Upper Chetco drainage, Oregon, by L. Ramp. 1975. 3730

B. 89 Geology and mineral resources of Deschutes County, Oregon, by N. V. Peterson and others. 1976. 3731

B. 90 Land-use geology of western Curry County, Oregon, by J. D. Beaulieu and P. W. Hughes. 1976. 3732

B. 91 Geologic hazards of parts of

northern Hood River, Wasco and Sherman Counties, Oregon, by J. D. Beaulieu. 1977. 3733

B. 92 Fossils in Oregon; a collection of reprints from The Ore Bin, by M. L. Steere. 1977. 3734

B. 93 Geology, mineral resources, and rock material of Curry County, Oregon, by L. Ramp, H. G. Schlicker, and J. J. Gray, 1977. 3735

B. 94 Land use geology of central Jackson County, Oregon, by J. D. Beaulieu and P. W. Hughes. 1977. 3736

B. 95 North American ophiolites, by R. G. Coleman and W. P. Irwin. 1977. 3737

B. 96 Magma genesis; proceedings of the American Geophysical Union Chapman Conference on Partial Melting in the Earth's Upper Mantle, Sept. 10-14, 1976, by H. J. B. Black. 1977. 3738

B. 97 Bibliography of the geology and mineral resources of Oregon, 1971-75, 6th supplement, by C. P. R. Hulick. 1978. 3739

B. 98 Geologic hazards of eastern Benton County, Oregon, by J. L. Bela. 1979. 3740

B. 99 Geology and geologic hazards of northwestern Clackamas County, Oregon. 1980. 3741

B. 100 Geology and mineral resources of Josephine County, Oregon, by L. Ramp and N. V. Peterson. 1979. 3742

B. 101 Geologic field trips in western Oregon and southwestern Washington, ed. by K. F. Oles and others. 1980. 3743

Miscellaneous Paper. 1950-

MP. 1 A description of some Oregon rocks and minerals. 1950. 3744

MP. 2 Key to Oregon mineral deposits

map. 1951. Rev. 1964. 3745

ι MP. 3 Facts about fossils; reprints of papers on paleontology published by the State Department of Geology and Mineral Industry. 1959. 3746

MP. 4 Laws relating to oil, gas, and geothermal exploration and development in Oregon. 3747

Pt. 1: Oil and natural gas rules and regulations. Rev. 1980. 3748

Pt. 2: Geothermal resources rules and regulations. Rev. 1979. 3749

MP. 5 Oregon's gold placers (reprints). 1954. 3750

MP. 6 Oil and gas exploration in Oregon, revised by V. C. Newton. 1965. 3751

MP. 7 Bibliography of theses and dissertations on Oregon geology from January 1, 1959, to December 31, 1965 (supplement), by M. Roberts. 1966. 3752

MP. 8 Well records on file of oil and gas exploration in Oregon, by V. C. Newton. 1960. Rev. 1980. 3753

MP. 9 Petroleum exploration in Oregon. 1962. 3754

MP. 10 Articles on recent vulcanism in Oregon. 1965. 7 vols. 3755

MP. 11 A collection of articles on meteorites from the Department's monthly publication, The Ore Bin. 1968. 3756

MP. 12 Index to published geologic mapping in Oregon, 1898-1967, by R. E. Corcoran. 1968. 3757

MP. 13 Index to The Ore Bin, 1950 to 1974, by M. C. Lewis and N. T. Mussotto. 1975. 3758

MP. 14 Thermal springs and wells in Oregon. 1975. 3759

MP. 15 Quicksilver deposits in

Oregon, by H. C. Brooks. 1971. 3760

MP. 16 Mosaic of Oregon from ERTS-1 imagery, by R. D. Lawrence and C. E. Poulton. 1973. 3761

MP. 17 Geologic hazards inventory of the Oregon coastal zone, by J. D. Beaulieu, P. W. Hughes, and R. L. Mathiot. 1974. 3762

MP. 18 Proceedings of the citizens' forum on potential future energy sources, held Jan. 17, 1974. 1975. 3763

MP. 19 Geothermal exploration studies in Oregon, by R. G. Bowen, D. D. Blackwell, and D. A. Hull. 1977. 3764

MP. 20 Investigations of nickel in Oregon, by L. Ramp. 1978. 3765

Short Paper. 1939-

SP. 24 The Almeda Mine, Josephine County, Oregon, by P. W. Libbey. 1967. 3766

SP. 25 Petrography of the Rattlesnake formation at the type area, central Oregon, by H. E. Enlows. 1976. 3767

SP. 26 Not published? 3768

SP. 27 Rock material resources of Benton County, Oregon, by H. G. Schlicker, J. J. Gray, and J. L. Bela. 1978. 3769

Special Paper. 1978-

S. 1 Mission, goals, and programs, 1979-1985; a six year plan for geologic investigation, research, and regulation. 1978. 3770

S. 2 Field geology of S. W. Broken Top quadrangle, Oregon, by E. M. Taylor. 1978. 3771

S. 3 Rock material resources of Clackamas, Columbia, Multnomah, and Washington Counties, Oregon, by

J. J. Gray, G. R. Allen, and G. S. Mack. 1978. 3772

S. 4 Heat flow of Oregon, by D. D. Blackwell and others. 1978. 3773

S. 5 Analysis and forecasts of the demand for rock materials in Oregon, by J. M. Friedman, E. G. Niems, and W. E. Whitelaw. 1979. 3774

S. 6 Geology of the La Grande area, Oregon, by W. Barrash and others. 1980. 3775

S. 7 Pluvial Fort Rock Lake, Lake County, Oregon, by I. S. Allison. 1979. 3776

S. 8 Geology and geochemistry of Mt. Hood volcano, by C. White. 1980. 3777

PENNSYLVANIA

Educational Series. 1962-

ES. 7 Coal in Pennsylvania, by
W. E. Edmunds and E. F. Koppe.
1968.   3778

ES. 8 Geology of Pennsylvania's
oil and gas, by W. R. Wagner and
W. S. Lytle. 1969.   3779

ES. 9 Geologic hazards in Pennsylvania, by J. P. Wilshusen. 1979.
                                         3780

Environmental Geology Report.
1972-

EGR. 1 Engineering characteristics of the rocks of Pennsylvania,
by W. G. McGlade and others.
1972.   3781

EGR. 2 Environmental geology for
land-use planning, by A. R. Geyer
and W. G. McGlade. 1972.   3782

EGR. 3 Subsurface liquid waste
disposal and its feasibility in Pennsylvania, by N. Rudd. 1972.
                                         3783

EGR. 4 Environmental geology of
the greater Harrisburg metropolitan
area, by W. G. McGlade and A. R.
Geyer. 1976.   3784

EGR. 5 Building stones of Pennsylvania's capital area, by A. R. Geyer.
1977.   3785

EGR. 6 Environmental geology of
the greater York area, York County,
by J. P. Wilshusen. 1979.   3786

EGR. 7 Outstanding scenic geological features of Pennsylvania, by
A. R. Geyer and W. H. Bolles.
1979.   3787

General Geology Report. 1931-

G. 38 Lithostratigraphy of the Middle
Ordovician Salona and Coburn forma-
tions in central Pennsylvania, by R. R.
Thompson. 1963.   3788

G. 39 Symposium on Middle and Upper
Devonian stratigraphy of Pennsylvania
and adjacent states, by V. C. Shepps,
ed. 1963.   3789

G. 40 Fossil collecting in Pennsylvania, by D. M. Hoskins. 1964. 3790

G. 41 Guidebook to the geology of the
Philadelphia area, by B. K. Goodwin.
1964.   3791

G. 42 Annotated bibliography of Pennsylvania geology; supplement to 1959,
by H. R. Cramer. 1965.   3792

G. 43 Provenance, dispersal, and
depositional environments of Triassic
sediments in the Newark-Gettysburg
Basin, by J. D. Glaeser. 1966. 3793

G. 44 Structure and stratigraphy of
the limestones and dolomites of Dauphin County, by D. B. MacLachlan.
1967.   3794

G. 45 Structure of the Jacksonburg
Formation in Northampton and Lehigh
Counties, by W. C. Sherwood. 1964.
                                         3795

G. 46 Geologic investigations of the
Pennsylvania Piedmont; collected reprints. 1964.   3796

G. 47 Stratigraphy of Lower Ordovician Nittany dolomite in central Pennsylvania, by A. R. Spelman. 1966.
                                         3797

G. 48 Stratigraphy and paleontology
of the Mahantango Formation in south-

central Pennsylvania, by R. L. Ellison. 1965. 3798

G. 49 Stratigraphy of the Cambrian to Middle Ordovician rocks of central and western Pennsylvania, by W. R. Wagner. 1966. 3799

G. 50 Guide to the Horseshoe Curve section between Altoona and Galitzin, central Pennsylvania, by F. M. Swartz. 1965. 3800

G. 51 Stratigraphy of the Devonian Trimmers rock in eastern Pennsylvania, by L. A. Frakes. 1967. 3801

G. 52 Stratigraphy of the Lower Ordovician Axemann limestone in Central Pennsylvania, by J. A. Lees. 1967. 3802

G. 53 A lithostratigraphic, petrographic, and chemical investigation of the lower Middle Ordovician carbonate rocks in central Pennsylvania, by M. Rones. 1969. 3803

G. 54 Stratigraphy, structure, and sedimentary patterns in the Upper Devonian of Bradford County, by D. L. Woodrow. 1968. 3804

G. 55 Pleistocene stratigraphy in northwestern Pennsylvania, by G. W. White and others. 1969. 3805

G. 56 Geology and land use in eastern Washington County (Hackett and Ellsworth quadrangles), by G. H. Kent and others. 1969. 3806

G. 56A Geology and land use in eastern Washington County (California and Monongahela quadrangles), by S. P. Schweinfurth and others. 1971. 3807

G. 57 Surficial geology of the Stroudsburg quadrangle, by J. B. Epstein. 1969. 3808

G. 58 Carbonates of the Lower and Middle Ordovician in central Pennsylvania, by H. S. Chafetz. 1969. 3809

G. 59 Geology of the Pittsburgh

area, by W. R. Wagner and others. 1970. 3810

G. 60 Pleistocene geology and unconsolidated deposits of the Delaware Valley, Matamoras to Shawnee on Delaware, by G. H. Crowl. 1971. 3811

G. 61 Annotated bibliography of Pennsylvania geology; supplement to 1969, by H. R. Cramer. 1972. 3812

G. 62 Upper Devonian marine-nonmarine transition, southern Pennsylvania, by R. D. Walker. 1972. 3813

G. 63 Upper Devonian stratigraphy and sedimentary environments in northeastern Pennsylvania, by J. D. Glaeser. 1974. 3814

G. 64 Pleistocene beach ridges of northwestern Pennsylvania, by E. E. Schooler. 1974. 3815

G. 65 Caves of southeastern Pennsylvania, comp. by J. R. Reich. 1974. 3816

G. 66 Geology and biology of Pennsylvania caves, ed. by W. B. White. 1976. 3817

G. 67 Caves of western Pennsylvania, comp. by W. B. White. 1976. 3818

G. 68 Caves of the Valley and Ridge province, by W. B. White. 1980. 3819

G. 69 Caves of the Great Valley, by W. B. White. 1980. 3820

G. 70 Middle Devonian stratigraphy in central Pennsylvania--a revision, by R. T. Faill and others. 1978. 3821

G. 71 Glacial border deposits of late Wisconsinan age in northeastern Pennsylvania, by G. H. Crowl and W. D. Sevon. 1980. 3822

Information Circular. 1951-

IC. 51 The mineral industry of Pennsylvania in 1962, by C. C. Yeloushan and others. 1964. 3823

IC. 52 Interpretation of an aeromagnetic survey in western Pennsylvania and parts of eastern Ohio,
northern West Virginia, and western Maryland, by M. E. Beck and
R. E. Mattick. 1964. 3824

IC. 53 The mineral industry of
Pennsylvania in 1963, by J. R. Kerr.
1965. 3825

IC. 54 Directory of the mineral industry in Pennsylvania, by B. J.
O'Neill. 3rd ed. 1977. 3826

IC. 55 The mineral industry of
Pennsylvania in 1964, by J. R. Kerr.
1966. 3827

IC. 56 The geography and geology
of Erie County, by J. C. Tomikel
and V. C. Shepps. 1967. 3828

IC. 57 A method for determining
post-depositional alteration in underclays and associated lithogies, by
M. G. Jaron. 1967. 3829

IC. 58 The mineral industry of
Pennsylvania in 1965, by C. C.
Yeloushan. 1967. 3830

IC. 59 The mineral industry of
Pennsylvania in 1966, by C. C.
Yeloushan. 1968. 3831

IC. 60 Devonian Tully limestone
in Pennsylvania and comparison to
type Tully limestone in New York,
by P. H. Heckel. 1969. 3832

IC. 61 Revised lithostratigraphic
nomenclature of the Pottsville and
Allegheny Groups (Pennsylvania),
Clearfield County, by W. E. Edmunds. 1969. 3833

IC. 62 The Precambrian in the subsurface of northwestern Pennsylvanian and adjoining states, by T. E.
Saylor. 1968. 3834

IC. 63 The mineral industry of
Pennsylvania in 1967, by C. C.
Yeloushan. 1969. 3835

IC. 64 The W. L. Newman phosphate
mine, Juniata County, by W. D.
Carter. 1969. 3836

IC. 65 Peat bog investigations in
northeastern Pennsylvania, by C. D.
Edgerton. 1969. 3837

IC. 66 Geology of flagstones in the
Endless Mountains region, northern
Pennsylvania, by J. D. Glaeser. 1969.
3838

IC. 67 Analyses and potential uses of
selected shales and clays, Greene
County, by J. B. Roen and J. W. Hosterman. 1969. 3839

IC. 68 Chemical analyses of three
Triassic (?) diabase dikes in Penn-
sylvania, by D. M. Lapham and T. E.
Saylor. 1970. 3840

IC. 69 The mineral industry of Pennsylvania in 1968, by C. C. Yeloushan.
1970. 3841

IC. 70 Summary of isotopic age determinations in Pennsylvania, by D. M.
Lapham and S. I. Root. 1971. 3842

IC. 71 The mineral industry of Pennsylvania in 1969, by C. G. Leaf. 1971.
3843

IC. 72 Coal reserves of Pennsylvania:
total, recoverable, and strippable
(January 1, 1970), by W. E. Edmunds.
1972. 3844

IC. 73 The mineral industry of Pennsylvania in 1970, by F. D. Cooper.
1972. 3845

IC. 74 The mineral industry of Pennsylvania in 1971, by F. D. Cooper.
1974. 3846

IC. 75 Geologic causes and possible
preventions of roof fall in room-and-
pillar coal mines, by B. H. Kent.
1974. 3847

IC. 76 Potential high-calcium limestone resources in the Mount Joy area,
Lancaster County, by B. J. O'Neill.
1975. 3848

IC. 77 Geologic interpretation of
aeromagnetic maps of southeastern
Pennsylvania, by A. A. Socolow. 1974.
3849

IC. 78 The mineral industry of Pennsylvania in 1972, by F. D. Cooper. 1975. 3850

IC. 79 Interpretation of K-Ar and Rb-Sr isotopic dates from a Precambrian basement core, Erie County, by D. M. Lapham. 1975. 3851

IC. 80 Greater Pittsburgh region revised surface structure and its relation to oil and gas fields, by W. R. Wagner and W. S. Lytle. 1976. 3852

IC. 81 The mineral industry of Pennsylvania in 1973, by F. D. Cooper. 1976. 3853

IC. 82 The mineral industry of Pennsylvania in 1974, by W. Kebbish. 1978. 3854

IC. 83 The Huntley Mountain Formation: Catskill-to-Burgoon transition in north-central Pennsylvania, by T. M. Berg and W. E. Edmunds. 1979. 3855

IC. 84 Geologic conditions affecting safe bituminous coal mining in Pennsylvania: selected papers, comp. by S. I. Root. 1979. 3856

IC. 85 The mineral industry of Pennsylvania in 1975, by W. Kebbish. 1979. 3857

IC. 86 Quality of gravel resources in northwestern Pennsylvania, by J. L. Craft. 1979. 3858

IC. 87 The mineral industry of Pennsylvania in 1976, by W. Kebbish. 1980. 3859

Mineral Resource Report. 1922-

MRR. 46 Underground gas storage in Pennsylvania, by W. S. Lytle. 1963. 3860

(MRR. 47 in base index)

MRR. 48 Petrography of the Upper Freeport coal--Harmar and Springdale Mines, Allegheny and West-

moreland Counties, by E. F. Koppe. 1963. 3861

MRR. 49 Mineral aspects of the Grace Mine magnetite deposit, by A. Tsusue. 1964. 3862

MRR. 50 Atlas of Pennsylvania's mineral resources. 3863

Pt. 1: Limestones and dolomites of Pennsylvania, by B. J. O'Neill. 1964. 3864

Pt. 1S: Supplement, limestones and dolomites of Pennsylvania, by G. F. Deasy and others. 1967. 3865

Pt. 2A: Clays and shales of Pennsylvania, by B. J. O'Neill and K. V. Hoover. n. d. 3866

Pt. 2B: Economics of Pennsylvania's clay and shale production, by G. F. Deasy and P. R. Griess. 1969. 3867

Pt. 3: Metal mines and occurrences in Pennsylvania, by A. W. Rose. 1970. 3868

Pt. 4: The distribution of limestones containing at least 90 percent $CaCO_3$ in Pennsylvania, by B. J. O'Neill. 1976. 3869

MRR. 51 Properties and uses of Pennsylvania shales and clays, by B. J. O'Neill and others. 1965. 3870

MRR. 52 Oil and gas geology of the Warren quadrangle, by W. S. Lytle. 1965. 3871

MRR. 53 Oil and gas geology of the Youngsville quadrangle, by W. G. McGlade. 1964. 3872

MRR. 54 Oil and gas geology of the Amity and Claysville quadrangles, by W. G. McGlade. 1967. 3873

MRR. 55 Petrography of coal in the Houtzdale quadrangle, Clearfield County, by E. F. Koppe. 1967. 3874

MRR. 56 Geology and origin of the Triassic magnetite deposit and diabase at Cornwall, by D. M. Lapham and Carlyle Gray. 1973. 3875

MRR. 57 Geology of the Red Valley sandstone in Forest and Venango Counties, by D. R. Kelley. 1967. 3876

MRR. 58 Salt deposits in the Salina Group in Pennsylvania, by W. B. Fergusson and B. A. Prather. 1968. 3877

MRR. 59 Geology of the Elk Run gas pool, Jefferson County, by L. Heyman. 1969. 3878

MRR. 60 Medina and Oriskany production along the shore of Lake Erie, Pierce field, Erie County, by D. A. Kelley and W. G. McGlade. 1969. 3879

MRR. 61 A summary of major geophysical logging methods, by D. R. Kelley. 1969. 3880

MRR. 62 Oil and gas geology of the Kinzua quadrangle, Warren and McKean Counties, by W. S. Lytle and J. H. Goth. 1968. 3881

MRR. 63 Properties and uses of Pennsylvania shales and clays, southeastern Pennsylvania, by K. V. Hoover and others. 1971. 3882

MRR. 64 Proceedings: fifth forum on geology of industrial minerals, ed. by K. V. Hoover. 1970. 3883

MRR. 65 The petroleum industry and the future petroleum province in Pennsylvania, 1970, by D. R. Kelley and others. 1970. 3884

MRR. 66 Analyses and measured sections of Pennsylvania bituminous coals, by R. D. Sponseller. 1973. 3885

MRR. 67 Greater Pittsburgh region construction aggregates, by B. J. O'Neill. 1974. 3886

MRR. 68 Bituminous coal resources in western Pennsylvania, comp. by M. A. Sholes and V. W. Skema. 1974. 3887

MRR. 69 Analyses and measured sections of Pennsylvania bituminous coals, Pt. II, by V. W. Skema and others. 1975. 3888

MRR. 70 Oil fields of the Greater Pittsburgh region, by W. S. Lytle. 1976. 3889

MRR. 71 Greater Pittsburgh region properties and uses of clays and shales, by B. J. O'Neill. 1976. 3890

MRR. 72 Zinc and lead occurrences in Pennsylvania, by R. C. Smith. 1977. 3891

MRR. 73 Tully (Middle Devonian) to Queenston (Upper Ordovician) correlations in the subsurface of western Pennsylvania, by L. Heyman. 1977. 3892

MRR. 74 Analyses and measured sections of Pennsylvania bituminous coals, Pt. III, by V. W. Skema and others. 1977. 3893

MRR. 75 The economic geology of the Upper Freeport coal in the New Stanton area of Westmoreland County, Pennsylvania: a model for coal exploration, by M. A. Sholes and others. 1979. 3894

MRR. 76 The economic geology of the Upper Freeport coal in northeastern Greene County, by V. W. Skema and others. 1980. 3895

MRR. 77 Properties and uses of shales and clays, southwestern Pennsylvania, by B. J. O'Neill and J. H. Barnes. 1979. 3896

MRR. 78 Mineralogy associated with burning anthracite deposits of eastern Pennsylvania, by D. M. Lapham and others. 1980. 3897

Park Guide. 1969-

PG. 1 Trough Creek State Park: ice mine and balanced rock, by J. P. Wilshusen. 1969. 3898

PG. 2 Hickory Run State Park: Hickory Run boulder field, by A. R. Geyer. 1969. 3899

PG. 3 Archbald Pothole State Park: Archbald pothole, by W. G. Mc-Glade. 1969. 3900

PG. 4 Moraine State Park, by W. S. Lytle. 1970. 3901

PG. 5 Leonard Harrison and Colter Point State Parks: Grand Canyon of Pennsylvania, by W. G. Mc-Glade. 1971. 3902

PG. 6 French Creek State Park: story of the rocks, by W. B. Fergusson. 1971. 3903

PG. 7 Ohiopyle State Park: geologic features of interest, by K. O. Bushnell. 1971. 3904

PG. 8 Valley Forge State Park: history of the rocks, by P. L. Samson. 1974. 3905

PG. 9 McConnells Mill State Park: Slippery Rock Creek gorge, by K. O. Bushnell. 1975. 3906

PG. 10 Gifford Pinchot State Park: diabase (molten liquid rock), by D. M. Hoskins. 1978. 3907

PG. 11 Ravensburg State Park, by A. R. Geyer and D. W. Royer. 1980. 3908

PG. 12 Worlds End State Park, by D. W. Royer. 1980. 3909

PG. 13 Ricketts Glen State Park, by J. D. Inners. 1980. 3910

PG. 14 Nockamixon State Park, by J. D. Inners. 1980. 3911

Water Resource Report. 1933-

WRR. 1 Ground water in southwestern Pennsylvania, by A. M. Piper. 1933. 3912

WRR. 2 Ground water in southeastern Pennsylvania, by G. M. Hall. 1934. 3913

WRR. 3 Ground water in northwestern Pennsylvania, by R. M. Leggette. 1936. 3914

WRR. 4 Ground water in northeastern Pennsylvania, by S. W. Lohman. 1937. 3915

WRR. 5 Ground water in south-central Pennsylvania, by S. W. Lohman. 1938. 3916

WRR. 6 Ground water in north-central Pennsylvania, by S. W. Lohman. 1939. 3917

WRR. 7 Ground-water resources of Pennsylvania, by S. W. Lohman. 1941. 3918

WRR. 8 Ground-water resources of the valley-fill deposits of Allegheny County, by J. H. Adamson and others. 1949. 3919

WRR. 9 Ground-water resources of Beaver County, by D. W. Van Tuyl and N. H. Klein. 1951. 3920

WRR. 10 Ground water for air conditioning at Pittsburgh, by D. W. Van-Tuyl. 1951. 3921

WRR. 11 Ground water resources of Bucks County, by D. W. Greenman. 1955. 3922

WRR. 12 Borehole geophysical methods for analyzing specific capacity of multi-aquifer wells, by G. D. Bennett and E. P. Patten. 1960. 3923

WRR. 13 Ground-water resources of the coastal plain area of southeastern Pennsylvania, by D. W. Greenman and others. 1961. 3924

WRR. 14 Geology and hydrology of the Stockton Formation in southeastern Pennsylvania, by D. R. Rima and others. 1962. 3925

WRR. 15 Geology and hydrology of the Neshannock quadrangle, Mercer and Lawrence Counties, by L. D. Carswell and G. D. Bennett. 1963. 3926

WRR. 16 Geology and hydrology of the Mercer quadrangle, Mercer, Lawrence, and Butler Counties, by C. W. Poth. 1963. 3927

WRR. 17 Methods of flow measure-

ment in well bores, by E. P. Patten and G. D. Bennett. 1962.
3928

WRR. 18 Hydrogeology of the carbonate rocks of the Lebanon Valley, by H. Meisler. 1963.
3929

WRR. 19 Application of electrical and radioactive well logging to groundwater hydrology, by E. P. Patten and G. D. Bennett. 1963.
3930

WRR. 20 The ground-water observation-well program in Pennsylvania, by C. W. Poth. 1963.
3931

WRR. 21 Hydrology of the New Oxford Formation in Adams and York Counties, by C. R. Wood and H. E. Johnston. 1964.
3932

WRR. 22 Ground-water resources of the Brunswick Formation in Montgomery and Berks Counties, by S. M. Longwill and C. R. Wood. 1965.
3933

WRR. 23 Hydrology of the New Oxford Formation in Lancaster County, by H. E. Johnston. 1966.
3934

WRR. 24 Geology and hydrology of the Martinsburg Formation in Dauphin County, by L. D. Carswell and others. 1968.
3935

WRR. 25 Hydrology of the metamorphic and igneous rocks of central Chester County, by C. W. Poth. 1968.
3936

WRR. 26 Hydrology of the carbonate rocks of the Lancaster 15-minute quadrangle, south-central Pennsylvania, by H. Meisler and A. E. Becher. 1971.
3937

WRR. 27 Ground-water resources of the Loysville and Mifflintown quadrangles in south-central Pennsylvania, by H. E. Johnston. 1970.
3938

WRR. 28 Hydrology of the Pleistocene sediments in the Wyoming Valley, Luzerne County, by J. R. Hollowell. 1971.
3939

WRR. 29 Ground-water resources of Montgomery County, by T. G. Newport. 1971.
3940

WRR. 30 Hydrology of the Martinsburg Formation in Lehigh and Northampton Counties, by C. W. Poth. 1972.
3941

WRR. 31 Water resources of Lehigh County, by C. R. Wood and others. 1972.
3942

WRR. 32 Summary ground-water resources of Clarion County, by T. G. Newport. 1973.
3943

WRR. 33 Geology and ground-water resources of northern Mercer County, by G. R. Schiner and G. E. Kimmel. 1976.
3944

WRR. 34 Summary ground-water resources of Armstrong County, by C. W. Poth. 1973.
3945

WRR. 35 Summary ground-water resources of Allegheny County, by J. T. Gallaher. 1973.
3946

WRR. 36 Summary ground-water resources of Butler County, by C. W. Poth. 1973.
3947

WRR. 37 Summary ground-water resources of Westmoreland County, by T. G. Newport. 1973.
3948

WRR. 38 Summary ground-water resources of Washington County, by T. G. Newport. 1973.
3949

WRR. 39 Summary ground-water resources of Beaver County, by C. W. Poth. 1973.
3950

WRR. 40 Summary ground-water resources of Luzerne County, by T. G. Newport. 1977.
3951

WRR. 41 Ground-water resources of Lackawanna County, by J. R. Hollowell and H. E. Koester. 1975. 3952

WRR. 42 Ground-water resources of central and southern York County, by O. B. Lloyd and D. J. Growitz. 1977.
3953

WRR. 43 Summary ground-water re-

sources of Lancaster County, by
C. W. Poth. 1977.   3954

WRR. 44 Geology and groundwater
resources of northern Berks County,
by C. R. Wood and D. B. MacLachlan. 1978.   3955

WRR. 45 Groundwater resources
of the DuBois area, Clearfield and
Jefferson Counties, by E. T. Shuster. 1979.   3956

WRR. 46 Geology and groundwater
resources of western Crawford
County, by G. R. Schiner and J. T.
Gallaher. 1979.   3957

WRR. 47 Geology and groundwater
resources of Monroe County, by
L. D. Carswell and O. B. Lloyd.
1979.   3958

WRR. 48 Summary groundwater
resources of Centre County, Pennsylvania, by C. R. Wood. 1980.
                                3959

WRR. 49 Groundwater resources of
the Gettysburg and Hammer Creek
Formations, southeastern Pennsylvania, by C. R. Wood. 1980.
                                3960

RHODE ISLAND

No activity.

Bulletin. 1904-

B. 1 A preliminary report on the clays of South Carolina, by E. Sloan. 1904. 3961

B. 2 Catalog of the mineral localities of South Carolina, by E. Sloan. 1908. 3962

(B. 3-14 are not related to geology)

B. 15 Ground-water investigations in South Carolina, by G. E. Siple. 1946. 3963

B. 16 Chemical character of the surface waters of South Carolina, 1945-47, by W. L. Lamar. 1948. 3964

B. 16A Chemical character of the surface waters of South Carolina, 1945-50, by F. H. Pacszek. 1951. 3965

B. 16B Chemical character of surface waters of South Carolina, 1945-55, by G. A. Billingsley. 1956. 3966

B. 16C Chemical character of the surface waters of South Carolina, 1945-60, by K. F. Harris. 1962. 3967

B. 17 Surface water supply of South Carolina. 1948. 3968

B. 18 Geology and preliminary ore dressing studies of the Carolina Barite belt, by E. C. Van Horn, J. R. LeGrand, and L. L. McMurray. 1949. 3969

B. 19 The distribution and properties of the shales of South Carolina, by B. F. Buie and G. C. Robinson. 1949. 3970

B. 20 South Carolina raw materials for light-colored face brick, by G. C. Robinson. 1949. 3971

(B. 21-22 are not related to geology)

B. 23 Silica for glass manufacture in South Carolina, by B. F. Buie and G. C. Robinson. 1958. 3972

B. 24 Guidebook for the South Carolina coastal plain field trip of the Carolina Geological Society, by G. E. Siple. 1959. 3973

B. 25 Common clays of the coastal plain of South Carolina and their use in structural clay products, by G. C. Robinson, B. F. Buie, and H. S. Johnson. 1961. 3974

B. 26 Exploration for heavy minerals on Hilton Head Island, South Carolina, by C. K. McCauley. 1960. 3975

B. 27 Barium resources of South Carolina, by C. K. McCauley. 1962. 3976

B. 28 Limestone resources of the coastal plain of South Carolina, by S. D. Heron. 1962. 3977

B. 29 Corundum resources of South Carolina, by C. K. McCauley and J. F. McCauley. 1964. 3978

B. 30 Gem stone resources of South Carolina, by C. K. McCauley. 1964. 3979

B. 31 Clays and opal-bearing claystones of the South Carolina coastal plain, by S. D. Heron, G. C. Robinson, and H. S. Johnson. 1965. 3980

B. 32 Gold resources of South Carolina, by C. K. McCauley and J. R. Butler. 1966. 3981

191

B. 33 Geology and mineral resources of York County, South Carolina, by J. R. Butler. 1966.
3982

B. 34 Geology and mineral resources of Oconee County, South Carolina, by C. J. Cazeau. 1967. 3983

B. 35 Heavy minerals in South Carolina, by L. Williams. 1967.
3984

B. 36 Ground-water resources of Orangeburg County, South Carolina, by G. E. Siple. 1975. 3985

B. 37 Geology and kyanite resources of Little Mountain, South Carolina, by J. C. McKenzie and J. F. Mc-Cauley. 1968. 3986

B. 38 Ground-water resources of Greenville County, South Carolina, by N. C. Koch. 1968. 3987

Circular. 19?-

C. 1 Catalog of geologic publications. Revised. 3988

C. 2 Catalog of South Carolina mineral producers. 5th ed. 1978-79. 3989

C. 3 Index of 7. 5-minute topographic maps for South Carolina and listing of available orthophotoquads. Updated every six months. 3990

Mineral Resources Series. 1973-

MRS. 1 Fluvial monazite deposits in the drainage basins of the Enoree, Tyger, and Pacolet Rivers, South Carolina, by D. W. Caldwell and A. M. White. 1973. 3991

MRS. 2 Fluvial monazite deposits in the drainage basins of the Savannah and Saluda Rivers, South Carolina, by N. P. Cuppels and A. M. White. 1973. 3992

MRS. 3 Radioactive mineral resources of South Carolina, by S. D.

Heron and H. S. Johnson. 1969. 3993

MRS. 4 Ground-water records of South Carolina, by G. W. Stock and G. E. Siple. 1969. 3994

MRS. 5 The granitic stone resources of South Carolina, by H. D. Wagener. 1977. 3995

SOUTH DAKOTA

Bulletin. 1894-

B. 14 A guide to the stratigraphy of South Dakota, by A. F. Agnew and P. C. Tychsen. 1965.   3996

(B. 15 is in the base index)

B. 16 Mineral and water resources of South Dakota, by the U. S. Geological Survey, the U. S. Bureau of Reclamation, and the South Dakota Geological Survey. 1964, 1975.   3997

B. 17 Geology and ground-water supplies in Sanborn County, South Dakota, by F. V. Steece and L. W. Howells.   3998

Pt. 1, 1965.   3999

Pt. 2, 1968 (Ground-water basic data).   4000

B. 18 Geology and water resources of Beadle County, South Dakota, by L. S. Hedges.   4001

Pt. I: Geology. 1968.   4002

Pt. II: Water resources. 1968.   4003

B. 19 Geology and hydrology of Clay County, South Dakota, by C. M. Christensen and J. C. Stephens.   4004

Pt. I: Geology. 1967.   4005

Pt. II: Water resources. 1968. 4006

Pt. III: Basic data. 1970.   4007

B. 20 Geology and water resources of Campbell County, South Dakota, by L. S. Hedges and N. C. Koch.   4008

Pt. I: Geology. 1972.   4009

Pt. II: Water resources. 1970. 4010

Pt. III: Basic data. 1970.   4011

B. 21 Geology and water resources of Bon Homme County, South Dakota, by C. M. Christensen and D. G. Jorgensen.   4012

Pt. I: Geology. 1974.   4013

Pt. II: Water resources. 1971. 4014

B. 22 Geology and water resources of Charles Mix and Douglas Counties, South Dakota, by L. S. Hedges and J. Kume.   4015

Pt. I: Geology. 1975.   4016

Pt. II: Water resources. 1977. 4017

B. 23 Geology and water resources of Marshall County, South Dakota, by N. C. Koch.   4018

Pt. I: Geology and water resources. 1975.   4019

B. 24 Geology and water resources of Day County, South Dakota, by D. Leap. n. d.   4020

B. 25 Geology and water resources of Brown County, South Dakota, by N. C. Koch and W. Bradford.   4021

Pt. I: Manuscript in review.   4022

Pt. II: Water resources. 1976. 4023

B. 26 Geology and water resources of McPherson, Edmunds, and Faulk Counties, South Dakota, by C. M. Christensen and L. J. Hamilton. 4024

Pt. I: Geology. 1977.   4025

Circular. 1917-

C. 32 List of Widco logs run by State Geological Survey before March 1, 1964, by L. S. Hedges. 1964. 4026

C. 33 Bibliography of reports containing maps on South Dakota geology published before January 1, 1959, by C. M. Christensen, A. F. Agnew, and M. J. Tipton. 1966. 4027

C. 34 List of oil and gas tests in South Dakota before July 1, 1964, by D. J. Hedges. 1964. 4028

C. 35 Selected formation tops in oil and gas tests in South Dakota drilled before January 1, 1965, by R. A. Schoon. 1965. 4029

C. 36 Selected formation tops in water wells logged by the South Dakota Geological Survey to January 1, 1968, by R. A. Schoon. 1968. 4030

C. 37 Mining laws of South Dakota, by E. Cox and D. J. McGregor. 1967. 4031

C. 38 Sand and gravel deposits in Campbell County, South Dakota, by L. S. Hedges. 1969. 4032

C. 39 Sand and gravel resources in Bon Homme County, South Dakota, by C. M. Christensen. 1970. 4033

C. 40 Recovering microvertebrates with acetic acid, by M. Green. 1970. 4034

C. 41 Results from drill stem tests of oil tests in South Dakota drilled before July 1, 1970, by R. A. Schoon. 1970. 4035

C. 42 Sand and gravel resources in Charles Mix and Douglas Counties, South Dakota, by L. S. Hedges. 1972. 4036

C. 43 A pump test in the Dakota sandstone at Wall, South Dakota, by J. P. Gries, P. H. Rahn, and R. K. Baker. 1976. 4037

Educational Series. 1962-

ES. 2 Record of life, by D. J. Mc-Gregor and B. C. Petsch. 1968. 4038

ES. 3 South Dakota's rock history, by B. C. Petsch and D. J. McGregor. 1969. 4039

ES. 4 A history of the South Dakota Geological Survey, by P. M. Vanorny. 1970. 4040

ES. 5 Minerals and rocks of South Dakota, by B. C. Petsch and D. J. McGregor. 1973. 4041

ES. 6 Not published? 4042

ES. 7 A tourist guide of the Black Hills, by B. C. Petsch. n. d. 4043

Guide Book Series. 1965-

GBS. 1 Reprint of South Dakota part of Inqua guidebook and supplemental data for field conference C, Upper Mississippi Valley, INQUA. 1965. 4044

GBS. 2 Guidebook to the major Cenozoic deposits of southwestern South Dakota, by J. C. Harksen and J. R. Macdonald. 1969. 4045

Report of Investigation. 1930-

RI. 92 Geology of the Bridger area, west-central South Dakota, by D. F. B. Black. 1964. 4046

(RI. 93 in base index)

RI. 94 Gravity survey of southwestern South Dakota, by E. L. Tullis. 1963. 4047

RI. 95 Evaluation of exploration methods for coarse aggregate in eastern South Dakota, by R. L. Bruce and B. E. Lundberg. 1964. 4048

RI. 96 Miocene Batesland Formation named in southwestern South Dakota, by J. C. Harksen and J. R. Macdonald. 1967. 4049

RI. 97 Rosebud Formation in South Dakota, by J. R. Macdonald and J. C. Harksen. 1968.   4050

RI. 98 Red Dog loess named in southwestern South Dakota, by J. C. Harksen. 1968.   4051

RI. 99 Type sections for the Chadron and Brule Formations of the White River Oligocene in the Big Badlands, South Dakota, by J. C. Harksen and J. R. Macdonald. 1969.   4052

RI. 100 Thin Elk Formation, Lower Pliocene, South Dakota, by J. C. Harksen and M. Green. 1971.   4053

RI. 101 Never published.   4054

RI. 102 Hydrology of Lake Poinsett, by A. Barari. 1971.   4055

RI. 103 Hydrology of Lake Kampeska, by A. Barari. 1971.   4056

RI. 104 Geology and hydrology of the Dakota Formation in South Dakota, by R. A. Schoon. 1971.   4057

RI. 105 Monroe Creek (Early Miocene) microfossils from the Wounded Knee area, South Dakota, by L. Macdonald. 1972.   4058

RI. 106 Review of oil possibilities in Harding and Butte Counties with emphasis on the Newcastle sandstone, by R. A. Schoon. 1972.   4059

RI. 107 Large springs in the Black Hills, South Dakota and Wyoming, by P. H. Rahn and J. P. Gries. 1973.   4060

RI. 108 Radiocarbon dating of terraces along Bear Creek, Pennington County, South Dakota, by J. C. Harksen. 1974.   4061

RI. 109 Ground-water resources of the western half of Fall River County, South Dakota, by J. R. Keene. 1973.   4062

RI. 110 Geothermal potentials in South Dakota, by R. A. Schoon and D. J. McGregor. 1974.   4063

RI. 111 Miocene channels in the Cedar Pass area, Jackson County, South Dakota, by J. C. Harksen. 1974.   4064

Special Report. 1959-

SR. 20 Water supply for the city of Scotland, by C. M. Christensen. 1963.   4065

SR. 21 Water supply for the city of Vermillion, by R. L. Bruce. 1963.   4066

SR. 22 Water supply for the city of Beresford, by G. K. Baker. 1963.   4067

SR. 23 Water supply for the city of Lesterville, by L. R. Rukstad. 1963.   4068

SR. 24 Water supply for the city of Redfield, by C. M. Christensen. 1963.   4069

SR. 25 Water supply for the city of Claremont, by G. K. Baker. 1963.   4070

SR. 26 Ground water supply for the city of Harrisburg, by J. A. McMeen. 1964.   4071

SR. 27 Ground water supply for the city of Marion, by J. A. McMeen. 1964.   4072

SR. 28 Ground water supply for the city of Watertown, by L. R. Rukstad and L. S. Hedges. 1964.   4073

SR. 29 Ground water supply for the city of Langford, by A. Wood and L. S. Hedges. 1964.   4074

SR. 30 Ground water supply for the city of Bowdle, by L. R. Rukstad and L. S. Hedges. 1964.   4075

SR. 31 Ground water supply for the city of Canton, by J. A. McMeen. 1965.   4076

SR. 32 Ground water supply for the city of Ipswich, by S. W. Pottratz. 1965. 4077

SR. 33 Ground water supply for the city of Britton, by J. A. McMeen. 1965. 4078

SR. 34 Ground water supply for the city of Lake Norden, by D. G. Jorgensen. 1965. 4079

SR. 35 Ground water supply for the city of Bryant, by D. G. Jorgensen. 1965. 4080

SR. 36 Ground water supply for the city of Winner, by A. Barari. 1965. 4081

SR. 37 Ground water supply for the city of Lake Andes, by G. W. Shurr. 1966. 4082

SR. 38 Ground water supply for the city of Wessington Springs, by F. V. Steece and G. W. Shurr. 1966. 4083

SR. 39 Ground water supply for the city of Dell Rapids, by A. Barari. 1967. 4084

SR. 40 Ground water supply for the city of Waubay, by J. D. Beffort and L. S. Hedges. 1967. 4085

SR. 41 Ground water supply for the city of Mission, by A. Barari. 1967. 4086

SR. 42 Ground water supply for the city of Mitchell, by A. Barari. 1967. 4087

SR. 43 Ground water supply for the city of Viborg, by J. D. Beffort and C. M. Christensen. 1968. 4088

SR. 44 Ground-water investigations for the city of Gettysburg, by J. D. Beffort. 1969. 4089

SR. 45 Ground water supply for the city of Brookings, by A. Barari. 1968. 4090

SR. 46 Ground-water investigation for the city of Lennox, by J. D.

Beffort. 1969. 4091

SR. 47 Ground-water investigation for the city of Howard, by A. Barari. 1972. 4092

SR. 48 Ground-water investigation for the city of Colome, by A. Barari. 1969. 4093

SR. 49 Ground-water investigation for the city of Gregory, by A. Barari. 1969. 4094

SR. 50 Ground-water investigations for the city of Columbia, by A. Barari and D. Brinkley. 1970. 4095

SR. 51 Ground-water investigation for the city of Volga, by A. Barari. 1971. 4096

SR. 52 Water investigation for the city of Pierre, by D. Brinkley. 1971. 4097

SR. 53 Ground-water investigation for the city of Hazel, by A. Barari. 1972. 4098

SR. 54 Ground-water investigation for the city of Spencer, by A. Barari. 1972. 4099

SR. 55 Ground-water investigation for the city of Parkston, by A. Barari. 1972. 4100

SR. 56 Ground-water investigation for the city of Baltic, by A. Barari. 1972. 4101

SR. 57 Ground-water investigation for the city of Webster, by A. Barari. 1974. 4102

SR. 58 Ground-water investigation for the city of Groton, by A. Barari. 1974. 4103

SR. 59 Ground-water investigation for the city of Peever, by A. Barari and D. Buehrer. 1974. 4104

SR. 60 Ground-water investigation for the city of Eureka, by A. Barari and L. J. Steffen. 1975. 4105

SR. 61 Ground-water investigation for

Bulletin. 1910-

B. 65 Limestone and dolomite resources of Tennessee, by R. E. Hershey and S. W. Maher. 1963. 4109

B. 66 Tennessee rock and mineral resources, by R. J. Floyd. 1965. 4110

B. 67 Annotated bibliography of the geology of Tennessee, January 1951 through December 1960, by C. W. Wilson. 1965. 4111

B. 68 Geology of the Wells Creeks structure, Tennessee, by C. W. Wilson and R. G. Stearns, assisted by H. A. Tiedemann, J. T. Wilcox, and P. S. Marsh. 1968. 4112

B. 69 Descriptions of Tennessee caves, by L. E. Matthews. 1971. 4113

B. 70 Geology of Knox County, Tennessee, by 20 contributors. 1973. 4114

B. 71 Annotated bibliography of the geology of Tennessee, January 1961 through December 1970, by C. W. Wilson. 1973. 4115

B. 72 Gold deposits of the Coker Creek District, Monroe County, Tennessee, by R. C. Hale. 1974. 4116

B. 73 Place names of Tennessee, by R. O. Fullerton. 1974. 4117

B. 74 The geologic history of Tennessee, by R. A. Miller. 1974. 4118

B. 75 Stratigraphy of the outcropping Upper Cretaceous, Paleocene, and Lower Eocene in western Tennessee (including descriptions of Younger fluvial deposits), by E. E. Russell and W. S. Parks. 1975. 4119

B. 76 Subsurface information catalog of Tennessee, 1866-1974, by A. Statler, P. Bloss, and R. Zurawski. 1975. 4120

B. 77 Vertebrate fossils of Tennessee, by J. X. Corgan. 1976. 4121

B. 78 Seismic investigations in eastern Tennessee, by E. R. Tegland. 1978. 4122

B. 79 Geology of Hamilton County, Tennessee, by 8 contributors. 1978. 4123

B. 80 Natural bridges of Tennessee, by J. X. Corgan and J. T. Parks. 1979. 4124

Environmental Geology Series. 1972-

EGS. 1 Geologic evaluation of sanitary landfill sites in Tennessee, by R. A. Miller and S. W. Maher. 1972. 4125

EGS. 2 Environmental geology summary of the Kingston Springs quadrangle, Tennessee, by R. A. Miller. 1973. 4126

EGS. 3 Landslides in the Nashville, Tennessee, area--winter 1975, by R. A. Miller and J. D. Wiethe. 1975. 4127

EGS. 4 Earthquake hazards in Tennessee, by R. G. Stearns and R. A. Miller. 1977. 4128

EGS. 5 Geologic hazards map of Tennessee, by R. A. Miller. 1978. 4129

EGS. 6 Sinkhole collapse in Montgomery County, Tennessee, by
P. R. Kemmerly. 1980.   4130

EGS. 7 Environmental geology summary of the Bellevue quadrangle, by R. A. Miller. 1980.   4131

EGS. 8 Earthquake data for Tennessee and surrounding areas, by T. R. Templeton and B. C. Spencer. 1980.   4132

Information Circular.   1953-

IC. 11 Monteagle limestone, Hartselle formation, and Bangor limestone--a new Mississippian nomenclature for use in middle Tennessee, with a history of its development, by R. G. Stearns. 1963.   4133

IC. 12 Iron, zinc, and barite deposits between Morristown and Etowah, Tennessee, by S. W. Maher. 1964.   4134

IC. 13 Investigations of miscellaneous mineral deposits in east Tennessee, by S. W. Maher and C. P. Finlayson. 1965.   4135

IC. 14 The copper-sulfuric acid industry in Tennessee, by S. W. Maher. 1966.   4136

IC. 15 Circumferential faulting around Wells Creek Basin, Houston and Stewart Counties, Tennessee-- a manuscript by J. M. Safford and W. T. Lander circa 1895, by C. W. Wilson and R. G. Stearns. 1966.   4137

IC. 16 Trace element content of some ore deposits in the southeastern states, by S. W. Maher and J. M. Fagan. 1970.   4138

IC. 17 Coal mining in Tennessee (as of November 1974), by A. R. Leamon and S. W. Maher. 1975.   4139

Report of Investigations. 1955-

RI. 19 The brown iron ores of east

Tennessee, by S. W. Maher. 1964.   4140

RI. 20 Well sample descriptions and drillers' logs, Morgan County, Tennessee, by H. B. Burwell. 1967.   4141

RI. 21 Well sample descriptions and drillers' logs, Scott County, Tennessee, by H. B. Burwell. 1967.   4142

RI. 22 The physiography of Sequatchie Valley and adjacent portions of the Cumberland Plateau, Tennessee, by R. C. Millici. 1968.   4143

RI. 23 Papers on the stratigraphy and mine geology of the Kingsport and Mascot Formations (Lower Ordovician) of east Tennessee. 1959.   4144

RI. 24 Stratigraphy of the Chickamauga super-group in its type area, by R. C. Milici and J. W. Smith. 1969.   4145

RI. 25 Ceramic evaluations of clays and shales in east Tennessee, by R. P. Hollenbeck and M. E. Tyrrell. n. d.   4146

RI. 26 Stratigraphy of the Fort Pillow test well, Lauderdale County, Tennessee, by G. K. Moore and D. L. Brown. 1969.   4147

RI. 27 Anomalous zinc in water wells, northeastern Henry County, Tennessee, by R. E. Hershey and J. M. Wilson. 1970.   4148

RI. 28 Barite resources of Tennessee, by S. W. Maher. 1970.   4149

RI. 29 Structure of the Dumplin Valley Fault Zone in east Tennessee, by R. D. Hatcher. 1970.   4150

RI. 30 Middle Ordovician stratigraphy in central Sequatchie Valley, Tennessee, by R. C. Milici. 1970.   4151

RI. 31 Preliminary investigations of heavy minerals in the McNairy sand of west Tennessee, by J. T. Wilcox. 1971.   4152

RI. 32 Stream sediment geochemical

studies in northwest Tennessee
and east Tennessee, by R. H. Carpenter, G. D. Robinson, and J. M.
Fagan. 1971.   4153

RI. 33 Carboniferous depositional
environments in the Cumberland
Plateau of southern Tennessee and
northern Alabama, by J. C. Ferm,
R. C. Milici, J. E. Eason, and
others. 1972.   4154

RI. 34 Strippable coal in the north-
ern Cumberland Plateau area of
Tennessee, by R. C. Johnson and
E. T. Luther. 1972.   4155

RI. 35 Analyses of geologic materials, Boone Lake area, Tennessee, by S. W. Maher. 1973. 4156

RI. 36 Field trips in west Tennessee, by 7 contributors. 1975.
                                4157

RI. 37 Field trips in the southern
Appalachians, by 8 contributors.
1978.   4158

RI. 38 Mississippian and Pennsylvanian section on Interstate 75,
south of Jellico, Campbell County,
Tennessee, by members of the
Sedimentation Seminar, University
of Cincinnati. 1981.   4159

RI. 39 Guide to the geology along
interstate highways in Tennessee,
by R. L. Wilson. 1981.   4160

State Park Series. 1980-

SPS. 1 Geology of Cedars of Lebanon State Park and forest and
vicinity in Wilson County, Tennessee, by C. W. Wilson. 1980.
                                4161

Geological Circular. 1965?-

GC. 65-1 Bloating characteristics of East Texas clays, by W. L. Fisher and L. E. Garner. 1965.
4162

GC. 65-2 Texas mineral resources: problems and predictions, by P. T. Flawn. 1965.
4163

GC. 65-3 A revision of Taylor nomenclature: Upper Cretaceous, central Texas, by K. P. Young. 1965.
4164

GC. 65-4 Texas minerals: trends in production, by W. L. Fisher. 1965.
4165

GC. 65-5 Geology in the State Government of Texas, by P. T. Flawn. 1965.
4166

GC. 67-1 Uranium in Texas--1967, by P. T. Flawn, 1967.
4167

GC. 67-2 Fluorspar in Brewster County, Texas, by W. N. McAnulty. 1967.
4168

GC. 67-3 History of geology at the University of Texas, by K. P. Young. 1967.
4169

GC. 67-4 Depositional systems in the Wilcox group of Texas and their relationship to occurrence of oil and gas, by W. L. Fisher and J. H. McGowen. 1967.
4170

GC. 68-1 Glen Rose cycles and facies, Paluxy River Valley, Somervell County, Texas, by J. S. Nagle. 1968.
4171

GC. 69-1 Edwards Formation (Lower Cretaceous), Texas: dolomitization in a carbonate platform

system, by W. L. Fisher and P. U. Rodda. 1969.
4172

GC. 69-2 Sulfur in west Texas: its geology and economics, by J. B. Zimmerman and E. Thomas. 1969.
4173

GC. 69-3 Virgil and Lower Wolfcamp repetitive environments and the depositional model, north-central Texas, by L. F. Brown. 1969.
4174

GC. 69-4 Geometry and distribution of fluvial and deltaic sandstones (Pennsylvanian and Permian), north-central Texas, by L. F. Brown. 1969.
4175

GC. 70-1 Mineral resources and conservation in Texas, by P. T. Flawn. 1970.
4176

GC. 70-2 Geological considerations in disposal of solid municipal wastes in Texas, by P. T. Flawn. L. J. Turk, and C. H. Leach. 1970.
4177

GC. 70-3 Effects of Hurricane Celia --a focus on environmental geologic problems of the Texas coastal zone, by J. H. McGowen, C. G. Groat, L. F. Brown, W. L. Fisher, and A. J. Scott. 1970.
4178

GC. 70-4 Depositional systems in the Jackson Group of Texas--their relationship to oil, gas, and uranium, by W. L. Fisher, C. V. Proctor, W. E. Galloway, and J. S. Nagle. 1970.
4179

GC. 71-1 Resource capability units --their utility in land-and-water-use management with examples from the Texas coastal zone, by L. F. Brown, W. L. Fisher, A. W. Erzleben, and J. H. McGowen. 1971.
4180

GC. 72-1 Mineral deposits in the

West Chinati Stock, Chinati Mountains, Presidio County, Texas, by W. N. McAnulty. 1972. 4181

GC. 72-2 Cretaceous paleogeography: implications of endemic ammonite faunas, by K. P. Young. 1972. 4182

GC. 72-3 Evaluation of sanitary landfill sites, Texas coastal zone --geologic and engineering criteria, by L. F. Brown, W. L. Fisher, and J. F. Malina. 1972. 4183

GC. 72-4 Depositional systems and oil-gas reservoirs in the Queen City Formation of Texas, by E. H. Guevara and R. Garcia. 1972. 4184

GC. 73-1 Asbestos in the Allamoore talc district, Hudspeth and Culberson Counties, by R. G. Rohrbacher. 1973. 4185

GC. 74-1 Depositional-episodes: their relationship to the Quaternary stratigraphic framework in the northwestern portion of the Gulf Basin, by D. E. Frazier. 1974. 4186

GC. 74-2 Shoreline changes on Galveston Island (Boliver roads to San Luis Pass); an analysis of historical shoreline changes of the Texas Gulf shoreline, by R. A. Morton. 1974. 4187

GC. 74-3 A numeric code for describing rocks in sedimentary basins, by E. G. Wermund and C. A. Caughey. 1974. 4188

GC. 74-4 Potential geothermal resources of Texas, by M. Dorfman and R. O. Kehle. 1974. 4189

GC. 75-1 Geothermal resources-- Frio formation, South Texas, by D. G. Bebout, M. H. Dorfman, and O. K. Agagu. 1975. 4190

GC. 75-2 Shoreline changes on Brazos Island and South Padre Island (Mansfield Channel to mouth of the Rio Grande); an analysis of historical changes of the Texas Gulf

shoreline, by R. A. Morton and M. J. Pieper. 1975. 4191

GC. 75-3 Upper Pennsylvanian limestone banks, north-central Texas, by E. G. Wermund. 1975. 4192

GC. 75-4 Shoreline changes in the vicinity of the Brazos River delta (San Luis Pass to Brown Cedar Cut); an analysis of historical changes of the Texas Gulf shoreline, by R. A. Morton and M. J. Pieper. 1975. 4193

GC. 75-5 Flood hazards along the Balcones Escarpment in central Texas; alternative approaches to their recognition, mapping, and management, by V. R. Baker. 1975. 4194

GC. 75-6 Shoreline changes between Sabine Pass and Bolivar Roads; an analysis of historical changes of the Texas Gulf shoreline, by R. A. Morton. 1975. 4195

GC. 75-7 Microrelief (Gilgai) structures on expansive clays of the Texas coastal plain--their recognition and significance in engineering construction, by T. C. Gustavson. 1975. 4196

GC. 75-8 Geothermal resources, Frio formation, Middle Texas Gulf Coast, by D. G. Bebout, O. K. Agagu, and M. H. Dorfman. 1975. 4197

GC. 75-9 Physiographic features and stratification types of coarse-grained point bars: modern and ancient examples, by J. H. McGowen and L. E. Garner. 1975. 4198

GC. 76-1 Hydrogeologic significance of depositional systems and facies in Lower Cretaceous sandstones, north-central Texas, by W. D. Hall. 1976. 4199

GC. 76-2 Land resources inventory of lignite strip-mining areas, East Texas--an application of environmental geology, by C. D. Henry. 1976. 4200

GC. 76-3 Geothermal resources-- Frio formation, upper Texas Gulf

Coast, by D. G. Bebout, R. G.
Loucks, S. C. Borsch, and M. H.
Dorfman. 1976.   4201

GC. 76-4 Shoreline changes on Matagorda Island and San Jose Island
(Pass Cavallo to Aransas Pass);
an analysis of historical changes
of the Texas Gulf shoreline, by
R. A. Morton and M. J. Pieper.
1976.   4202

GC. 76-5 Regional Tertiary cross
sections--Texas Gulf Coast, by
D. G. Bebout, P. E. Luttrell, and
J. H. Seo. 1976.   4203

GC. 76-6 Shoreline changes on
Matagorda Peninsula (Brown Cedar
Cut to Pass Cavallo); an analysis
of historical changes of the Texas
Gulf shoreline, by R. A. Morton,
M. J. Pieper, and J. H. McGowen.
1976.   4204

GC. 76-7 Geothermal resources of
the Texas Gulf Coast--environmental
concerns arising from the production and disposal of geothermal
waters, by T. C. Gustavson and
C. W. Kreitler. 1976.   4205

GC. 77-1 Shoreline changes on
Mustang Island and North Padre Island (Aransas Pass to Yarborough
Pass)--an analysis of historical
changes of the Texas Gulf shoreline, by R. A. Morton and M. J.
Pieper. 1977.   4206

GC. 77-2 Shoreline changes on
Central Padre Island (Yarborough
Pass to Mansfield Channel)--an
analysis of historical changes of
the Texas Gulf shoreline, by R. A.
Morton and M. J. Pieper. 1977.
                             4207

GC. 77-3 The Gulf shoreline of
Texas: processes, characteristics,
and factors in use, by J. H. Mc-
Gowen, L. E. Garner, and B. H.
Wilkinson. 1977.   4208

GC. 77-4 Hydrogeology of Gulf
Coast aquifers, Houston-Galveston
area, Texas, by C. W. Kreitler,
E. Guevara, G. Granata, and D.
McKalips. 1977.   4209

GC. 77-5 Relationship of porosity for-
mation and preservation to sandstone
consolidation history--Gulf Coast
Lower Tertiary Frio Formation, by
R. G. Loucks, D. G. Bebout, and W.
E. Galloway. 1977.   4210

GC. 77-6 Historical shoreline changes
and their causes, Texas Gulf coast,
by R. A. Morton. 1977.   4211

GC. 77-7 Depositional systems in the
Sparta Formation (Eocene), Gulf Coast
Basin of Texas, by J. U. Ricoy and
L. F. Brown. 1977.   4212

GC. 77-8 Depositional systems in the
Paluxy Formation (Lower Cretaceous),
northeast Texas--oil, gas, and ground-
water resources, by C. A. Caughey.
1977.   4213

GC. 78-1 Mineral lands in the City
of Dallas, by A. E. St. Clair. 1978.
                             4214

GC. 78-2 Regional distribution of
fractures in the southern Edwards
Plateau and their relationship to tectonics and caves, by E. G. Wermund,
J. C. Cepeda, and P. E. Luttrell.
1978.   4215

GC. 78-3 Electric power generation
from Texas lignite, by W. R. Kaiser.
1978.   4216

GC. 78-4 Sand-body geometry and the
occurrence of lignite in the Eocene
of Texas, by W. R. Kaiser, J. E.
Johnston, and W. N. Bach. 1978.
                             4217

GC. 78-5 Texas energy reserves and
resources, by W. L. Fisher. 1978.
                             4218

GC. 78-6 Identification of surface
faults by horizontal resistivity profiles, by C. W. Kreitler and D. G.
McKalips. 1978.   4219

GC. 79-1 Geology and geohydrology
of the Palo Duro Basin, Texas Panhandle; a report on the progress of
nuclear waste isolation feasibility
studies (1978), by S. P. Dutton, R. J.
Finley, W. E. Galloway, T. C. Gustavson, C. R. Handford, and M. W.

Presley. 1979. 4220

GC. 79-2 Geochemistry of bottom
sediments, Matagorda Bay system,
Texas, by J. H. McGowen, J. R.
Byrne, and B. H. Wilkinson. 1979.
4221

GC. 79-3 Precambrian rocks of
the southeastern Llano region,
Texas, by R. V. McGehee. 1979.
4222

GC. 79-4 Sandstone distribution
and potential for geopressured
geothermal energy production in
the Vicksburg Formation along the
Texas Gulf Coast, by R. G. Loucks.
1979. 4223

GC. 80-1 Quaternary faulting in
East Texas, by E. W. Collins, D. K.
Hobday, and C. W. Kreitler. 1980.
4224

GC. 80-2 Importance of secondary
leached porosity in Lower Tertiary
sandstone reservoirs along the
Texas Gulf Coast, by R. G. Loucks.
1980. 4225

GC. 80-3 Hydrology and water quality of the Eocene Wilcox Group:
significance for lignite development
in East Texas, by C. D. Henry,
J. M. Basciano, and T. W. Duex.
1980. 4226

GC. 80-4 The Queen City Formation in the East Texas embayment:
a depositional record of riverine,
tidal, and wave interactions, by
D. K. Hobday, R. A. Morton, and
E. D. Collins. 1980. 4227

GC. 80-5 Studies of the suitability
of salt domes in East Texas Basin
for geologic isolation of nuclear
wastes, by C. W. Kreitler. 1979.
4228

GC. 80-6 Distribution and significance of coarse biogenic and clastic
deposits on the Texas inner shelf,
by R. A. Morton and C. D. Winker.
1980. 4229

GC. 80-7 Geology and hydrology
of the Palo Duro Basin, Texas

Panhandle, a report on the progress
of nuclear waste isolation feasibility
studies (1979), by T. C. Gustavson,
M. W. Presley, C. R. Handford, R. J.
Finley, S. P. Dutton, R. W. Baumgardner, K. A. McGillis, and W. W.
Simpkins. 1980. 4230

GC. 80-8 Depositional systems and
hydrocarbon resource potential of the
Pennsylvanian system, Palo Duro and
Dalhart Basins, Texas Panhandle, by
S. P. Dutton. 1980. 4231

GC. 80-9 Facies patterns and depositional history of a Permian Sabkha
complex: Red Cave Formation, Texas
Panhandle, by C. R. Handford and P. E.
Fredericks. 1980. 4232

GC. 80-10 Petroleum source rock potential and thermal maturity, Palo
Duro Basin, Texas, by S. P. Dutton.
1980. 4233

GC. 80-11 Climatic controls on erosion in the rolling plains and along
the caprock escarpment of the Texas
Panhandle, by R. J. Finley and T. C.
Gustavson. 1980. 4234

GC. 80-12 Not yet published. 4234a

GC. 80-13 Structure of the Presidio
Bolson area, Texas, interpreted from
gravity data, by J. R. Mraz and G. R.
Keller. 1980. 4235

GC. 80-14 The Mississippian and
Pennsylvanian (Carboniferous) systems
in the United States--Texas, by R. S.
Kier, L. F. Brown, and E. F. Mc-
Bride. 1980. 4236

Guidebook. 1958-

GB. 4 The geologic story of Longhorn
Cavern, by W. H. Matthews. 1963.
4237

GB. 5 Field excursion: geology of
Llano Region and Austin area, by
V. E. Barnes, W. C. Bell, S. E. Bla-
baugh, P. E. Cloud, R. V. McGehee,
and K. P. Young. 1963. 4238

GB. 6 Texas rocks and minerals; an
amateur's guide, by R. M. Girard.

1967.   4239

GB. 7 The Big Bend of the Rio
Grande; a guide to the rocks, landscape, geologic history, and set-
tlers of the area of Big Bend National Park, by R. A. Maxwell.
1968.   4240

GB. 8 The geologic story of Palo
Duro Canyon, by W. H. Matthews.
1969.   4241

GB. 9 Field excursion, East Texas:
clay, glauconite, ironstone deposits,
by T. E. Brown, L. E. Newland,
D. H. Campbell, and A. J. Ehlmann.
1969.   4242

GB. 10 Geologic and historic guide
to the state parks of Texas, by
R. A. Maxwell, L. F. Brown, G. K.
Eifler, and L. E. Garner. 1970.
                                   4243

GB. 11 Recent sediments of southeast Texas--a field guide to the
Brazos alluvial and deltaic plains
and the Galveston Barrier Island
complex, by H. A. Bernard, C. F.
Major, B. S. Parrott, and R. J.
LeBlanc. 1970.   4244

Appendix: Resume of the Quater-
nary geology of the northwestern
Gulf of Mexico Province, by H. A.
Bernard and R. J. LeBlanc. 1970.
                                   4245

GB. 12 Uranium geology and mines,
South Texas, by D. H. Eargle, G. W.
Hinds, and A. M. D. Weeks. 1971.
                                   4246

GB. 13 Geology of the Llano Re-
gion and Austin area, by V. E.
Barnes, W. C. Bell, S. E. Clabaugh,
P. E. Cloud, R. V. McGehee, P. U.
Rodda, and K. P. Young. 1972.
                                   4247

GB. 14 Pennsylvanian depositional
systems in north-central Texas; a
guide for interpreting terrigenous
clastic facies in a cratonic basin,
by L. F. Brown, A. W. Cleaves,
and A. W. Erxleben. 1973. 4248

GB. 15 The Edwards reef complex

and associated sedimentation in Central Texas, by H. F. Nelson. 1973.
                                   4249

GB. 16 Guide to points of geologic
interest in Austin, by A. R. Trippet
and L. E. Garner. 1977.   4250

GB. 17 Padre Island National Sea-
shore--a guide to the geology, natural
environments, and history of a Texas
barrier island, by B. R. Weise and
W. A. White. 1980.   4251

GB. 18 South Texas uranium province:
geologic perspective, by W. E. Galloway, R. J. Finley, and C. D. Henry.
1979.   4252

GB. 19 Cenozoic geology of the Trans-
Pecos volcanic field of Texas, by
A. W. Walton and C. D. Henry, editors.
1979.   4253

GB. 20 Modern depositional environments of the Texas coast, by R. A.
Morton and J. H. McGowen. 1980.
                                   4254

Handbook. 1918-

H. 1 Aids to identification of geological formations, by J. A. Udden. 1918.
                                   4255

H. 2 Sulfur in Texas, by S. P. Ellison. 1971.   4256

H. 3 Fluorspar in Texas, by W. N.
McAnulty. 1974.   4257

H. 4 Bituminous coal in Texas, by
T. J. Evans. 1974.   4258

Mineral Resource Circular. 1930-

MRC. 45 The mineral industry of
Texas in 1962, by F. F. Netzeband,
T. R. Early, and R. M. Girard. 1963.
                                   4259

MRC. 46 The mineral industry of
Texas in 1963, by F. F. Netzeband
and R. M. Girard. 1964.   4260

MRC. 47 The mineral industry of
Texas in 1964, by F. F. Netzeband,

H. F. Pierce, and R. M. Girard.
1965.   4261

MRC. 48 The mineral industry of
Texas in 1965, by F. F. Netzeband
and R. M. Girard. 1966.   4262

MRC. 49 The mineral industry of
Texas in 1966, by F. F. Netzeband
and R. M. Girard. 1967.   4263

MRC. 50 The mineral industry of
Texas in 1967, by F. F. Netzeband
and R. M. Girard. 1968.   4264

MRC. 51 The mineral industry of
Texas in 1968, by F. F. Netzeband
and R. M. Girard. 1969.   4265

MRC. 52 The mineral industry of
Texas in 1969, by O. W. Jones,
F. F. Netzeband, and R. M. Girard.
1970.   4266

MRC. 53 The mineral industry of
Texas in 1970, by R. F. Zaffarano,
R. M. Girard, and E. R. Slatick.
1972.   4267

MRC. 54 The mineral industry of
Texas in 1971, by S. O. Wood and
R. M. Girard. 1973.   4268

MRC. 55 The mineral industry of
Texas in 1972, by S. O. Wood and
R. M. Girard. 1974.   4269

MRC. 56 Gold and silver in Texas,
by T. J. Evans. 1975.   4270

MRC. 57 Native bituminous materials in Texas, by T. J. Evans.
1975.   4271

MRC. 58 The mineral industry of
Texas in 1973, by C. J. Jirik and
R. M. Girard. 1976.   4272

MRC. 59 The mineral industry of
Texas in 1974, by M. E. Hawkins
and R. M. Girard. 1977.   4273

MRC. 60 The mineral industry of
Texas in 1975, by M. E. Hawkins
and T. J. Evans. 1979.   4274

MRC. 61 Coal problems and prospects, by W. C. J. van Rensburg,
H. B. H. Cooper, W. R. Kaiser, and

S. H. Spurr. 1979.   4275

MRC. 62 Coal gasification and liquefaction, by W. C. J. van Rensburg.
1979.   4276

MRC. 63 The future utilization of
Texas lignites: a review, by W. C. J.
van Rensburg. 1979.   4277

MRC. 64 Development of the mercury
mining industry: Trans-Pecos Texas,
by R. D. Sharpe. 1980.   4278

MRC. 65 The classification of coal
resources and reserves, by W. C. J.
van Rensburg. 1980.   4279

MRC. 66 The mineral industry of
Texas in 1976, by M. E. Hawkins and
T. J. Evans. 1976.   4280

Publication. 1901-

This series is a part of the consecu-
tively-numbered Publications series
of the University of Texas; those miss-
ing were not published by the Bureau
of Economic Geology. Series no
longer used by the Bureau.

P. 6304 Upper Cretaceous ammonites
from the Gulf Coast of the United
States, by K. P. Young. 1963. 4281

P. 6413 Evolution of Athleta petrosa
stock (Eocene, gastropoda) of Texas,
by W. L. Fisher, P. U. Rodda, and
J. W. Dietrich. 1964.   4282

P. 4283 Geology of the Big Bend National Park, Brewster County, Texas,
by R. A. Maxwell, J. T. Lonsdale,
R. T. Hazzard, and J. A. Wilson.
1967.   4283

Report of Investigations. 1946-

RI. 49 Pleistocene geology of Red
River Basin in Texas, by J. C. Frye
and A. B. Leonard. 1963.   4284

RI. 50 Lignites of the Texas Gulf
Coastal Plain, by W. L. Fisher.
1963.   4285

RI. 51 Relation of Ogallala Formation

to the Southern High Plains in Texas, by J. C. Frye and A. B. Leonard. 1964. 4286

RI. 52 Stratigraphy of the Fredericksburg Division, south-central Texas, by C. H. Moore. 1964. 4287

RI. 53 Potential low-grade iron ore and hydraulic-fracturing sand in Cambrian sandstones, northwestern Llano Region, Texas, by V. E. Barnes and D. A. Schofield. 1964. 4288

RI. 54 Rock and mineral resources of East Texas, by W. L. Fisher, C. R. Chelf, C. A. Shelby, L. E. Garner, D. E. Owen, and D. A. Schofield. 1965. 4289

RI. 55 Heavy minerals in the Wellborn Formation, Lee and Burleson Counties, Texas, by C. A. Shelby. 1965. 4290

RI. 56 Limestone and dolomite resources, Lower Cretaceous rocks, Texas, by P. U. Rodda, W. L. Fisher, W. R. Payne, and D. A. Schofield. 1966. 4291

RI. 57 Sedimentary petrology and history of the Haymond Formation (Pennsylvanian), Marathon Basin, Texas, by E. F. McBride. 1966. 4292

RI. 58 Nomenclature revision of basal Cretaceous rocks between the Colorado and Red Rivers, Texas, by W. L. Fisher and P. U. Rodda. 1966. 4293

RI. 59 Lower Cretaceous sands of Texas: stratigraphy and resources, by W. L. Fisher and P. U. Rodda. 1967. 4294

RI. 60 Sand resources of Texas Gulf Coast, by L. E. Garner. 1967. 4295

RI. 61 Hurricanes as geological agents: case studies of hurricanes Carla, 1961, and Cindy, 1963, by M. O. Hayes. 1967. 4296

RI. 62 Late Cambrian and Early

Ordovician faunas from Wilberns Formation of Central Texas, by D. Winston and H. Nicholls. 1967. 4297

RI. 63 Lithology and petrology of the Gueydan (Catahoula) Formation in South Texas, by E. F. McBride, W. L. Lindemann, and P. S. Freeman. 1968. 4298

RI. 64 Palynology of the Eddleman coal (Pennsylvanian) of north-central Texas, by J. F. Stone. 1969. 4299

RI. 65 Lower Cretaceous stratigraphy, northern Coahuila, Mexico, by C. I. Smith. 1970. 4300

RI. 66 Trilobites of the Upper Cambrian ptychaspid biomere, Wilberns Formation, Central Texas, by S. A. Longacre. 1970. 4301

RI. 67 Facies and genesis of a hurricane-washover fan, St. Joseph Island, Central Texas Coast, by P. B. Andrews. 1970. 4302

RI. 68 Pre-Chappel conodonts of the Llano Region, Texas, by G. Seddon. 1970. 4303

RI. 69 Gum Hollow fan delta, Nueces Bay, Texas, by J. H. McGowen. 1970. 4304

RI. 70 Correlation of Tertiary rock units, West Texas, by R. A. Maxwell and J. W. Dietrich. 1971. 4305

RI. 71 Stratigraphy of Lower Cretaceous Trinity deposits of Central Texas, by F. L. Stricklin, C. I. Smith, and F. E. Lozo. 1971. 4306

RI. 72 Van Horn sandstone, West Texas, an alluvial fan model for mineral exploration, by J. H. McGowen and C. G. Groat. 1971. 4307

RI. 73 Depositional systems in the Woodbine Formation (Upper Cretaceous), northeast Texas, by W. B. Oliver. 1971. 4308

RI. 74 Edwards Group, surface and subsurface, Central Texas, by P. R. Rose. 1972. 4309

RI. 75 Depositional systems and shelf-

slope relationships in Upper Pennsylvanian rocks, north-central Texas, by W. E. Galloway and L. F. Brown. 1972. 4310

RI. 76 Presidio Bolson, Trans-Pecos Texas, and adjacent Mexico: geology of a desert basin aquifer system, by C. G. Groat. 1972. 4311

RI. 77 Fossil vertebrates from the Late Pleistocene Ingleside fauna, San Patricio County, Texas, by E. L. Lundelius. 1972. 4312

RI. 78 Stuart City Trend, Lower Cretaceous, South Texas--a carbonate shelf-margin model for hydrocarbon exploration, by D. G. Bebout and R. G. Loucks. 1974. 4313

RI. 79 Texas lignite: near-surface and deep-basin resources, by W. R. Kaiser. 1974. 4314

RI. 80 Depositional systems, San Angelo Formation (Permian), north Texas--facies control of Red-Bed copper mineralization, by G. E. Smith. 1974. 4315

RI. 81 Approaches to environmental geology: a colloquium and workshop, by E. G. Wermund, editor. 1974. 4316

RI. 82 Depositional systems in the Canyon Group (Pennsylvanian system), north-central Texas, by A. W. Erxleben. 1975. 4317

RI. 83 Determining the source of nitrate in ground water by nitrogen isotope studies, by C. W. Kreitler. 1975. 4318

RI. 84 Land capability in the Lake Travis vicinity, Texas; a practical guide for the use of geologic and engineering data, by C. M. Woodruff. 1975. 4319

RI. 85 Lineations and faults in the Texas coastal zone, by C. W. Kreitler. 1976. 4320

RI. 86 Environmental geology of the Austin area: an aid to urban planning, by L. E. Garner and K. P. Young. 1976. 4321

RI. 87 Catahoula Formation of the Texas coastal plain--depositional systems, composition, structural development, ground-water flow history, and uranium distribution, by W. E. Galloway. 1977. 4322

RI. 88 The Moore Hollow Group of Central Texas, by V. E. Barnes and W. C. Bell. 1977. 4323

RI. 89 Cretaceous carbonates of Texas and Mexico--applications to subsurface exploration, by D. G. Debout and R. G. Loucks, editors. 1977. 4324

RI. 90 Proceedings, Gulf Coast Lignite Conference: geology, utilization, and environmental aspects, ed. by W. R. Kaiser. 1978. 4325

RI. 91 Frio sandstone reservoirs in the deep subsurface along the Texas Gulf Coast--their potential for the production of geopressured geothermal energy, by D. G. Bebout, R. G. Loucks, and A. R. Gregory. 1977. 4326

RI. 92 Land and water resources, historical changes, and dune criticality, Mustang and North Padre Islands, Texas, by W. A. White and R. A. Morton. 1978. 4327

RI. 93 Landstat analysis of the Texas Coastal Zone, by R. J. Finley. 1979. 4328

RI. 94 Depositional model for the Lower Cretaceous Washita Group, north-central Texas, by R. W. Scott. 1978. 4329

RI. 95 Land and water resources of the Corpus Christi area, Texas, by R. S. Kier and W. A. White. 1978. 4330

RI. 96 Geologic setting and geochemistry of thermal water and geothermal assessment, Trans-Pecos Texas and adjacent Mexico, with tectonic map of the Rio Grande area, Trans-Pecos Texas, by C. D. Henry. 1979. 4331

RI. 97 Depositional framework of the Lower Dockum Group (Triassic), Texas Panhandle, by J. H. McGowen, G. E. Granata, and S. J. Seni. 1979. 4332

RI. 98 Environmental geology of the Wilcox Group lignite belt, East Texas, by C. D. Henry and J. M. Basciano. 1979. 4333

RI. 99 Depositional patterns of Miocene facies, middle Texas coastal plain, by J. D. Doyle. 1979. 4334

RI. 100 Catahoula Formation of the Texas coastal plain: origin, geochemical evolution, and characteristics of uranium deposits, by W. E. Galloway and W. R. Kaiser. 1980. 4335

RI. 101 Trace elements in soils of the South Texas uranium district: concentrations, origin, and environmental significance, by C. D. Henry and R. R. Kapadia. 1980. 4336

RI. 102 Lower Permian facies of the Palo Duro Basin, Texas: depositional systems, shelf-margin evolution, paleogeography, and petroleum potential, by C. R. Handford and P. E. Fredericks. 1980. 4337

RI. 103 Molluscan distribution in Copano Bay, Texas, by T. R. Calnan. 1980. 4338

RI. 104 Lignite resources in Texas, by W. R. Kaiser, W. B. Ayers, and L. W. LaBrie. 1980. 4339

RI. 105 Sand-body geometry and depositional systems, Ogallala Formation, Texas, by S. J. Seni. 1980. 4340

Bulletin. 1948-

B. 49 Kanab, southern gateway to Utah, by H. E. Gregory. 1963.
4341

(B. 50-53 in base index)

B. 54 Oil and gas possibilities of Utah, re-evaluated, by A. L. Crawford. 1963.
4342

B. 54A Surface, structure, and stratigraphy of Utah, by A. L. Crawford. 1963.
4343

(B. 55-68 in base index)

B. 69 Geology of Salt Lake County, by A. L. Crawford. 1964.
4344

(B. 70-71 in base index)

B. 72 Geology of Uintah County, by G. E. and B. R. Untermann. 1964.
4345

B. 73 Mineral and water resources of Utah, ed. by L. S. Hilpert. 1964.
4346

B. 74 Drilling records for oil and gas in Utah, January 1, 1954-December 31, 1963, comp. by E. B. Heylmun, R. E. Cohenour, and R. B. Kayser. 1965.
4347

B. 75 Reconnaissance of the Tertiary sedimentary rocks in western Utah, by E. B. Heylmun. 1965.
4348

B. 76 Minerals of Utah, by K. C. Bullock. 1967.
4349

B. 77 Geology of the Deep Creek Mountains, Tooele and Juab Counties, Utah, by K. F. Bick. 1966.
4350

B. 78 Cordilleran miogeosyncline in Nevada and Utah, by R. L. Armstrong. 1968.
4351

B. 79 A directory of the mining industry of Utah, 1965, by R. F. Garvin and R. B. Kayser. 1965.
4352

B. 80 Central Utah coals: a guidebook prepared for the Geological Society of America and associated societies. 1966.
4353

B. 81 Radioactive and isotopic age determinations of Utah rocks, comp. by J. A. Whelan. 1970.
4354

B. 82 Guidebook of northern Utah, by M. L. Jensen. 1969.
4355

B. 83 Mineralization in the Gold Hill Mining District, Tooele County Utah, by H. M. El-Shatoury and J. A. Whelan. 1970.
4356

B. 84 A directory of the mining industry of Utah, 1967, comp. by N. S. Goeltz, N. Nielsen, and R. L. Graham. 1970.
4357

B. 85 Land and mineral resources of Sanpete County, Utah, by A. R. Pratt and E. Callaghan. 1970.
4358

B. 86 Abstracts of theses concerning the geology of Utah to 1966, comp. and ed. by B. S. Childers and B. Y. Smith. 1970.
4359

B. 87 Stratigraphy of San Rafael Group, southwest and south-central Utah, by A. E. Thompson and W. L. Stokes. 1970.
4360

B. 88 Iron deposits of Utah, by K. C. Bullock. 1970.
4361

B. 89 Palynology of the Upper Cretaceous Straight Cliffs sandstone,

Garfield County, Utah, by R. Orlansky. 1971. 4362

B. 90 Landslides of Utah, by J. F. Shroder. 1971. 4363

B. 91 Index to the Salt Lake Mining Review, 1899-1928, comp. by C. W. Warren. 1971. 4364

B. 92 Gravity base station network in Utah--1967, by K. L. Cook, T. H. Nilsen, and J. F. Lambert. 1971. 4365

B. 93 Geologic hazards in Morgan County, with applications to planning, by B. N. Kaliser. 1972. 4366

B. 94 Oil and gas production in Utah to 1970, comp. by C. H. Stowe. 1972. 4367

B. 95 Magnetic and gravity study of Desert Mountain, Juab County, Utah, by W. G. Calkins. 1972. 4368

B. 96 Environmental geology of Bear Lake area, Rich County, Utah, by B. N. Kaliser. 1972. 4369

B. 97 Stratigraphy of the Duchesne River Formation (Eocene-Oligocene?), northern Uinta Basin, northeastern Utah, by D. W. Andersen and M. D. Picard. 1972. 4370

B. 98 Analysis of gravity and aeromagnetic data, San Francisco Mountains and vicinity, southwestern Utah, by J. W. Schmoker. 1972. 4371

B. 99 Mineral deposits of the Deep Creek Mountains, Tooele and Juab Counties, Utah, by K. C. Thomson. 1973. 4372

B. 100 Petrology, geochemistry, and stratigraphy of Black Shale facies of Green River Formation (Eocene), Uinta Basin, Utah, by M. D. Picard, W. D. Thompson, and C. R. Williamson. 1973. 4373

B. 101 Utah mineral operator directory. 3rd, 1972/73, by C. H. Stowe. n. d. 4374

B. 102 Mineral resource potential of Piute County, Utah, and adjoining area, by E. Callaghan. 1973. 4375

B. 103 Bibliography of Utah geology, 1950 to 1970, by W. R. Buss and N. S. Geoltz. 1974. 4376

B. 104 Stratigraphic and depositional analysis of the Moenkopi Formation, southeastern Utah, by R. C. Blakey. 1974. 4377

B. 105 Utah's mineral activity: an operational and economic review, by C. H. Stowe. 1974. 4378

B. 106 Utah mineral industry statistics through 1973, comp. by C. H. Stowe. 1975. 4379

B. 107 Geology and mineral resources of Garfield County, Utah, by H. H. Doelling. 1975. 4380

B. 108 Utah mineral industry operator directory, 1975, comp. by C. H. Stowe. 1976. 4381

B. 109 Allosaurus Fragilis: a revised osteology, by J. H. Madsen. 1976. 4382

B. 110 Fluorite occurrences in Utah, by K. C. Bullock. 1976. 4383

B. 111 Utah mineral industry operator directory, 1977, comp. by C. H. Stowe. 1977. 4384

B. 112 Coal drilling at Trail Mountain, North Horn Mountain, and Johns Peak areas, Wasatch Plateau, Utah, by F. D. Davis and H. H. Doelling. 1977 4385

B. 113 The geology and uranium-vanadium deposits of the San Rafael River mining area, Emery County, Utah, by L. M. Trimble and H. H. Doelling. 1978. 4386

B. 114 Geology, ore deposits, and history of the Big Cottonwood mining

district, Salt Lake County, Utah,
by L. P. James. 1979.   4387

B. 115 Geology and mineral resources of Box Elder County, Utah,
by H. H. Doelling, J. A. Campbell,
J. W. Gwynn, and L. I. Perry.
1980.   4388

B. 116 Great Salt Lake; a scientific, historical, and economic over-
view, ed. by J. W. Gwynn. 1980.
                              4389

Monograph Series. 1972-

MS. 1 Southwestern Utah coal
fields: Alton, Kaiparowits Plateau
and Kolob-Harmony, by H. H. Doelling and R. L. Graham. 1972.
                              4390

MS. 2 Eastern and northern Utah
coal fields: Vernal, Henry Moun-
tains, Sego, La Sal-San Juan,
Tabby Mountain, Coalville, Henrys
Fork, Goose Creek, and Lost
Creek, by H. H. Doelling and R. L.
Graham. 1972.   4391

MS. 3 Central Utah coal fields:
Sevier-Sanpete, Wasatch Plateau,
Book Cliffs, and Emery, and H. H.
Doelling. 1972.   4392

Special Studies. 1962-

SS. 3 A reconnaissance survey of
the coal resources of southwestern
Utah, by R. A. Robison. 1963.
                              4393

SS. 4 Hydrothermal alteration in
the southeast part of the Frisco
quadrangle, Beaver County, Utah,
by B. Stringham. 1963.   4394

SS. 5 Oil seeps at Rozel Point, by
A. J. Eardley. 1963.   4395

SS. 6 Geology and hydrothermal
alteration in northwestern Black
Mountains and southern Shauntie
Hills, Beaver and Iron Counties,
Utah, by M. P. Erickson and E. J.
Dasch. 1963.   4396

SS. 7 Progress report on the coal
resources of southern Utah, 1963, by
R. A. Robison. 1964.   4397

SS. 8 Shallow oil and gas possibilities
in east and south-central Utah, by
E. B. Heylmun. 1964.   4398

SS. 9 Alteration area south of the
Horn Silver Mine, Beaver County,
Utah, by B. Stringham. 1964. 4399

SS. 10 Foundation characteristics of
sediments, Salt Lake metropolitan
area, by R. D. Bauman. 1965. 4400

SS. 11 Engineering implications and
geology, Hall of Justice excavation,
Salt Lake City, Utah, by J. C. Osmond, W. P. Hewitt, and R. Van
Horn. 1965.   4401

SS. 12 Hydrothermal alteration and
mineralization, Staats Mine and Blawn
Mountain areas, central Wah Wah
Range, Beaver County, Utah, by J. A.
Whelan. 1965.   4402

SS. 13 Concentrated subsurface brines
in the Moab region, Utah, by E. J.
Mayhew and E. B. Heylmun. 1965.
                              4403

SS. 14 Geothermal power potential in
Utah, by E. B. Heylmun. 1966. 4404

SS. 15 Review of the coal deposits of
eastern Sevier County, Utah, by R. E.
Maurer. 1966.   4405

SS. 16 Hydrothermal alteration near
the Horn Silver Mine, Beaver County,
Utah, by B. Stringham. 1967. 4406

SS. 17 Igneous complex at Wah Wah
Pass, Beaver County, Utah, by M. P.
Erickson. 1966.   4407

SS. 18 Geology and coal resources
of the tropic area, Garfield County,
Utah, by R. A. Robison. 1966. 4408

SS. 19 Bituminous sandstone deposits,
Asphalt Ridge; Uintah County, Utah,
by R. B. Kayser. 1966.   4409

SS. 20 Kaiparowits Plateau, Garfield
County, Utah, Escalante-Upper Valley

coal area, by H. H. Doelling. 1967. 4410

SS. 21 Rates of denudation as measured by bristlecone pines, Cedar Breaks, Utah, by A. J. Eardley and W. Viavant. 1967. 4411

SS. 22 Uranium deposits of Garfield County, Utah, by H. H. Doelling. 1967. 4412

SS. 23 Pyrophyllite-bearing clay in Clinton deposit, Utah County, Utah, by G. V. Henderson. 1968. 4413

SS. 24 Mineral resources, San Juan County, Utah, and adjacent areas. 4414

Pt. 1: Petroleum, potash, groundwater, and miscellaneous minerals, by H. R. Ritzma and H. H. Doelling. 1969. 4415

Pt. 2: Uranium and other metals in sedimentary host rocks, by H. H. Doelling. 1969. 4416

SS. 25 Carcass Canyon coal area, Kaiparowits Plateau, Garfield and Kane Counties, Utah, by H. H. Doelling. 1968. 4417

SS. 26 Determination of oil shale potential, Green River Formation, Uinta Basin, northeast Utah, by H. R. Ritzma and B. K. Seeley. 1969. 4418

SS. 27 Stratigraphy and economic significance, Currant Creek Formation, northwest Uinta Basin, Utah, by R. F. Garvin. 1969. 4419

SS. 28 Study of the Kennecott Copper Corporation, Great Salt Lake Authority tailings test, by H. S. Suekawa. 1970. 4420

SS. 29 Tertiary volcanic rocks of Needles Range, western Utah, by O. G. Conrad. 1969. 4421

SS. 30 Subsurface brines and soluble salts of subsurface sediments, Sevier Lake, Millard County, Utah, by J. A. Whelan. 1969. 4422

SS. 31 P. R. Spring oil-impregnated sandstone deposit, Uintah and Grand Counties, Utah, by W. D. Byrd. 1970. 4423

SS. 32 Governor's conference on geologic hazards in Utah, December 14, 1967. 1970. 4424

SS. 33 Sedimentology of oil-impregnated, Lacustrine and fluvial sandstone, P. R. Spring area, southeast Uinta Basin, Utah, by M. D. Picard and L. R. High. 1970. 4425

SS. 34 Origin of pyrophyllite-rectorite in shales of north central Utah, by G. V. Henderson. 1971. 4426

SS. 35 Clay mineralogy at the brine-sediment interface in the south arm of Great Salt Lake, Utah, by L. L. Hedburg and W. T. Parry. 1971. 4427

SS. 36 Petrographic criteria for recognition of La Custrine and fluvial sandstone, P. R. Spring oil-impregnated sandstone area, southeast Uinta Basin, Utah, by M. D. Picard. 1971. 4428

SS. 37 Instrumental analysis of tars and their correlation in oil-impregnated sandstone beds, Uintah and Grand Counties, Utah, by J. W. Gwynn. 1971. 4429

SS. 38 Engineering geology of the city and county building, Salt Lake City, Utah, by B. N. Kaliser. 1971. 4430

SS. 39 Analysis of oil extracted from oil-impregnated sandstone deposits in Utah, by R. E. Wood and H. R. Ritzma. 1972. 4431

SS. 40 Preliminary report on possible solutions to "fill effect" causing dilution of South Arm brines and concentration of North Arm brines, Great Salt Lake, Utah, by J. A. Whelan and N. Stauffer. 1972. 4432

SS. 41 A sulfur isotopic study of Uinta Basin hydrocarbons, by R. L. Mauger, R. B. Kayser, and J. W. Gwynn. 1973. 4433

SS. 42 Geology and diatremes of Desert Mountain, Utah, by D. C. Rees, M. P. Erickson, and J. A. Whelan. 1973. 4434

SS. 43 Geochemical reconnaissance at Mercur, Utah, by G. W. Lenzi. 1973. 4435

SS. 44 Lead and zinc resources in Utah, by A. H. James. 1973. 4436

SS. 45 Micropaleontology and paleoecology of the Tununk Member of the Mancos Shale, by R. H. Lessard. 1973. 4437

SS. 46 Geology and mineralogy of the Milford Flat quadrangle, Star District, Beaver County, Utah, by S. Abou-Zied and J. A. Whelan. 1973. 4438

SS. 47 Geology and mineralization of the Church Hills, Millard County, Utah, by R. L. Sayre. 1974. 4439

SS. 48 Petrology of the Morrison Formation, Dinosaur Quarry quadrangle, Utah, by S. A. Bilbey, R. L. Kerns, and J. T. Bowman. 1974. 4440

SS. 49 Coal studies. 1979. 4441

Methane content of Utah coals, by H. H. Doelling, A. D. Smith, and F. D. Davis. 4442

Sunnyside coal zone, by H. H. Doelling, A. D. Smith, F. D. Davis, and D. L. Hayhurst. 4443

Chemical analyses of coal from the Blackhawk Formation, by J. R. Hatch, R. H. Affolter, and F. D. Davis. 4444

SS. 50 Geology and petroleum resources of the major oil-impregnated sandstone deposits of Utah, by J. A. Campbell and H. R. Ritzma. 1979. 4445

SS. 51 Geology for assessment of seismic risk in the Tooele and Rush Valleys, Tooele County, Utah, by B. L. Everitt and B. N. Kaliser.

1980. 4446

SS. 52 Studies in Late Cenozoic volcanism in west-central Utah, by W. P. Nash, J. B. Peterson, and C. H. Turley. 1980. 4447

Water-Resources Bulletin. 1962-

WRB. 2 Ground-water conditions in the southern and central parts of the east shore area, Utah, 1953-1961, by R. E. Smith and J. S. Gates. 1963. 4448

WRB. 3 Pt. I: Dissolved-mineral inflow to Great Salt Lake and chemical characteristics of the Salt Lake brine: selected hydrologic data, by D. C. Hahl and C. G. Mitchell. 1963. 4449

Pt. II: Dissolved-mineral inflow to Great Salt Lake and chemical characteristics of the Salt Lake brine: technical report, by D. C. Hahl and R. H. Langford. 1964. 4450

WRB. 4 Hydrogeologic reconnaissance of part of the headwaters area of the Price River, Utah, by R. M. Cordova. 1964. 4451

WRB. 5 Reconnaissance of water resources of a part of western Kane County, Utah, by H. D. Goode. 1964. 4452

WRB. 6 Evaporation studies, Great Salt Lake. 1965. 4453

Pt. I: Evaporation and ground water, Great Salt Lake, by E. L. Peck and D. R. Dickson. 4454

Pt. II: Evaporation from the Great Salt Lake as computed from eddy flux techniques, by D. R. Dickson and C. McCullom. 4455

WRB. 7 Geology and ground-water resources of the Jordan Valley, Utah, by I. W. Marine and D. Price. 1964. 4456

WRB. 8 Second reconnaissance of water resources in western Kane County, Utah, by H. D. Goode. 1966. 4457

WRB. 9 Reconnaissance of the chemical quality of water in western Utah: Sink Valley area, drainage basins of Skull, Rush, and Government Creek Valleys and the Dugway Valley-Old River Bed areas, by K. M. Waddell. 1967. 4458

WRB. 10 Not published? 4459

WRB. 11 Reconnaissance appraisal of the water resources near Escalante, Garfield County, Utah, by H. D. Goode. 1969. 4460

WRB. 12 Great Salt Lake, Utah: chemical and physical variations of the brine, 1963-1966, by D. C. Hahl and A. H. Handy. 1969.
4461

WRB. 13 Not published? 4462

WRB. 14 Effects of a causeway on the chemistry of the brine in Great Salt Lake, Utah, by R. J. Madison. 1970. 4463

WRB. 15 Evaluation of eddy flux techniques in computing evaporation from the Great Salt Lake, by D. R. Dickson and A. E. Rickers. 1970.
4464

WRB. 16 Nonthermal springs of Utah, by J. C. Mundorff. 1971.
4465

WRB. 17 Great Salt Lake, Utah: chemical and physical variations of the brine, 1966-1972, by J. A. Whelan. 1973. 4466

WRB. 18 The effects of restricted circulation on the salt balance of Great Salt Lake, Utah, by K. M. Waddell and E. L. Bolke. 1973.
4467

WRB. 19 Hydrogeology of the Bonneville Salt Flats, Utah, by L. J. Turk. 1973. 4468

WRB. 20 Great Salt Lake, Utah: chemical and physical variations of the brine, water-year 1973, by J. A. Whelan and C. A. Peterson. 1975. 4469

WRB. 21 Model for evaluating the effects of dikes on the water and salt balance of Great Salt Lake, Utah, by K. M. Waddell and F. K. Fields. 1976. 4470

WRB. 22 Great Salt Lake, Utah: chemical and physical variations of the brine, water-years 1974 and 1975, by J. A. Whelan and C. A. Petersen. n. d. 4471

WRB. 23 Hydrogeology of Utah Lake with emphasis on Goshen Bay, by J. D. Dustin and L. B. Merritt. 1980.
4472

Bulletin. 1950-

B. 20 Geology of the Island Pond area, Vermont, by B. K. Goodwin. 1963. 4473

B. 21 Bedrock geology of the Randolph quadrangle, Vermont, by E. H. Ern. 1963. 4474

B. 22 Geology of the Lunenburg-Brunswick-Guildhall area, Vermont, by W. I. Johansson. 1963. 4475

B. 23 Geology of the Enosburg area, Vermont, by J. G. Dennis. 1964. 4476

B. 24 Geology of the Hardwick area, Vermont, by R. H. Konig and J. G. Dennis. 1964. 4477

B. 25 Stratigraphy and structure of a portion of the Castleton quadrangle, Vermont, by E-A. Zen. 1964. 4478

B. 26 Geology of the Milton quadrangle, Vermont, by S. W. Stone and J. G. Dennis. 1964. 4479

B. 27 Geology of the Vermont portion of the Averill quadrangle, by P. B. Myers. 1964. 4480

B. 28 Geology of the Burke quadrangle, Vermont, by B. G. Woodland. 1965. 4481

B. 29 Bedrock geology of the Woodstock quadrangle, Vermont, by P. H. Chang, E. H. Ern, and J. B. Thompson. 1965. 4482

B. 30 Bedrock geology of the Pawlet quadrangle, Vermont, by R. C. Shumaker and J. B. Thompson. 1967. 4483

B. 31 The surficial geology and Pleistocene history of Vermont, by D. P. Stewart and P. MacClintock. 1969. 4484

Economic Geology. 1966-

EG. 1 A report on magnetic surveys of ultramafic bodies in the Dover, Windham, and Ludlow areas, Vermont, by V. J. Murphy. 1966. 4485

EG. 2 Report on a resistivity survey of the Monkton kaolin deposit and drill hole exploration, by J. A. Wark. 1968. 4486

EG. 3 Geology and origin of the kaolin at East Monkton, Vermont, by D. G. Ogden. 1969. 4487

EG. 4 Report on the Cuttingsville pyrrhotite deposit, Cuttingsville, Vermont, by C. G. Doll. 1969. 4488

EG. 5 The geology of the Elizabeth Mine, Vermont, by P. F. Howard. 1969. 4489

EG. 6 Magnetic surveys of ultramafic bodies in the vicinity of Lowell, Vermont, by V. J. Murphy and A. V. Lacroix. 1969. 4490

EG. 7 Geochemical investigations in Essex and Caledonia Counties, Vermont, by R. W. Grant. 1970. 4491

EG. 8 Geochemical investigation of the Pomfret Dome, Vermont, by J. E. Thresher. 1972. 4492

Environmental Geology. 1971-

ENG. 1 Geology for environmental planning in the Barre-Montpelier re-

gion, Vermont, by D. P. Stewart.
1971.   4493

ENG. 2   Geology for environmental
planning in the Rutland-Brandon region, Vermont, by D. P. Stewart.
1972.   4494

ENG. 3   Geology for environmental
planning in the Burlington-Middle-
bury region, Vermont, by D. P.
Stewart.   1973.   4495

ENG. 4   Geology for environmental
planning in the Johnson-Hardwick
region, Vermont, by F. N. Wright.
1974.   4496

ENG. 5   Geology for environmental
planning in the Milton-St. Albans
region, Vermont, by D. P. Stewart.
1974.   4497

ENG. 6   Not published?   4497a

ENG. 7   Geology for environmental
planning in the Brattleboro-Windsor
region, Vermont, by D. P. Stewart.
1975.   4498

Special Publication.   1968-

SP. 2   Mineral collecting in Vermont, by R. W. Grant.   1968.
                                         4499

Studies in Vermont Geology.   1970-

SVG. 1   The morphometry and re-
cent sedimentation of Joe's Pond,
West Danville, Vermont, by J. S.
Moore and A. S. Hunt.   1970. 4500

SVG. 2   Surficial geology of the
Brandon-Ticonderoga 15 minute
quadrangles, Vermont, by G. G.
Connally.   1970.   4501

Bulletin. 1905-

B. 78 Geology and mineral resources of Greene and Madison Counties, by R. M. Allen. 1963.
4502

B. 79 Geology and mineral resources of Fluvanna County, by J. W. Smith, R. C. Millici, and S. S. Greenberg. 1964.
4503

B. 80 Geology and mineral resources of Frederick County, by C. Butts and R. S. Edmundson. 1966.
4504

B. 81 Geology and mineral resources of Page County, by R. M. Allen. 1967.
4505

B. 82 Post-Miocene stratigraphy and morphology, southeastern Virginia, by R. Q. Oaks and N. K. Coch. 1973.
4506

B. 83 Geologic studies, coastal plain of Virginia. 1973.
4507

Pt. 1: Stratigraphic units of the Lower Cretaceous through Miocene series, by R. H. Teifke.
4508

Pt. 2: Paleogeology of Early Cretaceous through Miocene time, by R. H. Tiefke.
4509

Pt. 3: Pleistocene-Holocene environmental geology, by E. Onuschak.
4510

B. 84 Stratigraphy and coal beds of Upper Mississippian and Lower Pennsylvanian rocks in southwestern Virginia, by M. S. Miller. 1974.
4511

B. 85 Descriptions of Virginia caves, by J. R. Holsinger. 1975.
4512

B. 86 Geology of the Shenandoah National Park, Virginia, by T. M. Gathright. 1976.
4513

Information Circular. 1959-

IC. 7 Guide to fossil collecting in Virginia, by E. K. Rader. 1964.
4514

IC. 8 Directory of the mineral industry in Virginia, by D. C. Le Van and R. F. Pharr. 1964.
4515

IC. 9 Geologic literature of the coastal plain of Virginia, 1783-1962, by J. L. Ruhle. 1965.
4516

IC. 10 Characteristics of the Everona Formation in Virginia, by T. Mack. 1965.
4517

IC. 11 Directory of the mineral industry in Virginia--1966, by D. C. Le Van and R. F. Pharr. 1966. 4518

IC. 12 Magnetic and radiometric data, southwest Piedmont of Virginia. 1966.
4519

IC. 13 Directory of the mineral industry in Virginia--1967, by D. C. Le Van. 1967.
4520

IC. 14 Bibliography of Virginia geology and mineral resources--1941-1949, by F. B. Hoffer. 1968.
4521

IC. 15 A computer-program system to grid and contour random data, by S. S. Johnson, C. L. Huxsaw, and D. R. Thomas. 1971.
4522

IC. 16 Field trip to the igneous rocks of Augusta, Rockingham, Highland, and Bath Counties, Virginia, by R. W. Johnson, C. Milton, and J. M. Dennison. 1971.
4523

IC. 17 Virginia gravity base net, by S. S. Johnson. 1972. 4524

IC. 18 Bibliography of published measured sections west of the Blue Ridge in Virginia, by H. W. Webb and W. E. Nunan. 1972. 4525

IC. 19 Bibliography of Virginia geology and mineral resources-- 1950-1959, by F. B. Hoffer. 1972. 4526

IC. 20 Geographic and cultural names in Virginia, by T. H. Biggs. 1974. 4527

Mineral Resources Report. 1960-

MRR. 5 Analyses of clay, shale, and related materials--west-central counties, by J. L. Calver, C. E. Smith, and D. C. Le Van. 1964. 4528

MRR. 6 Analyses of clay, shale, and related materials--southwestern counties, by S. S. Johnson, M. V. Denny, and D. C. Le Van. 1966. 4529

MRR. 7 Base- and precious-metal and related ore deposits of Virginia, by G. W. Luttrell. 1966. 4530

MRR. 8 Analyses of clay and related materials--eastern counties, by S. S. Johnson and M. E. Tyrrell. 1967. 4531

MRR. 9 Ground-water resources of Accomack and Northampton Counties, Virginia, by A. Sinnott and G. C. Tibbitts. 1968. 4532

MRR. 10 Development of ground-water supplies in Shenandoah National Park, Virginia, by R. H. De-Kay. 1972. 4533

MRR. 11 High-silica resources of Clarke, Frederick, Page, Rocking-ham, Shenandoah, and Warren Counties, Virginia, by W. B. Harris. 1972. 4534

MRR. 12 Analyses of clay, shale, and related materials--southern

counties, by P. C. Sweet. 1973. 4535

MRR. 13 Clay-material resources in Virginia, by P. C. Sweet. 1976. 4536

Publication. 1977-

P. 1 Bibliography of Virginia geology and mineral resources, 1960-1969, by F. B. Hoffer. 1977. 4537

P. 2 Geology of the Blairs, Mount Hermon, Danville, and Ringgold quadrangles, Virginia, by W. S. Henika. 1977. 4538

Triassic system, by P. A. Thayer. 1977. 4539

P. 3 Geology of the Waynesboro east and Waynesboro west quadrangles, Virginia, by T. M. Gathright, W. S. Henika, and J. L. Sullivan. 1977. 4540

P. 4 Geology of the Greenfield and Sherando quadrangles, Virginia, by M. J. Bartholomew. 1977. 4541

P. 5 Geology of the Omega, South Boston, Cluster Springs, and Virgilina quadrangles, by R. D. Kreisa. 1980. 4542

P. 6 Bouguer gravity in southwestern Virginia, by S. S. Johnson. 1977. 4543

P. 7 Contributions to Virginia geology --III. 1978. 4544

P. 8 Geology of the Norfolk North quadrangle, Virginia, by W. J. Barker and E. D. Bjorken. 1978. 4545

P. 9 Geology of the Norfolk South quadrangle, Virginia, by W. J. Barker and E. D. Bjorken. 1978. 4546

P. 10 Geology of the Grottoes quadrangle, Virginia, by T. M. Gathright, W. S. Henika, and J. L. Sullivan. 1978. 4547

P. 11 Geology of the Mount Sidney quadrangle, Virginia, by T. M. Gathright, W. S. Henika, and J. L. Sullivan. 1978. 4548

P. 12 Geology of the Fort Defiance quadrangle, Virginia, by T. M. Gathright, W. S. Henika, and J. L. Sullivan. 1978. 4549

P. 13 Geology of the Crimora quadrangle, Virginia, by T. M. Gathright, W. S. Henika, and J. L. Sullivan. 1978. 4550

P. 14 Geology of the Blacksburg quadrangle, Virginia, by M. J. Bartholomew and W. D. Lowry. 1979. 4551

P. 15 Geologic factors affecting land modification, Warren County, Virginia, by E. K. Rader and H. W. Webb. 1979. 4552

P. 16 Geology of the Abingdon, Wyndale, Holston Valley, and Shady Valley quadrangles, by C. S. Bartlett and T. H. Biggs. 1980. 4553

P. 17 Geology of the Massies Corner quadrangle, Virginia, by M. T. Lukert and C. R. Halladay. 1980. 4554

P. 18 Geology of the Bon Air quadrangle, Virginia, by B. K. Goodwin. 1980. 4555

P. 19 Gold in Virginia, by P. C. Sweet. 1980. 4556

P. 20 Geology of the Oak Grove Core. 1980. 4557

Report of Investigations. 1960-

RI. 4 A magnetic survey of the coastal plain in Virginia, by D. C. Le Van and R. F. Pharr. 1963. 4558

RI. 5 Geology of the Clinchport quadrangle, Virginia, by W. B. Brent. 1963. 4559

RI. 6 Triassic formations of the Danville Basin, by C. T. Meyertons. 1963. 4560

RI. 7 Geology of the Vesuvius quadrangle, Virginia, by H. J. Werner. 1966. 4561

RI. 8 Geology of the Millboro quadrangle, Virginia, by S. J. Kozak. 1965. 4562

RI. 9 Miocene and Pleistocene foraminifera and ostracoda of southeastern Virginia, by J. D. McLean. 1966. 4563

RI. 10 Geology of the Dillwyn quadrangle, Virginia, by W. R. Brown. 1969. 4564

RI. 11 Geology of the Leesburg quadrangle, Virginia, by E. C. Toewe. 1966. 4565

RI. 12 Geology of the Staunton, Churchville, Greenville, and Stuarts Draft quadrangles, Virginia, by E. K. Rader. 1967. 4566

RI. 13 Geology of the Natural Bridge, Sugarloaf Mountain, Buchanan, and Arnold Valley quadrangles, Virginia, by E. W. Spencer. 1968. 4567

RI. 14 Geologic investigations of the Lincoln and Bluemont quadrangles, Virginia, by P. E. Parker. 1968. 4568

RI. 15 Geology of the Buckingham quadrangle, Virginia, by E. H. Ern. 1968. 4569

RI. 16 Geology of the Martinsville West quadrangle, Virginia, by J. F. Conley and E. C. Toewe. 1968. 4570

RI. 17 Geology of the Benns Church, Smithfield, Windsor, and Chuckatuck quadrangles, Virginia, by N. K. Coch. 1968. 4571

RI. 18 Geology of the Williamsburg, Hog Island, and Bacons Castle quadrangles, Virginia, by K. F. Bick and N. K. Coch. 1969. 4572

RI. 19 Geology of the Stokesville and Parnassus quadrangles, Virginia, by E. K. Rader. 1969. 4573

RI. 20 Magnetic and gravity surveys of Albemarle and Fluvanna Counties, Virginia, by S. S. Johnson and P. C. Sweet. 1969. 4574

RI. 21 Geology of the Elliott Knob, Deerfield, Craigsville, and Augusta Springs quadrangles, Virginia, by S. J. Kozak. 1970. 4575

RI. 22 Geology of the Philpott Reservoir quadrangle, Virginia, by J. F. Conley and W. S. Henika. 1970. 4576

RI. 23 Geology of the Hylas and Midlothian quadrangles, Virginia, by B. K. Goodwin. 1970. 4577

RI. 24 Geology of the Eagle Rock, Strom, Oriskany, and Salisbury quadrangles, Virginia, by O. S. McGuire. 1970. 4578

RI. 25 Geology of the Bristol and Wallace quadrangles, Virginia, by C. S. Bartlett and H. W. Webb. 1971. 4579

RI. 26 Geology of the Bassett quadrangle, Virginia, by W. S. Henika. 1971. 4580

RI. 27 Bouguer gravity in Virginia, 36°30' to 39°30'N., 78°00' to 79°00'W., by S. S. Johnson. 1971. 4581

RI. 28 Geology of the Newport News South and Bowers Hill quadrangles, Virginia, by N. K. Coch. 1971. 4582

RI. 29 Bouguer gravity in Virginia, 36°30' to 38°52'30"N., 79°00' to 80°00'W., by S. S. Johnson. 1972. 4583

RI. 30 Geology of the Yorktown, Poquoson West, and Poquoson East quadrangles, Virginia, by G. H. Johnson. 1972. 4584

RI. 31 Geochemical reconnaissance for zinc, lead, and copper in the Staunton quadrangle, Virginia, by R. S. Good and G. C. Allen. 1972. 4585

RI. 32 Bouguer gravity, northeastern Virginia and the Eastern Shore Peninsula, by S. S. Johnson. 1973. 4586

RI. 33 Geology of the Snow Creek, Martinsville East, Price, and Spray quadrangles, Virginia, by J. F. Conley and W. S. Henika. 1973. 4587

RI. 34 Geology of the Berryville, Stephenson, and Boyce quadrangles, Virginia, by R. S. Edmundson and W. E. Nunan. 1973. 4588

RI. 35 Geology of the Woodstock, Wolf Gap, Conicville, and Edinburg quadrangles, Virginia, by R. S. Young and E. K. Rader. 1974. 4589

RI. 36 Geology of the Ashby Gap quadrangle, Virginia, by T. M. Gathright and P. G. Nystrom. 1974. 4590

RI. 37 Geology of the Salem quadrangle, Virginia, by R. V. Amato. 1974. 4591

RI. 38 Geology of the Studley, Yellow Tavern, Richmond, and Seven Pines quadrangles, Virginia, by P. A. Daniels and E. Onuschak. 1974. 4592

RI. 39 Bouguer gravity in southeastern Virginia, by S. S. Johnson. 1975. 4593

RI. 40 Geology of the Front Royal quadrangle, Virginia, by E. K. Rader, T. H. Biggs, and R. H. Dekay. 1975. 4594

RI. 41 Geology of the Mulberry Island, Newport News North, and Hampton quadrangles, Virginia, by G. H. Johnson. 1976. 4595

RI. 42 Geology of the Daleville quadrangle, Virginia. 1976. 4596

Economic geology, by J. M. Wunder. 4597

Road log, by E. K. Rader and J. M. Wunder. 4598

RI. 43 Bouguer gravity in Virginia, 36°30' to 38°00'N., 80°00'W., by S. S. Johnson. 1976. 4599

RI. 44 Geology of the Linden and Flint Hill quadrangles, Virginia,

by M. T. Lukert and E. B. Nuckols. 1976. 4600

Hydrogeology, by R. H. DeKay. 1976. 4601

RI. 45 Geology of the Strasburg and Toms Brook quadrangles, Virginia, by E. K. Rader and T. H. Biggs. 1976. 4602

Bulletin. 1910-

B. 51 Barite in Washington, by
W. S. Moen. 1964.   4603

B. 52 Limestone resources of
western Washington, by W. R. Danner, 1966.   4604

B. 53 Stratigraphy and foraminifera
of the Satsop River area, southern
Olympic Peninsula, Washington, by
W. W. Rau. 1966.   4605

B. 54 Geology and mineral resources of the Kelso-Cathlamet
area, Cowlitz and Wahkiakum Counties, Washington, by V. E. Livingston. 1966.   4606

B. 55 Building stone of Washington,
by W. S. Moen. 1967.   4607

B. 56 Geology of the Wynoochee
Valley quadrangle, Grays Harbor
County, Washington, by W. W. Rau.
1967.   4608

B. 57 Mines and mineral deposits
of Whatcom County, Washington,
by W. S. Moen. 1969.   4609

B. 58 Chemical and physical controls for base metal deposition in
the Cascade Range of Washington,
by A. R. Grant. 1969.   4610

B. 59 Bibliography and index of
the geology and mineral resources
of Washington, 1957-1962, by W. H.
Reichert. 1969.   4611

B. 60 Cenozoic volcanism in the
Cascade Mountains of southern
Washington, by W. S. Wise. 1970.
                                     4612

B. 61 Lead-zinc deposits in the
Kootenay arc, northeastern Wash-

ington and adjacent British Columbia,
by A. E. Weissenborn, F. C. Armstrong,
and J. T. Fyles. 1970.   4613

B. 62 Foraminifera, stratigraphy, and
paleoecology of the Quinault formation,
Point Grenville-Raft River coastal
area, Washington, by W. W. Rau.
1970.   4614

B. 63 Geology and mineral resources
of King County, Washington, by V. E.
Livingston. 1971.   4615

B. 64 Geology and mineral deposits
of the Loomis quadrangle, Okanogan
County, Washington, by C. D. Rinehart and K. F. Fox. 1972.   4616

B. 65 Distribution of copper and
other metals in gully sediments of
part of Okanogan County, Washington,
by K. F. Fox and C. D. Rinehart.
1972.   4617

B. 66 Geology of the Washington coast
between Point Grenville and the Hoh
River, by W. W. Rau. 1973.   4618

B. 67 Mining laws of the State of
Washington, by J. L. Neff and R. L.
Magnuson. 1974.   4619

B. 68 Geology of the Methow Valley,
Okanogan County, Washington, by
J. D. Barksdale. 1975.   4620

B. 69 Silver occurrences of Washington, by W. S. Moen. 1976.   4621

B. 70 Zinc and lead ore deposits in
carbonate rocks, Stevens County,
Washington, by J. W. Mills. 1977.
                                     4622

B. 71 Geology of parts of Grant,
Adams, and Franklin Counties, east-
central Washington, by M. J. Grolier
and J. W. Bingham. 1978.   4623

B. 72 Washington coastal geology between the Hoh and Quillayute Rivers, by W. W. Rau. 1980. 4624

Information Circular. 1938-

IC. 37 1962 directory of Washington mining operations, by G. W. Thorsen. 1963. 4625

IC. 38 A geologic trip along Snoqualmie, Swauk, and Stevens Pass highways, by University of Washington Geology Department staff, rev. by V. E. Livingston. 1963. 4626

IC. 39 Marketing of metallic and non-metallic minerals, by D. L. Anderson. 1963. 4627

IC. 40 Caves of Washington, by W. R. Halliday. 1963. 4628

IC. 41 Origin of Cascade landscapes, by J. H. Mackin and A. S. Cary. 1965. 4629

IC. 42 1964 directory of Washington mining operations by W. S. Moen and G. W. Thorsen. 1965. 4630

IC. 43 1965-1966 directory of Washington mining operations, by W. S. Moen. 1967. 4631

IC. 44 1967-68 directory of Washington mining operations, by W. S. Moen. 1969. 4632

IC. 45 Geologic history and rocks and minerals of Washington, by V. E. Livingston. 1969. 4633

IC. 46 1969-70 directory of Washington mining operations. 1971. 4634

IC. 47 Geology in land use planning --some guidelines for the Puget Lowland, by E. R. Artim. 1973. 4635

IC. 48 1971-72 directory of Washington mining operations, by J. E. Schuster. 1973. 4636

IC. 49 Conconully Mining District of Okanogan County, Washington, by W. S. Moen. 1973. 4637

IC. 50 Energy resources of Washington. 1974. 4638

Geothermal energy potential of Washington, by J. E. Schuster. 4639

Terrestrial heat flow and its implications on the location of geothermal reservoirs in Washington, by D. D. Blackwell. 4640

Coal in Washington, by V. E. Livingston. 4641

Oil and gas in Washington, by W. W. Rau and H. C. Wagner. 4642

Uranium in Washington, by A. E. Weissenborn and W. S. Moen. 4643

Electrical energy resources of Washington, by L. C. Buchanan. 4644

IC. 51 Piercement structure outcrops along the Washington coast, by W. W. Rau and G. R. Grocock. 1974. 4645

IC. 52 Landslides in Seattle, by D. W. Tubbs. 1974. 4646

IC. 53 Compilation of earthquake hypocenters in western Washington, July 1970-Dec. 1972, by R. S. Crosson. 1975. 4647

IC. 54 A geologic road log over Chinook, White Pass, and Ellensburg to Yakima Highways, by N. P. Campbell. 1975. 4648

IC. 55 Compilation of earthquake hypocenters in western Washington-- 1973, by R. S. Crosson. 1975. 4649

IC. 56 Compilation of earthquake hypocenters in western Washington--1974, by R. S. Crosson and R. C. Millard. 1975. 4650

IC. 57 Handbook for gold prospectors in Washington, by W. S. Moen and M. T. Huntting. 1975. 4651

IC. 58 Engineering geologic studies. 1976. 4652

Soil--what is it? by K. Othberg.
4653

The role of ground water in slope
stability, by W. D. Paterson. 4654

Potential land use problems of
Puget Sound shore bluffs, by D. W.
Mintz, R. S. Babcock, and T. A.
Terich. 4655

Seismic risk, by E. R. Artim.
4656

IC. 59 Washington gravity base
station network, by T. Nilsen.
1976. 4657

IC. 60 St. Helens and Washougal
Mining Districts of the southern
Cascades of Washington, by W. S.
Moen. 1977. 4658

IC. 61 Annotated guide to sources
of information on the geology, minerals, and ground-water resources
of the Puget Sound region, Washington, King County section, by
W. H. Reichert and D. D. Dethier.
1978. 4659

IC. 62 Heat flow studies in the
Steamboat Mountain-Lemei Rock
area, Skamania County, Washington,
by J. E. Schuster, D. D. Blackwell,
P. E. Hammond, and M. T. Huntting.
1978. 4660

IC. 63 Directory of Washington
mining operations, 1977, by P. C.
Milne and C. W. Walker. 1978.
4661

IC. 64 Compilation of earthquake
hypocenters in western Washington
--1975, by R. S. Crosson and L.
Noson. 1979. 4662

IC. 65 Compilation of earthquake
hypocenters in western Washington
--1976, by R. S. Crosson and L.
Noson. 1979. 4663

IC. 66 Compilation of earthquake
hypocenters in western Washington
--1977, by R. S. Crosson and L. J.
Noson. 1979. 4664

IC. 67 Oil and gas exploration in

Washington, 1900-1978, by C. R.
McFarland. 1979. 4665

IC. 68 Index to published geologic
mapping in Washington, 1854-1970,
by W. H. Reichert. 1979. 4666

IC. 69 Directory of Washington mining
operations, 1979, by C. R. McFarland
and others. 1979. 4667

IC. 70 Theses on Washington geology
--a comprehensive bibliography, 1901-
1979, by C. Manson. 1980. 4668

IC. 71 The 1980 eruption of Mount
St. Helens, Washington. Pt. 1: March
20-May 19, 1980, by M. A. Korosec,
J. G. Rigby, and K. L. Stoffel. 1980.
4669

Report of Investigations. 1926-1964

RI. 22 Tertiary geologic history of
western Oregon and Washington, by
P. D. Snavely and H. C. Wagner. 1963.
4670

RI. 23 Mineralogy of black sands at
Grays Harbor, Washington, by G. W.
Thorsen. 1964. 4671

Bulletin. 1901-

B. 24 Structure of Devonian strata along Allegheny front, by J. M. Dennison and O. D. Naegele. 1963. 4672

B. 25 Sulphate minerals in West Virginia, by J. H. C. Martens. 1963. 4673

B. 26 West Virginia's oil and gas --lubricants and fuels, by O. L. Haught. 1964. 4674

B. 27 Occurrence and availability of ground-water in Ohio County, West Virginia, by T. M. Robison. 1964. 4675

B. 28 Appalachian connate water, by E. T. Heck, C. E. Hare, and H. A. Hoskins. 1964. 4676

B. 29 Oil and gas report and map of Braxton and Clay Counties, West Virginia, by O. L. Haught and W. K. Overbey. 1964. 4677

B. 30 Lithification of sandstones in West Virginia, by M. T. Heald. 1965. 4678

B. 31 Oil and gas report and map of Barbour and Upshur Counties, West Virginia, by O. L. Haught. 1965. 4679

B. 32 Ground water in Mason and Putnam Counties, West Virginia, by B. M. Wilmoth. 1966. 4680

B. 33 Oil and gas report and map on Ohio, Brooke, and Hancock Counties, West Virginia, by O. L. Haught. 1968. 4681

B. 34 Geology of the Charleston

area, by O. L. Haught. 1968. 4682

B. 35 The Newburg of West Virginia, by D. H. Cardwell. 1971. 4683

B. 36 Hydrology of limestone karst, by W. K. Jones. 1973. 4684

Circular. 1965-

C. 1 Some low-alumina quartzitic sandstones in West Virginia, by P-F Chen, R. G. Hunter, and R. B. Erwin. 1965. 4685

C. 2 A simple technique for the determination of the weight per cent of calcite and dolomite in carbonate rocks, by J. J. Renton and R. G. Hunter. 1965. 4686

C. 3 Geology of oil and gas, by O. L. Haught. 1965. 4687

C. 4 Preliminary palynological and mineralogical analyses of a Lake Monongahela (Pleistocene) terrace deposit at Morgantown, West Virginia, by J. A. Clendening, J. J. Renton, and B. M. Parsons. 1967. 4688

C. 5 A pressure chamber for the impregnation of porous rock specimens, by J. J. Renton. 1967. 4689

C. 6 Newburg gas development in West Virginia, by D. G. Patchen. 1967. 4690

C. 7 Keefer sandstone gas development and potential in West Virginia, by D. G. Patchen. 1968. 4691

C. 8 A summary of Tuscarora sandstone (Clinton sand) and Pre-Silurian test wells in West Virginia, by D. G. Patchen. 1968. 4692

C. 9 Oriskany sandstone--Huntersville chert gas production in the eastern half of West Virginia, by D. G. Patchen. 1968. 4693

C. 10 Proceedings of the 19th annual highway geology symposium, May 16 and 17, 1968, ed. by R. B. Erwin. 1968. 4694

C. 10A Field trip guide for the 19th annual highway geology symposium. 1968. 4695

C. 11 Salty ground water in the Pocatalico River Basin, by G. L. Bain. 1970. 4696

C. 12 The use of pelletized samples for x-ray diffraction analysis of clay minerals in shales, by R. V. Hidalgo and J. J. Renton. 1970. 4697

C. 13 The Trenton Group of Nittany Anticlinorium, eastern West Virginia, by W. J. Perry. 1972. 4698

C. 14 Petrology of the Middle Silurian McKenzie Formation, Wayne County, West Virginia, by R. A. Smosna. 1974. 4699

C. 15 Computer methods for petroleum geologists, program and abstracts of the tenth annual Appalachian petroleum geology symposium. 1979. 4700

C. 16 The eleventh annual Appalachian petroleum geology symposium, "Current research and exploration in the Appalachian Basin." Program and abstracts, March 31-April 2, 1980, at Morgantown, West Virginia. 1980. 4701

Coal-Geology Bulletin. 1973-

CGB. 1 Suitability of West Virginia coals to coal-conversion processes, by S. P. Babu, J. A. Barlow, L. L. Craddock, R. V. Hidalgo, and E. Friel. 1973. 4702

CGB. 2 Coal and coal mining in West Virginia, by J. A. Barlow.

1974. 4703

CGB. 3 Palynological evidence for a Pennsylvanian age assignment of the Dunkard Group in the Appalachian Basin: Part II, by J. A. Clendening. 1974. (Part I will appear as a paper in the proceedings of the first I. C. White memorial symposium, The Age of the Dunkard.) 4704

CGB. 4 Some geochemical considerations of coal, by J. J. Renton and R. V. Hidalgo. 1975. 4705

CGB. 5 Coal analyses of McDowell County, West Virginia, by C. J. Smith, S. Vinton, G. Ahnell, and B. Blake. 1977. 4706

CGB. 6 Proceedings of the coal processing and conversion symposium, June 1-3, 1976, comp. by C. J. Smith and D. G. Nichols. 1979. 4707

Educational Series. 1960-

ES. 3 Plant fossils of West Virginia, by W. H. Gillespie, I. S. Latimer, and J. A. Clendening. 1966. 4708

ES. 3A Plant fossils of West Virginia, by W. H. Gillespie, J. A. Clendening, and H. W. Pfefferkorn, Rev. ed. 1978. 4709

ES. 4 History and bibliography of West Virginia paleobotany, by W. H. Gillespie and I. S. Latimer. 1961. 4710

ES. 5 Oil and gas in West Virginia, by O. L. Haught. 1964. 4711

ES. 6 Synopsis of drilling in West Virginia, by O. L. Haught. 1964. 4712

ES. 7 Common minerals and rocks of West Virginia. 5th ed. 1966. 4713

ES. 8 Minerals of West Virginia, by J. H. C. Martens. 1964. 4714

ES. 9 Coal and coal mining in West Virginia, by O. L. Haught. Rev. ed. 1964. 4715

ES. 10 Geologic history of West
Virginia, by D. H. Cardwell. 1975.
4716

ES. 11 A practical handbook for
individual water-supply systems
in West Virginia, by R. A. Landers.
1976.   4717

ES. 12 Coal blooms: description of
sulfate mineral efflorescences associated with coal, by E. B. Nuhfer.
1976.   4718

Environmental Geology Bulletin.
197?-

EGB. 1 Geological considerations
of sanitary landfill site evaluations,
by P. Lessing and R. S. Reppert.
3rd ed. 1973.   4719

EGB. 2 Sanitary landfill sites in
the eastern panhandle, by R. S.
Reppert and P. Lessing. 1971.
4720

EGB. 3 Sanitary landfill sites in
south-eastern West Virginia, by P.
Lessing and R. S. Reppert. 1971.
4721

EGB. 4 Sanitary landfill sites in
south-western West Virginia, by
R. S. Reppert and P. Lessing.
1971.   4722

EGB. 5 Sanitary landfill sites in
central West Virginia, by P. Lessing and R. S. Reppert. 1972. 4723

EGB. 6 Sanitary landfill sites in
north-western West Virginia, by
R. S. Reppert and P. Lessing.
1972.   4724

EGB. 7 Sanitary landfill sites in
northern West Virginia, by P.
Lessing and R. S. Reppert. 1972.
4725

EGB. 8 Bibliography of environmental geology in West Virginia,
by R. A. Landers and P. Lessing.
1973.   4726

EGB. 9 The waste of our fuel resources, by I. C. White. 1972.
4727

EGB. 10 Geology underlies it all, by
R. B. Erwin. n. d.   4728

EGB. 11 Relative acid-producing potential of coal, by J. J. Renton, R. V.
Hidalgo, and D. L. Streib. 1973.
4729

EGB. 12 Earthquake history of West
Virginia, by P. Lessing. 1974. 4730

EGB. 13 Ground-water hydrology of
Berkeley County, West Virginia, by
W. A. Hobba. 1976.   4731

EGB. 14 Aerial and satellite imagery
of West Virginia, by S. M. Woodring.
1977.   4732

EGB. 15 West Virginia landslides and
slide-prone areas, by P. Lessing,
B. R. Kulander, B. D. Wilson, S. L.
Dean, and S. M. Woodring. 1976.
4733

EGB. 16 Not yet published. 4734

EGB. 17 Karst subsidence and linear
features, Greenbrier and Monroe
Counties, West Virginia, by P. Lessing, S. L. Dean, B. R. Kulander, and
J. H. Reynolds. 1979.   4735

EGB. 18 Land use statistics for West
Virginia. Pt. 1. 1979.   4736

Pt. 2. 1980.   4737

Report of Investigations. 1947-

RI. 22 Cone-in-cone in coal, by P. H.
Price and B. M. Shaub. 1963. 4738

RI. 23 Stratigraphy and petrography
of the Williamsport sandstone, West
Virginia, by D. G. Patchen. 1974.
4739

RI. 24 Determination of calcite/dolo-
mite ratios by infrared spectroscopy
in the 750 to 200 cm^{-1} region and
comparison with x-ray diffraction
analysis, by J. J. Renton and J. Kovach. 1974.   4740

RI. 25 Stratigraphy and petrology of
Middle Silurian McKenzie Formation in
West Virginia, by D. G. Patchen and

R. A. Smosna. 1976.   4741

RI. 26-1 Lower Paleozoic stratigraphy, tectonics, paleogeography,
and oil/gas possibilities in the
Central Appalachians (West Virginia
and adjacent states). Part I:
Stratigraphic maps, by P-F Chen.
1977.   4742

RI. 27 Gravity, magnetics, and
structure: Allegheny Plateau/West-
ern Valley and Ridge in West Virginia and adjacent states, by B. R.
Kulander and S. L. Dean. 1978.
    4743

RI. 28 Relationships between depositional environments, Tonoloway
limestone, and distribution of evaporites in the Salina Formation,
West Virginia, by R. A. Smosna,
D. G. Patchen, S. M. Warshauer,
and W. J. Perry. 1978.   4744

RI. 29 Silurian evolution of the
Central Appalachian Basin, by R. A.
Smosna and D. G. Patchen. 1978.
    4745

RI. 30 A very Early Devonian patch
reef and its ecological setting, by
R. A. Smosna and S. M. Warshauer.
1979.   4746

RI. 31 A scheme for multivariate
analysis in carbonate petrology with
an example from the Silurian Tonoloway limestone, by R. A. Smosna
and S. M. Warshauer. 1979. 4747

RI. 32 The Wills Mountain anticline;
a study in complex folding and faulting in eastern West Virginia, by
W. J. Perry. 1978.   4748

River Basin Basic Data Report
1968-

RBB. 1 Records of wells, springs,
and test borings, chemical analyses
of ground water and selected driller's logs from the Monongahela
River Basin in West Virginia, by
P. E. Ward and B. M. Wilmoth.
1968.   4749

RBB. 2 Records of wells, springs,

and test borings, chemical analyses
of water, sediment analyses, standard
streamflow data summaries, and selected driller's logs from the Little
Kanawha River Basin in West Virginia,
by E. A. Friel and G. L. Bain. 1971.
    4750

RBB. 3 Records of wells, springs,
and streams in the Potomac River
Basin, West Virginia, by E. A. Friel,
W. A. Hobba, and J. L. Chisholm.
1975.   4751

RBB. 4 Records of wells, springs,
chemical analyses of water, biological
analyses of water and standard streamflow data summaries from the Upper
New River Basin in West Virginia,
by J. L. Chisholm and P. M. Frye.
1975.   4752

RBB. 5 Hydrologic data for the Coal
River Basin, West Virginia, by F. O.
Morris, J. S. Bader, J. L. Chisholm,
and S. C. Downs. 1976.   4753

River Basin Bulletin. 1968-

RIB. 1 Ground-water hydrology of the
Monongahela River Basin in West
Virginia, by P. E. Ward and B. M.
Wilmoth. 1968.   4754

RIB. 2 Water resources of the Little
Kanawha River Basin, West Virginia,
by G. L. Bain and E. A. Friel. 1972.
    4755

RIB. 3 Water resources of the Poto-
mac River Basin, West Virginia, by
W. A. Hobba, E. A. Friel, and J. L.
Chisholm. 1972.   4756

State Park Bulletin. 1952-

SPB. 6 Blackwater Falls State Park
and Canaan Valley State Park--re-
sources, geology, and recreation, by
J. C. Ludlum and T. Arkle. Rev.
ed. 1971.   4757

Volume. 1899-

V. 23 West Virginia geological and
economic survey--its accomplishments

and outlook, ed. by I. S. Latimer,
J. C. Ludlum, J. C. Welden, and
R. C. Tucker.　1963.　　　4758

Bulletin. 1898-

B. 75 Soils of Dodge County, Wisconsin, by F. D. Hole, G. B. Lee, and E. A. Brickbauer. 1953. 4759

B. 76 Not published?   4760

(B. 77-84 in base index)

B. 85 Soil resources and forest ecology of Menominee County, Wisconsin, by C. J. Milfred, G. W. Olson, F. D. Hole, F. P. Baxter, F. G. Goff, W. A. Creed, and F. Stearns. 1967.   4761

B. 86 Soils of Jefferson County, Wisconsin, by C. J. Milfred and F. D. Hole. 1970.   4762

B. 87 Soils of Wisconsin, by F. D. Hole. 1976.   4763

Educational Series. 197?-

ES. 1 Stones used in state capitol, by E. F. Bean and C. A. Halbert. n. d.   4764

ES. 2 The mineral industry of Wisconsin, by R. C. Briggs and M. E. Ostrom. 1974.   4765

The mineral industry of Wisconsin, by R. C. Briggs and M. E. Ostrom. 1975.   4766

ES. 3 Mineral and rock collecting in Wisconsin, by M. E. Ostrom and G. F. Hanson. 1961.   4767

ES. 4 Fossil collecting in Wisconsin, by M. E. Ostrom. 1962.   4768

ES. 5 The water resources of Wisconsin, by C. L. R. Holt and K. B.

Young. 1964.   4769

ES. 6 The soils of Wisconsin, by M. T. Beatty, I. O. Hembre, F. D. Hole, L. R. Massie, and A. E. Peterson. 1964.   4770

ES. 7 The mineral resources of Wisconsin, by G. F. Hanson. 1967. 4771

ES. 8 Exploration for copper in Wisconsin, by M. E. Ostrom. 1973.   4772

ES. 9 Wisconsin's ground water: an invaluable resource, by D. Stephenson and J. W. Clark. 1974.   4773

ES. 10 List of high points in Wisconsin, by M. E. Ostrom. Rev. 1977.   4774

ES. 11 Not published?   4775

ES. 12 Soil surveys for town and country, by F. D. Hole. 1975. 4776

ES. 13 Mineral and water resources of Wisconsin. Report prepared by the U. S. Geological Survey and the Wisconsin Geological and Natural History Survey. 1976.   4777

ES. 14 The geology of Wisconsin. 1977.   4778

ES. 15 The geology of Wisconsin. Part II. 1977.   4779

ES. 16 The geology of Wisconsin. Part IV. 1978.   4780

ES. 17 Copper and zinc mining in Wisconsin, by M. G. Mudrey. 1978.   4781

ES. 18 Mines and minerals in Wisconsin, by M. Roshardt and T. J. Evans. 1978.   4782

ES. 19 Wisconsin lake levels--
their ups and downs, by R. P.
Novitski and R. W. Devaul. 1978.
4783

ES. 20 Minerals in the economy
of Wisconsin, by R. C. Briggs and
T. J. Evans. n. d. 4784

ES. 21 Geology in land use, by
M. L. Czechanski and R. G. Hennings.
1979. 4785

Field Trip Guide Book Series.
1978-

FTGB. 1 Upper Mississippi Valley
base-metal district, ed. by M. G.
Mudrey. 1978. 4786

FTGB. 2 Precambrian inliers in
south-central Wisconsin, ed. by
E. I. Smith, R. A. Paull, and M. G.
Mudrey. 1978. 4787

FTGB. 3 Lithostratigraphy, petrology, and sedimentology of Late
Cambrian-Early Ordovician rocks
near Madison, Wisconsin, by I. E.
Odom, M. E. Ostrom, C. W. Byers,
R. C. Morris, and R. A. Adams.
1978. 4788

Information Circular. 1955-

IC. 6 Cambro-Ordovician stratigraphy of southwest Wisconsin, by
M. E. Ostrom. 1965. 4789

IC. 7 Cambrian stratigraphy in
western Wisconsin, by M. E. Ostrom.
1966. 4790

IC. 8 Paleozoic stratigraphic nomenclature for Wisconsin, by M. E.
Ostrom. 1967. 4791

IC. 9 Trends in ground-water
levels in Wisconsin through 1966,
by R. W. Devaul. 1967. 4792

IC. 10 Geochemical prospecting by
spring sampling in the southwest
Wisconsin zinc mining area, by J.
De Geoffroy. 1969. 4793

IC. 11 Field trip guidebook for

Cambrian-Ordovician geology of western Wisconsin, by M. E. Ostrom,
R. A. Davis, and L. M. Cline. 1970.
4794

Special paper: Lithologic cycles in
Lower Paleozoic rocks of western Wisconsin, by M. E. Ostrom. 4795

Special paper: Lithostratigraphy of
the Prairie du Chien Group, by R. A.
Davis. 4796

IC. 12 Directory of Wisconsin mineral
producers, 1968, by M. E. Ostrom.
1970. 4797

IC. 13 Glacial geology of Two Creeks
Forest Bed, Valderan type locality
and northern Kettle Moraine State
Forest, by R. F. Black. 1970. 4798

IC. 14 Geology of the Baraboo District, Wisconsin; a description and
field guide incorporating structural
analysis of the Precambrian rocks
and sedimentologic studies of the Paleozoic strata, by I. W. D. Dalziel and
R. H. Dott. 1970. 4799

Glacial geology summary, by R. F.
Black. 4800

Plant ecology of the Baraboo Hills,
by J. H. Zimmerman. 4801

IC. 15 Pleistocene geology of southern
Wisconsin; field trip guide with spe-
cial papers by R. F. Black, N. K.
Bleuer, F. D. Hole, N. P. Lasca, and
L. J. Maher. 1970. 4802

IC. 16 Guidebook to the Upper Mississippi Valley base metal district,
by A. V. Heyl, W. A. Broughton, and
W. S. West. 1970. 4803

Geology of the Upper Mississippi
Valley base-metal district, by A. V.
Heyl, W. A. Broughton, and W. S.
West. 1978. 4804

IC. 17 Field trip guidebook to the
hydrogeology of the Rock-Fox River
Basin of southeastern Wisconsin, by
C. L. R. Holt, R. D. Cotter, J. H.
Green, and P. G. Olcott. 1970. 4805

IC. 18 Preliminary report on results

of physical and chemical tests of Wisconsin silica sandstones, by M. E. Ostrom. 1971. 4806

IC. 19 Conodonts and biostratigraphy of the Wisconsin Paleozoic, by D. L. Clark, ed. 1971. 4807

IC. 20 Soil absorption of septic tank effluent; a field study of some major soils in Wisconsin, by J. Bouma and others. 1972. 4808

IC. 21 Trends in ground-water levels in Wisconsin, 1967-71, by R. M. Erickson. 1972. 4809

IC. 22 Ground-water quality in Wisconsin through 1972, by C. L. R. Holt and E. L. Skinner. 1973. 4810

IC. 23 Mineral prospecting and mining transactions, by P. E. McKeever and J. Preston. 1975. 4811

IC. 24 Model mineral reservation and mine zoning ordinance, by J. Preston, E. Strauss, and T. O. Friz. 1974. 4812

IC. 25 Not published? 4813

IC. 26 Mineral resources, mining, and land-use planning in Wisconsin, by T. O. Friz. 1975. 4814

IC. 27 Measurement of water movement in soil pedons above the water table, by J. Bouma, F. G. Baker, and P. L. M. Veneman. 1974. 4815

IC. 28 A digital-computer model for estimating drawdowns in the sandstone aquifer in Dane County, Wisconsin, by R. S. McLeod. 1975. 4816

IC. 29 Ground-water resources of Waukesha County, Wisconsin, by J. B. Gontheir. 1975. 4817

IC. 30 A digital-computer model for estimating hydrologic changes in the aquifer system in Dane County, Wisconsin, by R. S. McLeod. 1975. 4818

IC. 31 The availability of ground water for irrigation in the Rice Lake-Eau Claire area, Wisconsin, by E. A. Bell and S. M. Hindall. 1975. 4819

IC. 32 Ground-water resources and geology of St. Croix County, Wisconsin, by R. G. Borman. 1976. 4820

IC. 33 Ground-water resources and geology of Jefferson County, Wisconsin, by R. G. Borman and L. C. Trotta. 1975. 4821

IC. 34 Ground-water resources and geology of Walworth County, Wisconsin, by R. G. Borman. 1976. 4822

IC. 35 Leachate attenuation in the unsaturated zone beneath three sanitary landfills in Wisconsin, by R. A. Gerhardt. 1977. 4823

IC. 36 Effects of irrigation on water quality in the sand plain of central Wisconsin, by S. M. Hindall. 1978. 4824

IC. 37 Ground-water resources and geology of Columbia County, Wisconsin, by C. A. Harr, L. C. Trotta, and R. G. Borman. 1978. 4825

IC. 38 Ground-water resources and geology of Washington and Ozaukee Counties, Wisconsin, by H. L. Young and W. G. Batten. 1980. 4826

Special Report. 1967-

SR. 1 Preliminary report on the irrigation potential of Dunn County, Wisconsin, by P. G. Olcott, F. D. Hole, and G. F. Hanson. 1967. 4827

SR. 2 Bibliography and index of Wisconsin ground-water, 1851-1972, by A. Zaporozec. 1974. 4828

SR. 3 Mining on your land? by D. Barrows, M. E. Ostrom, and J. Preston. 1975. 4829

SR. 4 Ground-water hydrology and geology near the July 1974 phenol spill at Lake Beulah, Walworth County, Wisconsin, by R. G. Borman. 1975. 4830

Bulletin. 1911-

B. 51 A field guide to the rocks and minerals of Wyoming, by W. H. Wilson. 1965. 4833

B. 52 Measured sections of Devonian rocks in northern Wyoming, by C. A. Sandberg. 1967. 4834

B. 53 Bibliography of Wyoming geology, 1917-1945, by M. L. Troyer. 1969. 4835

B. 54 Fossils of Wyoming, by M. W. Hager. 1970. 4836

B. 55 Traveler's guide to the geology of Wyoming, by D. L. Blackstone. 1971. 4837

B. 56 Minerals and rocks of Wyoming, by F. K. Root. 1977. 4838

B. 57 Bibliography of Wyoming geology, 1945-1949, by J. M. Love. 1973. 4839

B. 58 Bibliography of Wyoming coal, by G. B. Glass and R. W. Jones. 1974. 4840

B. 59 Caves of Wyoming, by C. Hill and others. 1976. 4841

B. 60 Thermal springs of Wyoming, by R. M. Breckenridge and B. S. Hinckley. 1978. 4842

Chapter on flora, by T. T. Terrell. 1978. 4843

B. 61 Not yet published. 4844

B. 62 Bibliography of Wyoming geology, 1950-1959, by H. L. Nace. 1979. 4845

B. 63 Paleontology of the Green

River Formation, with a review of the fish fauna, by L. Grande. 1980. 4846

Memoir. 1968-

M. 1 A regional study of rocks of Precambrian age in that part of the Medicine Bow Mountains lying in southeastern Wyoming, by R. S. Houston and others. 1968. 4847

M. 2 Geology and mammalian paleontology of the Sand Creek facies, lower Willwood Formation (Lower Eocene), Washakie County, Wyoming, by T. M. Bown. 1979. 4848

Preliminary Report. 1961-

PR. 2 The Kirwin mineralized area, Park County, Wyoming, by W. H. Wilson. 1964. 4849

PR. 3 The Keystone gold-copper prospect area, Albany County, Wyoming, by D. R. Currey. 1965. 4850

PR. 4 Bentonite deposits of the Clay Spur district, Crook and Weston Counties, Wyoming, by J. C. Davis. 1965. 4851

PR. 5 The Haystack Range, Goshen and Platte Counties, Wyoming, by M. L. Millgate. 1965. 4852

PR. 6 Gravity thrusting in the Bradley Peak-Seminoe Dam quadrangles, Carbon County, Wyoming and the relationship to the Seminoe iron deposits, by D. L. Blackstone. 1965. 4853

PR. 7 The Centennial Ridge gold-platinum district, Albany County, Wyoming, by M. E. McCallum. 1968. 4854

PR. 8 Not published?   4855

PR. 9 Gypsum deposits in the Cody
area, Park County, Wyoming, by
J. M. Bullock and W. H. Wilson.
1969.   4856

PR. 10 Taconite in the Wind River
Mountains, Sublette County, Wyoming, by R. G. Worl. 1968. 4857

PR. 11 Structural geology of the
Rex Lake quadrangle, Laramie
Basin, Wyoming, by D. L. Black-
stone. 1969.   4858

PR. 12 The Phosphoria and Goose
Egg Formations in Wyoming, by
D. W. Lane. 1973.   4859

PR. 13 Structural geology of the
eastern half of the Morgan quadrangle, the Strouss Hill quadrangle,
and the James Lake quadrangle,
Albany and Carbon Counties, Wyoming, by D. L. Blackstone. 1973.
                                    4860

PR. 14 Geology and mineral deposits of the Silver Crown Mining
District, Laramie County, Wyoming, by T. Klein. 1974. 4861

PR. 15 Structural geology of the
Arlington-Wagonhound Creek area;
a revision of previous mapping, by
D. L. Blackstone. 1976.   4862

PR. 16 Late Cretaceous and Early
Tertiary provenance and sediment
dispersal, Hanna and Carbon Basins, Carbon County, Wyoming, by
J. D. Ryan. 1977.   4863

PR. 17 Petrography of selected
rock samples and a discussion of
structural fabric, northern Salt
River Range, Lincoln County, Wyoming, by D. R. Lageson. 1978.
                                    4864

Public Information Circular. 1976-

PIC. 1 Geothermal resources, present and future demand for power and
legislation in the State of Wyoming,
by E. R. Decker. 1976.   4865

PIC. 2 State-owned coal lands in Wyoming, by G. B. Glass. 1976. 4866

PIC. 3 Wyoming coal directory, by
G. B. Glass. n. d.   4867

PIC. 4 Not published?   4868

PIC. 5 Not published?   4869

PIC. 6 Directory of sources of Wyoming geological information. 1977.
                                    4870

PIC. 7 Occurrences of uranium in
Precambrian and younger rocks of
Wyoming and adjacent areas: ab-
stracts, ed. by D. R. Lageson and
W. D. Hausel. 1978.   4871

PIC. 8 The Wyoming mineral industry:
a summary by the Staff of the Geological Survey of Wyoming. 1978.
                                    4872

PIC. 9 Wyoming coal fields 1978, by
G. B. Glass. 1978.   4873

PIC. 10 Update on the Wyoming-
Idaho-Utah thrust belt: abstracts,
comp. by D. R. Lageson. 1979. 4874

PIC. 11 The overthrust belt: an over-
view of an important new oil and gas
province, by A. J. Ver Ploeg. 1979.
                                    4875

PIC. 12 Wyoming coal production and
summary of coal contracts, by G. B.
Glass. 1980.   4876

PIC. 13 Rocky Mountain foreland
basement tectonics, by D. R. Lageson.
1980.   4877

PIC. 14 Guidebook to the coal geology
of the Powder River coal basin, ed.
by G. B. Glass. 1980.   4878

Report of Investigations. 1939-

RI. 9 Analyses of rock and stream
sediment samples, Teton Corridor and
contiguous areas, Teton County, northwestern Wyoming, by J. D. Love and
J. C. Antweiler. 1974.   4879

RI. 10 Applied geology and archaeology: the Holocene history of Wyoming, by M. Wilson, ed. 1974. 4880

RI. 11 Analyses and measured sections of 54 Wyoming coal samples (collected in 1974), by G. B. Glass. 1975. 4881

RI. 12 Diamond in state-line kimberlite diatremes, Albany County, Wyoming, and Larimer County, Colorado, by M. E. McCallum and C. D. Mabarak. 1976. 4882

RI. 13 Stratigraphy and uranium potential of Early Proterozoic metasedimentary rocks in the Medicine Bow Mountains, Wyoming, by K. E. Karlstrom and R. S. Houston. 1979. 4883

RI. 14 Not yet published. 4884

RI. 15 Not yet published. 4885

RI. 16 Coal analyses and lithologic descriptions of five core holes drilled in the Carbon Basin of southcentral Wyoming, by G. B. Glass. 1978. 4886

RI. 17 Remaining strippable coal resources and strippable reserve base of the Hanna coal field in southcentral Wyoming, by G. B. Glass and J. T. Roberts. 1979. 4887

RI. 18 Geometry of the Prospect-Darby and LaBarge faults at their junction with the LaBarge platform, Lincoln and Sublette Counties, Wyoming, by D. L. Blackstone. 1979. 4888

RI. 19 Exploration for diamond-bearing kimberlite in Colorado and Wyoming; an evaluation of exploration techniques, by W. D. Hausel, M. E. McCallum, and T. L. Woodzick. 1979. 4889

RI. 20 A stratigraphic evaluation of the Eocene rocks of southwestern Wyoming, by R. Sullivan. 1980. 4890

RI. 21 Not yet published. 4891

RI. 22 Coals and coal-bearing rocks of the Hanna coal field, Wyoming, by G. B. Glass and J. T. Roberts. 1980. 4892

RI. 23 Gold districts of Wyoming, by W. D. Hausel. 1980. 4893

Abdulkareem, T. F. 1586
Abegglen, D. E. 1168
Abou-Zied, S. 4438
Abrams, C. E. 917
Achmad, G. 2311
Ackerman, D. J. 3395
Ackroyd, E. A. 2417
Adams, B. J. 3541
Adams, D. B. 1832
Adams, J. W. 424
Adams, R. A. 4788
Adams, S. S. 3104
Adamson, J. H. 3919
Addicott, W. O. 424
Adkison, W. L. 3690
Affolter, R. H. 4444
Agagu, O. K. 4190, 4197
Agnew, A. F. 3996, 4027
Ahnell, G. 4706
Albers, J. P. 415, 424, 502, 2879
Alberstadt, L. P. 3644
Albin, D. R. 403
Alenius, E. M. J. 349
Alewine, R. W. 430
Alexander, C. P. 649
Alexander, J. D. 2027
Aley, T. J. 2556
Alfonsi, P. P. 3647
Alfors, J. T. 424, 436
Allen, A. T. 898
Allen, C. R. 431
Allen, E. P. 3283, 3287
Allen, G. C. 4585
Allen, J. E. 3273
Allen, J. R. 3030
Allen, R. M. 4502, 4505
Allen, W. B. 2368
Allgaier, G. J. 1345
Allison, E. C. 424
Allison, I. S. 3776
Alptekin, O. S. 3174
Alverson, R. M. 116
Amato, R. V. 4591
American Geological Institute 591, 1779
Amos, D. H. 2627
Amsden, T. W. 3616, 3617, 3618,
3627, 3646, 3648, 3652, 3656
Anderson, A. L. 1156
Anderson, C. 3465
Anderson, D. L. 4627
Anderson, D. W. 4370
Anderson, E. G. 2148
Anderson, K. H. 2604, 2632, 2635
Anderson, P. W. 705
Anderson, R. C. 1223, 1237, 1270, 1302
Anderson, R. E. 227, 269, 2626
Anderson, R. R. 1654
Anderson, S. B. 3427, 3436, 3467, 3472, 3473, 3476, 3478, 3498
Anderson, T. P. 482, 580, 584
Anderson, W. 757, 763, 799, 837, 841, 868
Andrews, H. N. 2218
Andrews, P. B. 4302
Angino, E. E. 1811, 1872, 1893
Angurarohita, R. 2526
Ansari, M. B. 2926
Anthony, J. W. 376, 377
Antweiler, J. C. 4879
Apfel, E. T. 1621
Applin, E. R. 893
Applin, P. L. 893
Archbold, N. L. 2875, 2883, 2911
Arkle, T. 4757
Armstrong, A. K. 3212, 3245
Armstrong, C. A. 3326, 3327, 3335, 3352, 3353, 3375, 3380, 3381, 3389, 3391
Armstrong, C. F. 456, 457, 561
Armstrong, F. C. 4613
Armstrong, R. L. 699, 4351
Arndt, B. M. 3369, 3458, 3480, 3484, 3490
Arnold, E. C. 3118, 3202, 3221, 3226
Arrigo, J. L. 2858
Artim, E. R. 4635, 4656
Asher, R. R. 226, 230, 231, 1155, 1158
Ashley, R. P. 2920
Ashmead, L. 683
Asquith, G. B. 3195, 3204, 3211
Atherton, E. 1197, 1224, 1238

Atkins, R. L.   918, 919, 920
Au, A. K.   1455, 1462
Ault, C. H.   1509, 1511
Aune, Q. A.   424
Austin, G. S.   1513, 1537, 1567,
    2390, 2403, 2412, 2413, 2415,
    2421, 2423, 3213
Austin, H. A. R.   1061
Austin, S. R.   1147
Avcin, M. J.   1376, 1472, 1663
Avila, J.   2052
Avrett, J. R.   80, 94, 95, 171
Ayer, N. J.   1406, 1411
Ayers, W. B.   4339
Ayres, Lewis, Norris, and May,
    Inc.   2362

Babcock, C.   758, 761, 765, 770,
    771, 776, 780, 782, 788, 797
Babcock, R. S.   4655
Babu, S. P.   4702
Bach, W. N.   4217
Bacon, C. F.   486, 563
Bader, J. S.   4753
Bailey, B. L.   134, 180
Bailey, E. H.   406, 413, 424
Bain, G. L.   4696, 4750, 4755
Baird, A. K.   430, 519
Baker, A.   2883
Baker, A. A.   2021
Baker, C. H.   3328, 3329, 3330
Baker, D. H.   3168, 3172, 3192
Baker, F. G.   4815
Baker, G. K.   4067, 4070
Baker, L. S.   3047
Baker, R. K.   4037
Baker, V. R.   4194
Baker, W. W.   711
Baldridge, H. L.   2107
Baldwin, A. D.   1708
Baldwin, B.   3271, 3277
Baldwin, E. M.   3722, 3725
Balek, J.   2995
Ball, M. M.   1680
Ballard, F. V.   3312
Balster, C. A.   2817, 2818, 2823,
    2828
Baltosser, W. W.   3168
Barari, A.   4055, 4056, 4081,
    4084, 4086, 4087, 4090, 4092,
    4093, 4094, 4095, 4096, 4098,
    4099, 4100, 4101, 4102, 4103,
    4104, 4105, 4106, 4108
Barker, J. M.   2930
Barker, W. J.   4545, 4546
Barks, J. H.   2675, 2677
Barksdale, J. D.   4620
Barlow, J. A.   4702, 4703

Barnes, J.   2960
Barnes, J. H.   3896
Barnes, V. E.   4238, 4247, 4288,
    4323
Barnes, W. C.   2685
Barnett, S. G.   3088
Barraclough, J. T.   366, 827
Barrash, W.   3775
Barrett, J. K.   593, 612, 614
Barrows, A. G.   430, 450, 555, 577
Barrows, D.   4832
Bartholomew, M. J.   4541, 4551
Bartlett, C. S.   4553, 4579
Barton, J. M.   699
Basciano, J. M.   4226, 4333
Bashe, R. W.   2055
Bass, C.   1834
Bassett, A. M.   500
Bastian, T.   2244
Bateman, P. C.   417, 424
Bateman, R. L.   2927
Bateman, S. J.   2081
Batten, W. G.   4826
Baughman, W. T.   2520, 2527, 2529,
    2541, 2544
Bauleke, M.   1792
Bauleke, M. P.   1685, 1857, 1901
Bauman, R. D.   4400
Baumgardner, R. W.   4230
Baxter, F. P.   4761
Baxter, J. W.   1198, 1241, 1246,
    1269, 1304, 1328, 1473
Bayer, T. N.   2388
Bayne, C. K.   1723, 1752, 1768, 1775,
    1787, 1793, 1820, 1891, 1898
Beal, L. H.   2862, 2896
Bean, E. F.   4764
Beard, J. G.   2132
Beard, M. E.   844
Bearden, H. W.   794, 849
Beattie, B.   2952
Beatty, M. T.   4770
Beaulieu, J. D.   3712, 3715, 3719,
    3721, 3722, 3724, 3726, 3729,
    3732, 3733, 3736, 3762
Beaumont, E. C.   3168, 3208, 3257
Bebout, D. G.   2608, 4190, 4197,
    4201, 4203, 4210, 4313, 4324,
    4326
Becher, A. E.   3937
Beck, B. F.   921
Beck, F. M.   2213
Beck, M. E.   3824
Becker, G. W.   3171, 3191
Becker, L. E.   1524, 1531, 1574,
    1579, 1585
Bedrossian, T. L.   583
Beeler, C. W.   1252
Been, W.   2320

Beene, D. L.   1690, 1703, 1714, 1722, 1732, 1796, 1798, 1800, 1802, 1804, 1806, 1859, 1879, 1888, 1892, 1897, 1902
Beffort, J. D.   4085, 4088-9, 4091
Bela, J. L.   3740, 3769
Bell, A. H.   1224
Bell, E. A.   4819
Bell, W. C.   4238, 4247, 4323
Bellinger, F.   963
Bellis, W. H.   3677
Belt, Collins and Associates   1094, 1120
Beltrame, R. J.   2409, 2410, 2411
Benedek, C.   2953
Bengal, L. E.   1360
Bennett, E. H.   1182, 1183, 1186, 1187
Bennett, G. D.   3923, 3926, 3928, 3930
Benson, D. J.   3608
Benson, R. N.   734
Berberick, J.   1889, 1901
Berendsen, P.   1830
Berg, M. A.   160
Berg, R. B.   2728, 2733, 2740, 2752, 2758, 2781, 2789, 2790, 2809, 2813, 2814, 2816, 2819, 2821, 2822, 2826, 2831, 2835, 2839, 2840, 2841, 2844
Berg, R. C.   1452
Berg, T. M.   3855
Berggren, D. J.   1204, 1376
Bergstrom, R. E.   1272, 1282, 1346, 1353, 1396, 1412, 1422, 1484
Berkley, J. L.   2431, 2432
Berkson, J. M.   1450
Berkstresser, C. F.   3229
Bermes, B. J.   753, 819
Bernard, H. A.   4244, 4245
Berquist, S. G.   2313
Berrien, P. L.   657
Berry, D. W.   1764
Bertalan, D. T.   2334
Bertram, N.   2581
Berube, S. C.   2102
Besancon, J. R.   699
Best, D. M.   3309
Beveridge, T. R.   2549C, 2572, 2573, 2584, 2588
Bhappu, R. B.   3119, 3120, 3124, 3127, 3133, 3135, 3136, 3143, 3144, 3151, 3153, 3159, 3160
Bhatia, D. M. S.   2640
Bick, K. F.   4350, 4572
Bicker, A. R.   2491, 2494, 2495, 2499, 2509, 2517,

2522, 2542, 2545
Bickley, W. B.   3464
Bieber, P. P.   2195
Bieberman, R. A.   3126, 3142, 3150, 3163
Biehler, S.   430
Biggs, T. H.   4527, 4553, 4594, 4602
Bilbey, S. A.   4440
Billingsley, G. A.   3966
Bingham, J. W.   4623
Bingham, R. H.   72
Bingler, E. C.   2924, 2934, 3091, 3102, 3122
Bishop, C. C.   458, 545, 558
Bishop, D. T.   1160
Bishop, G. A.   2824
Bittner, E.   1144
Bjorken, E. D.   4545, 4546
Bjorlie, P. F.   3497
Black, D. F. B.   4046
Black, H. J. B.   3738
Black, H. N.   2090
Black, R. F.   4798, 4800, 4802
Blackstone, D. L.   2832, 4837, 4853, 4858, 4860, 4862, 4888
Blackwell, D. D.   3764, 3773, 4640, 4660
Blake, B.   4706
Blake, M. C.   2886
Blakely, R. F.   1565, 1568, 1569, 1571, 1577, 1578, 1581
Blakey, R. C.   4377
Blanc, R. P.   534
Blanchard, H. E.   957
Blanchard, R.   2865
Bleuer, N. K.   1563, 1566, 1596, 1604, 1611, 4802
Block, D. A.   3319
Bloom, A. L.   665, 2206
Bloom, P. A.   3130
Bloss, P.   4120
Bluemle, J. P.   3322, 3337, 3343, 3358, 3365, 3373, 3388, 3390, 3398, 3400, 3401, 3403, 3404, 3405, 3406, 3407, 3408, 3409, 3441, 3443, 3452, 3499
Bluemle, M. E.   3399, 3450
Blumer, J. W.   2727, 2754
Blyth, C. R.   1264
Boellstorff, J.   1820
Boettcher, A. J.   2771, 2773
Boggess, D. H.   778, 779, 792, 856, 2281
Bograd, M.   2537, 2540, 2543
Bolander, K.   3005
Bolke, E. L.   4467
Bolles, W. H.   3787
Bolt, B. A.   431
Bond, D. C.   1326, 1993, 2061

Bond, E. W.   2800, 2830
Bond, G. C.   280
Bond, J. G.   1148
Bondarenko, W.   179
Bondurant, K. T.   2730
Bonem, R. M.   3165
Bonham, H. F.   2869, 2893, 2919, 2934, 2935
Bonilla, M. G.   423
Bonini, W. E.   714, 1152, 2825, 3086
Bonneville, J. W.   3425
Bonnichsen, B.   2407
Bookstrom, A. A.   2887
Boone, G. M.   2218, 2219
Boone, P. A.   187
Borchardt, G. A.   574
Boreck, D. L.   615, 632
Borman, R. G.   4820, 4821, 4822, 4825, 4830
Borton, R. L.   3176
Bosch, S. C.   4201
Botoman, G.   3542, 3605
Bottino, M. L.   2218
Botz, M. K.   2734, 2800, 2811, 2812
Boucot, A. J.   1927, 2210, 2213, 2216, 2218
Bouma, J.   4808, 4815
Bowen, O. E.   424, 427, 501
Bowen, R. G.   3764
Bowie, J. E.   2665
Bowman, J. T.   4440
Bown, T. M.   4848
Bowyer, B.   2861
Boyd, E. L.   596
Boyd, W. T.   2014
Brabb, E. E.   423, 515
Bradbury, J. C.   1201, 1219, 1235, 1238, 1276, 1473
Bradford, B. B.   2074
Bradford, W.   4021, 4023
Brady, L.   1834, 1835
Brady, L. L.   1832, 1899
Braidech, L. L.   3583
Brandvold, L. A.   3178, 3196, 3216
Branson, C. C.   3621, 3685
Branson, E. R.   1927, 1931, 1933
Breckenridge, R. M.   4842
Bremser, S. M.   1388
Brennan, D. J.   352
Brenner, G. J.   2233
Brent, W. B.   4109, 4559
Breslauer, S.   2969
Brett, C. E.   179
Bretz, J. H.   2605
Brickbauer, E. A.   4759

Bricker, D. M.   2331
Bridge, J.   2569
Bridges, K. H. P.   1614, 1615
Bridges, W. C.   796
Brietkrietz, A.   2685, 2699, 2707, 2726, 2741
Briggs, G.   3676
Briggs, R. L.   4765, 4766, 4784
Brinkley, D.   4095, 4097
Bristol, H. M.   1280, 1314, 1365
Broderick, G. N.   424
Broeker, M. E.   1691, 1704, 1712, 1721, 1814, 1817
Brookhyser, C. S.   3720
Brookins, D. G.   699, 1747, 1748, 1749, 1751
Brooks, F. L.   2498
Brooks, H. C.   3697, 3703, 3760
Brophy, J. A.   3463
Brostuen, E. A.   3410, 3492
Broughton, W. A.   4803, 4804,
Broussard, W. L.   2171, 2196, 2198, 2199, 2200
Brower, E.   3170
Brower, J. C.   3679
Brown, C. N.   1624
Brown, D. C.   1474
Brown, D. L.   3308, 4147
Brown, D. P.   855
Brown, G. A.   430
Brown, J. A.   2035
Brown, L. F.   4174, 4175, 4178, 4180, 4183, 4212, 4236, 4243, 4248, 4310
Brown, P. M.   3284, 3308
Brown, S. L.   1727
Brown, T. E.   4242
Brown, W. R.   4564
Browning, J. S.   1953
Bruce, R. L.   4048, 4066
Brueckmann, J. E.   1232, 1484
Bruer, W. G.   434
Brumbach, J. J.   655
Buchanan, L. C.   4644
Buck, L.   2988
Buckler, W. R.   2364
Budding, A. J.   3174, 3268
Buehrer, D.   4104
Buffington, J. D.   3070
Bufvers, J.   327
Buhle, M. B.   1232
Buie, B. F.   3970, 3972, 3974
Bukry, D.   539
Bullock, K. C.   4349, 4361, 4383
Bundtzen, T. K.   283, 325
Bundy, J. L.   3298
Bunker, B. J.   1649, 1661, 1677
Burand, W. M.   213, 218, 219, 220, 221
Burchett, R. R.   2846, 2848, 2850,

2853, 2855, 2856, 2858
Burchfiel, B. C. 535
Burford, R. O. 430
Burger, A. M. 1536
Burger, H. R. 698, 2786
Burnett, J. L. 424, 436, 476, 507
Burr, J. H. 2439
Burt, D. M. 785
Burt, E. R. 3301
Burt, R. J. 1114
Burton, L. C. 3670
Burwell, H. B. 4141, 4142
Busch, W. L. 1213, 1229, 1248, 1263, 1274, 1291, 1303, 1311, 1325, 1333
Buschbach, T. C. 1197, 1224, 1250, 1347, 1361, 1428, 1481
Buschman, W. J. 3557
Bush, W. V. 358, 400
Bushnell, K. 2866
Bushnell, K. O. 3904, 3906
Bushnell, M. M. 584
Buss, W. R. 4376
Butcher, W. S. 3037
Butler, A. P. 424
Butler, E. A. 2146, 2149, 2158
Butler, J. 2983
Butler, J. N. 2898
Butler, J. R. 3981, 3982
Butrico, F. A. 3069
Butts, C. 4504
Buturla, F. 3344, 3345
Byerlay, J. R. 2382
Byers, C. W. 4788
Byrd, W. D. 4423
Byrne, J. R. 4221

Cagle, J. W. 21, 759, 820, 822, 1667, 1668, 1670
Cahill, R. A. 1355
Cain, D. 596
Calandro, A. J. 2166, 2174, 2177, 2198, 2199, 2200
Caldwell, D. W. 2205, 2207, 3991
Calkins, W. G. 4368
Callaghan, E. 4358, 4375
Callender, J. F. 3207
Calnan, T. R. 4338
Calver, J. L. 4528
Calvert, W. L. 3538, 3548, 3549, 3550, 3551, 3553
Cameron Engineers, Inc. 2829
Cammack, J. H. 127
Campbell, C. V. 3218
Campbell, D. D. 3097
Campbell, D. H. 4242
Campbell, I. 412, 424

Campbell, J. A. 4388, 4445
Campbell, K. S. W. 3642, 3650
Campbell, N. P. 2772, 4648
Campbell, R. B. 1645
Campbell, R. H. 577
Caplenor, C. D. 2516
Cardwell, D. H. 4683, 4716
Cardwell, G. T. 2172, 2193
Carillo, F. V. 2725, 2731, 2737, 2743
Carlson, C. G. 3355, 3362, 3376, 3429, 3434, 3436, 3438, 3444, 3471, 3473
Carlson, M. P. 2846, 2847, 2851, 2855, 2857
Carlton, R. W. 3590, 3598
Carmalt, S. W. 699
Carman, M. F. 498
Carmichael, I. S. E. 538
Caroon, L. E. 80, 171
Carpenter, G. L. 1515, 1538, 1539, 1540, 1541, 1542, 1543, 1544, 1545, 1546, 1547, 1548, 1549, 1550, 1551, 1552, 1553, 1554, 1555, 1592, 1984, 2002, 2011, 2015, 2136
Carpenter, P. A. 3290, 3294, 3298, 3299, 3301, 3303
Carpenter, R. H. 948/9, 950, 4153
Carr, B. L. 1784, 1786, 1788
Carr, D. D. 1505, 1507, 1509, 1511, 1514, 1520, 1522, 1526, 1595, 1605
Carroll, D. L. 1368
Carroll, W. K. 3496
Carswell, L. D. 3926, 3935, 3958
Carten, R. B. 3212
Carter, C. H. 3523, 3534, 3600, 3603
Carter, L. M. 633
Carter, R. F. 892, 927
Carter, R. S. 928
Carter, W. D. 3836
Cartright, C. J. 432
Cartwright, K. 1289, 1380, 1387, 1391, 1403, 1454, 1456, 1457, 1461
Cary, A. S. 4629
Cash, D. J. 3156, 3173
Caswell, W. B. 2221
Cathcart, J. B. 969
Cathey, J. B. 1914, 1915, 1918
Cathey, T. A. 3689
Caughey, C. A. 4188, 4213
Causey, L. V. 20, 26, 28
Causseaux, K. W. 2
Cavanaugh, T. D. 3309
Cazeau, C. J. 3983
Cepeda, J. C. 4215

Chadwick, R. A. 2693
Chafetz, H. S. 3809
Chaffin, H. S. 137, 156, 202, 203, 204, 206
Chamberlain, C. K. 2891
Chamberlain, V. D. 2316
Chang, P. H. 4482
Chapman, M. J. 2361
Chapman, R. H. 422, 430, 506, 510, 557, 558
Chapman, R. W. 697
Chase, G. B. 459, 460
Chase, G. W. 430
Chase, R. B. 2788
Chatel, B. 2998
Chelf, C. R. 4289
Chelikowsk, J. R. 1763, 1774
Chelini, J. M. 2679, 2694, 2695, 2708, 2709, 2718, 2792, 2797
Chen, C. S. 739
Chen, P. F. 4685, 4742
Chen, P. Y. 1576
Cheney, E. S. 2214
Chermock, R. L. 63, 182, 187, 192
Cherry, J. A. 3493
Cherry, R. N. 165, 829, 843, 855
Chesterman, C. W. 423, 424, 518
Chian, E. S. K. 1462
Childers, B. S. 4359
Childress, S. C. 2532, 2538, 2540
Chipp, E. R. 233, 277
Chisholm, J. L. 4751, 4752, 4753, 4756
Christensen, C. M. 4004, 4005, 4006, 4007, 4012, 4013, 4014, 4024, 4025, 4027, 4033, 4065, 4069, 4088
Christensen, M. N. 419
Christensen, N. I. 673
Christian, R. 154
Christiansen, P. W. 3278, 3280, 3282
Christiansen, R. L. 2917
Chronic, H. 586
Chronic, J. 586
Ciardy, B. F. 400
Ciesielski, P. F. 58
Cilweck, B. A. 453
Clabaugh, S. E. 4238, 4247
Clare, P. H. 3659
Clark, D. K. 516
Clark, D. L. 4807
Clark, J. G. 1890
Clark, J. W. 4773
Clark, K. F. 3105
Clark, L. D. 423, 428, 2348

Clark, M. W. 752
Clark, N. J. 3169
Clark, R. R. 1462
Clark, W. B. 417, 424, 426, 447, 471, 580, 584
Clark, W. E. 759, 820, 822, 825
Clark, W. Z. 916
Clarke, O. M. 55, 74, 75, 76, 77, 78, 124, 136, 139, 149, 152, 158
Clarke, T. G. 3304
Claus, G. 3005, 3019, 3060
Clayton, L. 3354, 3422, 3440, 3462, 3463, 3464, 3474, 3489, 3499
Cleaves, A. W. 4248
Cleaves, E. T. 2276, 2286, 2303
Clebsch, A. 3233
Clegg, K. E. 1236, 1329
Clements, T. H. 119, 121, 146
Clemons, R. E. 3108, 3111, 3113, 3187, 3223, 3280
Clendening, J. A. 4688, 4704, 4708, 4709
Cleveland, G. B. 424, 453, 534, 541, 560, 575
Clifford, M. J. 3517, 3538, 3591
Cline, L. M. 4794
Cloud, P. E. 4238, 4247
Cloud, W. K. 431
Clough, R. W. 432
Coash, J. R. 2867
Cobb, G. D. 3058
Cobban, W. A. 3270
Coble, R. W. 1666, 1669
Coch, N. K. 3045, 4506, 4571, 4572, 4582
Cocke, J. M. 3673
Coe, B. A. 617, 623
Coe, C. 873
Coe, W. D. 57
Coffey, J. C. 3300
Coffin, D. L. 2741
Coffroth, J. M. 2269
Cohen, B. 704
Cohenour, R. E. 4347
Colbert, E. H. 652
Cole, R. E. 1424
Cole, V. B. 1686, 1698, 1863, 1904, 1905
Coleman, D. D. 3198
Coleman, R. G. 3737
Collins, D. 1833
Collins, D. B. 594
Collins, E. W. 4224, 4227
Collins, H. R. 3509, 3513, 3520, 3539, 3582
Collins, R. P. 2703, 2714, 2725
Collins, W. B. 167
Collins, W. E. 2592
Collinson, C. 1197, 1199, 1211, 1244,

1371, 1406, 1432, 1452, 1468, 1472, 1474
Collinson, C.W. 1369, 1465
Compton, R.R. 419
Condie, K.C. 3268
Conklin, J.E. 58
Conley, J.F. 4570, 4576, 4587
Connally, G.G. 4501
Conover, C.S. 880
Conrad, O.G. 4421
Conrad, S.G. 3283, 3294
Conrey, B.L. 523
Conwell, C.N. 333, 334
Cook, C.W. 3482
Cook, D.R. 2868
Cook, E.F. 649, 2905
Cook, K.L. 4365
Cook, R.B. 911
Cook, T.E. 168
Cooper, F.D. 3845, 3846, 3850, 3853
Cooper, H.B.H. 4275
Cooper, J.B. 3228
Cooper, J.D. 915
Cooper, J.F. 518
Copeland, C.W. 73, 105, 141, 177
Corbaley, R. 584
Corcoran, R.E. 3757
Cordova, R.M. 4451
Corgon, J.X. 4121, 4124
Cornwall, H.R. 2878
Coskery, O.J. 705
Costa, J.E. 2276
Cotter, R.D. 4805
Couchot, M.L. 3515, 3518, 3606
Cox, D.C. 988
Cox, D.P. 522
Cox, E. 4031
Craddock, C. 339, 340, 2422, 3093
Craddock, L.L. 4702
Craft, J.L. 3858
Craig, R.J. 700
Crain, L.J. 867
Cramer, H.R. 898, 903, 909, 3792, 3812
Crawford, A.L. 4342, 4343, 4344
Crawford, W.A. 735
Creath, W.B. 2498
Creed, W.A. 4761
Creely, R.S. 407
Cressler, C.W. 910, 932, 933, 934, 935, 945, 954, 955, 957
Cressman, E.R. 1964
Crisler, K. 1811
Croft, M.G. 3356, 3357, 3377, 3378
Cronoble, W.R. 3629

Crooks, J.W. 755
Crossfield, J.A. 1754
Crosson, R.S. 4647, 4649, 4650, 4662, 4663, 4664
Crowell, J.C. 559
Crowl, G.H. 3811, 3822
Crowley, D.J. 1744
Crowley, F.A. 2682
Crowley, W.P. 680, 699, 2244, 2305
Croy, R.L. 3678
Cruickshank, G. 3002, 3038
Cuffy, R.J. 1782
Culver, L.G. 3141
Culver, T.L. 2321
Cunningham, J.E. 3185
Cuppels, N.P. 3992
Curray, J.R. 420
Currens, J.C. 1950, 2124, 2126, 2138
Currie, W.W. 2333
Currier, J.D. 3653
Curry, D.R. 4850
Curtis, N.M. 3685
Cushman, R.V. 1947, 2077
Cvancara, A. 3421, 3485, 3487
Czechanski, M.L. 4785
Czehura, S.J. 2840
Czirjak, L. 2955

Dahl, G.G. 2727, 2802, 2806
Dahlem, D.H. 2764
Dalton, R. 2948
Dalton, R.C. 3647
Dalziel, I.W.D. 4799
Damberger, H.H. 1301, 1349
Dane, C.H. 3154
Daniel, T.W. 51, 56, 57, 84, 96, 104, 124, 136, 149
Daniels, P.A. 4592
Danner, W.R. 4604
Darko, D.A. 1755
Darrow, R.L. 495
Dasch, E.J. 4396
Davie, W. 1700
Davies, D.A. 1005
Davis, D. 3065
Davis, D.A. 988, 1018, 1036, 1037, 1108
Davis, D.R. 796
Davis, F.D. 4385, 4442, 4443, 4444
Davis, F.F. 419, 424, 475
Davis, G.A. 415
Davis, J.C. 1864, 4851
Davis, J.F. 486, 489
Davis, L.L. 431
Davis, L.V. 3228, 3658

Davis, M. E.   180
Davis, R. A.   4794, 4796
Davis, R. W.   1961, 2097
Davis, S. N.   423, 1891
Dawson, L. C.   626
Dawson, T. A.   1515, 1531, 1592, 2092
Day, R. L.   1175
Deacon, R. J.   3702
Deadmore, D. L.   1206
Deal, D. E.   3361
Dean, L.   184
Dean, S. L.   4733, 4735, 4743
Dean, T. J.   2675
Deane, R. E.   676
Deasy, G. F.   3865, 3867
De Benedicty, G.   3056
Deboo, P. B.   27
De Brosse, T. A.   3557
Decker, E. R.   4865
Decker, J. E.   294
Degen, I.   2999
De Geoffroy, J.   4793
Deju, R. A.   3133, 3136, 3143, 3144, 3151
DeKay, R. H.   4533, 4594, 4601
Dellwig, L. F.   1708
DeLong, R. M.   3501, 3520, 3556, 3558, 3564, 3566, 3569, 3572, 3585, 3588, 3607
Denison, R. E.   3619
Dennis, J. G.   4476, 4477
Dennison, J. M.   4523, 4672
Denny, M. V.   4529
Desborough, G. A.   1240, 1241, 1269, 2627
Dethier, D. D.   4659
Devaul, R. W.   4783, 4792
Dever, G. R.   1929, 1930, 1949, 1954, 1987, 2006, 2016, 2137
DeVries, D. A.   2450
DeWitt, W.   3519
Dial, D. C.   2197, 2201
Dibblee, T. W.   409, 411, 421, 424
Dickerson, D. R.   1357, 1364
Dickert, P. F.   419
Dickey, H. P.   1786, 1788, 1789
Dickinson, W. R.   505, 3700
Dickson, D. R.   4454, 4455, 4464
Dietrich, J. W.   4282, 4305
Dietz, W. P.   1644
Dills, G. G.   138
Dilts, R. L.   1795, 1868, 1870, 1874, 1884, 1887, 1895, 1900
Dimmitt, M. N.   2278
Dingwell, A. G.   1360
Dinkins, T. H.   2500, 2505, 2507, 2511, 2523

Dinwiddie, G. A.   3233
Dion, N. P.   769
Dixon, L. H.   2143, 2144, 2145, 2152, 2157
Dockery, D. T.   2534, 2536
Dodge, M. M.   4225
Dodson, C. L.   23
Doelling, H. H.   4380, 4385, 4386, 4388, 4390, 4391, 4392, 4410, 4412, 4415, 4416, 4417, 4442, 4443
Doheny, E. J.   1527
Dole, H. M.   3704
Doll, C. G.   4488
Domokos, M.   3006
Doney, H. H.   3103
Doonan, C. J.   2370, 2376, 2378, 2379, 2380, 2381, 2382
Dora, T.   2987
Doraibabu, P.   3286
Dorfman, M.   4189, 4190, 4197, 4201
Dorheim, F. H.   1630, 1636, 1646
Dorman, L. M.   905
Dott, R. H.   3711, 4799
Dover, J. H.   1162
Dow, W. G.   3346
Dowdall, W. L.   668
Dowhan, J. J.   700, 702
Downey, J. S.   3359, 3360, 3374, 3375
Downs, S. C.   4753
Doyle, F. L.   8, 1198, 1251
Doyle, J. D.   4334
Doyle, R. G.   2213
Dracup, J. F.   3054
Drahovzal, J. A.   177
Dreeszen, V. H.   2849
Dreher, G. B.   1355, 1458
Dreschhoff, G.   1811
Drew, I. M.   537
Dros, M. O.   1690
Droste, J. B.   1527, 1560, 1570, 1579
Duarte-Vivas, A.   3647
DuBois, S. M.   1809
Duchscherer, W. M.   2034
Duckstein, L.   3065
Duex, T. W.   4226
Duke, C. M.   432, 470
Dumontelle, P. B.   1307, 1407, 1424, 1474
Duncan, J. M.   1111
Dunlop, D. V.   3062
Dunning, G.   518
Durham, C. O.   2147, 2151
Durrell, C.   417, 424
Dustin, J. D.   4472
Dutcher, R. R.   1477, 1478
Dutton, C. E.   2346
Dutton, S. P.   4220, 4230, 4231, 4233
Dyman, T.   4108

Eakins, G. R.   225, 228, 273, 276, 279, 329, 330, 338
Eakle, H.   2019
Eardley, A. J.   4395, 4411
Eargle, D. H.   4246
Earll, F. N.   2689, 2691, 2745, 2764, 2784
Early, T. R.   4259
Eason, J. E.   4154
Eaton, J. P.   422
Ebanks, W. J.   1904
Eckard, W. E.   2068
Eddy, G. E.   2323
Edgerton, C. D.   424, 2254, 2255, 2257, 3837
Edgington, W. J.   550, 568
Edmonds, C. S.   1085
Edmunds, W. E.   3778, 3833, 3844, 3855
Edmundson, R. S.   4504, 4588
Edwards, C. L.   3205
Edwards, J.   2229, 2250, 2258, 2269, 2308
Ehlig, P. L.   430
Ehlmann, A. J.   4242
Ehrlich, R.   87
Ehrlinger, H. P.   1300
Eier, D. D.   1163
Eifler, G. K.   4243
Eilertsen, N. A.   2248, 2249
Ekblaw, G. E.   1215, 1368
Elias, M. K.   3693
Ellard, J. S.   41, 133, 201
Ellis, C. W.   665, 695
Ellis, W. C.   650
Ellison, R. L.   3798
Ellison, S. P.   4256
Ells, G. D.   2343, 2363
Ellsworth, G. W.   2016
Ellsworth, L. E.   1172
Elowski, R. C.   2366
El-Shatoury, H. M.   4356
Elston, W. E.   3092
Elzeftawy, A.   1461
Emblen, D. J.   2689, 2764
Emerson, D. O.   423
Emery, P. A.   1710
Emirmen, F. D.   1853
Emmett, L. F.   2678
Emplaincourt, J. L.   186
Emrich, G. H.   1259
Engelhardt, D. W.   2489
Enlows, H. E.   3714, 3767
Epstein, S. B.   3808
Erickson, J. M.   3466
Erickson, K.   4832
Erickson, M. P.   4396, 4407, 4434
Erickson, R. M.   4809

Ern, E. H.   4474, 4482, 4569
Ervin, C. P.   1350
Erwin, J. W.   2925
Erwin, R. B.   4685, 4694, 4728
Erxleben, A. W.   4180, 4248, 4317
Escalera, S. J.   3160
Esogbue, A.   2965
Espenshade, G. H.   424
Essig, C. F.   795
Evans, D. R.   1709
Evans, F. E.   189
Evans, J. R.   424, 427, 430, 482, 496, 566, 580
Evans, T. J.   4258, 4270, 4271, 4274, 4280, 4782, 4784
Evenson, R. E.   418
Everitt, B. L.   4446

F. Beach Leighton and Associates   430
F. M. Fox and Associates   598, 599, 600, 604
Fabbi, B. P.   2900
Fabianek, J.   3001
Fader, S. W.   1692, 1730, 1767, 1816, 1826, 1848, 1871
Fagan, J. M.   4138, 4153
Faill, R. T.   3821
Fairchild, R. W.   846
Fairley, W. M.   894
Fallgren, R. B.   430
Farrand, W. R.   2315
Farvolden, R. N.   1402, 1421
Fay, R. D.   3620, 3628, 3639, 3640
Feder, G. L.   2668, 2673
Feldmann, R. M.   3368, 3524
Fellows, L. D.   2617, 2625, 3664
Feltis, R. D.   2749, 2767
Fender, H. B.   595
Fenstermaker, C. D.   2084
Fenton, T. E.   1644
Ferguson, W. B.   3877, 3903
Ferm, J. C.   4154
Ferrell, V. M.   282
Ficke, J. F.   3070
Field, M. T.   2210
Fields, F. K.   4470
Fies, M. H.   51
Fife, D. L.   454, 551, 554, 580
File, L. A.   3128, 3131, 3138, 3168
Fillo, P. V.   3130
Finch, W. I.   1943
Finger, G. C.   1276, 1357
Finlayson, C. P.   4135
Finley, R. J.   4220, 4230, 4234, 4252, 4328
Fischer, R. P.   424
Fish, S. E.   2091

Fisher, G.W. 2246, 2284, 2310
Fisher, W. L. 4162, 4165, 4170,
4172, 4178, 4179, 4180, 4183,
4218, 4282, 4285, 4289, 4291,
4293, 4294
Fisher, W.W. 380, 3166
Fisk, H. G. 2810
Fitak, M. R. 3540
Flanagan, M. A. 1788
Flawn, P. T. 4163, 4166, 4167,
4176, 4177
Fletcher, R. 2210
Flint, R. F. 670, 674, 679, 684,
687, 691, 692
Flippo, H. N. 840
Florida Bureau of Geology 883,
886
Florida Geological Survey 813,
878
Flower, R. H. 3236, 3237, 3243,
3244, 3246, 3247, 3248, 3249,
3250, 3251, 3252, 3253, 3260,
3261, 3265
Floyd, R. J. 4110
Flueckinger, L. A. 1893
Foerste, A. F. 2569
Fogel, M. 3065
Follmer, L. R. 1341, 1342, 1353,
1471, 1425
Folsom, C. B. 3428, 3432, 3435,
3437, 3439, 3446, 3448, 3451,
3455, 3456, 3461, 3468
Forbes, M. J. 2171, 2172, 2185,
2193, 2195
Forbes, R. B. 338
Forbes, W. H. 2210, 2218
Ford, J. A. 2139, 2140, 2141
Ford, J. P. 3584, 3594
Forsyth, J. L. 3560, 3567, 3570
Foster, J. B. 773, 784, 791,
828
Foster, R.W. 3096, 3099, 3141,
3155, 3182, 3190, 3220, 3274,
3277
Fouts, J. A. 897
Fox, K. F. 4616, 4617
Fox, R. D. 2706
Fox, W. T. 1850
Fox and Associates, F.M. see
F. M. Fox and Associates
Frakes, L. A. 3801
Franklin, M. A. 910
Franks, P. C. 1752
Franz, R. 2241
Franzoni, R. A. 2079
Fraser, G. S. 1299, 1322, 1324,
1439, 1443
Fraunfelter, G. H. 2612
Frazier, D. E. 4186

Fredericks, P. E. 4232, 4337
Frederickson, E. A. 3661
Freeman, A. C. 197
Freeman, L. B. 1907
Freeman, P. S. 4298
Freeman, T. 387
Freers, T. F. 3334, 3351, 3362,
3429, 3440
French, B. M. 2384
French, R. R. 1497, 1502, 1509,
1511
Frenier, W.W. 2279
Frey, L. H. 2012
Frezon, S. E. 390
Friberg, J. F. 2094
Friedman, I. 1249, 1253
Friedman, J. M. 3774
Friedman, S. A. 2001
Friel, E. 4702, 4750, 4751, 4755,
4756
Fritts, C. E. 272, 274
Friz, T. O. *4812, 4814
Froelich, A. J. 1960
Fronczek, C. J. 3054
Frost, R. R. 1431, 1444, 1455
Frund, E. 1242
Frye, C. I. 3350
Frye, J. C. 1196, 1197, 1203, 1220,
1256, 1283, 1293, 1296, 1317,
1323, 1327, 1341, 1342, 1394,
1418, 1436, 1446, 1467, 2903,
3193, 3198, 3199, 3214, 3215,
3263, 4284, 4286
Frye, P. M. 4752
Fuerstenau, M. C. 3119, 3124, 3127
Fulkerson, F. B. 2681, 2688
Fullagar, P. D. 2218
Fuller, D. L. 2645, 2647, 2648,
2649, 2650, 2651, 2652, 2653,
2654, 2655, 2656, 2657, 2658,
2659
Fuller, J. O. 3555
Fullerton, R. O. 4117
Furcron, A. S. 963
Furlow, J.W. 901
Furnish, W. M. 3666
Fyles, J. T. 4613

Gage, J. E. 3211
Galbraith, F.W. 352, 368, 377
Galbraith, J. 1136
Galbraith, J. H. 1182, 1183
Gallaher, J. T. 3946, 3957
Galle, O. K. 1701, 1716, 1784, 1786,
1788, 1789
Galloway, A. J. 423, 440
Galloway, M. J. 645
Galloway, W. E. 4179, 4210, 4220,

4225, 4252, 4310, 4322, 4335
Gann, E. E.   2665
Gannon, W. B.   928
Ganser, R. W.   3653
Garcia, R.   3048, 4184
Gardner, R. A.   2228, 2295
Gardund, H.   3048
Garner, L. E.   4162, 4198, 4208, 4243, 4250, 4289, 4295, 4321
Garrick, B. J.   1111
Garrity, T. A.   3149
Garside, L. J.   2882, 2892, 2893, 2915, 2919, 2923, 2935
Garvin, R. F.   4419
Gastil, R. G.   424
Gaston, M. P.   1142
Gates, J. S.   4448
Gates, R. M.   673, 678, 688, 694
Gathright, T. M.   4513, 4540, 4547, 4548, 4549, 4550, 4590
Gay, T. E.   416, 424, 436
Gaydos, M. W.   2172, 2174, 2193
Gazzier, C. A.   150
Geach, R. D.   2679, 2786, 2696, 2701, 2712, 2723, 2746, 2804
Gehr, J. B.   2022
Gentile, R. J.   2609, 2611, 2620, 2639
Geological Society of America   4353
George, J. V.   2028
Georgia Department of Natural Resources   960, 962
Geraghty and Miller, Inc.   2964
Gerdemann, P. E.   2635
Gerhard, L. C.   3467
Gerhardt, R. A.   4823
German, E. R.   180
Gernant, R. E.   2245, 2290
Gernazian, A.   958
Geyer, A. R.   3782, 3784, 3785, 3787, 3899, 3908
Gibson, A. M.   3692
Gibson, T. G.   2245
Gilbert, J. L.   3204
Gilbert, O. E.   135, 151, 183
Gilbert, W. G.   282, 285, 293, 294
Gilbreath, L. B.   398
Gilchrist, J. M.   668
Gilkeson, R. H.   1454
Gillerman, E.   3094
Gillespie, W. H.   4708, 4709, 4710
Gilman, R. A.   2213, 2218
Gilmore, C.   472
Gilmore, E. H.   2684, 2732, 2801, 2802, 2806
Gilmore, J. L.   1648

Gimlett, J. I.   2912
Girard, R. M.   4239, 4259, 4260, 4261, 4262, 4263, 4264, 4265, 4266, 4268, 4269, 4272, 4273, 4274
Giroux, P. R.   2374, 2377
Glaeser, J. D.   3793, 3814, 3838
Glaser, J. D.   2242, 2243, 2289, 2293
Glass, B. P.   3692
Glass, G. B.   4840, 4866, 4867, 4873, 4876, 4878, 4881, 4886, 4887, 4892
Glass, H. D.   1203, 1220, 1256, 1283, 1293, 1315, 1327, 1341, 1342, 3193, 3198, 3199, 3214
Glawe, L. N.   39, 953
Glenn, R. C.   2530
Gluskoter, H. J.   1228, 1288, 1332, 1355, 1419, 1437, 1448
Gockel, D. J.   1652
Godfrey, A. E.   2266
Goebel, E. D.   1683, 1689, 1690, 1703, 1736
Goebel, J. E.   2426
Goebel, L. M.   3162
Goeltz, N. S.   4357, 4376
Goff, F. G.   4761
Golde, M. V.   1526
Goldman, D.   1178
Goldman, H. B.   424, 527, 528, 529
Goldthwait, R. P.   3612
Gonthier, J. B.   4817
Good, D.   1842
Good, D. I.   1697, 1852, 1866
Good, R. S.   4585
Goode, H. D.   4452, 4457, 4460
Gooding, A. M.   1523
Goodman, A.   3010
Goodwin, B. K.   3791, 4473, 4555, 4577
Goodwin, R. W.   3044
Goolsby, D. A.   774, 791, 799
Goolsby, S. M.   630
Gopalakrishnan, B. S.   431
Gordon, D. L.   1630
Gosling, A. W.   2771
Goth, J. H.   3881
Gower, H. D.   424
Graeff, G. D.   2163
Graf, D. L.   1249, 1253, 1264
Graham, J. B.   1622
Graham, J. J.   516
Graham, R. L.   4357, 4390, 4391
Granata, G. E.   4209, 4332
Grande, L.   4846
Grandjean, M. A.   3126
Grannell, R. B.   430
Grant, A. R.   4610
Grant, D. L.   1178
Grant, R. W.   4491, 4499
Grantham, R. G.   91, 766, 827, 832, 838

Gray, C. 3875
Gray, C. H. 424, 427, 526
Gray, H. H. 1535, 1556, 1557, 1580, 1590, 1593, 1594, 1598, 1603, 1613
Gray, J. J. 3735, 3769
Gray, L. M. 1617
Greeley, R. 1180, 3713
Green, J. C. 2425, 2449
Green, J. E. 2079
Green, J. H. 4805
Green, J. W. 2537, 2538
Green, M. 154, 4034, 4053
Greenberg, S. S. 4503
Greenman, D. W. 3922, 3924
Greensfelder, R. W. 430, 459
Greenwood, W. R. 1126, 1127, 1130, 1132, 1144, 1174
Gregg, D. O. 924, 926
Gregory, A. R. 4326
Gregory, H. E. 4341
Grevlich, G. 3075
Gries, J. P. 4037, 4060
Griess, P. R. 3867
Griffin, G. M. 888
Griffin, M. M. 918, 919
Griffin, R. A. 1454, 1455, 1462
Griffin, W. L. 2444
Griffith, H. D. 2093
Griffiths, J. C. 1876
Griggs, J. H. 197
Grisafe, D. A. 1792, 1829, 1834
Grislom, A. 422
Griswold, G. B. 3121, 3123, 3168
Groat, C. G. 4178, 4307, 4311
Grocock, G. R. 4645
Groenewald, G. H. 3465, 3493, 3494, 3500, 3536
Groff, S. L. 2794, 2798, 2799
Grogan, R. M. 2115
Groh, E. A. 3699
Grolier, M. J. 4623
Groot, J. J. 711
Gross, D. L. 1315, 1352, 1406, 1409, 1411, 1423, 1430, 1434, 1445, 1450, 1460, 1471
Gross, E. B. 424, 518
Growitz, D. J. 3953
Grunewald, C. 1364
Grybeck, D. 336
Guevara, E. H. 4184, 4209
Guidotti, C. V. 696, 2211
Gundersen, J. N. 451
Gustavson, T. C. 4196, 4205, 4220, 4230, 4234
Gutentag, E. D. 1692, 1815, 1823
Gutjahr, A. L. 3220
Gwinn, J. E. 2023
Gwynn, J. W. 4388, 4389,

4429, 4433

Habashi, F. 2705, 2715, 2815
Hackel, O. 418
Hackett, J. E. 1294, 1377, 1379, 1384, 1398
Hackett, S. W. 284, 291
Hadley, D. W. 4831
Hager, M. W. 4836
Hahl, D. C. 4449, 4450, 4461
Hahn, W. F. 730, 731
Hains, C. F. 90, 115
Hakes, W. 1811
Halasi-Kun, G. J. 2944, 2949, 2957, 2958/59, 2963, 2991, 2994, 3012, 3017, 3035, 3042, 3046, 3049, 3054, 3057, 3059, 3068, 3073, 3080
Halbert, C. A. 4764
Hale, R. C. 4116
Hale, W. N. 2698, 2703, 2714, 2725, 2731, 2737, 2743
Haley, B. R. 388, 389, 390, 392, 394, 395, 397
Hall, B. A. 2213, 2217
Hall, D. C. 590, 596
Hall, G. F. 1644
Hall, G. M. 3913
Hall, P. F. 199
Hall, W. B. 1177, 2530
Hall, W. D. 4199
Hall, W. E. 490
Halladay, C. R. 4554
Hallberg, G. R. 1633, 1650, 1653, 1657, 1658
Halliday, W. R. 4628
Halloran, T. 2073
Ham, W. E. 3619, 3663, 3685, 3694
Hambleton, W. W. 1686, 1698, 1863, 1869
Hamblin, W. K. 1707, 1740
Hamilton, J. L. 597
Hamilton, L. J. 4024, 4025
Hamilton, W. B. 424
Hamlin, H. P. 762, 2282
Hammond, P. E. 4660
Hammond, S. V. 190
Hamon, J. H. 738
Hampton, E. R. 847
Hanby, K. P. 61, 197
Hancock, J. H. 3667
Handford, C. R. 4220, 4230, 4232, 4337
Handy, A. H. 4461
Haney, G. 2075
Hankins, W. T. 2472
Hanks, T. C. 431
Hansen, A. J. 2097

Hansen, D. E. 3316, 3325, 3347
Hansen, H. J. 2226, 2251, 2267,
 2280, 2285, 2299, 2307, 2308
Hansen, J. P. 92, 107
Hansen, M. 2736, 2742, 3411,
 3433
Hansen, M. C. 3512, 3514, 3516,
 3526
Hansen, R. E. 1674, 1675
Hansmann, E. W. 662
Hanson, G. F. 4767, 4771, 4827
Hanson, G. N. 2419
Hanson, L. G. 237
Haraldson, H. C. 3478
Harbaugh, J. M. 1650
Harbaugh, J. W. 1700, 1702,
 1841, 1847, 1849, 1853, 1862
Hardee, J. 842
Harder, A. H. 2170, 2181, 2189,
 2196
Hardie, L. A. 2247
Hardin, J. D. 61, 186
Hardy, R. G. 1728, 1750, 1793,
 1828, 1834, 1839, 1855, 1857,
 1872, 1873, 1878, 1901
Hare, C. E. 4676
Harksen, J. C. 4045, 4049, 4050,
 4051, 4052, 4053, 4054, 4061,
 4064
Harlow, F. H. 3259
Harnish, D. H. 2793
Harper, D. P. 3032, 3043
Harr, C. A. 4825
Harris, A. J. 36
Harris, B. 2550, 2638
Harris, H. B. 162, 163
Harris, K. F. 3967
Harris, K. L. 3462
Harris, L. D. 2054
Harris, S. E. 1638
Harris, W. B. 4534
Harris, W. F. 23, 25, 52, 162,
 164
Harris, W. L. 2710
Harrison, J. A. 1222
Harrison, J. L. 1491
Harrison, S. F. 37
Harrison, S. S. 3409, 3445
Harrison, W. 1488, 1533
Hart, D. L. 3641
Hart, E. W. 418, 435, 437,
 444, 457, 474, 569
Hart, O. D. 3625
Hart, S. S. 603, 646
Hart, T. A. 3653
Hartke, E. J. 1564, 1572, 1602,
 1605, 1610
Hartley, M. E. 904
Hartley, R. P. 3554, 3559

Hartman, D. C. 331
Hartman, H. 3205
Hartman, S. 2996
Hartwell, J. H. 834
Harvey, R. D. 1201, 1226, 1260,
 1271, 1324, 1397, 1414, 1426,
 1431, 1444
Hastings, E. L. 84, 85, 104, 166,
 170
Hatch, J. R. 1663, 2643, 4444
Hatcher, D. A. 1710
Hatcher, R. D. 902, 4150
Hatchett, J. L. 3318
Hatfield, C. B. 1514
Hathaway, L. R. 1784, 1786, 1788,
 1789
Hattin, D. E. 1713, 1755, 1771, 1822
Hauge, C. J. 471, 565
Haught, O. L. 4674, 4677, 4679,
 4681, 4682, 4687, 4711, 4712,
 4715
Hausel, W. D. 4871, 4889, 4893
Havenor, K. C. 3147
Hawaii Water Authority 989, 992
Hawkins, D. B. 332, 335, 337
Hawkins, M. E. 4273, 4280
Hawks, W. L. 3164
Hawley, J. W. 3108, 3112, 3113,
 3201, 3217, 3231
Hayes, E. 2982
Hayes, E. C. 68
Hayes, L. D. 3304
Hayes, M. O. 4296
Hayes, W. C. 2549A, 2578, 2582, 2583,
 2584, 2589
Hayhurst, D. L. 4443
Hazzard, R. T. 4283
Heald, M. T. 4678
Healy, H. G. 764, 768, 777, 785,
 790, 798, 802
Heath, E. W. 2216
Heath, R. C. 786
Heaton, M. J. 3654
Hebrank, A. W. 2602
Heck, D. B. 1417
Heck, E. T. 4676
Heckel, P. H. 1772, 1821, 3832
Hedberg, M. R. 1130
Hedburg, L. L. 4427
Hedges, L. S. 4001, 4002, 4003,
 4008, 4009, 4010, 4011, 4015,
 4016, 4017, 4026, 4028, 4036,
 4073, 4074, 4075, 4085
Hedges, R. B. 2762, 2768
Hedlund, R. W. 3636
Heidari, M. 1350
Heidel, S. G. 2279, 2281, 2287, 2291
Heigold, P. C. 1210, 1225, 1306,
 1324, 1344, 1350, 1351, 1363,

1400, 1435
Heim, G. E. 1428, 2622, 2660, 2661, 2662
Heimlich, R. A. 3524
Heineman, R. E. 365, 370
Heinitz, A. J. 1670
Heinrich, E. W. 634, 2317, 2362
Helfinstine, R. J. 1231, 1252, 1268, 1279, 1290, 1305, 1309, 1318, 1340, 1410
Helgerson, R. 4107
Heller, C. A. 1882
Helton, W. L. 1940, 2032, 2138
Hembre, I. O. 4770
Hemingway, E. L. 2253
Hemish, L. 3465
Henderson, G. W. 4413, 4426
Hendricks, E. L. 2950
Hendricks, T. A. 392, 395
Hendrickson, G. E. 2314, 2370, 2376, 2378, 2379, 2380
Hendry, C. W. 741, 744, 746, 748, 806, 807, 808, 809
Henika, W. S. 4538, 4540, 4547, 4548, 4549, 4550, 4576, 4580, 4587
Hennessy, G. J. 2026
Henning, M. W. 326
Hennings, R. G. 4785
Henry, C. D. 4200, 4226, 4252, 4253, 4331, 4333, 4336
Henry, H. R. 179
Herbert, C. F. 209, 214
Herdendorf, C. E. 3529, 3559, 3561, 3575, 3577, 3580, 3583
Hermel, M. 3060
Heron, S. D. 3977, 3980, 3993
Herpers, H. 3083
Herreid, G. 232, 235, 236, 240, 240, 242, 243, 244, 245, 246, 247, 252, 257, 259, 261, 262, 264, 270, 271, 275, 283
Herrick, S. M. 931, 937, 947, 974
Herring, R. R. 795
Herschler, L. H. 1067
Hershey, H. G. 1624, 1673
Hershey, R. E. 4109, 4148
Herz, N. 424
Hess, W. R. 1857
Hester, N. C. 1308, 1320, 1424, 1439, 1443, 1473
Hester, W. G. 910, 957
Hetland, D. L. 1147
Hewett, D. F. 424
Hewitt, W. P. 4401
Heyl, A. V. 1982, 2119, 2234, 4803, 4804
Heylmun, E. B. 4347, 4348,

4398, 4403, 4404
Heyman, L. 3878, 3892
Hickey, H. 1834
Hickman, R. C. 762
Hickman, R. G. 339
Hidalgo, R. V. 4697, 4702, 4705, 4729
Higgins, C. T. 576
Higgins, M. W. 896, 2244, 2310
High, L. R. 4425
Highland, J. D. 1657
Hill, C. 4841
Hill, J. M. 3226
Hill, J. R. 1564, 1567, 1572, 1573, 1599, 1602, 1605, 1618
Hill, W. E. 1687, 1689, 1701, 1709
Hillard, P. D. 3157
Hiller, D. 596
Hills, D. L. 2417
Hilpert, L. S. 4346
Hilpman, P. L. 1690, 1703
Himmelberg, G. R. 2431, 2432, 2441
Hinckley, B. S. 4842
Hindall, S. M. 4819, 4824
Hinds, G. W. 4246
Hine, G. T. 2129
Hines, M. S. 401
Hinkle, M. E. 2252
Hinkley, K. C. 1405, 1433
Hinman, E. E. 1625, 1643
Hintzman, D. E. 2690
Hinze, W. J. 2342, 2353, 2355
Hirashima, G. T. 1011, 1012, 1088, 1089
Hirsch, A. 3027
Hiss, W. L. 3194
Hitchcock, S. W. 663
Hladik, W. B. 1687, 1701, 1857
Ho, A. M. 1030
Hoare, R. D. 3506, 3510
Hobba, W. A. 4731, 4751, 4756
Hobday, D. K. 4224, 4227
Hodges, A. L. 2187
Hodgson, H. E. 627
Hodson, W. G. 1678, 1705, 1719, 1742
Hoffard, S. 1087, 1091
Hoffer, F. B. 4521, 4526, 4537
Hoffer, J. M. 3203
Hoffman, D. 2549B, 2597
Hoffman, J. F. 1133
Hoffman, J. P. 3174
Hofstra, W. E. 590
Hogberg, R. K. 2385, 2388, 2389, 2398, 2401, 2402, 2414
Hoggatt, W. C. 3212
Holdway, K. 1781, 1811
Hole, F. D. 4759, 4761, 4762, 4763, 4770, 4776, 4802, 4827

Holland, F. D. 3414, 3415, 3416, 3417, 3418, 3419, 3421, 3424
Hollenbeck, R. P. 1953, 4146
Hollowell, J. R. 3939, 3952
Hollyday, E. F. 2298
Holsinger, J. R. 4512
Holt, C. L. R. 4769, 4805, 4810
Holt, R. D. 588, 637, 644, 3188
Holtz, W. G. 646
Holtzman, R. C. 3495
Holy, Y. 4832
Honjo, S. 1688
Hook, J. W. 2121
Hook, S. C. 3265, 3270
Hoover, K. V. 3866, 3882, 3883
Hope, R. C. 3305
Hopkins, J. T. 2085
Hopkins, M. E. 1197, 1287, 1318, 1340, 1345, 1410, 1469, 1470
Horick, P. J. 1625
Hornbaker, A. L. 594, 637, 644, 1839, 1855, 1857, 1873, 1878, 1889
Horowitz, A. S. 1486, 1494, 1530
Horst, W. E. 3159
Horton, R. C. 2897, 2901
Hose, R. K. 2886
Hosking, J. S. 1261
Hoskins, D. M. 3790, 3907
Hoskins, H. A. 4676
Hosman, R. L. 2178, 2198, 2200
Hosterman, J. W. 1972, 2282, 3839
Hotz, P. E. 424
Houston, R. S. 4847, 4883
How, K. T. S. 1118
Howard, P. F. 4489
Howard, R. H. 1205, 1225, 1233, 1314
Howe, H. V. 2488
Howe, R. C. 2150
Howe, W. B. 1891, 2621, 2622, 2632, 2660, 2661, 2662
Howell, P. 2029
Howells, L. W. 3998, 3999, 4000
Howland, M. D. 342
Hoyer, B. E. 1632, 1633, 1634
Hreha, A. J. 1531
Hsu, K. J. 467
Hubert, J. F. 668
Hubricht, L. 2457, 2458, 2515
Hudson, D. E. 431
Huff, W. J. 2519
Huffman, G. G. 3647, 3653, 3667, 3689
Huffman, M. E. 455, 459, 561
Hughes, G. H. 835, 860, 861, 867, 868
Hughes, G. M. 1262, 1377, 1393,

1402, 1421, 1427, 1454, 1456
Hughes, P. W. 3726, 3729, 3732, 3736, 3762
Hughes, T. H. 143
Hulick, C. P. R. 3739
Hull, D. A. 3764
Hull, J. E. 775, 853
Humphrey, J. E. 3689
Hunt, A. S. 4500
Hunt, C. S. 1459
Hunt, J. B. 3469, 3470
Hunter, C. D. 1914, 1912, 1916, 2020
Hunter, R. E. 1237, 1247, 1255, 1258, 1273, 1284
Hunter, R. G. 4685, 4686
Huntting, M. T. 4651, 4660
Hurdon, R. 3008
Hurst, V. J. 895, 941
Hussey, A. M. 2209, 2218
Husted, J. E. 963
Hutchins, W. A. 993
Hutchinson, R. D. 3342, 3370, 3371, 3374, 3393
Hutchison, H. C. 1528, 1529
Huxel, C. J. 1042, 1052, 3314, 3315
Huxsaw, C. L. 4522
Hyde, L. W. 6, 40, 41, 122, 126, 129
Hynes, J. L. 622

Idaho Bureau of Mines and Geology 1188
Idaho Department of Highways 1188
Idaho Department of Reclamation 1188
Ikola, R. J. 2415
Imbrie, J. 1851
Inners, J. D. 3910, 3911
Interstate Oil Compact Commission 2591
Irwin, W. P. 406, 415, 424, 3337
Ives, W. 1752
Ivey, J. B. 605

Jack, R. N. 538
Jackman, H. W. 1222, 1231, 1268, 1279, 1290, 1305, 1309
Jacob, A. F. 3401, 3483, 3488
Jacobeen, F. H. 2260
Jacobs, A. M. 1298, 1321, 1471
James, A. H. 4436
James, A. T. 1299
James, H. L. 2345, 2347, 2348, 3280, 3281
James, L. P. 4387
Jamhindikar, S. 1729
Jamieson, W. 3020, 3025

Janssens, A. 3507, 3519, 3537, 3571, 3579, 3601
Jarjur, S. Z. 1761
Jaron, M. G. 3829
Jasper, M. W. 212, 215, 222, 223
Jayaprakash, G. 1811
Jefferson, P. O. 52
Jeffries, E. L. 3647
Jelgersma, S. 2447
Jenkins, E. B. 2043
Jenkins, E. D. 1825
Jenkins, G. M. 2156
Jenkins, O. P. 466
Jennings, A. R. 2090
Jennings, C. W. 414, 424, 452, 491
Jensen, H. M. 3338, 3339
Jensen, M. L. 4355
Jerome, S. E. 2868, 3097
Jewell, J. W. 42
Jewett, J. M. 1710
Jirik, C. J. 4272
Johansson, W. I. 4475
Johns, W. M. 2684, 2692, 2700, 2717, 2739
Johnson, A. 2359
Johnson, A. C. 2108
Johnson, A. M. F. 927
Johnson, G. H. 4584, 4595
Johnson, H. S. 3974, 3980, 3993
Johnson, K. S. 3663, 3678, 3680, 3682, 3684, 3685, 3686, 3687
Johnson, M. G. 2890
Johnson, P. H. 3153
Johnson, R. C. 4155
Johnson, R. G. 1470
Johnson, R. W. 4523
Johnson, S. H. 2422
Johnson, S. S. 4522, 4524, 4529, 4531, 4543, 4574, 4581, 4583, 4586, 4593, 4599
Johnson, W. H. 1234, 1313, 1315, 1317, 1471
Johnston, H. E. 3932, 3934, 3938
Johnston, J. E. 4217
Johnston, R. H. 708, 709, 727
Joiner, T. J. 35, 43, 46, 100, 101, 112, 115, 118, 123, 127, 179
Jolly, J. L. 1982
Jones, D. C. 595, 609, 610, 611, 615
Jones, D. J. 1910, 1919, 1920
Jones, D. L. 406, 424
Jones, F. P. 2709, 2728, 2740, 2752
Jones, J. 2976
Jones, M. L. 4109

Jones, O. W. 4266
Jones, R. C. 916
Jones, R. W. 1137, 1173, 4840
Jones, W. B. 110
Jones, W. K. 4684
Jordan, L. 3624
Jordan, P. A. 701
Jordan, R. R. 706, 711, 721
Jorgensen, D. G. 4012, 4017, 4014, 4079, 4080
Joyce, L. G. 920
Joyner, B. F. 757, 763, 772, 837, 840, 858
Jungmann, W. L. 1720

Kaesler, R. L. 1842
Kahle, J. E. 430, 450, 453, 577
Kaiser, W. R. 4216, 4217, 4275, 4314, 4325, 4335, 4339
Kalb, G. W. 3586
Kaliser, B. N. 4366, 4369, 4430, 4446
Kanivetsky, R. 2435
Kapadia, R. R. 4336
Kapple, G. W. 2267
Karklins, O. L. 2442
Karlstrom, K. E. 4883
Karner, F. R. 3447, 3479
Kasper, A. E. 2218
Kauffman, A. J. 2688
Kauffman, M. E. 2783
Kaufman, A. 1919
Kaufman, M. A. 239, 242
Kaufman, M. I. 769
Kaufmann, R. F. 2922
Kay, G. F. 1620, 1621, 1622
Kayser, R. B. 4347, 4352, 4409, 4433
Keady, D. M. 2546
Kebblish, W. 3854, 3857, 3859
Keene, J. R. 4062
Keene, K. 1834
Keene, K. M. 1787
Keener, M. J. 7
Kehle, R. O. 4189
Kehn, T. M. 2137
Keith, B. D. 1555
Keith, S. B. 353, 354, 358, 360, 362, 363
Keller, G. R. 4235
Keller, M. 3559
Keller, S. J. 1515, 1538, 1543, 1544, 1545, 1546, 1547, 1548, 1549, 1550, 1551, 1552, 1553, 1554, 1555, 1574, 1585, 1586
Keller, W. D. 2551
Keller, W. T. 657
Kelley, D. A. 3879

Kelley, D.R. 3876, 3880, 3884
Kelley, F.R. 424, 575
Kelley, R.W. 2315, 2318, 2322, 2337, 2339
Kelley, V.C. 3109, 3207, 3210, 3239, 3256, 3262, 3266, 3279
Kellogg, J.A. 1760
Kelly, F.H. 2689, 2764
Kelly, J.C. 44
Kelly, T.E. 3319, 3320, 3321, 3348, 3349
Kemmerly, P.R. 4130
Kemp, W.R. 2934
Kempton, J.P. 1210, 1212, 1273, 1293, 1312, 1404, 1405, 1433, 1459
Kendall, T.A. 1958
Kennedy, E.J. 1413, 1419, 1420
Kennedy, M.P. 438, 539, 564, 572, 577
Kenner, W.E. 755, 756
Kent, B.H. 3806, 3847
Kerhin, R.T. 2270
Kern, M.K. 2453, 2455, 2477, 2487
Kerns, R.L. 4440
Kerr, J.R. 3825, 3827
Kerr, P.F. 537
Khawaja, I.U. 1600
Kidd, J.T. 10
Kidd, R.E. 59, 132, 142, 146
Kiechel, W. 2979
Kier, R.S. 4236, 4330
Kiessling, E.F. 453
Kiessling, E.W. 543, 571
Kilburn, C. 2170
Kim, S.M. 1404
Kimmel, G.E. 3944
Kimrey, J.O. 3285
Kindt, E.A. 3510
King, J.E. 1352, 1475
King, J.S. 1180
King, P.B. 424
King, R.U. 424
King, W.E. 3185, 3231, 3258
Kinkel, A.R. 424
Kinoshita, W.T. 430
Kirkby, E.A. 2318, 2319, 2322, 2324, 2325
Kirkham, R.M. 607, 608, 620, 621, 628
Kirkpatrick, G.A. 1951
Kirstein, D.S. 3304
Kisvarsanyi, E.B. 2593, 2623, 2631, 2636, 2641
Kitts, D.B. 3668
Klapper, G. 1756
Klasner, J.S. 2365
Klausing, R.L. 3331, 3332,

3333, 3339, 3366, 3367, 3386, 3387, 3397
Klein, H. 824, 847
Klein, N.H. 3920
Klein, T. 4861
Kleinschmidt, M.K. 1785
Klett, M.C. 3466
Klingman, C.L. 2262, 2264, 2265
Knaebel, J. 229
Knapp, N.F. 3604
Knechtel, M.M. 2282
Knight, A.L. 191
Knight, R.D. 2577
Knochenmus, D.D. 844, 859
Knostman, R.W. 2681, 2688, 2698, 2703, 2714
Knowles, C.R. 1135
Knowles, D.B. 17
Knox, L.M. 4109
Knox, R.D. 458
Knutilla, R.L. 2367, 2370, 2371
Koch, D.L. 1642, 1646, 1647
Koch, N.C. 3987, 4008, 4009, 4010, 4011, 4018, 4019, 4021, 4023
Koehn, H.H. 3110, 3161
Koehn, M.A. 3110
Koenig, J.W. 2577, 2579, 2586, 2606, 2613
Koester, H.E. 3952
Kohlenberger, L.B. 1442
Kohler, S. 584
Kohout, D.L. 3582
Kohout, F.A. 834
Kohut, J.J. 1996
Kolata, D.R. 1347, 1358, 1361
Konizeski, R.L. 2699, 2726
Koontz, T. 3635
Kopp, R.S. 1183
Koppe, E.F. 3778, 3861, 3874
Korosec, M.A. 4669
Korsok, A.J. 3508
Kottlowski, F.E. 3090, 3125, 3155, 3158, 3168, 3201, 3254, 3257, 3271, 3273, 3278
Koufman, M.I. 836
Kovach, J. 4740
Kovacs, G. 3004
Kozak, S.J. 4562, 4575
Kraatz, P. 1262, 1391
Krause, R.E. 924, 952
Krauser, R.F. 3045
Kreisa, R.D. 4542
Kreitler, C.W. 4205, 4209, 4219, 4224, 4228, 4318, 4320
Krieger, R.A. 2077
Krohn, D.H. 2838
Kroll, R.L. 690
Krombach, J. 2978
Krumlauf, H.E. 343

Kudo, A. M. 3207, 3210
Kuff, K. R. 2277
Kuhn, J. K. 1252, 1355, 1415, 1442
Kuiper, L. K. 1651
Kulander, B. R. 4723, 4735, 4743
Kumar, M. B. 2153, 2154
Kume, J. 3311, 3316, 3347, 3442, 3452, 4015, 4016, 4017
Kun, P. 1176
Kupfer, D. H. 500
Kurtz, V. E. 2632, 2635
Kwon, B. D. 1578, 1581
Kyriazis, M. K. 1399

Labotka, T. C. 1320
LaBrie, L. W. 4339
Lacroix, A. V. 4490
Ladwig, L. R. 608, 628
Lageson, D. R. 4864, 4871, 4874, 4877
Laird, W. M. 3411, 3412, 3413, 3420, 3434
Lamar, D. L. 542
Lamar, J. E. 1193, 1215, 1373
Lamar, W. L. 3964
Lamb, G. M. 61, 140
Lambert, J. F. 4365
Lambert, T. W. 1947, 1963, 2097
Lamont, W. E. 85
Lamoreaux, P. E. 1, 43, 47, 48, 49, 60, 98, 114
Land, L. F. 854
Lander, W. T. 4137
Landers, R. A. 4717, 4726
Landis, E. R. 399, 424, 1662, 3154, 3372
Landon, R. A. 1262, 1312, 1390, 1402, 1421
Landreth, J. O. 2807
Landrum, J. R. 970
Lane, C. W. 1711, 1860
Lane, D. W. 4859
Lane, E. 890
Lane, N. G. 1699
Lang, M. 2986
Langenheim, R. L. 2872
Langenheim, V. A. 1194
Langford, R. H. 4450
Langton, J. M. 3667
Lapham, D. M. 3840, 3842, 3851, 3875, 3897
La Roque, A. 3502, 3503, 3504, 3505
Larsen, J. I. 1337, 1378, 1379
Larson, E. R. 2872

Lasca, N. P. 4802
Laswell, T. J. 2546
Latimer, I. S. 4708, 4710, 4758
Lauffer, J. B. 2030
Laughlin, C. P. 2224, 2225, 2236
Laughlin, D. J. 1680
Lawrence, E. F. 2860
Lawrence, R. D. 3761
Lawson, D. C. 2695, 2700, 2717, 2728, 2730, 2747, 2750, 2755, 2759, 2765, 2740, 2752, 2769, 2774, 2777, 2780, 2842
Lawton, D. E. 951
Layton, D. W. 1764
Leach, C. H. 4176, 4177
Leach, S. D. 832, 847
Leaf, C. G. 3843
Leahy, P. P. 729, 732
Leamnson, R. N. 1300
Leamon, A. R. 4139
Learned, E. M. 412
Leap, D. 4020
Leary, R. L. 1356
Le Blanc, R. J. 4244, 4245
Lee, C. H. 530
Lee, G. B. 4759
Lee, M. C. 1462
Lee, T. C. 1584
Leeds, D. J. 470
Lees, J. A. 3802
Lees, J. H. 1619, 1620
Leggette, R. M. 3914
Le Grand, H. E. 3000
Legrand, H. E. 47, 49, 60
LeGrand, J. R. 3969
Lehner, R. E. 2863
Leighton, F. B. 2913
Leighton and Associates, F. Beach see F. Beach Leighton and Associates
Leininger, R. K. 1526
Leland, H. V. 1406, 1408, 1417
Lemmon, D. M. 424
Lemone, D. V. 3158, 3185
Lenzi, G. W. 4435
Leo, G. W. 514
Leonard, A. B. 1317, 1323, 1327, 1342, 1343, 3193, 3198, 3199, 3214, 3215, 3263, 4284, 4286
Leonard, R. B. 1783, 1785, 1885
Lessard, R. H. 4437
Lesser, A. 2974
Lessing, R. S. 4719, 4720, 4721, 4722, 4723, 4724, 4725, 4726, 4730, 4733, 4735
Lester, J. G. 898
Lesure, F. G. 424
Le Van, D. C. 4515, 4518, 4520,

4528, 4529, 4558
Leve, G. W.   753, 774, 781, 819,
  830, 839
Lewis, J. D.   2328
Lewis, M. C.   3758
Libbey, F. W.   3766
Lichtler, W. F.   757, 763, 824,
  837, 848
Lidiak, E. G.   2852
Lieberman, N. I.   2109
Lilienthal, R. T.   2356, 2360
Lincoln, F. C.   2901
Lindemann, W. L.   4298
Lindorff, D. E.   1457
Lineback, J. A.   1197, 1242, 1257,
  1281, 1298, 1330, 1331, 1352,
  1406, 1411, 1423, 1430, 1434,
  1445, 1449, 1450, 1460, 1474,
  1475, 1518
Lineback, N. G.   193, 194
Lipp, R. L.   63, 187, 201
Lipscomb, R. G.   2673
Listerud, W. H.   2428
Livingston, N. D.   1832
Livingston, V. E.   4606, 4615,
  4626, 4633, 4641
Lloyd, N. A.   137
Lloyd, O. B.   3953, 3958
Lloyd, R. C.   584
Lobmeyer, D. H.   1692, 1827
Logan, M. H.   1083
Lohman, S. W.   3915, 3916, 3917,
  3918
Long, R. A.   2167, 2190
Longacre, S. A.   4301
Longwell, C. R.   2861
Longwill, S. M.   3933
Lonsdale, J. T.   4283
LoPinto, J.   3013
LoPinto, R. W.   3013, 3031
Lorenz, J. S.   2321
Los Angeles County Earthquake
  Commission   433
Loucks, D. P.   2954, 2989
Loucks, R. G.   4201, 4210, 4223,
  4225, 4313, 4324, 4326
Love, J. D.   4879
Love, J. M.   4839
Lovejoy, E. M. P.   3264
Lowe, E. N.   2468
Lowell, B. H.   1886
Lowenstam, H. A.   1463, 1466
Lowry, W. D.   4551
Lozo, F. E.   4306
Lubke, R. R.   1001
Lucas, R. C.   2230, 2232
Luce, P. B.   3141
Ludington, S.   1144
Ludlum, J. C.   4757, 4758

Ludvigson, G. A.   1649, 1661
Lukert, M. T.   4554, 4600
Lum, D.   1010, 1045, 1054, 1055,
  1056, 1057, 1083, 1092
Lum Associates, Inc., Walter see
  Walter Lum Associates, Inc.
Lund, C. R.   1378, 1382, 1383, 1385,
  1386
Lund, K.   1144
Lundberg, B. E.   4048
Lundelius, E. L.   4312
Lundgren, L.   669, 671, 675, 677,
  683, 685, 693
Lundin, R. F.   50, 3630, 3643
Luper, E. E.   2525
Lusk, T. W.   2459, 2461
Luther, E. T.   4155
Lutsey, I. A.   2935, 2937
Luttrell, G. W.   4530
Luttrell, P. E.   4203, 4215
Luttrell, S. P.   3391
Lutz-Garihan, A. B.   1782
Lutzen, E. E.   2552, 2553, 2555
Luza, K. V.   3682, 3683
Lydon, P. A.   445, 513
Lyle, W. M.   297
Lynch, B. M.   1194
Lynch, R. E.   143
Lynch-Blosse, M. A.   3024
Lytle, W. S.   3779, 3852, 3860, 3871,
  3881, 3889, 3901

Mabarak, C. D.   4882
McAnulty, W. N.   3267, 4168, 4181,
  4257
McBirney, A. R.   3707
McBride, E. F.   4236, 4292, 4298
McCaleb, J. A.   3666
McCallum, M. E.   4854, 4882, 4889
McCarthy, L. T.   704
McCarty, J. E.   2529
McCarville, M. A.   1908, 1911
McCauley, C. K.   3975, 3976, 3978,
  3979, 3981
McCauley, J. F.   3978, 3986
McClain, T.   1793, 1808
McClernan, H. G.   2763, 2827, 2833
MacClintock, P.   4484
McCollum, M. J.   939
McComas, M. R.   1278, 1294, 1405,
  1433
McComb, M. A.   2731, 2737, 2743
McConnell, K. I.   917, 959
McCoy, H. J.   767, 842
McCoy, J.   850
McCracken, E.   2557
McCracken, M. H.   2629
McCrone, A. W.   1681

McCullom, C. 4455
McCutcheon, T.E. 2463, 2496, 2502, 2504, 2512, 2514, 2521
McDaniel, R.D. 3301
McDermott, J.J. 2768
MacDonald, G.A. 416, 424, 980, 981, 982, 984, 985, 986, 987, 988
MacDonald, J.R. 4045, 4049, 4050, 4052
MacDonald, L. 4058
McElroy, M.N. 1696
MacElvain, R.C. 89, 109
McFarlan, A.C. 1921, 1922
McFarland, C.R. 4665, 4667
McGee, D.L. 289, 331
McGehee, R.V. 4222, 4238, 4247
McGillis, K.A. 4230
McGinnis, L.D. 1200, 1210, 1219, 1225, 1254, 1344, 1350, 1482
McGlade, W.G. 3781, 3782, 3784, 3872, 3873, 3879, 3900, 3902
McGovern, H.E. 1876
McGowen, J.H. 4170, 4178, 4180, 4198, 4202, 4208, 4221, 4254, 4304, 4307, 4332
McGrain, P. 1923, 1928, 1929, 1930, 1932, 1934, 1935, 1936, 1937, 1940, 1941, 1954, 1958, 1965, 1966, 1968, 1970, 1971, 1974, 1976, 1978, 1980, 1983, 1985, 1987, 1988, 1989, 1991, 1994, 1998, 1999, 2004, 2006, 2007, 2010, 2012, 2014, 2016, 2045, 2046, 2047, 2050, 2123, 2124, 2125, 2126, 2131, 2133, 2135, 2138
McGregor, D.J. 1487, 4031, 4038, 4063
McGregor, M.E. 2541
McGuire, O.S. 4578, 4596
McGuire, W.H. 2029, 2038, 2040, 2067, 2072
Machamer, J.F. 2438
Machin, J.S. 1206
McIntyre, D.B. 519
McIntyre, D.H. 1171
McIntyre, H.J. 1817
McIntyre, J.R. 3708
McJunkin, R.D. 462, 464, 585
Mack, F.K. 2235, 2237, 2295, 2300, 2306
Mack, T. 4517
McKalips, D.G. 4209, 4219
McKay, E.D. 1461, 1475, 1476
McKee, E.H. 2889, 2918, 2921
McKee, J.W. 153
McKeever, P.E. 4811

McKenzie, B.J. 3302
McKenzie, J.C. 3986
McKenzie, S.W. 737, 2298
Mackenzie, W.B. 2887
Mackey, H.L. 1875
MacKichan, K.A. 880
Mackin, J.H. 4629
McKinney, F.K. 53
McKinney, T.F. 3063
MacLachlan, D.B. 3794, 3955
MacLaughlin, R.E. 946
McLean, J.D. 4563
McLendon, J.T. 137
McLeod, C.A. 2484
McLeod, R.S. 4816, 4818
McLeroy, D.F. 3175
McMahon, T. 2993
McMannis, W.J. 2693
McMaster, W.M. 18, 25, 164
McMath, V.E. 417
McMeen, J.A. 4071, 4072, 4076, 4078
McMillen, J.R. 2646, 2648
McMurray, L.L. 3969
McMurtrey, R.G. 2699, 2711, 2721, 2726, 2729, 2735, 2741
McNellis, J.M. 1684, 1769, 1814, 1817, 1877, 1880, 1881, 1883, 1886
McNitt, J.R. 424
MacRoberts, P.B. 1762
McVay, T.L. 1953
McWilliams, R.G. 112, 117, 118, 120
Maddock, M.E. 423
Madison, R.A. 1951
Madison, R.J. 4463
Madri, P.P. 2990, 3005
Madsen, J.H. 4382
Magnuson, L.M. 1784, 1786, 1788
Magnuson, R.L. 4619
Maher, J.C. 3691
Maher, L.J. 4802
Maher, S.W. 4109, 4125, 4134, 4135, 4136, 4138, 4139, 4140, 4149, 4156
Major, C.F. 4244
Major, R.L. 1276, 1375, 1392
Malatino, A.M. 134
Malina, J.F. 4183
Malone, W.T. 2090
Malott, C.A. 1583, 1966
Mandle, R.J. 2306
Mangan, G.F. 3036
Manger, W.L. 3695, 3696
Maniak, U. 2961, 2984
Mankin, C.J. 3629
Mann, J.A. 793, 843
Mann, V.I. 3296, 3304

Manson, C. 4668
Manson, M. W. 580
Manson, V. 1851
Marafi, H. 3457
Maras, B. B. 3141
Marble, J. C. 2540
Marcher, M. V. 3685
Marie, J. R. 2175
Marine, I. W. 4456
Markofsky, J. 3076
Maroney, D. G. 1562
Marsalis, W. E. 944
Marsh, O. T. 740
Marsh, P. S. 4112
Marshall, R. A. 2615
Martens, J. H. C. 4673, 4714
Martin, C. C. 3127
Martin, C. W. 678, 681
Martin, H. M. 2313
Martin, J. A. 2560, 2614, 2615, 2660, 2661, 2662
Masingill, J. H. 199
Mason, J. E. 2116
Massello, J. W. 2556
Mast, R. F. 1277, 1307, 1339, 1441
Masters, J. M. 1341, 1359
Mathiot, R. K. 3726, 3762
Matias, J. 3076
Matson, R. E. 2727, 2738, 2744, 2754, 2804
Matthes, W. J. 1656
Matthews, L. E. 4113
Matthews, R. A. 468
Matthews, W. H. 4237, 4241
Matthiesen, R. B. 464
Mattick, R. E. 3824
Mattson, C. 3013, 3031
Maud, R. L. 580
Mauger, R. L. 4433
Maurer, R. E. 4405
Maxey, G. B. 2922
Maxwell, E. L. 783
Maxwell, R. A. 4240, 4243, 4283, 4305
May, H. G. 2165
May, J. H. 2528
Meade, B. K. 421
Mears, A. I. 592, 616, 642
Medlin, J. H. 941
Meehan, J. F. 432
Meents, W. F. 1243, 1245, 1249, 1253, 1339
Mehl, M. G. 2615
Meinere, D. G. 2428
Meisel, S. L. 3033
Meisler, H. 3929, 3937
Mellen, F. F. 2470, 2473, 2490, 2491, 2493

Meng, H. 1462
Menke, C. G. 759, 760, 820, 822
Merenyi, M. 2987
Merewether, E. A. 391, 394, 396
Meridith, E. W. 760
Mero, J. L. 420
Merriam, D. F. 1679, 1686, 1693, 1698, 1846, 1849, 1861, 1863
Merrill, G. K. 1472, 3525
Merrill, R. K. 384
Merritt, C. A. 3619
Merritt, L. B. 4472
Meshref, W. M. 2353
Mettee, M. F. 70, 184
Metzger, W. J. 86, 88
Metzler, K. J. 703
Meyer, F. W. 775, 829, 845, 853, 862
Meyer, J. A. 3182
Meyer, R. F. 3241
Meyer, R. P. 1423, 1430, 1445
Meyer, R. R. 2192
Meyer, W. R. 1692, 1856
Meyertons, C. T. 4560
Michna, L. 3040
Mifflin, M. D. 2895
Miles, P. M. 2105
Milfred, C. J. 4761, 4762
Milhous, H. C. 1986
Milici, R. C. 4143, 4145, 4151, 4154, 4503
Millard, R. C. 4650
Miller, B. 1820
Miller, D. 3066
Miller, D. E. 1711, 1718, 1741, 1860, 2674
Miller, E. T. 179
Miller, F. 1901
Miller, G. A. 1657
Miller, J. C. 720, 721, 773, 2670
Miller, J. P. 3235
Miller, J. W. 3014
Miller, M. L. 1462
Miller, M. S. 4511
Miller, N. G. 1475
Miller, P. J. B. 1138
Miller, R. A. 4118, 4125, 4126, 4127, 4128, 4129, 4131
Miller, R. V. 453, 454, 567, 584
Miller, W. G. 1355
Miller, W. R. 2776
Millgate, M. L. 4852
Mills, J. W. 4622
Mills, M. F. 453
Milne, C. 4661
Milner, S. 3219
Miloradov, M. 3064
Milton, C. 386, 750, 895, 4523
Mink, L. L. 1169, 1184

Mintz, D. W.   4655
Minyard, J. P.   2531
Misaqi, F. L.   3146, 3148
Miser, H. D.   386
Missaghi, F.   3121, 3139
Missouri Geological Survey   2595,
   2596, 2600, 2601, 2607, 2637
Missouri Oil and Gas Council   2598
Mitchell, C. G.   4449
Mitchell, D. W.   3140
Moen, W. S.   4603, 4607, 4609,
   4621, 4630, 4631, 4632, 4637,
   4643, 4651, 4658
Moffett, T. B.   161
Moftah, I.   147
Montana Bureau of Mines and
   Geology   2791, 2805, 2843
Montana State Planning Board
   2805
Montgomery, A.   3235, 3276
Montgomery, G.   2114
Montgomery, K. M.   2863
Moodie, F. B.   2123
Mooney, H. M.   2422, 2436
Moore, C. H.   4287
Moore, D. B.   35, 54, 93, 100,
   128
Moore, G. K.   163, 4147
Moore, J. D.   47, 48
Moore, J. G.   2876
Moore, J. S.   4500
Moore, L.   359, 424
Moore, M. C.   1516, 1559, 1604
Moore, R. T.   344, 348, 357, 381
Moore, W. H.   2452, 2457, 2464,
   2487, 2492, 2495, 2533
Moore, W. L.   3479, 3486
Moran, D. F.   432
Moran, S. R.   1315, 3454, 3462,
   3463, 3464, 3465, 3484, 3491,
   3493, 3499
Moravec, G. F.   48
Morey, G. B.   2387, 2408, 2418,
   2424, 2425, 2427, 2434, 2443,
   2444
Morgan, C. O.   1877, 1880, 1881,
   1883, 1885, 1886, 2162
Morgan, J. H.   1942
Morgan, T.   2680, 2687, 2697,
   2702, 2712, 2724
Morris, F. O.   4753
Morris, R. C.   4768
Morrison, D. A.   1127, 1174
Morrison, R. B.   2903
Morrissey, F. R.   2914
Morrow, E.   913
Morse, H. M.   2451, 2469
Morton, D. M.   425, 430, 453,
   519, 546, 577

Morton, P. K.   424, 446, 449, 454,
   508, 552, 553
Morton, R. A   4187, 4191, 4193, 4195,
   4202, 4204, 4206, 4207, 4211, 4227,
   4229, 4254, 4237
Moser, P. H.   3, 4, 5, 6, 7, 161,
   184
Mossler, J. H.   1766, 2406, 2429,
   2430
Mote, R. H.   1919
Motz, L. H.   869
Motzer, W. E.   1144
Mound, M. C.   1498
Mourant, W. A.   3229, 3230
Moye, F.   908
Moyer, R. E.   107
Mozola, A. J.   2344, 2354
Mraz, J. R.   4235
Muchmore, C. B.   1458
Mudrey, M. G.   4781, 4786, 4787
Muehlberger, W. R.   3100, 3277
Muilenburg, G.   1839, 1843
Muilenburg, G. A.   2574, 2575
Mull, D. S.   1947, 1955
Mullins, A. T.   1938, 2069
Mullis, C. H.   179
Multer, H. G.   3547, 3562
Mundorff, J. C.   4465
Mundt, H. W.   2568
Murdoch, J.   412
Murphy, M. A.   425
Murphy, V. J.   4485
Murray, D. K.   595, 610, 611, 615,
   624, 626, 630, 644, 648
Murray, H. H.   1491
Musgrove, R. H.   759, 766, 773, 820,
   822, 827, 828
Mussotto, N. T.   3758
Myers, A. J.   3692
Myers, P. B.   4480
Myers, R. E.   1672, 2601

Nace, H. L.   4845
Naegele, O. D.   4672
Naff, R. L.   2922
Nagle, J. S.   4171, 4179
Nance, R. B.   1470
Nash, W. P.   4447
Nault, M. J.   2160
Naylor, R. S.   2210
Nazalevicz, S. K.   3053
Neathery, T. L.   38, 64, 96, 99,
   108, 139
Neff, J. L.   4619
Nelson, H. F.   4249
Nelson-Moore, J. L.   594
Netzeband, F. F.   4259, 4260, 4261,
   4262, 4263, 4264, 4265, 4266

Neuman, R. W.   2142
Nevada Bureau of Mines   2864
Newcome, R.   2163, 2164, 2168,
   2186
Newell, M. L.   1159
Newhouse, C. J.   3699
Newland, L. E.   4242
Newport, T. G.   3940, 3943, 3949,
   3951
Newton, G. D.   50
Newton, J. G.   21, 126, 141
Newton, V. C.   3751, 3753
Nicholls, H.   4297
Nichols, D. G.   4707
Nicol, D.   872
Nicoll, R. S.   1495, 1500, 1525
Nielsen, N.   4357
Niems, E. G.   3774
Nilsen, T.   4657
Nilson, T. H.   4365
Nitzschke, E. T.   3149
Noble, A. G.   3508
Noble, D. C.   2917
Nodine-Zeller, D. E.   1776
Noel, J. A.   1608
Noger, M. C.   1964
Noland, A. V.   1607
Norby, R. D.   1472
Nordstrom, K. F.   3030
Norling, D. L.   3557
Normark, W. R.   578
Norris, R. L.   2025
Norris, S. E.   3597, 3611
North, W. G.   1297
Northrop, S. A.   3138, 3262
Northup, R. C.   1624
Noson, L.   4662, 4663, 4664
Nosow, E.   1925, 1969, 1973,
   1975, 1977, 1979, 1981, 1990,
   2005, 2013, 2015
Novitski, R. P.   4783
Nowell, W. R.   649
Nuckols, E. B.   4600
Nuhfer, E. B.   4718
Nunan, W. E.   4525, 4588
Nuttall, B. D.   2018
Nutter, L. J.   2231, 2238, 2263,
   2288, 2297, 2305
Nuzman, C. E.   1848, 1856, 1867
Nystrom, P. G.   4590

Oakes, M. C.   3634, 3649
Oakeshott, G. B.   421, 424, 429,
   430, 434, 528
Oaks, R. Q.   4506
O'Brien, J. C.   443, 445
O'Brien, N. R.   1227
O'Brien, W. J.   1762

O'Bryan, F. L.   2147
O'Connor, H. G.   1758, 1818, 1891
O'Connor, M. P.   3289
Odom, I. E.   1285, 4788
Ogawa, J.   2447
Ogden, D. G.   4487
Oglesby, W. R.   743, 805
Ojakangas, R. W.   2424, 2425, 2428
Olcott, P. G.   4805, 4827
Olea, R. A.   1837, 1838
O'Leary, W. J.   620, 621
Oles, K. F.   3714, 3743
Oliver, H. W.   430
Oliver, W. B.   4308
Olsen, J. M.   3190
Olsen, S.   2244
Olsen, S. J.   879
Olson, E. C.   3669, 3675
Olson, G. W.   1770, 1778, 4761
Olson, L. J.   3653
Olson, N. K.   966
O'Neill, B. J.   3826, 3848, 3864,
   3866, 3869, 3870, 3886, 3890,
   3896
O'Neill, J. F.   125
Onuschak, E.   4510, 4592
O'Rear, D. M.   81
Orlansky, R.   4362
Oros, M. O.   1703, 1714, 1722,
   1732, 1797, 1799, 1801, 1803,
   1805, 1807, 1859, 1893
Orr, R. W.   1217, 1519, 1560, 1562
Orville, P. M.   666
Osberg, P. H.   2215
Osborne, R. H.   3578, 3595
Osmond, J. C.   4401
Ostrom, M. E.   4765, 4766, 4767,
   4768, 4772, 4774, 4788, 4789,
   4790, 4791, 4794, 4797, 4806,
   4829, 4832
Othberg, K.   4653
Otton, E. G.   2236, 2288, 2292, 2296
Overbey, W. K.   4677
Owen, D. E.   4289
Owen, V.   930
Owens, G. L.   3565, 3576
Owens, W. G.   597
Oxley, M. L.   2508

Pabst, M. E.   1825
Paczek, F. H.   3965
Paden, J. W.   424
Page, B. M.   419
Page, L. V.   2163, 2164, 2165
Page, W. D.   2684
Pallister, H. D.   84
Palmer, J. E.   1477, 1478
Palmer, M. V.   1511

Pampeyan, E. H. 497, 2861, 2874
Pankiwskyj, K. A. 2212, 2218,
2222
Pansze, A. J. 1181
Papke, K. G. 2870, 2877, 2880,
2885, 2888, 2894, 2929, 2940
Parham, W. E. 1208, 1261, 1285,
2414, 2421, 2422, 2446
Parizek, E. J. 2611
Parke, D. L. 484
Parker, J. M. 3292
Parker, M. C. 1628, 1638, 1641,
1646
Parker, P. E. 4568
Parker, R. B. 493
Parker, R. L. 424
Parks, J. T. 4124
Parks, W. S. 2462, 2473, 2476,
2483, 2487, 2495, 2497, 4119
Parrott, B. S. 4244
Parry, W. T. 4427
Parsons, B. M. 4688
Parsons Brinckerhoff-Hirota Asso-
ciates 1093, 1096, 1113
Pascale, C. A. 784, 795, 863,
864
Pasini, J. 2083
Patchen, D. G. 2082, 4690, 4691,
4692, 4693, 4739, 4741, 4744,
4745
Paterson, W. D. 4654
Patrick, C. R. 3692
Patten, E. P. 3923, 3928, 3930
Patterson, J. L. 402
Patterson, O. F. 3305
Patterson, S. H. 947, 969, 1006,
1972
Patton, J. B. 1561, 2092
Paul, R. R. 2875
Paull, R. A. 4787
Paulson, Q. F. 3330, 3349, 3426
Pavlides, L. 2210
Payne, W. R. 4291
Peace, R. R. 19, 24, 162
Pearl, R. H. 587, 589, 593, 612,
614, 629, 636, 638
Pearre, N. C. 2234
Pearson, E. F. 540
Peck, D. L. 423
Peck, E. L. 4454
Pederson, D. T. 3449
Peirce, H. W. 347, 353, 355,
356, 382
Peirce, L. B. 29, 34, 178, 194
Pennington, A. J. 612
Penzien, J. 432
Peppers, R. A. 1192, 1195, 1301,
1319, 1336
Perhac, R. M. 3106

Perkins, J. H. 2101
Perkins, P. 2065
Perry, L. I. 4388
Perry, T. G. 1486, 1493
Perry, W. J. 4698, 4744, 4748
Pessel, G. H. 326, 331
Petersen, M. S. 2663
Petersen, R. C. 659
Peterson, A. E. 4770
Peterson, C. A. 4469, 4471
Peterson, D. H. 571
Peterson, G. L. 424, 438
Peterson, J. B. 4447
Peterson, N. V. 3699, 3708, 3731,
3742
Peterson, R. D. 1159
Petri, L. R. 3314, 3315
Petsch, B. C. 4038, 4039, 4041,
4043
Pettijohn, F. J. 2348, 2349, 2350,
2351
Pettyjohn, W. A. 3341, 3342
Pewe, T. L. 384, 385
Pfefferkorn, H. W. 4709
Pfefferkorn, R. A. 1319, 1336, 1348,
1356
Pflieger, W. L. 2673
Pharr, R. F. 4515, 4518, 4558
Philip, J. 3007
Phillips, T. L. 1319, 1336, 1376
Phinney, W. C. 2420
Picard, M. D. 4370, 4373, 4425, 4428
Pickard, G. W. 2117
Pickering, R. J. 1955
Pickering, S. M. 900, 915, 970
Pickett, T. E. 717, 721, 724
Pieper, M. J. 4191, 4193, 4202,
4204, 4206, 4207
Pierce, H. F. 4261
Pierce, J. W. 1866
Pierce, M. G. 951
Pinchock, J. M. 2834
Pinsak, A. P. 1492
Piper, A. M. 3912
Pirkle, E. C. 746, 871
Pirkle, W. A. 871
Piskin, K. 1272, 1346, 1353
Pitt, W. D. 3184, 3688
Pitts, L. P. 2543
Plarke, D. L. 576
Platt, P. A. 912, 956
Plebuch, R. O. 404, 1961
Plechaty, M. 3077
Plummer, N. 1685
Poe, R. M. 198
Pogge, E. C. 1762
Poindexter, O. F. 2313
Poland, J. F. 418
Pollard, D. D. 430

Pollard, L. D.   955
Pollock, C. A.   1607
Pope, D. E.   2156
Porcella, R. L.   464
Porro, A. A.   3079
Porter, L. D.   461, 462, 585
Poth, C. W.   3927, 3931, 3936, 3941, 3945, 3947, 3950, 3954
Potter, P. E.   1198, 1240, 1480, 1924
Pottratz, S. W.   4077
Poulton, C. E.   3761
Pounder, J. A.   2053
Powell, J. D.   1134
Powell, R. L.   1534, 1535, 1597, 1603
Powell, W. J.   80, 114, 177, 180
Power, B. A.   2951
Praszker, M.   530
Prather, B. A.   3877
Prather, P.   948
Pratt, A. R.   4358
Prescott, R.   1280
Presley, M. W.   4220, 4230
Preston, F. W.   1842, 1862
Preston, J.   4811, 4812, 4829
Price, D.   4456
Price, P. H.   4738
Price, W. E.   1951
Pride, R. W.   800, 829, 880
Prior, J. C.   1625
Proctor, C. V.   4179
Proctor, P. D.   3186
Proctor, R. J.   524
Proffett, B. H.   2933
Proffett, J. M.   2933
Prouty, D.   647
Psuty, N. P.   3030
Purandare, Y.   2980
Puri, H. S.   743, 884, 887
Putnam, A. L.   866
Putnam, G. B.   2089

Quilici, F. J.   2753
Quinn, J. H.   3666
Quinn, M. J.   3612

R. M. Towill Corporation   1084
Race, W. H.   209, 214, 216
Rader, E. K.   4514, 4552, 4566, 4573, 4589, 4594, 4597, 4602
Ragsdale, J. T.   462, 464, 585
Rahme, H. S.   3071
Rahn, P. H.   682, 4037, 4060
Rainwater, E. H.   2460, 2479, 2480
Ralston, D. R.   1132, 1140, 1178, 1185
Ramp, L.   3703, 3717, 3722, 3730, 3735, 3742, 3765
Randazzo, A. F.   749, 872, 3307
Randich, P. G.   3317, 3318, 3363, 3364, 3383, 3384
Rao, C. P.   1332
Rapp, G. R.   2386
Rapp, J. S.   485, 562, 566
Rath, D. L.   2594
Rau, W. W.   4605, 4608, 4614, 4618, 4624, 4642, 4645
Rautio, S. A.   2779
Ravn, R. L.   1664, 1665
Ray, E. O.   1915, 1918, 1920, 1925, 2063, 2100
Ray, T.   3101
Raymond, D.   43
Read, C. B.   3105
Reade, H. L.   17
Reade, N. S.   630
Real, C. R.   484, 488
Redey, G.   2975, 3061
Redman, E.   285
Redman, R. H.   3661
Reece, D. R.   1185
Reed, A. A.   668
Reed, A. H.   1921, 1922, 1968, 1971, 1974, 1976
Reed, E. C.   2849
Reed, P. C.   1429, 1447, 1451, 1453
Reed, R. C.   2329
Reed, T. E.   2711, 2721, 2729, 2735
Rees, D. C.   4434
Rees, O. W.   1221, 1228, 1252, 1483
Rehn, E. F.   2060
Reich, J. R.   3817
Reichert, W. H.   4611, 4659, 4666
Reid, J. R.   3449
Reid, R. R.   1044, 1149
Reidel, S. P.   3596
Reinertsen, D. L.   1230
Reinhardt, J.   2247, 2301
Reishus, M.   3477
Reiter, M.   3205
Rember, W. C.   1143
Remec, P. P.   3034
Renault, J. R.   3165, 3167, 3170, 3181
Renne, R.   2971
Renton, J. J.   4686, 4688, 4689, 4697, 4705, 4729, 4740
Reppert, R. S.   4719, 4720, 4721, 4722, 4723, 4724, 4725
Reshkin, M.   1602
Rexroad, C. B.   1211, 1244, 1490, 1495, 1496, 1500, 1501, 1525, 1532, 1536, 1559, 1605, 1607, 1612, 1616, 1617, 1927
Reynolds, D. H.   3120, 3135

Reynolds, D. W.   2051
Reynolds, J. H.   4735
Reynolds, J. W.   64
Reynolds, S. J.   1139
Ricci, R. D.   3039
Riccio, J. F.   61, 129, 150
Rice, C. L.   2137
Rice, S.   488
Rice, S. J.   424, 460, 465
Rich, E. I.   423
Richards, H. G.   710, 975
Richter, D. H.   210, 238, 241,
   243, 251, 256, 260, 265, 266
Richter, K. E.   186, 200
Rickers, A. E.   4464
Ricoy, J. U.   4212
Riese, R.   3165
Rigby, J. G.   4669
Riggs, S. R.   3289
Riley, C. S.   2070
Riley, H. L.   1978, 1980, 1983,
   1985, 1989, 1991, 1994, 1999,
   2004, 2007, 2010
Rima, D. R.   705, 3925
Rinehart, C. D.   4616, 4617
Ringler, R. W.   1363
Ripley, E. M.   2431, 2433
Ripman, H.   3009
Risser, H. E.   1214, 1286, 1375,
   1381, 1392, 1416, 1425, 1438,
   1440
Rittschoff, W. F.   3023
Ritzma, H. R.   4415, 4418, 4431,
   4445
Rivers, M. E.   1968, 1971, 1976,
   1978, 1980
Robbins, S. L.   430
Roberts, J. F.   3682, 3684, 3685
Roberts, J. T.   4887, 4892
Roberts, M.   3752
Roberts, M. S.   3720
Roberts, R. J.   2861, 2863
Roberts, R. S.   1719
Robertson, A. F.   851
Robertson, C. E.   2618, 2628,
   2634
Robertson, F.   2624
Robertson, J. M.   3114
Robin, G. W.   3021, 3067
Robinson, G. C.   3970, 3971,
   3972, 3974, 3980
Robinson, G. D.   1455, 4153
Robinson, R. A.   4393, 4397, 4408
Robison, T. M.   4675
Rockaway, J. D.   2553, 2555
Rodda, P. U.   425, 4172, 4247,
   4282, 4291, 4293, 4294
Rodgers, D. A.   441
Rodis, H. G.   854

Roen, J. B.   3839
Rogers, D. T.   138
Rogers, J. E.   2166, 2174
Rogers, S. M.   31, 2170, 2189
Rogers, T. H.   456, 457, 548
Rogers, W. P.   602, 606, 641
Rohrbacher, R. G.   4185
Roland, H. L.   2155, 2156
Rolfe, D.   1367
Rollo, J. R.   2161, 2169, 2173, 2192
Roman, R. J.   3135, 3166, 3171,
   3172, 3191
Romero, G.   1835
Rones, M.   3803
Roney, J. R.   3028
Rooney, L. F.   1502, 1503, 1504,
   1511, 1512, 1520, 1587
Root, F. K.   4838
Root, S. I.   3842, 3856
Ropes, L. H.   1877
Rose, A. W.   216, 248, 249, 250,
   252, 253, 254, 255, 263, 266,
   267, 3868
Rose, K. E.   1728
Rose, P. R.   4309
Rose, R. L.   2871
Rose, W. D.   1926, 2036
Rosengreen, T. E.   3593
Roseveare, G. H.   346, 375
Roshardt, M.   4782
Ross, C. A.   1207, 1216
Ross, C. P.   1150, 1151
Ross, D. C.   424
Ross, F. I.   433
Ross, S. H.   1124, 1125, 1128, 1170,
   1173
Roux, P. H.   3066
Rowett, C. L.   3676, 3672
Rowland, T. L.   3627, 3677
Roy, D. C.   2218
Royer, D. W.   3908, 3909
Royse, C. F.   355, 3453, 3475
Rozov, W. C.   914
Ruch, R. R.   1339, 1355, 1404, 1413,
   1417, 1419, 1420, 1437, 1441,
   1448, 1454
Rudd, N.   3783
Rudman, A. J.   1565, 1568, 1569,
   1577, 1578, 1581, 1584
Rudy, S.   380
Rueff, A. W.   2640
Ruhe, R. V.   1644, 3242
Ruhle, J. L.   4516
Rukstad, L. R.   4068, 4073, 4075
Ruotsala, A. P.   2399
Rush, C. A.   3174
Rusnak, G. A.   420
Russell, E. E.   2546, 4119
Russell, J. A.   3117, 3680

Russell, W. B. 2576, 2580, 2645, 2646, 2647
Russo, T. N. 857
Ryan, J. D. 4863
Ryder, P. D. 1945

Sable, E. G. 2137
Sackreiter, D. K. 3481
Sadove, R. 2968
Safford, F. B. 3530
Safford, J. M. 4137
Sahinen, U. M. 2689, 2700, 2717
Saile, D. K. 1797, 1799, 1801, 1805
St. Clair, A. E. 4214
Sajenko, V. 3001
Sakin, M. J. 1861
Salotti, C. A. 897
Sampair, J. L. 3291, 3308
Sampson, R. J. 1836, 1864
Sams, R. H. 923
Samson, P. L. 3905
Sandberg, C. A. 4834
Sandberg, J. P. 502
Sanders, J. E. 667, 695, 696
Sandhaus, E. H. 2667
Sandy, J. 119, 121
Sanford, A. R. 3132, 3145, 3156, 3174, 3180, 3197
Sanford, T. H. 38, 82, 2179
Sapp, C. D. 186, 200
Satterfield, I. R. 2642
Sauer, V. B. 2196
Saul, R. B. 430, 450
Saunders, R. H. 211, 217, 220, 248, 328
Savage, C. N. 1122, 1123, 1124, 1128, 1129, 1154, 1167
Savage, J. C. 430
Saylor, T. E. 3834, 3840
Sayre, R. L. 4439
Scarbrough, L. 61, 142
Scarbrough, L. W. 141
Scarbrough, W. L. 46, 100, 115
Schader, E. E. 432
Schambeck, F. J. 424
Schatz, H. L. 1178
Scheibach, R. B. 2927
Schell, E. M. 2683
Schilling, F. A. 3116
Schilling, J. H. 2892, 2904, 2915, 2929, 2936, 3089, 3168, 3272, 3275
Schiner, G. R. 3944, 3957
Schleicher, J. A. 1415, 1417, 1456
Schlicker, H. G. 3702, 3716, 3723, 3727, 3735, 3769

Schlobaum, D. 1793
Schlocker, J. 423
Schmidt, P. G. 3093
Schmidt, W. 752, 873, 875
Schmoker, J. W. 4371
Schneeflock, R. D. 127
Schneider, A. F. 1516, 1536, 1594
Schneider, C. W. 664
Schneider, J. J. 854
Schnitker, D. 2220, 3306
Schoewe, W. H. 1723
Schofield, D. A. 4288, 4289, 4291
Schooler, E. E. 3815
Schoon, R. A. 4029, 4030, 4035, 4057, 4059, 4063
Schroeder, M. C. 824
Schuetz, J. R. 1656
Schultz, C. B. 2854
Schumacher, O. L. 3183
Schuster, J. E. 4636, 4639, 4660
Schwab, D. A. 3135
Schwab, G. E. 3178
Schwalb, H. R. 1934, 1944, 1946, 1956, 1962, 2080, 2103, 2110
Schwartz, G. M. 2437
Schwarz, C. 468
Schweinfurth, S. P. 3807
Schwochow, S. D. 625, 631, 639, 640
Scott, A. J. 4178
Scott, J. C. 11, 12, 17, 22
Scott, M. W. 3459, 3463
Scott, R. B. 686
Scott, R. W. 1696, 4329
Scott, T. M. 877
Scrocco, V. M. 3022
Scurlock, J. R. 356
Seager, W. R. 3108, 3111, 3112, 3113, 3187, 3201
Searight, T. K. 1295
Seddon, G. 4303
Seed, H. B. 531
Seeger, C. R. 1995
Seeley, B. K. 4418
Seely, D. R. 3623
Seevers, W. J. 1737, 1810
Seifert, W. F. 2037
Seni, S. J. 4332, 4340
Seo, J. H. 4203
Sever, C. W. 936, 938, 940, 942, 969
Severson, R. F. 1266
Sevon, W. D. 3822
Shaak, C. 891
Shabica, C. W. 1470
Shah, S. M. I. 1137
Shamburger, V. M. 41
Shampine, W. J. 2177
Shannon, S. S. 1139, 1166

Shannon, S. W. 10
Shapiro, E. 710
Sharp, B. J. 1147
Sharp, R. V. 430
Sharpe, R. D. 4278
Shaub, B. M. 4738
Shaver, R. H. 1492, 1517, 1527, 1536, 1537, 1558, 1570, 1582, 1591, 1606
Shaw, C. W. 1269
Shelby, C. A. 4289, 4290
Sheldon, M. G. 3690
Shelton, D. C. 602, 606, 643, 647
Shelton, J. W. 3645, 3655
Shepps, V. C. 3789, 3828
Sherburne, R. W. 565
Sheridan, M. F. 355
Sherman, F. B. 1403
Sherman, R. 1803, 1805
Sherrod, N. 3431
Sherwood, C. B. 838, 852
Sherwood, K. W. 340
Sherwood, W. C. 3795
Shetron, S. G. 2358
Shiley, R. H. 1357, 1364
Shimp, N. F. 1252, 1253, 1277, 1318, 1340, 1408, 1410, 1413, 1417, 1420, 1432, 1436, 1437, 1442, 1448, 1454, 1455
Sholes, M. A. 3887, 3894
Shomaker, J. W. 3188, 3194, 3208, 3209, 3230, 3257
Shotts, R. Q. 130, 131
Shows, T. N. 2510, 2518, 2524, 2541
Shroba, R. R. 639, 640
Shroder, J. F. 4363
Shufflebarger, T. E. 3308
Shumaker, R. C. 4483
Shurr, G. W. 4082, 4083
Shuster, E. T. 3956
Siemers, C. T. 1822
Silberman, J. D. 2104
Silberman, M. L. 2921, 3212
Silver, E. A. 578
Silver, M. L. 1331
Silverman, A. J. 2710
Simanek, D. E. 661
Simkins, W. W. 4230
Simmons, L. A. 2142
Simmons, W. B. 634
Simo, F. G. 3048
Simon, J. A. 1194, 1197, 1222, 1288, 1318, 1340, 1410, 1469
Simpson, D. R. 504
Simpson, T. A. 79, 96, 168
Sims, P. K. 2404, 2405, 2412, 2415, 2387, 2437

Sinclair, W. C. 803
Sinnott, A. 4532
Siple, G. E. 3963, 3973, 3985, 3994
Sjoreen, A. 1584
Skartvedt, R. 1371
Skelton, J. 2664, 2666, 2667, 2669, 2671, 2672, 2676
Skema, V. W. 3887, 3888, 3893, 3895
Skinner, E. L. 4810
Slagle, S. E. 1815, 1824
Slatick, E. R. 4267
Slaughter, T. H. 2225, 2236, 2239
Slemmons, D. B. 417
Slifer, D. 2241
Sloan, E. 3961, 3962
Sloan, R. E. 2385, 2416
Sloss, R. 2164, 2168
Slosson, J. E. 430
Slote, L. 2967, 2977
Small, J. B. 421
Smigaj, M. J. 2231
Smiley, C. J. 1137, 1143
Smith, A. D. 4442, 4443
Smith, A. G. 2684, 3665
Smith, A. R. 424, 512, 544
Smith, B. E. 3509, 3573
Smith, B. Y. 4359
Smith, C. E. 4528
Smith, C. I. 4300, 4306
Smith, C. J. 4706, 4707
Smith, C. T. 3129
Smith, D. L. 888, 3676
Smith, E. 170
Smith, E. A. 2467
Smith, E. I. 4787
Smith, G. E. 1267, 1934, 1950, 1952, 4315
Smith, G. I. 424, 526
Smith, H. R. 701
Smith, J. G. 1946
Smith, J. L. 430
Smith, J. W. 970, 4145, 4503
Smith, M. B. 424
Smith, M. O. 1927, 2127
Smith, N. M. 1589
Smith, R. B. 2825
Smith, R. C. 3891
Smith, R. E. 4448
Smith, R. I. 2689, 2695, 2708, 2728, 2740, 2752, 2764
Smith, R. L. 1762
Smith, R. M. 2886
Smith, S. G. 1591
Smith, T. C. 459
Smith, T. E. 234, 324, 325, 340
Smith, T. K. 2147
Smith, W. C. 424, 1389, 1395, 1401

Smith, W. E.   135, 139, 149, 181, 183
Smith, W. H.   224, 1204, 1267, 1275, 1295, 1470
Smith, W. W.   2062
Smosna, R. A.   4699, 4741, 4744, 4745, 4746, 4747
Smyth, P.   3527, 3531, 3532, 3543
Snavely, P. D.   4670
Sneach, P. H. A.   1861
Snell, L. J.   841, 867
Snider, D. W.   2357
Snider, J. L.   2177, 2185, 2188
Snowden, J. O.   2513
Snyder, F. G.   2610
Snyder, J.   2945
Socolow, A. A.   3849
Sokol, D.   1157
Soliman, S. M.   408
Solley, W. B.   2259
Sonderegger, J. L.   44, 45, 955, 2779, 2842
Sorensen, E. F.   3192
Sorensen, H.   2359
Soule, J. M.   606, 613, 3179
South Dakota Geological Survey   3997
Southard, R. B.   3050
Southwick, D. L.   2284
Spaine, S. I.   113
Spangle, L. B.   2074
Speed, R. C.   2884
Spelman, A. R.   3797
Spencer, B. C.   4132
Spencer, E. W.   4567
Spoljaric, N.   707, 713, 715, 721, 722, 735, 736
Sponseller, R. D.   3885
Springfield, V. T.   3000
Sproul, C. R.   741, 792
Spurr, S. H.   4275
Stack, W.   2039
Stafford, L. P.   906
Stager, H. K.   424, 2889
Stahmann, W. S.   3120
Staley, W. W.   1145
Stanfield, C. P.   2159
Stanley, R. S.   672, 689
Stanonis, F. L.   2059, 2071
Stanton, R. J.   520
Staplin, F.   1206
Statler, A.   4120
Statler, A. T.   2008, 2015
Stauffer, N.   4432
Stearns, F.   4761
Stearns, H. T.   976, 977, 979, 980, 981, 982, 983, 984, 986, 987, 1058, 1061, 1156

Stearns, N. D.   978
Stearns, R. G.   4112, 4128, 4133
Steele, F. V.   3998, 3999, 4000, 4083
Steele, J. D.   1454
Steere, M. L.   3720, 3734
Steffen, L. J.   4105, 4106
Stein, D. R.   234, 324
Steinbrugge, K. V.   432, 532
Steinhilber, W. L.   1667, 1673, 1675
Steinmetz, J. C.   1322, 1426
Stem, J. E.   3047
Stemmler, R. S.   1264
Stephens, E.   488
Stephens, H. G.   490
Stephens, J. C.   4004, 4005, 4006, 4007
Stephenson, D.   4773
Stephenson, D. A.   1265, 1380
Stevens, D. J.   2579
Stevenson, D. L.   1218, 1239, 1292, 1310, 1338
Stewart, D. P.   4484, 4493, 4494, 4495, 4497, 4498
Stewart, H. G.   754, 831
Stewart, J. H.   424, 2879, 2889
Stewart, J. W.   843, 861, 922, 974
Stewart, R. M.   424, 472
Stewart, W. J.   3254, 3255
Stieglitz, R. D.   3515, 3605, 3609
Stinson, M. C.   424
Stith, D. A.   3521, 3533, 3535, 3542, 3574, 3587, 3609, 3614
Stitt, J. H.   3632, 3651, 3694
Stock, G. W.   3994
Stoffel, K. L.   4669
Stokes, W. L.   4360
Stoll, W. J.   2883
Stone, A.   653
Stone, J. F.   4299
Stone, S. W.   4479
Stone, W. J.   3208
Stotelmeyer, R. B.   3192
Stout, J. H.   281
Stout, K. S.   2689, 2719, 2761, 2764, 2782
Stout, L. N.   2549B, 2562, 2638
Stout, T. M.   2854
Stover, D. W.   1458
Stowasser, W. F.   801, 805, 806
Stowe, C. H.   4367, 4374, 4378, 4379, 4381, 4384
Stramel, G. J.   1726
Strand, R. G.   491
Strasma, J.   4832
Stratton, J. F.   1530
Strauss, E.   4812
Straw, W. T.   1603
Streib, D. L.   4729
Strever, M. T.   632

Stricklin, F. L.   4306
Strimple, H. L.   1642, 3622
Stringer, R. P.   3071
Stringfield, V. T.   47, 49, 60
Stringham, B.   4394, 4399, 4406
Struble, R. A.   3520, 3582
Stuckey, J. L.   3286, 3293
Stullken, L. E.   1815, 1816, 1823, 1826
Sturgeon, M. T.   3506, 3510
Sublett, R. B.   2049
Suekawa, H. S.   4420
Sullivan, D. M.   1521
Sullivan, G. A.   2074
Sullivan, J. L.   4540, 4547, 4548, 4549, 4550
Sullivan, R.   4890
Summers, W. K.   3134, 3137, 3152, 3169, 3178, 3234
Summerson, C.   1927
Sunderland, G. L.   3374
Sunderman, J. A.   1499, 1511, 1587
Sunn, Low, Tom, and Hara, Inc.   1085
Sutcliffe, H.   772, 865
Sutherland, P. K.   3235, 3259, 3276, 3626, 3631, 3695, 3696
Sutton, C. J.   622, 623
Sutton, D. G.   1936
Sutton, J. A.   2271, 2272, 2278
Suyemoto, M. M.   664
Swain, F. M.   2439, 3284
Swain, L. A.   1109
Swann, D. H.   1224, 1242, 1245, 1307, 1465, 1466, 1479
Swanson, D. E.   958
Swanson, V. E.   399
Swartz, F. M.   3800
Swartz, J. H.   980, 981
Sweeney, J. W.   789, 807, 808, 809, 885, 889
Sweet, P. C.   4535, 4536, 4556, 4574
Sweet, W. C.   1996
Swindel, G. W.   2187
Swineford, A.   1791
Swinehart, R. P.   3498
Sylvester, A.   1179
Sylvester, A. G.   430
Synder, G. L.   683
Szabo, M. W.   97, 102, 124, 136, 139, 144, 148, 149, 160, 202, 203, 204, 205, 206, 207, 208
Szebenyi, A.   3011
Szidarovszky, F.   3072

Tabet, D. E.   3222
Takasaki, K. J.   1001, 1030,
1049, 1051, 1053
Taliaferro, W. J.   1068, 1069, 1072
Talley, J. H.   725, 726, 731, 733
Tan, S. S.   567, 568
Tanaka, H. H.   3658
Tanzer, M. S.   3060
Taranik, J. V.   1633, 1634
Taruski, A. G.   661
Tarver, G. R.   753, 819, 823
Taylor, A. M.   3231
Taylor, G. C.   476, 556, 566
Taylor, K. R.   2256, 2259
Taylor, O. J.   2722, 2785
Taylor, R. B.   2383
Tegland, E. R.   4122
Teifke, R. H.   4508, 4509
Teller, J. T.   3568
Templeton, J. S.   1191, 1464
Templeton, T. R.   4132
Tennissen, A. C.   2585, 2616
Terich, T. A.   4655
Terrell, T. T.   4843
Textoris, D. A.   3310
Thacker, J. L.   2635, 2642
Thayer, P. A.   3310, 4539
Thayer, T. P.   424
Thomas, D. R.   4522
Thomas, E.   4173
Thomas, G. R.   2030
Thomas, J.   1300, 1349, 1431, 1444
Thomas, J. D.   2283, 2287
Thomas, N. O.   2077
Thomas, R. L.   1460
Thomas, R. N.   1909
Thomas, W. A.   195
Thomas, W. O.   2237
Thompson, A. E.   4360
Thompson, A. J.   371
Thompson, J. B.   4482, 4483
Thompson, R. R.   3788
Thompson, S.   3206
Thompson, T. B.   3177
Thompson, T. L.   1683, 1736, 1776, 2604, 2619, 2625, 2630
Thompson, W. D.   4373
Thomson, K. C.   4372
Thomson, K. S.   661
Thomson, M. T.   892
Thorsen, G. W.   4625, 4630, 4671
Thresher, J. E.   4492
Thurlow, C.   3052
Thurrell, R. F.   685
Tibbitts, G. C.   4532
Tibbs, J. S.   2122
Tiedemann, H. A.   4112
Timmons, B. J.   885
Ting, F. T. C.   3460
Tipton, M. J.   4027
Tirabassi, M.   2973

Titus, F. B.   3227
Todd, V. R.   3212
Toewe, E. C.   4565, 4570
Toler, L. G.   773, 828, 833
Tolman, C.   2624
Tolson, J. S.   47, 198, 199
Tolsted, L. L.   1791
Tomikel, J. C.   3828
Toomey, D. F.   1865, 3665
Toppozada, T. R.   484, 576, 3174, 3180, 3197
Torries, T. F.   2481
Toth, M.   1144
Totten, S. M.   3589, 3613
Toulmin, L. D.   196
Towill Corporation, R. M.   see R. M. Towill Corporation
Townsend, P. S.   2042
Trace, R. D.   2120
Trapp, H.   3323, 3324
Trauger, F. D.   3232
Traylor, C. T.   193
Treworgy, C. G.   1360
Treworgy, J. D.   1361, 1365, 1476
Trexler, B. D.   1185
Trexler, D. T.   2934
Trible, T. C.   234, 235, 324, 325
Trihey, E. W.   1140
Trimble, L. M.   4386
Triplehorn, D. M.   287, 334
Trippet, A. R.   4250
Troell, A. R.   1743
Trotta, L. C.   4821, 4825
Trowbridge, A. C.   1639
Troxel, B.   451
Troxel, B. W.   424, 521, 526, 547
Troyer, M. L.   4835
Tschanz, C. M.   2874
Tschudy, R. H.   3224
Tsusue, A.   3862
Tubbs, D. W.   4646
Tucker, R. C.   4758
Tucker, W. E.   59, 101, 169
Tufford, S. P.   2385, 2399, 2406
Tullis, E. L.   4047
Turcan, A. N.   2194
Turk, L. J.   4177, 4468
Turley, C. H.   4447
Turnage, N. E.   193, 194
Turner, D. L.   282, 283, 336
Turner, M. P.   168
Turner, W. D.   2568
Tuthill, S. J.   1630, 3423, 3430
Twenter, F. R.   1666
Tychsen, P. C.   3996
Tyrrell, M. E.   152, 158, 2144, 2145, 4146, 4531

Udden, J. A.   4255
Ulmer, J. H.   3481
Ulrich, E. O.   2569
Ulteig, J. R.   3552
Unger, W. L.   1423
U. S. Bureau of Mines   2748, 2751, 2756, 2760, 2766, 2770, 2775, 2778, 2796, 4784
U. S. Bureau of Reclamation   351, 2829, 3997
U. S. Geological Survey   351, 813, 878, 1188, 2607, 2796, 2843, 2864, 3706, 3997, 4777
University of Washington Geology Department   4626
Untermann, B. R.   4345
Untermann, G. E.   4345
Updike, R. G.   342
Upshaw, C. F.   2498
Utgaard, J.   1493

Vaksvik, K. N.   976, 979
Valadares, L. T.   3081
Valenciano, S.   1030
Valkenburg, N.   154
Vallely, J. L.   125
Vallery, J. F.   179
Van Couvering, J. A.   1939
Van Den Berg, J.   1992, 1997, 2000, 2009, 2015, 2134
Van Dyke, L. H.   1307
Van Eck, O. J.   1624, 1627, 1637, 1645, 1662
Van Horn, E. C.   3969
Van Horn, R.   4401
Van Horn, R. G.   3515, 3599, 3610
Vanlier, K. E.   2372, 2373, 2375
Vanorny, P. M.   4040
Van Rensburg, W. C. J.   4275, 4276, 4277, 4279
Van Tuyl, D. W.   3920, 3921
Van Voast, W. A.   2757, 2762, 2768, 2820
Varga, R. J.   383
Varma, M. M.   1571
Varnedoe, W. W.   110
Varrin, R. D.   711
Veneman, P. L. M.   4815
Ventress, W. P. S.   3616
Verity, V. H.   3115
Vernon, R. O.   787
Ver Planck, W. E.   410, 501
Ver Ploeg, A. J.   4875
Vestal, F. E.   2471
Viavant, W.   4411
Vigrass, L. W.   3700

Vincent, J. K.   2051
Vincent, J. W.   2130
Vine, W. A.   2689, 2764
Vineyard, J. D.   2549A, 2558,
  2559, 2561, 2563, 2564, 2565,
  2579, 2606, 2617, 2638, 2673
Vinton, S.   4706
Vitz, H. E.   3097
Volborth, A.   2881, 2899, 2900
Volz, G.   2422
Von Huene, R.   526
Vondra, C. F.   2854
Vorhis, R. C.   931
Vosburg, D. L.   3624
Voskuil, W. H.   1214, 2906
Vuich, J. S.   361, 379

Waddell, D. E.   3637
Waddell, K. M.   4458, 4467, 4470
Waddington, J. L. B.   2447
Wade, W. J.   4109
Wagener, H. D.   3995
Wagner, H. C.   4642, 4670
Wagner, R. E.   2623
Wagner, W. K.   3168
Wagner, W. R.   3779, 3799, 3810,
  3852
Wahl, K. D.   30, 33, 52, 83,
  1671, 1673, 1676, 1677
Wahrhaftig, C.   417, 423
Wait, R. L.   926, 929
Waldron, G.   1793
Waldron, G. A.   1777
Wali, M. K.   3402
Walker, C. W.   3181, 4661
Walker, F. H.   1913, 1914, 1915,
  1917, 1918, 1920, 2041, 2064
Walker, G. W.   424
Walker, P. N.   2240, 2294
Walker, R. D.   3813
Wallace, A. T.   1163, 1165, 1168,
  1169
Wallace, D. T.   2386
Wallace, R. E.   423
Wallin, H.   2335
Walsh, T. H.   1177
Walter Lum Associates, Inc.   1117
Walters, J. V.   179
Walters, R. A.   2931
Walters, R. F.   1780, 1889
Walton, A. W.   4253
Walton, W. C.   2417
Ward, C. R.   1354
Ward, J. R.   1735, 1768, 1819
Ward, P. E.   4749, 4754
Ward, W. E.   189
Warford, J.   2985
Wark, J. A.   4486

Warman, J. L.   100, 115, 123
Warner, G. H.   3220
Warner, J. W.   620
Warren, C. W.   4364
Warren, W. M.   47, 159
Warshauer, S. M.   4744, 4746, 4747
Wartman, F. S.   371
Washington Division of Geology and
  Earth Resources   4638, 4652
Wasson, B. E.   2465
Watson, T.   925
Watson, W. T.   36, 37, 185
Watts, N. A.   2447
Waugh, T. C.   1717, 1784, 1786, 1788,
  1789
Waugh, W. N.   1687, 1701, 1716
Wayne, W. J.   1485, 1506, 1508,
  1588, 1601
Weakley, E. C.   1815, 1824, 1827
Weaver, K. N.   2269
Webb, E. J.   2688, 2106
Webb, H. W.   4525, 4552, 4579
Webb, R. W.   412
Webb, W. E.   2227, 2228, 2291, 2295
Webb, W. M.   1505
Webendorfer, B.   4832
Weber, F. H.   424, 430, 442, 450,
  453, 494
Weber, W. M.   2787
Webers, G. F.   2440
Wedge, W. K.   2640, 2643
Weed, W. H.   661
Weeks, A. M. D.   4246
Wehr, A. G.   2485
Weidener, J. P.   3051, 3055
Weidman, C.   3205
Weigle, J. M.   2227, 2228, 2237,
  2302, 2311
Weiler, C. T.   424
Weimer, B. S.   2935
Weir, K. L.   2352
Weisbord, N. E.   747, 751
Weise, B. R.   4251
Weissenborn, A. E.   4613, 4643
Weist, M. M.   2838
Welday, E. E.   519
Welden, J. C.   4758
Weller, S.   1366
Wells, J.   2557, 2599
Wells, J. S.   2614
Wells, M. W.   1121
Welter, B.   1793
Wentworth, C. K.   1073
Wermund, E. G.   4188, 4192, 4215,
  4316
Werner, H. J.   4561
West, J. M.   961
West, L. R.   163, 431
West, W. S.   4803, 4804

West Virginia Geological Survey 4736, 4737
Westfall, I. 2404, 2405
Westheimer, J. M. 3662
Wetterhall, W. S. 760, 821, 826
Wharton, H. M. 2633, 2638, 2644
Wheat, M. M. 2895
Whelan, J. A. 4354, 4356, 4402, 4422, 4432, 4434, 4438, 4466, 4469, 4471
Whipple, W. 2966
Whitcomb, J. H. 431
White, A. M. 3991, 3992
White, C. 3777
White, D. E. 424
White, D. H. 139
White, G. W. 3501, 3522, 3563, 3581, 3592, 3602, 3613, 3805
White, H. G. 101, 103, 106, 111, 169
White, I. C. 4727
White, W. A. 745, 1208, 1209, 1227, 1261, 1388, 1399, 1406, 1408, 1454, 4251, 4327, 4330
White, W. B. 3817, 3818, 3819, 3820
White, W. S. 2147
Whitelaw, W. E. 3774
Whitesides, D. V. 1945, 1948
Whitfield, J. W. 2554
Whitfield, M. S. 2180
Whiting, L. L. 1239
Whitman, C. D. 2176, 2199
Whitman, H. M. 1961, 2170, 2189, 2191
Whitmore, F. C. 2245
Whitmore, S. 3142
Whitworth, W. R. 657
Whyte, M. R. 3209
Wickham, J. T. 1362, 1460, 1476
Wicklein, P. C. 639, 640
Widmer, K. 2949, 3015, 3084, 3085, 3087
Wieland, D. R. 2111
Wielchowsky, C. C. 179
Wier, C. E. 1510, 1529
Wiethe, J. D. 4127
Wilcox, J. T. 4112, 4152
Wilde, P. 696
Wilke, K. R. 2773
Wilkins, C. H. T. 92, 107
Wilkinson, B. H. 4208, 4221
Willard, M. E. 3238
Willaw, J. R. 1364
Willden, R. 2859, 2884
Willey, G. R. 2141
Williams, C. H. 2501, 2503
Williams, F. E. 3130

Williams, J. F. 2309
Williams, J. H. 2556, 2610, 2615, 2666, 2675
Williams, J. W. 457, 548
Williams, K. E. 872
Williams, L. 3297, 3984
Williams, L. A. 2732
Williams, R. B. 1790
Williams, R. E. 1161, 1163, 1164, 1165, 1168, 1169, 1184, 1185
Williamson, A. D. 2132
Williamson, C. R. 4373
Willman, H. B. 1191, 1194, 1196, 1197, 1203, 1220, 1256, 1283, 1293, 1296, 1316, 1323, 1327, 1334, 1335, 1341, 1342, 1358, 1363, 1464, 1465, 1466, 1467, 1475
Willoughby, J. 1889
Wills, B. 2579, 2581
Wilmoth, B. M. 4680, 4749, 4754
Wilshusen, J. P. 3780, 3786, 3898
Wilson, B. D. 4733
Wilson, C. W. 4111, 4112, 4115, 4161
Wilson, E. D. 344, 345, 350, 364, 366, 367, 369, 372, 374, 375, 378
Wilson, E. N. 1939, 2003, 2056, 2066
Wilson, F. H. 336
Wilson, F. W. 1794, 1809
Wilson, G. V. 123, 132, 145, 198
Wilson, J. A. 4283
Wilson, J. L. 2099, 2930
Wilson, J. M. 4148
Wilson, L. 2972
Wilson, M. 4880
Wilson, M. E. 2567
Wilson, R. L. 4160
Wilson, R. P. 3231
Wilson, S. E. 2340, 2341
Wilson, T. A. 420
Wilson, W. E. 870
Wilson, W. F. 3283, 3287, 3288, 3294, 3295, 3301, 3302, 3303
Wilson, W. H. 4833, 4849, 4856
Wilt, J. C. 353, 361
Wimberly, E. T. 786
Windham, S. R. 889
Wingate, F. H. 3657
Wininger, D. C. 2261
Winkler, C. D. 4229
Winner, M. D. 2171
Winslow, J. D. 1691, 1704, 1712, 1721, 1757, 1848, 1867, 1875
Winston, D. 4297
Winston, G. O. 882, 887
Winter, T. C. 2406, 2447
Winters, A. S. 2720
Winters, H. A. 3313
Wiram, V. P. 1575

Wisconsin Geological and Natural
  History Survey   4777, 4784
Wise, W. S.   4612
Witherspoon, P. A.   1277
Withington, C. F.   424
Witinok, P. M.   1650
Wolfe, H.   1835
Wood, A.   4074
Wood, C. R.   3932, 3933, 3942,
  3955, 3959, 3960
Wood, E. B.   1906, 1908, 1911
Wood, J. M.   1489
Wood, P. R.   3670
Wood, R. E.   4431
Wood, S. O.   4268, 4269
Woodard, H. J.   792
Woodland, B. G.   4481
Woodring, S. M.   4732, 4733
Woodrow, D. L.   3804
Woodruff, C. M.   4319
Woodruff, K. D.   707, 712, 716,
  718, 719, 721, 728
Woods, C.   1803
Woods, M. J.   1747, 1748
Woodward, L. A.   3183, 3207
Woodzick, T. L.   4889
Wootton, T. M.   487
Workman, L. E.   1463
Worl, R. G.   2118, 4857
Worzel, L.   2962
Wray, J. L.   1695
Wright, A. P.   886
Wright, F. N.   4496
Wright, H. E.   2447
Wright, H. T.   2223
Wright, J. R.   3117
Wright, J. S.   3097
Wright, L. A.   424, 525, 547
Wright, R. P.   1609
Wright, T. J.   3283
Wrobel, W.   2969
Wunder, J. M.   4597, 4598
Wyoming Geological Survey   4872

Yamanaga, G.   1001, 1005,
  1018, 1036, 1037, 1042, 1051,
  1052, 1108
Yarger, H.   1761
Yarish, C.   664
Yaseen, A.   2997
Yasso, W. E.   2956, 3018, 3041
Yeloushan, C. C.   3823, 3830,
  3831, 3835, 3841
Yerkes, R. F.   577
Yoho, W. H.   746, 871, 1640
Yon, J. W.   742, 743, 744, 748,
  881
Young, D. M.   1912

Young, H. L.   4826
Young, K. B.   4769
Young, K. P.   4164, 4169, 4182,
  4238, 4247, 4281, 4321
Young, R. J.   3115
Young, R. N.   2102
Young, R. S.   4589
Youngblood, I.   3161
Yung-Yao, C.   3029
Yunghans, R.   3078

Zack, A. L.   2202
Zaffarand, R. F.   4267
Zaporozec, A.   3074, 4828
Zarley, C.   1833, 1834
Zeller, D. E.   1725, 1731, 1739, 1746
Zeller, E. J.   1811
Zeller, R. A.   3095, 3107, 3200, 3240
Zen, E. -A.   4478
Zenger, D. H.   540
Zidek, J.   3189
Ziegler, R.   1577
Ziegler, R. E.   905
Zietz, I.   2310
Zimmerman, E. A.   2704, 2716
Zimmerman, J. B.   4173
Zimmerman, J. H.   4801
Zisa, A. C.   916
Zones, C. P.   1007
Zurawski, R.   4120

Abingdon Quadrangle, Virginia
  Geology 4553
Accomack County, Virginia
  Groundwater 4532
Accretion
  Gleys--Illinois--Central Western
    1467
Acetic Acid 4034
Acid Mine Conditions
  Indiana
    --Greene County 1575
    --Latta 1575
Acid Mine Drainage
  Alabama
    --Cane Creek Basin 122
    --Oakman 122
    --Walker County 122
Acid Solutions 3140, 3166
Acid Water
  Maryland--Western 2298
Acidization 2111
Acker Lake, Mississippi
  Landslides 2546
Actinoceroids 3243, 3260
Adair County, Kentucky
  Economics 2101
  Geology 2101
Adair County, Oklahoma
  Geology 3667
Adams County, Illinois
  Coal 1230
Adams County, Mississippi
  Geography 2540
  Geology 2540
  Land Use 2540
  Man 2540
  Physiography 2540
  Recreation 2540
  Transportation 2540
  Wastes 2540
  Water 2540
Adams County, North Dakota
  Aquifers 3378
  Geology 3376
  Groundwater 3377, 3378
  Water 3377
  Wells 3377
Adams County, Ohio

Bedrock Geology 3596
Adams County, Pennsylvania
  Hydrology 3932
  New Oxford Formation 3932
Adams County, Washington
  Geology 4623
Addyston Quadrangle, Ohio
  Bedrock Geology 3584
Admiralty Island, Alaska
  Geochemistry 216
  Geology 216
Aeration
  New Jersey--Passaic River 2966
Aerial Imagery 1177
  Alabama 179
  Iowa 1634
  West Virginia 4732
Aeromagnetic Maps
  Alaska--Brooks Range 291
  Appalachians 2310
Aeromagnetic Surveys
  Alaska 317
  Illinois
    --Hardin County 1219
    --Southwestern 1351
  Maryland--Western 3824
  Michigan
    --Dickinson County 2357
    --Upper Peninsula 2353
  Ohio--Eastern 3824
  Pennsylvania
    --Southeastern 3849
    --Western 3824
  Utah
    --San Francisco Mountains
      4371
    --Southwestern 4371
  West Virginia--Northern 3824
Afterbay Reservoir Site, Hawaii
  Geology 1010
Aftershocks
  California--San Fernando Valley
    431
Aggregate 2144
  Alabama 84
    --Porters Creek 158
  California--Los Angeles 580
  Colorado--Front Range 639, 640

Aggregate (Cont.)
   Indiana 1509
   --Northwestern 1587
   Maine--Farmington 2205
   Mississippi 2482
   New Mexico 3099
   Ohio 3574
   Pennsylvania--Pittsburgh 3886
   South Dakota 4048
Ahuamda Camp, Hawaii
   Water Supply 1074
Air Pollution 2977, 2988, 2997
   Connecticut 703
Alabama River, Alabama
   Induced Infiltration 142
   Photogeology 142
Alabaster Cavern, Oklahoma
   Guide 3692
Alachla County, Florida
   Ceramic Clay 762
   Geology 872
   Water Resources 759, 822
Alaeloa, Hawaii
   Drilling Logs 1026
   Pumping Tests 1026
Alafia River Basin, Florida
   Fluoride 833
   Hydrology 836
Alameda County, California
   Geology 408, 488
Alapaha, Georgia
   Aquifers 942
Alaska Peninsula
   Coal 334
   Mesozoic 297
   Mineralization 330
   Petroleum Reservoirs 297
   Rock 297
   Tertiary 297
Alaska Range
   Bedrock Geology 237
   Formations 282
   Geology 280, 281
   Mineral Deposits 340
   Mount Galen 294
   Paleocene 282
   Paleozoic 280
   Petrology 280
   Rocks 280
   Sedimentology 280
   Teklanika Formation 282
   Tertiary 280, 294
   Volcanic Formations 282, 294
Albany County, Wyoming
   Diamond 4882
   Gold 4854
   Keystone Gold-Copper Prospect
      4850
   Kimberlite Diatremes 4882

   Platinum 4854
   Structural Geology 4860
Albemarle County, Virginia
   Gravity Surveys 4574
   Magnetic Data 4574
Albemarle Sound
   Oyster Shells 3291
Albian
   Oklahoma
      --Southern 3657
Albuquerque, New Mexico 3279
Albuquerque Basin, New Mexico 3207
   Basalt 3210
   Geology 3266
   Volcanics 3210
Alcorn College, Mississippi
   Water Supply 2544
Alexander County, Illinois
   Geology 1461
   Groundwater 1461
Alexandria, Louisiana
   Water 2186
Algae
   Kansas
      --Southeastern 1695
Algal Banks
   Kansas
      --Eastern 1744
Alger County, Michigan
   Groundwater 2372
Alkaline Environment 1324
Allamoore Talc District, Texas
   Asbestos 4185
Allegheny County, Pennsylvania
   Groundwater 3919, 3946
   Petrography 3861
Allegheny Front, West Virginia
   Devonian Rocks 4672
Allegheny Group
   Pennsylvania
      --Clearfield County 3833
Allegheny Plateau, West Virginia
   Gravity 4743
   Magnetics 4743
   Structure 4743
Allen County, Indiana
   Geology 1604
Allen County, Kansas
   Geology 1741
   Groundwater 1741
Allen County, Kentucky
   Economic Geology 1931
Allen County, Ohio
   Limestone 3521
   Tymochtee Dolomite 3521
   Upper Silurian 3521
Allosaurus Fragilis 4382
Alluvial Aquifers
   Illinois 1289

Alluvial Groundwater
Iowa
--Floyd River Basin 1676
Alluvial Intervals
New Mexico
--Northeastern 3204
Alluvial Plains
Texas
--Southeast 4244
Alluvium
Kentucky 1945
Almeda Mine, Oregon 3766
Alpine County, California
Mineral Resources 447
Mines 447
Alquist-Priolo Special Zones Act
474, 479
Alto Formation
Illinois
--Southern 1217
Alton, Utah
Coal 4390
Alumina
Kansas 1701
West Virginia 4685
Aluminum
California 424
Alunite 365
Amador County, California
Geophysics 558
Ambler River Quadrangle, Alaska
Geochemistry 274
Geology 274
Amboy Crater, California
Volcanism 493
Americus Limestone
Kansas 1853
Oklahoma 1853
Amity Quadrangle, Pennsylvania
Gas 3873
Oil 3873
Ammonite 4182
Gulf Coast 4281
Western U.S. 3270
Ammonoids 3666
Amphitheatre Mountains, Alaska
Geology 254
Anadarko Basin, Oklahoma and
Texas
Early Devonian 3648
Evaporites 3624
Hunton Group 3648
Late Ordovician 3648
Permian Salt 3624
Silurian 3648
Anchorage, Alaska
Geology 240
Andalusite
California 424

Andesite 3704, 3707
Andrew County, Missouri
Water 2659
Anhydrite
California 424
Kentucky
--Northwestern 1940
Animas Mountains, New Mexico
Structural Geology 3179
Annapolis, Maryland
Aquifers 2300
Anne Arundel County, Maryland
Groundwater 2232
Wells 2232
Anomalies 1578
California
--San Fernando 430
Idaho 1152
Illinois
--Northern 1210
Iowa
--Fayette County 1648
Kentucky
--Davies County 2025
--Western 2038
Maryland
--Chesapeake Bay 2308
Michigan
--Dickinson County 2357
--Southern Peninsula 2342,
2342
Minnesota
--Lake Washington 2415
--Meeker County 2415
New Jersey 3086
New Mexico
--Philmont Ranch Region
3146
North Dakota
--Pembina County 3479
Anomalous Zinc
Tennessee
--Henry County 4148
Ansonia Quadrangle, Connecticut
Surficial Geology 679
Anthophyllite
Alabama
--Chambers County 38
--Tallapoosa County 38, 107
North Carolina 3283
Anthracite
Pennsylvania
--Eastern 3897
Anticlines
Georgia
--Southwestern 947
Montana
--Cedar Creek Anticline 2785,
2824

Anticlinoria
  Maryland
    --South Mountain 2266
  Maine 2217
  West Virginia
    --Eastern 4698, 4748
    --Pendleton County 4698
Antietam Creek, Maryland
  Soluble Dye 2259
Antimony
  California 424
  Nevada 2860
Anton Larsen Bay, Alaska
  Geochemistry 266
  Geology 266
  Stream Sediments 266
Antrim Shale
  Michigan
    --Michigan Basin 2363
Apache Warm Springs, New Mexico
  Beryllium 3157
  Geology 3157
Apalachicola Embayment
  Georgia
    --Southwestern 947
Apollo 9
  Alabama 177
Appalachian Basin
  Cambrian 2088
  Central 4745
  Dunkard Group 4704
  Geology 2088
Appalachian Logging Services 2022
Appalachian Piedmont
  Glenarm Series 2284
Appalachians 899
  Connate Water 4676
  Devonian 1909
  Field Trips 4158
  Gas 1909, 4742
  Limestone 1929
  Lower Paleozoic 4742
  Oil 4742
  Shale 1909
Aquatic-Biotic Community 138
Aqueducts
  Maryland
    --Susquehanna Aqueduct 2286
Aquia Formation
  Maryland
    --Coastal Plain 2288
    --Digital Simulation Model
      2267
Aquifers 3923
  Alabama
    --Stevenson Area 134
  Delaware 708, 716, 718, 723
    --Central 709
    --Dover 729

    --Laird Tract Well Field 731
    --Newark 731
    --Piney Point 729, 732
    --Southeastern 709
  Florida 764, 768, 777, 785, 790,
    802
    --Broward County 823
    --Duvall County 846
    --Hillsboro County 803
    --Miami 802
    --Orange County 859
  Georgia
    --Alapaha 942
    --Coastal 924
    --Coolidge 942
    --Fitzgerald 942
    --Glynn County 926
    --Montezuma 942
    --Thomasville 942
  Illinois 1289
    --Illinois Basin 1326
    --Northeastern 1262
  Kentucky
    --Mayfield 1942
    --Muhlenberg County 1939
    --New Cypress Pool 1939
    --Western Coal Field Region
      1961
  Louisiana
    --Baton Rouge 2162, 2173
    --Coastal Area 2183
    --Lake Charles 2184
    --Southwestern 2180, 2189
  Maryland 2263
    --Annapolis 2300
    --Aquia Aquifer 2267
    --Piney Point Aquifer 2309
    --Southern 2267, 2306
  Montana
    --Eastern 2830
  North Carolina 3310
  North Dakota
    --Adams County 3378
    --Benson County 3364
    --Bowman County 3378
    --Burke County 3353
    --Grand Forks County 3347,
      3348, 3349
    --McLean County 3367
    --Mountrail County 3353
    --Nelson County 3360
    --Pierce County 3364
    --Renville County 3342
    --Stutsman County 3315
    --Traill County 3339
    --Walsh County 3360
    --Ward County 3342
    --Wells County 3345
    --Williams County 3336

Ohio
--Marion 3611
Texas
--Galveston 4209
--Houston 4209
--Trans-Pecos 4311
Wisconsin
--Dane County 4816, 4818
Aquilapollenites
Western U.S. 3224
Arbuckle Mountains, Oklahoma
Arbuckle Group 3632
Articulate Brachiopods 3644
Chitons 3665
Conodonts 3693
Early Devonian 3650
Geology 3694
Kindblade Formation 3665
Late Cambrian 3632
Ordovician 3632, 3644, 3665
Timbered Hills Group 3632
Trilobites 3632, 3650
Viola Formation 3644
Archaeolithophyllum
Kansas
--Southeastern 1695
Archbald Pothole State Park,
Pennsylvania
Guide 3900
Archeology
Maryland
--Middle Chesapeake 2223
Wyoming 4880
Arctic Alaska
Rock 233
Stream Sediments 233
Ardmore Basin, Oklahoma
Pennsylvanian Fusulinids 3637
Areal Geology
Alabama
--Jefferson County 15
--Valley and Ridge Province
10
Illinois
--Illinois Flurospar District
1198, 1241, 1269
Missouri 2562
Arenaceous Foraminifers
Alabama
--Gipsy Limestone County
58
Indiana
--Northern 1498
Arid Regions 60
Arkansas Valley Coal Field,
Arkansas
Coal 397, 398
Arkoma Basin, Oklahoma
Early Devonian 3656

Hunton Group 3656
Late Ordovician 3656
Silurian 3656
Arlington, Wyoming
Structural Geology 4802
Armstrong County, Pennsylvania
Groundwater 3945
Arnold Valley Quadrangle, Virginia
Geology 4567
Aromatic Fluorine
Illinois 1357
Arroyo Penasio Formation
New Mexico
--North Central 3245
Arsenic
California 424
--Plumas County 544
Artesian Aquifers
Florida 764, 768, 777, 785, 790,
792, 802
--Miami 862
Georgia
--Coastal 924
--Glynn County 926
Artesian Basins
Minnesota
--Minneapolis 2397
--St. Paul 2397
New Mexico
--Chaves County 3147
--Roswell Artesian Basin 3147
Artesian Pressures
Montana 2711, 2721, 2729, 2735
Artesian Water
Southeast 3000
Articulates
Illinois
--Southwestern 3646
Missouri
--Eastern 3466
Oklahoma 3616, 3617, 3646
--Arbuckle Mountains 3644
--Eastern 3652
Artisophyton
Illinois 1348
Asbestos 2996
California 424
North Carolina 3283
Texas
--Allamoore Talc District 4185
--Culberson County 4185
--Hudspeth County 4185
Ascension Parish, Louisiana
Groundwater 2167
Ash 1252
Colorado 615
Illinois 1221
Kansas 1857

Ash (Cont.)
  Missouri
    --Southeast 2626
  Montana 2753
  Nevada
    --Yerington District 2933
Ashby Gap Quadrangle, Virginia
  Geology 4590
Ashland County, Ohio
  Glacial Geology 3602
Ashmore Gas Area, Illinois 1243
Ashtabula County, Ohio
  Glacial Geology 3613
Asphalt
  California 424
  Kentucky 1965
Atascadero, California
  K-Feldspar 569
  Mesozoic 569
  Sandstone 569
Atchison County, Missouri
  Groundwater 2662
Athens, Alabama
  Geology 18
  Groundwater 18
Athleta Petrosa
  Texas 4282
Atkinson County, Georgia
  Phosphate 967
Atlanta, Georgia
  Brevard Lineament 896
  Geology 896
  Water Supply 974
Atlantic Seashore
  Delaware 720
Atomic Absorption 3196
Attala County, Mississippi
  Ceramic Tests 2463
  Geology 2464
  Mineral Resources 2462
  Subsurface 2464
  Water Resources 2465
Au Sable River, Michigan 2314
Auburn Damsite, California
  Seismic Safety 486
Augusta County, Virginia
  Igneous Rocks 4523
Augusta Springs, Virginia
  Geology 4575
Austin, Texas
  Geology 4238, 4247, 4250
Authors' Guide 890
Automation 2109
Avalanches
  Colorado 592, 642
    --Crested Butte 613
    --Gunnison County 613
Averill Quadrangle, Vermont
  Geology 4480

Avoyelles Parish, Louisiana 2175
Axemann Limestone
  Pennsylvania
    --Central 3802
Azusa Quadrangle, California
  Geology 546

Backbone Anticline, Arkansas
  Geology 390
Bacons Castle Quadrangle, Virginia
  Geology 4572
Bacteria
  Alabama 154
Bainbridge, Georgia
  Groundwater 938
Baker County, Florida
  Groundwater 839
Bakken Formation
  North Dakota 3311
  South Dakota 3311
Balcones Escarpment, Texas 4194
Baldwin County, Alabama
  Disposal Wells 116
  Gravity Surveys 118
Baltic, South Dakota
  Groundwater 4101
Baltimore Canyon Trough
  Hydrocarbon Resources 734
Baltimore County, Maryland
  Rocks 2305
  Springs 2224
  Wells 2224
Bangor Limestone
  Alabama 53
  Tennessee
    --Middle 4133
Baraboo District, Wisconsin
  Glacial Geology 4800
  Plant Ecology 4801
  Precambrian Rocks 4799
  Sedimentology 4799
Baraga County, Michigan
  Geology 2382
  Groundwater 2382
Barber Quadrangle, Arkansas
  Geology 389
Barbour County, Alabama
  Iron Ore 166
  Mineral Resources 173
  Water Resources 173
Barbour County, West Virginia
  Oil and Gas 4679
Barite
  Alabama 143
  California 424
  Maryland
    --Piedmont Upland 2234
  Mississippi Valley 2119

Missouri 2575
--Washington County 2633
Nevada 2897
New Mexico 3130
South Carolina
--Carolina Barite Belt 3969
Tennessee 4134, 4149
Washington 4603
Barium 3170
South Carolina 3976
Barkley Lake, Kentucky
Limestone 1941
Barnes County, North Dakota
Geology 3319, 3320, 3321
Glaciation 3319
Groundwater 3319, 3320, 3321
Water 3320
Barre, Vermont
Geology 4493
Barrier Dunes
Michigan
--Lake Michigan 2364
Barrier Islands
Texas
--Galveston Island 4244
--Padre Island National Sea-
shore 4251
Barry County, Michigan
Geology 2356
Subsurface 2356
Barton County, Kansas
Geology 1775
Structure 1775
Barton County, Missouri
Tar Sand 2599
Bartow County, Georgia
Geohydrology 957
Basal Castaic Formation
California
--Los Angeles County 520
Basalt
Connecticut
--Central 697
Idaho
--Benewah County 1160
Minnesota 2419
New Mexico
--Albuquerque Basin 3210
--Carrizozo Basalt Field
3167
--McCartys Basalt Field
3167
--Potrillo Basalt Field
3167, 3203
Base Metals
Alabama 113
New Mexico
--Luna County 3186
Virginia 4530

Washington
--Cascade Range 4610
Wisconsin
--Upper Mississippi Valley
4786, 4803, 4804
Basement
Illinois 1238
--Northeastern 1482
Iowa 1640
Kansas 1686
Maryland
--Salisbury Embayment 2307
Oklahoma
--Southern 3619
Pennsylvania
--Erie County 3851
Rocky Mountains 4877
Basement Rocks
Florida 750
Georgia 895
Kansas 1698
Missouri
--St. Francois Mountains 2631
Basins
Minnesota 2417
Bassett Quadrangle, Virginia
Geology 4580
Batch Grinding 3171
Batchelon, Louisiana
Water Resources 2196
Bates County, Missouri
Geology 2639
Mineral Resources 2639
Rocks 2639
Batesland Formation
South Dakota
--Southwestern 4049
Batesville Formation
Missouri
--Southwestern 2630
Bath County, Virginia
Igneous Rocks 4523
Batholiths
California
--Lakeview Mountains 519
--Riverside 519
--Sierra Nevada 423
Idaho 1144
Montana
--Bitterroot Range 2788
Bathymetric Changes
Alabama 186
Bathythermograph Recordings
Ohio
--Lake Erie 3529
Baton Rouge, Louisiana
Aquifers 2162
Groundwater 2162
Salt Water Encroachment 2192

Battle Creek, Michigan
  Groundwater 2375
Bauxite
  Alabama
    --Eufaula Bauxite District
      55
  Hawaii
    --Kauai 1006
    --Maui 1006
  Mississippi 2455
Baxter State Park, Maine
  Geology 2207
Bay Area, California
  Geology 423
  Hydrogeology 423
Bay County, Florida
  Geology 752
Bayou Carlin, Louisiana
  Miocene 2153
Beach Erosion
  Connecticut
    --Coast 695
Beach Ridges 1439, 1443
  Pennsylvania
    --Northwestern 3815
Beaches
  Hawaii 1117
  Massachusetts
    --Cape Cod Bay 3041
  New Jersey
    --Sandy Hook 3018, 3030
  New Mexico
    --Northwestern 3218
  New York
    --Long Island 2980
Beadle County, South Dakota
  Geology 4001, 4002
  Water 4001, 4003
Bear Creek, Alaska
  Geology 247
Bear Creek, South Dakota
  Radiation Dating 4061
  Terraces 4061
Bear Island, Maryland
  Rocks 2246
Bear Lake, Utah
  Environmental Geology 4369
Bearmouth, Montana
  Geology 2783
  Metallic Resources 2784
Beaufort County, North Carolina
  Pungo River Formation 3285
Beaver County, Pennsylvania
  Groundwater 3920, 3950
Beaver County, Utah
  Alteration Area 4399
  Geology 4396, 4438
  Hydrothermal Alteration 4394,
    4396, 4402, 4406
  Igneous Complex 4407
  Mineralization 4407
  Mineralogy 4438
Beaverdam Mountains, Arizona
  Geology 357
Beaverhead County, Montana
  Minerals 2746
  Mines 2746
  Thorium 2804
Beavers Bend State Park, Oklahoma
  Guidebook 3688
Bedford, Indiana
  Glacial Lake Sediments 1556
Bedrock
  Alaska
    --Alaska Range 237
    --Rainbow Mountain 237
  Colorado
    --Denver 597
  Connecticut
    --Bridgeport Quadrangle 680
    --Clinton Quadrangle 685
    --Colchester Quadrangle 683
    --Collinsville Quadrangle 672
    --Deep River Quadrangle 669
    --Essex Quadrangle 671
    --Haddam Quadrangle 693
    --Hamburg Quadrangle 675
    --Long Hill Quadrangle 680
    --Moodus Quadrangle 683
    --Newtown Quadrangle 689
    --Norwalk North Quadrangle
      690
    --Norwalk South Quadrangle
      690
    --Old Lyme Quadrangle 677
    --Sharon Quadrangle 694
    --South Canaan Quadrangle 688
    --Southbury Quadrangle 686
    --Torrington Quadrangle 681
    --Waterbury Quadrangle 678
    --West Torrington Quadrangle
      673
  Idaho
    --Blaine County 1162
    --Central 1162
    --Custer County 1162
    --Pioneer Mountains 1162
  Illinois
    --North-eastern 1262
    --Northern 1210
  Maine
    --Presque Isle Quadrangle 2210
  Maryland 2263
  Missouri
    --St. Charles County 2595
  Montana
    --Madison County 2786
    --Sheridan District 2786

North Dakota
--Cavalier County 3443
Ohio
--Adams County 3596
--Caldwell North Quadrangle
3573
--Carroll County 3566, 3572,
3588
--Carrollton Quadrangle
3572
--Clermont County 3578
--Columbiana County 3566
--Garrettsville Quadrangle
3555
--Guernsey County 3573
--Hamilton County 3578,
3584, 3594, 3595
--Highland County 3596
--Hocking County 3564
--Jefferson County 3588
--Licking County 3585
--Madeira Quadrangle 3578
--Minerva Quadrangle 3566
--Muskingum County 3585
--Noble County 3573
--Northwestern 3533
--South Bloomingville Quad-
rangle 3564
--Stark County 3501, 3566
--Vinton County 3564
--Wayne County 3562
Oregon
--Wheeler County 3714
Vermont
--Pawlet Quadrangle 4483
--Randolph Quadrangle 4474
--Woodstock Quadrangle 4482
Bedrock Wells
Kentucky 1948
Bellevue Quadrangle, Tennessee
Environmental Geology 4131
Belmont, Louisiana
Water Resources 2200
Belmont Field, Ohio
Coal 3606, 3607
Belt Creek, Alaska
Geochemistry 230
Geology 230
Belton, Missouri
Gas and Oil 2567
Beluga Basin, Alaska
Gravity Survey 284
Ben Hill County, Georgia
Geohydrology 925
Phosphate 967
Ben Lomond Mountain, California
Rock 514
Bendeleben Quadrangle, Alaska
Geology 246

Benewah County, Idaho
Basalt 1160
Benns Church Quadrangle, Virginia
Geology 4571
Benson County, North Dakota
Aquifers 3364
Geology 3362
Glaciation 3362, 3364
Groundwater 3363, 3364
Water 3363, 3364
Wells 3363
Benson Sand
West Virginia
--Barbour County 4679
--Upshur County 4679
Benton County, Oregon
Geologic Hazards 3740
Rocks 3769
Bentonite
Louisiana
--Claiborne Parish 2147
Mississippi
--West Point Quadrangle 2481
Montana 2733
--Ingomar 2814
--Rosebud County 2814
--Treasure County 2814
--Vananda 2814
Nevada 2877
Wyoming
--Clayspur District 4851
--Crook County 4851
--Weston County 4851
Berea Formation
Kentucky
--Pike County 2030
West Virginia
--Brooke County 4681
--Hancock County 4681
--Ohio County 4681
Beresford, South Dakota
Water 4067
Bergen County, New Jersey
Water 3087
Bergholz Quadrangle, Ohio
Bedrock Geology 3588
Berkeley County, West Virginia
Groundwater Hydrology 4731
Berks County, Pennsylvania
Brunswick Formation 3933
Geology 3955
Groundwater 3937, 3955
Berlin Heights Quadrangle, Ohio
Geology 3561
Bernalillo County, New Mexico
Clay 3164
Minerals 3092
Bernheim Forest, Kentucky
Geology 2047

Berrien County, Georgia
  Phosphate 967
Berryville Quadrangle, Virginia
  Geology 4588
Beryl 3119
  North Carolina 3295
Beryllium
  California 424
  New Mexico
  --Apache Warm Springs 3157
  --Socorro County 3157
Bethel Channel
  Kentucky 2051
Bethel Sandstone
  Indiana
  --South-Central 1957, 2094
  Kentucky
  --Western 1957, 2094
Bethesda Church, Georgia
  Geology 941
  Mineral Resources 941
Bettles Quadrangle, Alaska
  Geochemistry 270
  Geology 270
Bibb County, Alabama
  Coal 130, 131
Bibliographies
  Carbonate Rocks 43, 47
  Coal
  --Alabama 135
  --Colorado 588, 595
  --Kentucky 2096
  --Montana 2799, 2802, 2834
  --Wyoming 4840
  Coasts
  --Alabama 63
  Estuaries
  --Alabama 127
  Geology
  --Arizona 344
  --Georgia 898, 903, 909
  --Hawaii 985, 978
  --Illinois 1194
  --Iowa 1623, 1625
  --Kansas 1779
  --Louisiana 2160
  --Minnesota 2409
  --Missouri 2558, 2559, 2561,
    2563, 2564, 2565, 2606
  --Nevada 2902, 2926, 2937
  --New Mexico 3101, 3110,
    3117
  --North Carolina 3288, 3289
  --North Dakota 3459
  --Ohio 3527, 3531, 3532,
    3543
  --Oregon 3709, 3720, 3739,
    3752
  --Pennsylvania 3792, 3812

  --South Dakota 4027
  --Tennessee 4111, 4115
  --Utah 4376
  --Virginia 4521, 4526, 4537
  --Washington 4611, 4668
  --West Virginia 4726
  --Wyoming 4835, 4839, 4845
  Geophysics
  --New Mexico 3173
Geothermal Resources
  --Montana 2779
Groundwater
  --Florida 814
  --Hawaii 985
  --Kansas 1719
  --Montana 2800
  --Wisconsin 4828
Hydrogeology
  --Colorado 587
Indians
  --Louisiana 2142
Karst Geology
  --Kentucky 2138
Measured Sections
  --Virginia 4525
Metallic Mineral Resources
  --Montana 2833
Metallurgy
  --Minnesota 2409
Mineral Technology
  --New Mexico 3101, 3110, 3117
Minerals
  --Alabama 151
  --Arizona 344, 361
  --Kentucky 2124
  --Oregon 3709, 3720, 3739
  --Virginia 4521, 4526, 4537
  --Washington 4611
Mining
  --Minnesota 2409
  --Nevada 2926
Offshore Areas
  --Alabama 127
Paleobotany 4710
Pleistocene
  --Iowa 1623
Precambrian
  --Kansas 1846
  --Missouri 2581, 2593
  --New Mexico 3114
Radioactive Minerals
  --Colorado 594
Sandstone
  --New Mexico 3176
Stratigraphy
  --Montana 2801
Uranium
  --New Mexico 3116
Water

--Hawaii 978
Big Badlands, South Dakota
  Formations 4052
  White River Oligocene 4052
Big Bend National Park, Texas
  Geology 4283
  Guide 4240
Big Cottonwood Mining District,
  Utah
  Geology 4387
  History 4387
  Ore 4387
Big Delta Quadrangle, Alaska
  Geochemistry 211
Big Hatchet Mountains, New Mexico
  Stratigraphy 3240
Big Hatchet Peak Quadrangle,
  New Mexico
  Structural Geology 3200
Big Pine Fault, California
  Geology 458
Big Sinking Field, Kentucky
  Water Flooding 2039
Big Thompson Flood, Colorado
  606
Bighorn Crags, Idaho
  Geochemistry 1135
Bioacoustics 3067
Bioassays
  New Jersey
  --Hackensack Meadowlands
    3031
Biogeochemistry
  New Mexico
  --Eagle Nest Quadrangle
    3148
Biological Testing 3077
Biology
  Pennsylvania 3817
Biomere
  Texas
  --Central 4301
Biostratigraphy
  Kansas 1681
  Missouri
  --Southwestern 2625, 2630
  Montana
  --Dawson County 2824
  New Mexico
  --North Central 3245
  --Sangre De Cristo Mountains 3259
  Oklahoma 3626
  Wisconsin 4807
Biothermal Facies
  Iowa
  --Eastern 1643
Biotic-Aquatic Community 138

Biotite
  California 424
Biphenyls 1462
Birchdale, Minnesota
  Geochemistry 2428
  Geology 2428
  Sulfide Mineralization 2428
Birds
  Florida
  --Reddick 738
Birmingham, Alabama
  Floods 2
  Forest Resources 192
  Sinkholes 126
Birmingham County, Alabama
  Fish 70
Birmingham Red Iron District, Alabama
  Structural Geology 79
Birmingham Valley, Alabama
  Groundwater 161
Biscayne Aquifer
  Florida
  --Broward County 823
Bismarck, North Dakota
  Geology 3414, 3500
  Hydrogeology 3500
  Land Use 3500
Bismuth
  California 424
  --Plumas County 544
Bitterroot Range, Montana
  Batholiths 2788
  Glaciation 2787
  Petrology 2788
  Pleistocene 2787
Bituminous Coal 1252, 3885
  Arkansas
  --Arkansas Valley Coal Field
    397
  Pennsylvania 3856, 3888, 3893
  --Western 3887
  Texas 4258
Bituminous Materials
  Texas . 4271
Bituminous Rocks
  California 424
Bituminous Sandstone
  Utah
  --Uintah County 4409
Bituminous Shale
  Kentucky
  --Eastern 1912
Bivalvia 3510
Biwabik Iron Formation
  Minnesota
  --Mesabi Range 2384
Black Hills, South Dakota
  Guide 4043

Black Hills, South Dakota (Cont.)
  Springs 4060
Black Mesa Coal Field, Arizona
  Coal Samples 381
Black Mountain Volcanic Center,
  Nevada
  Road Log 2917
  Trip Guide 2917
Black Mountains, Utah
  Geology 4396
  Hydrothermal Alteration 4396
Black River Group
  Ohio
  --Southwestern 3614
Black Sands
  Washington
  --Grays Harbor 4671
Black Shale
  Arkansas
  --North Central 399
  Utah
  --Uinta Basin 4373
Black Warrior Basin, Alabama
  Drainage 12
  Fossils 153
  Pennsylvanian 153
  Sediments 153
Blackbird Mountain, Idaho
  Geochemistry 1187
  Geology 1187
Blackhawk Formation
  Utah 4444
Blacksburg Quadrangle, Virginia
  Geology 4551
Blackwater Falls State Park,
  West Virginia 4757
Blaine County, Idaho
  Bedrock Geology 1162
  Geochemistry 1166
  Geology 1166
Blaine Formation
  Kansas
  --Southern 3620
  Oklahoma
  --Northwestern 3620
Blairs Quadrangle, Virginia
  Geology 4538
Blawn Mountain, Utah
  Hydrothermal Alteration 4402
  Mineralization 4402
Blepharoceridae
  Connecticut 649
Blowouts 2334
Blue Crabs 3022
Blue Grass Region, Kentucky
  Farm Ponds 2127
  Geology 2127
Blue Ridge, Virginia
  Measured Sections 4525

Blue River Group
  Indiana 1526
Bluemont Quadrangle, Virginia
  Geology 4568
Bluff Area, Alaska
  Geology 245
Bluffs
  Illinois
  --Lake Michigan 1452
  North Dakota
  --Riverdale 3481
  Washington
  --Puget Sound 4655
Bodega Head, California
  Geology 459
Bogs
  Illinois
  --Lake County 1433
  --Northeastern 1343
  Iowa
  --Summer Bog 1653
  Pennsylvania
  --Northeastern 3837
Bois D'Arc Formation
  Oklahoma
  --Arbuckle Mountains 3650
Bokan Mountain, Alaska
  Geobotany 276
  Prospecting 276
  Uranium 276
Bokchito Formation
  Oklahoma
  --Southern 3657
Bolivar County, Mississippi
  Geology 2467
Bolivar Roads, Texas
  Shorelines 4195
Bollinger County, Missouri
  Kaolin 2585
Bolsons
  Texas 4311
Bon Air Quadrangle, Virginia
  Geology 4555
Bon Harbor Gas Pool, Kentucky 2025
Bon Homme County, South Dakota
  Geology 4012, 4013
  Gravel 4033
  Sand 4033
  Water 4012, 4014
Bonito Mining District, New Mexico
  Minerals 3177
Bonner County, Idaho
  Geology 1123
  Mineral Resources 1123
Bonneville Salt Flats, Utah
  Hydrogeology 4468
Books Cliff Coal Field, Utah 4392
Boone County, Illinois
  Dolomite 1235

Drilling 1453
Gravel 1273
Sand 1273
Boone County, Kentucky
High Bridge Group 1949
Middle Ordovician 1949
Rock 1949
Boone Lake, Tennessee
Geology 4156
Borate
Death Valley 2930
Borax
California 424
Borden Siltstone
Illinois 1281
--Southern 1257
--Southwestern 1242
Borders 3158
Boron Compounds
California 424
Bossier Parish, Louisiana
Water Resources 2165
Botryceras
New Mexico 3246
Bottomlands
Illinois 1363
Bottomless Lakes State Park, New
Mexico
Geology 3273
Bouguer Anomaly 1578
Bouguer Gravity
Montana 2825
New Jersey 3086
Virginia 4581, 4583, 4599
--Eastern Shore 4586
--Northeastern 4586
--Southeastern 4593
--Southwestern 4543
Boulder County, Colorado
Geology 596
Water 596
Boulder-Weld Coal Field, Colorado
Land Use 605
Subsidence 605
Boundaries
New Jersey 2945
Bowdle, South Dakota
Groundwater 4075
Bowers Hill Quadrangle, Virginia
Geology 4582
Bowling Green, Kentucky
Limestone Terrane 1963
Water 1963
Bowman County, North Dakota
Aquifers 3378
Geology 3376
Groundwater 3377, 3378
Water 3377
Wells 3377

Box Elder County, Utah
Geology 4388
Minerals 4388
Boyce Quadrangle, Virginia
Geology 4588
Brachiopoda
Illinois
--Southwestern 3646
New Mexico
--Sangre De Cristo Mountains
3259
Ohio 3506
Oklahoma 3615, 3616, 3617, 3618,
3646
--Arbuckle Mountains 3644
--Eastern 3652
Missouri
--Eastern 3646
Brackish Water
Louisiana
--Lake Charles 2184
Maryland 2291
Bradford County, Florida
Water Resources 759, 822
Bradford County, Pennsylvania
Upper Devonian 3804
Bradley County, Arkansas
Groundwater 403
Bradley Peak Quadrangle, Wyoming
Gravity 4853
Braided Alluvial Intervals
New Mexico
--Northeastern 3204
Branch County, Michigan
Water 2377
Brandon, Vermont
Geology 4494
Brandon Quadrangle, Vermont
Surface Geology 4501
Branford Quadrangle, Connecticut
Surficial Geology 670
Brassfield Formation
Indiana
--Cincinnati Arch 1496
--Southeastern 1500
Kentucky 1500
Ohio
--Southwestern 3609
Brattleboro, Vermont
Geology 4498
Braxton County, West Virginia
Oil and Gas 4677
Brazos Alluvial Plains, Texas 4244
Brazos Island, Texas
Shorelines 4191
Brazos River Delta, Texas
Shorelines 4193
Brevard Lineament

Brevard Lineament (Cont.)
  Georgia
    --Atlanta 896
Brewster County, Texas
  Flurospar 4168
  Geology 4283
Bricks
  Illinois 1261
  South Carolina 3971
Bridgeport Quadrangle, Connecticut
  Bedrock Geology 680
Bridger, South Dakota
  Geology 4046
Bridges
  California
    --San Fernando Valley 432
Brigantine Inlet, New Jersey
  Currents 3024
  Facies 3045
  Morphology 3023
  Sediments 3024, 3045
Brine
  Nevada 2888
  Utah
    --Great Salt Lake 4427,
      4432, 4449, 4450, 4461,
      4463, 4466, 4467, 4469,
      4471
    --Millard County 4422
    --Moab 4403
    --Sevier Lake 4422
Bristlecone Pines
  Utah
    --Cedar Breaks 4411
Bristol Quadrangle, Virginia
  Geology 4579
Britton, South Dakota
  Groundwater 4078
Broadwater County, Montana
  Economic Geology 2692
  Ore 2683
  Winston Mining District 2691
Broken Top Quadrangle, Oregon
  Geology 3771
Bromide Formation
  Oklahoma
    --Southern 3679
Bromine
  California 424
  New Mexico
    --Carlsbad Potash District
      3104
Brooke County, West Virginia
  Oil and Gas 4681
Brookings, South Dakota
  Groundwater 4090
Brooklyn Lake, Florida
  Hydrology 820
Brooks County, Georgia

  Phosphate 965
  Sand 965
Brooks Range, Alaska
  Aeromagnetic Map 291
  Geochemistry 277
  Geology 277, 293
Broward County, Florida
  Biscayne Aquifer 823
  Organic Contamination 857
  Sea Water Intrusion 814
  Water 838, 852
  Water Control 814
Brown County, Illinois
  Coal 1230
Brown County, Kansas
  Geology 1723
  Groundwater 1723
Brown County, Minnesota
  Kaolin Clay 2414
Brown County, South Dakota
  Geology 4021
  Water 4021, 4023
Brown Iron Ore
  Missouri
    --Howell County 2570
    --Oregon County 2571
    --Wayne County 2578
  Tennessee
    --East 4140
Brown Shale
  Kentucky
    --Eastern 2099
Brownsville Paleovalley
  Kentucky
    --Edmonson County 1967
    --Hart County 1967
Brownsville Quadrangle, Mississippi
  Environmental Geology 2537
Brule Formation
  South Dakota
    --Big Badlands 4052
Brunswick, Vermont
  Geology 4475
Brunswick Formation
  Pennsylvania
    --Berks County 3933
    --Montgomery County 3933
Bryan County, Oklahoma
  Cenomanian 3636
  Geology 3653
  Minerals 3653
  Palynology 3636
  Red Branch Member 3636
  Woodbine Formation 3636
Bryant, South Dakota
  Groundwater 4080
Bryant Pond Quadrangle, Maine
  Geology 2211
Bryantsville Quadrangle, Kentucky

Fractures 2129
Groundwater 2129
Joints 2129
Bryce State Hospital, Alabama
  Groundwater 83
Bryozoa
  Alabama 53
  Indiana
  --Eastern 1493
  --Southern 1486
  Kentucky 1486
  Minnesota 2442
  Ohio
  --Western 1493
Buchanan County, Missouri
  Water 2658
Buchanan Quadrangle, Virginia
  Geology 4567
Buckingham Quadrangle, Virginia
  Geology 4569
Bucks County, Pennsylvania
  Groundwater 3522
Buff-Burning Clay
  Illinois
  --Southern 1208
  --Southwestern 1208
  --Western 1209
Buffalo Wallow Group
  Indiana
  --Southern 1580
Building Stone 1561
  Montana 2758
  Pennsylvania 3785
  Washington 4607
Buildings
  California
  --San Fernando Valley 432
  --San Francisco Bay 532
Bull Run Watershed, Oregon
  Geological Hazards 3724
Bullion Creek Formation
  North Dakota 3489
Bunkerville Mining District, Nevada
  Geology 2862
  Minerals 2862
Bureau County, Illinois
  Coal 1204, 1275
Burial Mounds
  Louisiana
  --La Salle Parish 2141
Burke County, North Dakota
  Aquifers 3353
  Geology 3351, 3352
  Glaciation 3351
  Groundwater 3352, 3353
  Wells 3352
Burke Quadrangle, Vermont
  Geology 4481
Burleigh County, North Dakota

Geology 3316, 3317, 3318, 3452
Glaciation 3316
Groundwater 3316, 3317, 3318
Water 3317
Burleson County, Texas
  Heavy Minerals 4290
  Wellborn Formation 4290
Burlington, Vermont
  Geology 4495
Burlington Limestone
  Missouri
  --Southwestern 2619
Burlington Quadrangle, Ohio
  Bedrock Geology 3584
Butler County, Alabama
  Iron Ore 170
  Mineral Resources 172
  Water Resources 172
Butler County, Kentucky
  Coal 1938
  Gas and Oil 1962, 2018
  Petroleum Exploration 2110
Butler County, Pennsylvania
  Geology 3927
  Groundwater 3947
  Hydrology 3927
Butte County, California
  Geology 407
Butte County, South Dakota
  Newcastle Sandstone 4059
  Oil 4059

Cabaniss Formation
  Missouri
  --Western 2616
Cabaniss Shale
  Kansas 1728
Caddo County, Oklahoma
  Groundwater 3658
  Rush Springs Sandstone 3658
Caddo Parish, Louisiana
  Water Resources 2165
Cadmium
  California 424
  Missouri
  --Southeast 2638
Cahaba Valley, Alabama
  Groundwater 161
Calaveras County, California
  Geology 428
  Geophysics 558
Calaveras Fault, California 474
Calcareous Algae
  Kansas
  --Southeastern 1695
Calcareous Rock
  Kansas
  --Rawlins County 1700

Calcareous Rock (Cont.)
  Kansas (Cont.)
    --Stafford County 1700
Calcasieu Parish, Louisiana
  Groundwater 2187
Calceocrinids
  Oklahoma
    --Southern 3679
Calcite 4740
  California 424
Calcium Chloride 1253
  California 424
Calcium Sulfate 3586
Calhoun County, Alabama
  Mineral Resources 139
  Surface Water 91
Calderas
  Nevada
    --Northumberland Caldera
      2918
Caldwell North Quadrangle, Ohio
  Bedrock Geology 3573
Caledonia County, Vermont
  Geochemistry 4491
Calhoun County, Arkansas
  Groundwater 403
Calhoun County, Illinois
  Coal 1230
Calibration Base Lines
  New Jersey 3054
Caliche
  New Mexico
    --Central-Eastern 3198
Callinectes Sapidus 3022
Calloway County, Kentucky
  Economic Geology 1932
Caloosahatchee River, Florida
  Salt Water Contamination 779
Calvert City Quadrangle, Kentucky
  Engineering Geology 1943
Camas County, Idaho
  Geochemistry 1166
  Geology 1166
Cambrian
  Appalachian Basin 2088
  California
    --Death Valley 521
    --Inyo County 521
    --Salt Spring Hills 521
  Illinois 1259
    --Northeastern 1481
  Indiana 1531
  Kansas 1905
  Missouri 2569
    --St. Francois Mountains
      2621, 2635
    --Southern 2635
    --Washington County 2623
  Ohio 3507

Oklahoma
  --Arbuckle Mountains 3632
  --Delaware County 2635
  --Murray County 3632
  --Wichita Mountains 3651
Ontario
  --Southwestern 2053
Ozarks 2632
Pennsylvania
  --Central 3799
  --Western 3799
Texas
  --Central 4297, 4301
  --Llano Region 4288
Virginia
  --Lee County 2054
Wisconsin
  --Madison 4788
  --Western 4790, 4794, 4795,
    4796
Cambridge, Maryland
  Groundwater 2295
Cambro-Ordovician 2104
  Kentucky 2067
    --Western 2080
  Maryland
    --Frederick Valley 2301
  Wisconsin
    --Southwest 4789
Camden County, Georgia
  Phosphate 965
  Sand 965
Camp Breckinridge, Kentucky
  Geology 2052
Camp Dodge, Iowa 1619
Campbell County, South Dakota 4010
  Geology 4008, 4009
  Water 4008, 4010
Campbell County, Tennessee
  Mississippian 4159
  Pennsylvanian 4159
Camptonite
  New Mexico
    --Sacramento Mountains 3195
Canaan Valley State Park, West Vir-
    ginia 4757
Canada Gobernadora Quadrangle, Cal-
    ifornia
  Geology 552
Canadian
  Alaska 3248
  Southwest 3265
Canals
  Florida 786
    --Broward County 857
    --Plantation Canal 857
    --St. Lucie County 849
    --Tampa Bypass Canal System
      869

Candida Albincans 2990
Candle Quadrangle, Alaska
  Geology 247
Cane Creek Basin, Alabama
  Acid Mine Drainage 122
Cane Valley Limestone
  Kentucky
  --South-Central 1959
Cannonball Formation
  North Dakota 3421
  --Williston Basin 3487
Canton, South Dakota
  Groundwater 4076
Cantwell, Alaska
  Mineral Occurrences 339
Canyon Group
  Texas
  --North-Central 4317
Cape Cod Bay, Massachusetts
  Beaches 3041
  Shorelines 3041
Capitan, New Mexico
  Geology 3273
Caprock Escarpment
  Texas
  --Panhandle 4234
Captorhinid Reptiles
  Oklahoma 3654
  Texas 3654
Capulin Mountain, New Mexico
  Geology 3277
Carbon Basin, Wyoming
  Coal 4886
  Early Tertiary 4863
  Late Cretaceous 4863
  Sediments 4863
Carbon County, Montana
  Geology 2832
Carbon County, Wyoming
  Early Tertiary 4863
  Gravity 4853
  Iron 4853
  Late Cretaceous 4863
  Sediments 4863
  Structural Geology 4860
Carbon Dioxide
  California 424
Carbonate Facies
  Kansas
  --Southeast 1766
  New Mexico
  --North-Central 3245
Carbonate Reservoirs 2111
Carbonate Rock 43, 47
  Alabama 60
  Southwestern States 49
Carbonate Terranes 47, 60
Carbonates 1264, 1397, 1414,
  1426, 1431, 1444, 3546, 4686,
  4747

California 427
  Idaho 1122
  --Lewiston 1146
  --Snake River 1146
  Indiana 1497
  Iowa
  --Cedar Rapids 1677
  Kentucky 1964
  Maryland 2247
  --Frederick Valley 2297
  --Hagerstown Valley 2297
  Mexico 4324
  Missouri
  --St. Francois Mountains 2621
  North Carolina 3310
  North Dakota 3436
  --Eastern 3478
  Oklahoma
  --Osage County 3677
  Pennsylvania
  --Central 3803, 3809
  --Lebanon Valley 3929
  --South-Central 3937
  Texas 4172, 4324
  --South 4313
  Washington
  --Stevens County 4622
Carbondale Formation
  Illinois
  --Illinois Basin 1195
  --Northern 1470
  --Western 1470
  Kentucky
  --Western 1952
Carboniferous System
  Alabama
  --Northern 4154
  Kentucky 2137
  Tennessee
  --Southern 4154
  Texas 4236
Carcass Canyon Coal Area, Utah 4417
Caribou County, Idaho
  Groundwater 1179
  Precipitation 1140
Carlsbad Potash District, New Mexico
  Bromine 3104
Carolina Barite Belt, South Carolina
  Geology 3969
  Ore 3969
Carolina Slate Belt, North Carolina
  Minerals 3290
  Petrography 3307
  Stratigraphy 3307
Carper Sand
  Illinois
  --Fayette County 1218
Carrington Shale
  North Dakota
  --Central 3497

Carrizozo Basalt Field, New Mexico 3167
Carroll County, Missouri
  Water 2657
Carroll County, Ohio
  Bedrock Geology 3566, 3572, 3588
  Glacial Geology 3568
Carrollton Quadrangle, Ohio
  Bedrock Geology 3572
Carter Cave, Kentucky
  Geology 2046
Carter County, Oklahoma
  Pennsylvanian Fusulinids 3637
Carter County, Tennessee
  Geology 4156
Cartesian Coordinates 1697
Carterville Formation
  Missouri
    --Southwestern 2630
Cascade, Montana
  Geology 2706
  Groundwater 2706
Cascade Cave, Kentucky
  Geology 2046
Cascade Mountains, California
  Geology 416, 424
  Minerals 416
Cascade Mountains, Washington
  Base Metals 4610
  Cenozoic Volcanism 4612
  Landscape 4629
  St. Helens Mining District 4658
  Washougal Mining District 4658
Caseyville Sedimentation
  Illinois
    --Illinois Basin 1240
Casing 2023
Cass County, North Dakota
  Geology 3331, 3332, 3333
  Glaciation 3331
  Groundwater 3331, 3332, 3333
  Land Use 3484
  Water 3332, 3333
Cassinian
  New Mexico 3253
Castaic Formation
  California
    --Los Angeles 520
Castle Mountains Mining District, Montana
  Geology 2720
  Ore 2720
Castleton Quadrangle, Vermont
  Stratigraphy 4478
  Structure 4478
Catahoula Formation
  Texas
    --Coastal Plain 4322, 4335

South 4298
Cataract Formation
  Indiana
    --Northeastern 1532
Cathlamet, Washington
  Geology 4606
  Minerals 4606
Catlin, Indiana
  Geology 1528
Catoosa County, Georgia
  Geology 934
  Groundwater 934
Catron County, New Mexico
  Petroleum 3096
  Stratigraphy 3096
Caulopteris
  Illinois 1348
Causeways
  Utah
    --Great Salt Lake 4463
Cavalier County, North Dakota
  Bedrock 3443
  Geology 3369
  Glaciation 3369
  Groundwater 3370, 3371
  Springs 3370
  Water 3370
  Wells 3370
Caves
  Alabama 57
    --Madison County 110
  Georgia 921
  Idaho 1125
  Kentucky
    --Carter Cave 2046
    --Cascade Cave 2046
  Maryland 2241
  Minnesota 2388
  Missouri 2579
    --Highway 65, 2617
  Montana 2772
  New Jersey 2948
  Oklahoma
    --Alabaster Cavern 3692
  Pennsylvania 3817
    --Great Valley 3820
    --Southeastern 3816
    --Valley and Ridge Province 3819
    --Western 3818
  Tennessee 4113
    --Knox County 4114
  Texas
    --Longhorn Cavern 4237
    --Southern Edwards Plateau 4215
  Virginia 4512
  Washington 4628
  Wyoming 4841

Caving
  Georgia 921
  Idaho 1125
Cayugan
  Ohio
    --Northeastern 3552
Cebolla Quadrangle, New Mexico
  Geology 3103
Cedar Bluff Irrigation District,
  Kansas
  Water 1783
Cedar Breaks, Utah
  Denudation 4411
Cedar County, Missouri
  Tar Sand 2599
Cedar Creek Anticline, Montana
  Groundwater 2785, 2824
Cedar Hills, New Mexico
  Tertiary Geology 3187
Cedar Pass, South Dakota
  Miocene Channels 4064
Cedar Rapids, Iowa
  Carbonate Units 1677
  Devonian 1677
  Geohydrology 1677
  Silurian 1677
Cedar Valley Formation
  Iowa
    --Northern 2429
  Minnesota 2429
Cedars of Lebanon State Park,
  Tennessee
  Geology 4161
Cement 2023
  Indiana 1512
  Kentucky 1987
Cement Rock
  North Dakota
    --Eastern 3471, 3478
Cement Slurries 2074
Cenomanian
  Oklahoma
    --Bryan County 3636
Cenozoic
  Arizona
    --White Mountains 384
  California
    --Sierra Nevada 417
  Idaho
    --Owyhee County 1171
    --Reynold Creek Experimental
      Watershed 1171
  Louisiana
    --Central 2152
  New Mexico
    --Northeastern 3214
    --Southeastern 3199
  North Dakota
    --Mercer County 3355

    --Oliver County 3355
    --Riverdale 3481
  Oklahoma
    --Ellis County 3668
  South Dakota
    --Southwestern 4045
  Texas
    --Trans-Pecos 4253
  Utah
    --West-Central 4447
  Washington
    --Cascade Mountains 4612
Centennial Ridge Gold-Platinum District, Wyoming 4854
Centre County, Pennsylvania
  Groundwater 3959
Cephalopod 3236, 3244
  Alaska 3249
  Southwest 3265
Ceramic Block
  Illinois 1227
Ceramic Clay
  Florida
    --Alachua County 762
    --Clay County 762
    --Putnam County 762
Ceramic Decorations
  Louisiana
    --Sicily Island 2139
Ceramic Investigations
  Alabama
    --Eufaula Bauxite District 152
Ceramic Materials
  Kentucky 1988
Ceramic Properties
  Missouri
    --Western 2616
Ceramic Tests
  Mississippi
    --Attala County 2463
Cerro de Cristo Rey Uplift
  Geology 3264
Cerro Gordo County, Iowa
  Geology 1673
  Groundwater 1673
Cesium
  California 424
Chadron Formation
  South Dakota
    --Big Badlands 4052
Chalcocite 3166
Chalcopyrite 3160
Chalk 1444
  Alabama
    --West-Central 160
  Kansas 1755
Chama Quadrangle, New Mexico
  Geology 3100

Chambers County, Alabama
  Anthophyllite 38
  Talc 38
Champaign County, Illinois
  Glacial Geology 1362
Champaign County, Ohio
  Glacial Geology 3612
Champlainian Series
  Illinois 1191
Chandalar Area, Alaska
  Geochemistry 277
  Geology 277
  Channels 3064
  Delaware 714
  --New Castle County 713
  Kentucky
  --Henderson County 2132
  --Webster County 2132
  --Western 2051
  Maryland
  --Salisbury 2237
  South Dakota
  --Cedar Pass 4064
  --Jackson County 4064
Chapel Hill Quadrangle, North
  Carolina
  Geology 3304
Chappel Formation
  Texas
  --Llano 4303
Chariton County, Missouri
  Water 2656
Charles County, Maryland
  Faulting 2260
  Groundwater 2236
  Springs 2225
  Wells 2225
Charles Mix County, South Dakota
  Geology 4015, 4016
  Gravel 4036
  Sand 4036
  Water 4015, 4017
Charleston, West Virginia
  Geology 4682
Charlotte County, Florida
  Groundwater 769
  Water Resources 865
Charlton County, Georgia
  Phosphorite 966
Chatanika, Alaska
  Geochemistry 213
Chatham County, Georgia
  Economic Geology 901
  Phosphate 901, 971
  Stratigraphy 901
Chattahoochee Anticline, Georgia
  947
Chattooga County, Georgia
  Geology 933

  Groundwater 933
  Paleozoic 933
  Rock 933
Chaves County, New Mexico
  Glorieta Sandstone 3176
  Hydrogeology 3147
  Stratigraphy 3147
  Structure 3147
Chazyan 3261
  Alaska 3249
Chelating Properties 1689
Chemical Mining 3153
Chemical Weathering
  Maryland
  --Piedmont 2303
Cherokee County, Alabama
  Geology 26
  Groundwater 26
Cherokee County, Georgia
  Geohydrology 957
Cherokee County, Kansas
  Conodonts 1736
  Endothyrid Foraminifera 1776
  Keokuk Limestone 1717
  Subsurface Beds 1776
  Surficial Rocks 1810
Cherokee Group
  Iowa 1664
Chert 1324
  Iowa
  --Laporte City 1641
  Kentucky 1998
  Missouri
  --Southwestern 2618
  West Virginia
  --Eastern 4693
Chesapeake Bay, Maryland
  Magnetic Anomaly 2308
Chesapeake Region, Maryland
  Archaeology 2223
Chester County, Pennsylvania
  Hydrology 3936
  Igneous Rocks 3936
  Metamorphic Rocks 3936
Chester Group
  Alabama 53
  Arkansas 2630
  --Northwestern 3695
  Illinois 1479
  --Illinois Basin 1277, 1314
  --Southern 2630
  Illinois Basin 1495, 1924
  Indiana 1557
  --Southern 1486, 1494, 1580
  Kentucky 1486, 1494
  Missouri
  --Southwestern 2630
  Oklahoma 2630
  --Northeastern 3695

Texas 2630
Cheyenne Basin, Colorado
 Coal 628
 Energy Resources 608
 Hydrogeology 620
 Stratigraphic Data 620
Cheyenne Bottoms, Kansas
 Geology 1775
 Structure 1775
Chicago, Illinois
 Geology 1316, 1474
 Niagaran Reefs 1463
 Sedimentology 1474
Chicago Metropolitan Area, Illinois
 Flood Control 1428
 Geology 1384, 1428
 Land Use 1384
 Pollution Control 1428
Chickamauga Super-Group
 Tennessee 4145
Chickasha Formation
 Oklahoma 3669
Chicot Aquifer, Louisiana
 Groundwater 2183
Chignik, Alaska
 Coal 334
Chinati Mountains, Texas
 Minerals 4181
Chinook Highway, Washington
 Geology 4648
Chipola Formation
 Florida 747
Chippewa County, Michigan
 Drill Cores 2359
 Fiborn Limestone 2359
Chistochina River, Alaska
 Geology 263
Chitinozoa
 Indiana 1609
Chitons
 Oklahoma
 --Arbuckle Mountains 3665
Chittenden, California
 Earthquakes 515
Chloride Contamination
 Georgia
 --Glynn County 926
Chloride Flat, New Mexico
 Pennsylvanian 3185
Chlorine
 Illinois 1228
Chlorite
 Montana 2790
 Nevada 2885
Choctaw County, Oklahoma
 Geology 3647
 Minerals 3647
Choctawhatchee River Basin, Alabama

Drainage 13
Choptank Formation
 Maryland 2290
 Virginia 2290
Chromite
 Alaska
 --Anchorage Quadrangle 253
 --Eklutna 253
 California 424
Chronology
 Illinois
 --Northeastern 1343
Chuckatuck Quadrangle, Virginia 4571
Chugach Mountains, Alaska
 Cooper 326
 Geochemistry 260
 Geology 260
 Gold 326
 Lead 326
 Silver 326
Church Creek, California
 Stratigraphy 505
 Tertiary 505
Church Hills, Utah
 Geology 4439
 Mineralization 4439
Churchill Butte Quadrangle, Nevada
 Geology 2871
Churchill County, Nevada
 Geology 2884
 Minerals 2884
Churchville Quadrangle, Virginia
 Geology 4566
Cincinnati Arch, Indiana
 Brassfield Series 1496
 Conodonts 1496
 Paleontology 1496
 Silurian 1496
 Stratigraphy 1496
Cincinnati Arch Province
 Petroleum 2092
Cincinnati East Quadrangle, Ohio
 Bedrock Geology 3595
Cincinnati Group
 Indiana
 --Eastern 1493
 Ohio
 --Western 1493
Cincinnati West Quadrangle, Ohio
 Bedrock Geology 3594
Circle Hot Springs, Alaska
 Geochemistry 213
Circleville, Ohio
 Groundwater 3597
Civil Boundaries
 New Jersey 2945
Clackamas County, Oregon
 Geologic Hazards 3724, 3741
 Rocks 3772

Claiborne County, Mississippi
  Geology 2499, 2501
  Mineral Resources 2459
  Subsurface Stratigraphy 2500
Claiborne Parish, Louisiana
  Bentonite 2147
Claims
  Alaska 321
    --Native Lands 316
    --Recording 321
    --Staking 298
Claremont, South Dakota
  Water 4070
Clarion County, Pennsylvania
  Groundwater 3943
Clark County, Illinois
  Coal 1236
  Geology 1236
  Pennsylvanian 1236
Clark County, Indiana
  Conodonts 1607
  Louisville Limestone 1607
  Silurian 1607
  Wabash Formation 1607
Clark County, Kentucky
  Geology 2029
  Sandstone 2029
Clark County, Nevada
  Geology 2861
  Gold Butte Vermiculite 2913
  Granite Complex 2881
  Minerals 2861, 2862
Clark Fork Basin, Montana
  Water 2771
Clarke County, Alabama
  Gravity Surveys 112
Clarke County, Mississippi
  Invertebrate Macropaleontology
    2536
Clarke County, Virginia
  Silica 4534
Clarksdale, Mississippi
  Desiccation Sinking 2459
Clarksville Quadrangle, Arkansas
  Geology 394
Clatsop County, Oregon
  Environmental Geology 3716,
    3721
Clay 1576, 1970, 2143, 2144,
  2145, 4697
  Alabama 84
    --Northeastern 76
    --Northwestern 75
    --Piedmont Province 74
    --Southeastern 97
    --Southwestern 27, 78
    --Warrior Basin 86
  Alaska
    --Healy 287

California 424, 574
  --Los Angeles County 537
  --Portuguese Bend 537
Delaware 717
Florida
  --Alachua County 762
  --Clay County 762
  --Putnam County 762
Georgia
  --Echols County 964
Hawaii
  --Kauai 1006
Illinois 1227
  --Southern 1208
  --Southwestern 1208
  --Western 1209
Indiana 1491, 1513
Kansas 1687, 1701, 1792
  --Cloud County 1685
  --Ellsworth County 1685
Kentucky 1958
  --Olive Hill District 1972
Maine
  --Farmington 2205
Maryland
  --Southern 2282
Minnesota 2446
  --Brown County 2414
  --Goodhue County 2413
  --Minnesota River Valley 2414
  --Redwood County 2414
  --Renville County 2414
  --Southeastern 2421
  --Wabasha County 2413
Mississippi 2455
  --Claiborne County 2502
  --Copiah County 2512
  --George County 2504
  --Hinds County 2496
  --Rankin County 2532
  --Smith County 2526
  --Wayne County 2530
Missouri
  --Macon County 2620
  --Putnam County 2609
  --Randolph County 2620
  --Western 2616
Montana 2695, 2709, 2728, 2740,
  2752
New Mexico
  --Bernalillo County 3164
  --Central-Eastern 3198
  --East-Central 3193
  --Los Alamos County 3164
  --Northeastern 3214
  --Sandoval County 3164
  --Santa Fe County 3164
Ohio 3590, 3598
Pennsylvania 3866, 3867, 3870

--Greene County 3839
--Pittsburgh 3890
--Southeastern 3882
--Southwestern 3896
South Carolina 3961
--Coastal Plain 3974, 3980
Tennessee
--East 4146
Texas
--Coast 4196
--East 4162, 4242
Utah
--Great Salt Lake 4427
--Utah County 4413
Virginia 4536
--Eastern 4531
--Southern 4535
--Southwestern 4529
--West-Central 4528
Clay County, Alabama
Copper-Bearing Pyrite Ore 85
Sulfide Mineralization 108
Clay County, Florida
Ceramic Clay 762
Water Resources 759, 822
Clay County, Georgia
Rock 904
Clay County, Minnesota
Land Use 3484
Clay County, South Dakota 4007
Geology 4004, 4005
Hydrology 4004
Water 4006
Clay County, West Virginia
Oil and Gas 4677
Clay Minerals 1454, 1455
Clay Products 2145
Illinois 1261
South Carolina
--Coastal Plain 3974
Clay Spur District, Wyoming
Bentonite 4851
Claystones
South Carolina
--Coastal Plain 3980
Claysville Quadrangle, Pennsylvania
Oil and Gas 3873
Clayton, New Mexico
Geology 3277
Clear Lake Region, California
Geophysics 557
Clearfield County, Pennsylvania
Allegheny Group 3833
Coal Petrography 3874
Groundwater 3956
Lithostratigraphy 3833
Pennsylvanian 3833
Pottsville Group 3873
Clearwater, Idaho

Geology 1148
Cleburne County, Alabama
Kyanite 99
Clermont County, Ohio
Bedrock Geology 3578
Cleveland County, Oklahoma
Groundwater 3670
Clifty Falls State Park, Indiana
Geology 1612
Climate
Connecticut 655
Hawaii 1103
Idaho 1124
Nevada 2895, 2942
North Dakota 3464
--Southeastern 3423
Tennessee 4118
Texas
--Panhandle 4234
West Virginia
--Charleston 4682
Clinch County, Georgia
Phosphate 965
Phosphorite 966
Sand 965
Clinch Port Quadrangle, Virginia
Geology 4559
Clinton County, Ohio
Glacial Geology 3568
Clinton Deposits
Utah
--Utah County 4413
Clinton Mining District, Montana
Geology 2650
Ore 2650
Clinton Quadrangle, Connecticut
Bedrock Geology 685
Surficial Geology 684
Clinton Quadrangle, Mississippi
Environmental Geology 2537
Clinton Sand
West Virginia 4692
Cloud County, Kansas
Clay 1685
Groundwater 1730
Cluster Springs Quadrangle, Virginia
Geology 4542
Coachella Valley, California
Geology 524
Coahoma County, Mississippi
Geology 2467
Coal 1252, 1279, 1286, 1290, 1305,
1309, 1355, 1437, 1442, 1448,
1483, 3513, 3885, 4275, 4276,
4279, 4705, 4707, 4718, 4729,
4738
Alabama 135, 189
--Bibb County 130, 131
--Fabins 51

Coal (Cont.)
  Alabama (Cont.)
    --Jackson County 51
  Alaska
    --Alaska Peninsula 334
    --Chignik 334
    --Healy 287
    --Herendeen Bay 334
  Arizona 353
    --Black Mesa Field 381
  Arkansas
    --Arkansas Valley Coal Field
      397, 398
  California 424
  Colorado 588, 595, 609, 615,
    624, 626, 627, 630, 632, 633,
    637, 644, 648
    --Boulder-Weld Coal Field
      605
    --Cheyenne Basin 628
    --Denver Basin 621, 628
    --Mesa County 625
    --Raton Mesa 630
    --San Juan 630
    --San Juan Basin 3257
    --Uinta 630
    --Ute Indian Reservation 3188
  Illinois 1206, 1221, 1222, 1228,
    1268, 1288, 1301, 1318, 1332,
    1340, 1349, 1360, 1410, 1419
    --Adams County 1230
    --Brown County 1230
    --Bureau County 1204, 1275
    --Calhoun County 1230
    --Clark County 1236
    --De Witt County 1329
    --Edgar County 1236
    --Fairfield Basin 1345
    --Fulton County 1204
    --Grundy County 1275
    --Hancock County 1230
    --Henderson County 1230,
      1295
    --Henry County 1204, 1295
    --Illinois Basin 1195, 1354,
      1376
    --Kankakee County 1275
    --Knox County 1204
    --La Salle County 1275
    --Livingston County 1275
    --McDonough County 1230
    --McLean County 1329
    --Marshall County 1204,
      1275
    --Mercer County 1204, 1295
    --Peoria County 1204
    --Piatt County 1329
    --Pike County 1230
    --Putnam County 1275
    --Rock Island County 1295
    --Schuyler County 1230
    --Southeastern 1287, 1345
    --Stark County 1204
    --Tazewell County 1204
    --Warren County 1204, 1230,
      1295
    --Will County 1275
  Indiana 1510, 1529, 1597
  Iowa 1662, 1663, 1664
    --Southern 1665
  Kansas 1832
  Kentucky 2096
    --Butler County 1938
    --Edmonson County 1938
    --Grayson County 1938
    --Muhlenberg County 1938
    --Ohio County 1938
    --Warren County 1938
    --Western 1952
  Maryland 2269
  Missouri 2603, 2628, 2634, 2640,
    2643
    --Macon County 2620
    --Putnam County 2609
    --Randolph County 2620
  Montana 2680, 2687, 2697, 2702,
    2712, 2724, 2730, 2736, 2742,
    2747, 2750, 2755, 2759, 2765,
    2769, 2799, 2802, 2806, 2829,
    2834
    --Central 2710
    --Decker 2757, 2762
    --Eastern 2732, 2843
    --Fort Union Coal Region 2843
    --Foster Creek 2732
    --McCone County 2738
    --Moorhead Coal Field 2744
    --Powder River County 2727
    --Southeastern 2754, 2757
  New Mexico 3125
    --Jornada Del Muerto Coal Field
      3222
    --Northwest 3208
    --Rio Arriba County 3154
    --San Juan Basin 3209, 3257
    --Tierra Amarilla Coal Field
      3154
    --Ute Indian Reservation 3188
  North Dakota
    --West-Central 3493
  Ohio 3539, 3542, 3604
    --Belmont Field 3606, 3607
    --Monroe County 3520
    --Noble County 3520
    --Pawpaw Creek Watershed
      3520
    --Southeastern 3582
    --Washington County 3520

Pennsylvania 3778, 3844, 3856, 3888, 3893
--Allegheny County 3861
--Clearfield County 3874
--Eastern 3897
--Western 3887
--Westmoreland County 3861, 3894
Tennessee 4139
--Cumberland Plateau 4155
--Hamilton County 4123
Texas 4258
--North-Central 4299
Utah 4441, 4442, 4444
--Alton 4390
--Books Cliff 4392
--Carcass Canyon 4417
--Central 4353, 4392
--Coalville 4391
--Eastern 4391
--Emery 4392
--Escalante-Upper Valley 4410
--Garfield County 4408, 4410, 4417
--Goose Creek 4391
--Henry Mountains 4391
--Henrys Fork 4391
--Kaiparowits Plateau 4390, 4410
--Kane County 4417
--Kolob-Harmony 4390
--La Salsan Juan 4391
--Lost Creek 4391
--Northern 4391
--Sego 4391
--Sevier County 4405
--Sevier-Sanpete 4391
--Southern 4397
--Southwestern 4390, 4393
--Sunnyside Coal Zone 4443
--Tabby Mountain 4391
--Vernal 4391
--Wasatch Plateau 4385, 4392
Virginia
--Southwestern 4511
Washington 4641
West Virginia 4702, 4703, 4715
--McDowell County 4706
Wyoming 4840, 4866, 4867, 4873, 4876, 4881
--Carbon Basin 4886
--Hanna Coal Field 4887, 4892
--Powder River Coal Basin 4878
--South-central 4886, 4887
Coal Fields

Colorado 609, 637
Coal Gasification 4276
Coal Hill Quadrangle, Arkansas
Geology 394
Coal Liquefaction 4276
Coal Mines 3847
Arkansas
--Arkansas Valley Coal Field 398
Colorado 609, 610, 626, 627
Maryland
--Western 2298
Coal Mining
Indiana 1597
Coal Production
Kentucky 1950
Coal Property 2091
Coal River Basin, West Virginia
Hydrology 4753
Coalville Coal Field, Utah 4391
Coastal Plains
Alabama 105
Alaska 236
Arkansas 404
California 419, 421, 423, 424, 435
Delaware 727, 728
Georgia 893, 906, 931, 937, 946, 962, 970, 975
Maryland
--Charles County 2260
--Facies 2299
--Geology 2243, 2286
--Prince George's County 2260
--Subsurface 2299
Middle Atlantic
--Cretaceous 2289
--Petrology 2289
--Sediments 2289
South Carolina
--Clay 3974, 3980
--Geology 3973
--Limestone 3977
--Opals 3980
Virginia
--Geology 4507, 4508, 4509, 4510, 4516
--Magnetic Data 4558
Coastal Waters
Hawaii 1117
Coasts
Alabama 63, 186
Connecticut 665
New Jersey 3023
North Carolina
--Cretaceous 3308
--Geology 3289
--Paleocene 3308
--Rocks 3308

Coasts (Cont.)
  Oregon
    --Geologic Hazards 3762
    --Geology 3711, 3716
  Texas 4229, 4328
  Washington
    --Geology 4618, 4624
    --Piercement Structure
        Outcrop 4645
    --Quinault Formation 4614
Cobalt
  California 424
  Maryland
    --Piedmont Upland 2234
Coburn Formation
  Pennsylvania
    --Central 3788
Coccoliths
  California
    --San Diego County 539
  Kansas 1755
Cochise County, Arizona
  Mining Properties 358
Cochran Quadrangle, Georgia
  Economic Geology 900
  Paleontology 900
  Stratigraphy 900
Cockfield Formation
  Mississippi
    --Hinds County 2489
Cody, Wyoming
  Gypsum 4856
Coeur D'Alene River Basin, Idaho
  1153
  Community Perception 1172
  Effluents 1169
  Water Quality 1169
Coffee County, Georgia
  Phosphate 967
Coke 1231, 1309
  Illinois 1222
Coker Creek District, Tennessee
  Gold 4116
Coking 1268, 1279, 1290, 1305
Coking Coal
  Colorado 630
    --Raton Mesa 630
    --San Juan 630
    --Uinta 630
Colbert County, Alabama
  Environmental Geology 6
  Geology 163, 205, 207, 208
  Groundwater 163
  Mineral Resources 205, 207,
    208
Colchester Quadrangle, Connecticut
  Bedrock Geology 683
Colemanite
  California

--Kramer Junction 482
    --San Bernardino County 482
  Nevada 2931
Coles County, Illinois 1243
Collier County, Florida
  Groundwater 767
  Hydrology 850
Collignoniceras Woollgari Woollgari
  Western United States 3270
Collinsville Quadrangle, Connecticut
  Bedrock Geology 672
Colome, South Dakota
  Groundwater 4093
Colorado River Basin 3037
Colstrip, Montana
  Hydrogeology 2768
  Mining 2768
Colter Point State Park, Pennsylvania
  Guide 3902
Columbia, South Dakota
  Groundwater 4095
Columbia County, Oregon
  Rocks 3772
Columbia County, Wisconsin
  Geology 4825
  Groundwater 4825
Columbia Sediments
  Delaware 706, 708
    --Middletown 707
    --Odessa 707
Columbiana, Alabama
  Limestone Terranes 114
  Subsidence 114
Columbiana County, Ohio
  Bedrock Geology 3566
Columbus, New Mexico 3280
Columbus, Ohio
  Geology 3516
Combustion 2027
Companies
  Alaska 304, 314
Compressed Air 343
Computer Models
  Drawdowns
    --Wisconsin 4816
  Hydrologic Changes
    --Wisconsin 4818
Computer Programming 3072
Computer Systems
  Montana 2812
Computers 2070, 2093, 2965, 3165
Concentrates 3196
Concunully Mining District, Washington 4637
Concrete 3061
Condensation 2962
Conglomerate
  Missouri
    --Macon County 2620

--Putnam County 2609
--Randolph County 2620
Conicville Quadrangle, Virginia
  Geology 4589
Connate Water
  Appalachians 4676
Connecticut Valley
  Fossils 652
Conococheague Creek, Maryland
  Soluble Dye 2259
Conodonts 1199
  Illinois 1211, 1244
  --La Salle County 1472
  --Southern 1217
  Illinois Basin 1495
  Indiana 1211, 1490, 1525, 1996
  --Cincinnati Arch 1496
  --Clark County 1607
  --Northeastern 1532
  --Southern 1501
  --Southeastern 1500
  Kansas 1736
  --Cherokee County 1776
  --Western 1683, 1756
  Kentucky 1500, 1525, 1996
  --Jefferson County 1607
  Michigan Basin 1519
  Minnesota 2440
  Missouri 1211
  --Southwestern 2619, 2625,
    2630
  Nebraska
  --Southwestern 1756
  Ohio 1996
  --Northeastern 3525
  Oklahoma
  --Arbuckle Mountains 3693
  --Ouachita Mountains 3693
  Texas
  --Llano 4303
  Wisconsin 4807
Conservation 2041, 2064
  Alabama 185, 197
  Hawaii 990
  Michigan 2319
  Missouri 2588
Construction
  Colorado 646, 647
Construction Aggregates
  Pennsylvania
  --Pittsburgh 3886
Construction Materials
  Idaho 1155
  Iowa
  --Southwestern 1624
Consultants
  Alaska 305
Contamination 1457
  Delaware 723

Florida
  --Broward County 857
  --Caloosahatchee River 779
  --Plantation Canal 857
Georgia
  --Glynn County 926
  Missouri 2556
Continental Margin
  California 420
Continental Shelf 3027, 3033
  North Carolina 3289
Contra Costa County, California
  Geology 458, 497
  Geophysics 458
  Mineral Resources 497
Conulata
  Indiana
  --Delphi 1562
Cook County, Georgia
  Phosphate 967
Cook County, Illinois
  Drilling 1385, 1386
Cook County, Minnesota
  Geology 2395
  Precambrian 2395
Cook Inlet, Alaska
  Gravity Survey 284
  Stratigraphy 331
Cook Inlet Basin, Alaska
  Salinity Study 289
Cook Mountain Formation
  Louisiana 2150
  --Winn Parish 2150
Cookeite
  Arkansas
  --Jeffery Quarry 386
  --North Little Rock 386
  --Pulaski County 386
Coolidge, Georgia
  Aquifers 942
Coos County, Oregon
  Environmental Geology 3729
  Fuel Resources 3722
  Geography 3722
  Geology 3722
  Minerals 3722
Copano Bay, Texas
  Mollusks 4338
Copiah County, Mississippi
  Clay 2512
  Geology 2509
  Mineral Resources 2509
  Subsurface Stratigraphy 2511
  Water Resources 2510
Copper 2358, 3140
  Alabama
  --Clay County 85
  --Pyriton 85

Copper (Cont.)
  Alaska
    --Chugach Mountains 326
    --Nelchina Glacier 326
    --Spirit Mountain 275
    --Talkeetna Mountains 325
  California 424
    --Plumas County 544
  Georgia
    --Lincoln County 950
    --Metasville Quadrangle 950
    --Wilkes County 950
  Maryland
    --Piedmont Upland 2234
  Midcontinent 3678
  Minnesota
    --Duluth Complex 2407
  Missouri
    --Southeast 2638
  Oklahoma
    --Jackson County 3663
  Tennessee 4136
  Texas
    --North 4315
  Utah
    --Kennecott Copper Company
      4420
  Virginia
    --Staunton Quadrangle 4585
  Washington
    --Okanogan County 4617
  Wisconsin 4772, 4781
  Wyoming
    --Alabany County 4850
Copper River Basin, Alaska
  Geochemistry 248
  Geology 248
Copper Sulfide
  Arizona 380
Coquina
  Florida 875
Coral
  Florida 747, 751
  Missouri
    --Central 2612
  Oklahoma 3626, 3631, 3671,
    3672, 3673
Cordilleran Miogeosynclines
  Nevada 4351
  Utah 4351
Cores
  Georgia
    --Coastal Plain 946, 962
  Illinois
    --Illinois Basin 1267
  Michigan
    --Chippewa County 2359
    --Mackinac County 2359
    --Schoolcraft County 2359

Nevada 2936
  Wyoming
    --Carbon Basin 4886
    --South-central 4886
Cornwall, Pennsylvania
  Diabase 3875
  Geology 3875
  Magnetite 3875
  Triassic 3875
Corpus Christi, Texas
  Land 4330
  Water 4330
Corrosion 2107
Corundum
  South Carolina 3978
Cosmos Hills, Alaska
  Geochemistry 272, 274
  Geology 272, 274
Council Grove Group
  Kansas 1699
Covington Quadrangle, Ohio
  Bedrock Geology 3594
Cowlitz County, Washington
  Geology 4606
  Minerals 4606
Coxville Sandstone
  Indiana
    --Greene County 1575
    --Latta 1575
Coyote Lake, California
  Earthquakes 464
Crabs 3022
  North Dakota 3421
Cragford District, Alabama
  Sulfide Mineralization 108
Craig County, Oklahoma
  Oil and Gas 3621
  Geology 3621
Craig Quadrangle, Alaska
  Geochemistry 283
  Geology 283
  Rock 232
  Stream Sediments 232, 235
Craigsville Quadrangle, Virginia
  Geology 4575
Cratonic Basins
  Texas
    --North-Central 4248
Crawford County, Pennsylvania
  Geology 3957
  Groundwater 3957
Crawford Upland, Indiana
  Geomorphology 1535
  Groundwater 1535
  Hydrology 1535
Crawfordsville Moraine, Indiana 1588
Crenshaw County, Alabama
  Brown Iron Ore 92, 170
Crescent City, Illinois

Geology 1412
Crested Butte, Colorado
Geologic Hazards 613
Creta, Oklahoma
Copper 3663
Flowerpot Shale 3663
Permian 3663
Cretaceous 4182
California
--San Diego County 539
--San Joaquin County 545
--Stanislaus County 545
Delaware 710, 724
--Greenwood 726
Florida 882
--Peninsular 776
Gulf Coast 4281
Illinois
--Southern 1284
--Western 1220
Kansas 1755, 1771, 1787
--Central 1713
--Western 1822
Maryland
--Salisbury Embayment 2307
--Southern 2285
Mexico 4324
--Coahuila 4300
Middle Atlantic Coastal Plain 2289
Minnesota 2416
Montana
--Southeastern 2823
New Mexico
--Northwestern 3218
--Todilto Park 3238
North Carolina
--Coastal Plain 3308
North Dakota 3368, 3466
--Burke County 3353
--Mercer County 3355
--Mountrail County 3353
--Oliver County 3355
--Southwestern 3486
--Williston Basin 3485
Oklahoma
--Southern 3657
South Dakota 3466
Tennessee
--Western 4119, 4157
Texas 4172, 4291, 4293, 4294, 4324
--Central 4164, 4306
--North-Central 4199, 4329
--Northeast 4213, 4308
--South 4313
Utah
--Garfield County 4362
Virginia

--Coastal Plain 4508, 4509
Western U.S. 3270
Wyoming
--Carbon County 4863
Creve Coeur Quadrangle, Missouri
Engineering Geology 2553
Crimora Quadrangle, Virginia
Geology 4550
Crinoidea
Indiana
--Southern 1494
Kentucky 1494
Oklahoma 3622
Crittenden County, Kentucky
Fluorspar 2123
Crook County, Oregon
Geology 3700
Crook County, Wyoming
Bentonite 4851
Crooks Site, Louisiana 2141
Cross-Bedding
Illinois Basin 1924
Crude Oil
Illinois 1339, 1441
--Illinois Basin 1277
North Dakota 3428, 3432, 3435, 3437, 3439, 3446, 3448, 3451, 3455, 3456, 3461
Crushed Stone
California 424
Indiana 1497, 1509, 1526
Crustal Movement
California 430
Mississippi Valley 1200
Crustal Tectonics
Illinois
--Northeastern 1482
Cryptosome Bryozoa
Minnesota 2442
Crystal Valley, Colorado
Geology 604
Groundwater 604
Mineral Resources 604
Crystall Falls, Michigan
Geology 2348, 2349, 2350, 2352
Magnetic Data 2348, 2349, 2350, 2352
Stratigraphy 2349, 2350
Crystalline Rocks
Appalachians 2310
Georgia
--Dawson County 936
Maryland
--Baltimore 2305
--Bear Island 2246
--Piedmont 2305
--Potomac River 2246

Ctenoconularia Delphiensis
  Indiana
    --Delphi 1562
Culberson County, Texas
  Asbestos 4185
Culture
  Virginia 4527
Cumberland Falls State Park,
  Kentucky
  Geology 2045
Cumberland Plateau, Alabama
  Carboniferous Deposits 4154
Cumberland Plateau, Tennessee
  Carboniferous Deposits 4154
  Coal 4155
  Physiography 4143
Cumulus Mineralogy
  Minnesota
    --Itasca County 2432
Currant Creek Formation
  Utah
    --Uinta Basin 4419
Currents
  New Jersey
    --Brigantine Inlet 3024
Curry County, Oregon
  Geology 3735
  Land Use Geology 3732
  Minerals 3735
  Rocks 3735
Custer County, Idaho
  Bedrock Geology 1162
Custer County, Oklahoma
  Geology 3638, 3639, 3640
  Groundwater 3641
  Minerals 3638
  Stratigraphy 3639
Cut Bank, Montana
  Water 2716
Cuttingsville, Vermont
  Pyrrhotite 4488
Cuyuna Range, Minnesota
  Geology 2409, 2410
  Metallurgy 2409
  Mining 2409, 2410
Cyanide Solution 2715
Cyclothemic Limestone Facies
  Kansas
    --Eastern 1821
Cyclothems
  Illinois
    --Illinois Basin 1192
  Kansas 1681
Cystoids 1642

Dade County, Florida
  Flood Control 834
  Salinity 853

Salt Water 832
Dade County, Georgia
  Geology 932
  Groundwater 932
Dade County, Missouri
  Tar Sand 2599
Dahlen Esker
  North Dakota
    --Grand Forks 3442
    --Walsh County 3442
Dakota Formation
  Kansas
    --Ford County 1827
    --Hodgeman County 1827
  South Dakota 4057
Dakota Group
  North Dakota
    --Williams County 3336
Dakota Sandstone
  New Mexico
    --Northeastern 3204
  South Dakota
    --Wall 4037
Daleville Quadrangle, Virginia
  Geology 4596, 4597, 4598
  Road Log 4598
Dalhart Basin, Texas
  Pennsylvanian 4231
Dallas, Texas
  Minerals 4214
Dams 1389, 3053
  Illinois
    --McHenry County 1401
Dana Point Quadrangle, California
  Geology 550
Dane County, Wisconsin
  Aquifers 4816, 4818
  Drawdowns 4816
  Hydrologic Changes 4818
Danian
  Maryland
    --Salisbury Embayment 2307
Danville, Illinois
  Glacial Drift 1313
Danville Basin, Virginia
  Triassic Formations 4560
Danville Quadrangle, Virginia
  Geology 4538
Data Processing
  Montana 2812
Dating Methods
  Utah 4354
Dauphin County, Pennsylvania
  Dolomites 3794
  Geology 3935
  Hydrology 3935
  Limestone 3794
Davies County, Missouri
  Water 2653

Daviess County, Kentucky
  Gas Pool 2025
Dawson County, Georgia
  Crystalline Rock 936
  Geology 936
  Groundwater 936
Dawson County, Montana
  Biostratigraphy 2824
  Cedar Creek Anticline 2824
  Geology 2824
  Stratigraphy 2824
Day County, South Dakota
  Geology 4020
  Water 4020
Dead Mountains, Nevada
  Geology 2881
  Granite Complex 2881
Death Valley
  Borate 2930
  Cambrian 521
  Geologic Features 547
  Geology 508
  Precambrian 521
  Sedimentary Rock 521
  Talc 525
Death Valley National Monument,
  California
  Mineral Deposits 566
  Mines 566
De Baca County, New Mexico 3230
Debris-Flow Hazards
  Colorado
  --Glenwood Springs 616
Decatur, Illinois
  Geology 1353
Decatur County, Iowa
  Groundwater 1667
Decatur County, Kansas
  Geology 1742
  Groundwater 1742
Decker, Montana
  Coal 2762
  Hydrogeology 2762
  Strip Mining 2757
Decorah Formation
  Minnesota
  --Southeastern 2421
Decorah Shale
  Minnesota 2442
Deep Creek Mountains, Utah
  Geology 4350
  Minerals 4372
Deep River Basin, North Carolina
  Triassic Flora 3305
Deep River Quadrangle, Connecticut
  Bedrock Geology 669
Deep-Well Disposal
  Alabama 59
  --Baldwin County 116

--Escambia County 116
--Mobile County 116
Hawaii
  --Oahu 1045, 1092
  --Waimanaco 1045, 1092
Deer Lake Complex, Minnesota
  Cumulus Mineralogy 2432
  Geology 2431
  Petrology 2432
  Sulfide Minerals 2433
Deerfield Quadrangle, Virginia
  Geology 4575
Deformation Intensity 1126
De Kalb County, Illinois
  Dolomite 1235
  Drilling 1382, 1453
  Geology 1409
  Gravel 1223
  Sand 1223
Dekalb County, Missouri
  Water 2652
Del Mar Quadrangle, California
  Geology 438
Delamar, Idaho
  Geology 1181
  Ore Deposits 1181
Delaware Basin, New Mexico
  Geology 3213
  Minerals 3213
  Ochoan Rocks 3213
Delaware County, Oklahoma
  Late Cambrian 2635
  Traverse 2635
Delaware Estuary
  Salinity 704
Delaware Quadrangle, Arkansas
  Geology 396
Delaware Valley, Pennsylvania
  Pleistocene Geology 3811
Dell Rapids, South Dakota
  Groundwater 4084
Dellroy Quadrangle, Ohio
  Glacial Geology 3569
Delphi, Indiana
  Conulata 1562
Deltas
  Illinois
  --Lake Douglas 1322
  --Southern 1257
  --Southwestern 1242
  Texas
  --Brazos River 4193
  --North-Central 4175
  --Nueces Bay 4304
  --Southeast 4244
Deming, New Mexico
  Geology 3280
Denali, Alaska
  Geology 239

Denali, Alaska (Cont.)
  Mineral Deposits 239
Denton Shale
  Oklahoma
    --Southern 3657
Denudation
  Utah
    --Cedar Breaks 4411
Denver, Colorado
  Bedrock 597
  Geology 597
  Soil 597
Denver Basin, Colorado
  Coal 628
  Energy Resources 608
  Water Wells 621
Denver Formation
  Colorado
    --Cheyenne Basin 628
    --Denver Basin 628
Deposits
  Texas
    --Coast 4254
    --Dalhart Basin 4231
    --Palo Duro Basin 4231
    --Texas Inner Shelf 4229
    --Texas Panhandle 4231
Derbyshire Fluorspar Deposits
  United Kingdom 2116
Desalination 2960
  Cuba
    --Guantanamo Naval Base
      2953
  Hawaii 1111
Deschutes County, Oregon
  Geology 3731
  Minerals 3731
Desert Hot Springs, California
  Geology 524
Desert Mountain, Utah
  Diatremes 4434
  Geology 4434
  Gravity Data 4368
  Magnetic Data 4368
Desiccation Sinking
  Mississippi
    --Clarksdale 2459
Des Moines Series
  Iowa
    --Southern 1665
Desmoinesian Series
  Missouri 2608
    --Central 2612
DeSoto County, Florida
  Groundwater 769, 870
Desulfurization 3044
Detergents 1388, 2073
Detroit River Formation
  Indiana 1560

    --Northern 1527
Deuterophlebiidae
  Connecticut 649
Devils Backbone Area, Alabama
  Gold 125
Devils Lake, North Dakota
  Geology 3413
Devonian 1642, 4746
  Arkansas
    --North-Central 399
  Central Appalachian Area 1909
  Illinois 1301, 1339
    --Southern 1217, 1245, 1297
  Indiana 1497, 1558, 1609
    --Cincinnati Arch 1524
    --Delphi 1562
    --Northern 1527
    --Southeastern 1530
    --Southwestern 1579
  Iowa 1646
    --Cedar Rapids 1677
    --Laporte City 1641
    --North-Central 1647
    --Northern 2429
  Kentucky 1907, 1944
    --Eastern 1912, 2020
    --Pulaski County 2128
  Michigan
    --Michigan Basin 2363
  Michigan Basin 1519
  Minnesota 2429
  Nevada 2891
  North Dakota 3427
  Ohio 3519, 3576
    --Erie County 3571
    --Logan County 3571
    --Northwestern 3579, 3587
  Oklahoma 3615, 3616, 3617,
    3618, 3622, 3643
    --Anadarko Basin 3648
    --Arkoma Basin 3656
  Pennsylvania 3789, 3832
    --Bradford County 3804
    --Central 3821
    --Eastern 3801
    --Northeastern 3814
    --Southern 3813
    --Western 3892
  Tennessee
    --West 4157
  West Virginia
    --Allegheny Front 4672
  Wyoming
    --Northern 4834
Dewey Formation
  Oklahoma 3673
De Witt County, Illinois
  Coal 1329
  Geology 1225, 1329, 1459

Gravity Field 1225
Niagran Reefs 1233
Oil 1273
Pennsylvanian 1329
Diabase
Pennsylvania 3840
--Cornwall 3875
--Gifford Pinchot State Park
  3907
Diabase Dikes
North Carolina
--Eastern Piedmont 3301
Diamond
Colorado 4889
--Larimer County 4882
Wyoming 4889
--Albany County 4882
Diana Lakes Area, Alaska
Geochemistry 269
Geology 269
Diaspores
Missouri 2551
Diatomaceous Earth 364
California 541
Diatomite
California 424
Diatoms
Connecticut
--Eastern 662
Diatremes
Colorado
--Larimer County 4882
Utah
--Desert Mountain 4434
Wyoming
--Albany County 4882
Dibunophyllid Coral
Missouri
--Central 2612
Dibunophyllum
Oklahoma 3673
Dickey County, North Dakota
Geology 3390
Geophysics 3391
Glaciation 3390
Groundwater 3391
Springs 3391
Water 3391
Dickinson, North Dakota
Geology 3415
Dickinson County, Michigan
Aeromagnetic Anomaly 2357
Geophysics 2357
Groundwater 2376
Mineral Resources 2357
Digital Simulation
Aquia Aquifer 2267
Water Levels 2306
Dikes

North Carolina
--Eastern Piedmont 3301
Pennsylvania 3840
Utah
--Great Salt Lake 4470
Dillwyn Quadrangle, Virginia
Geology 4564
Dinosaur Quarry Quadrangle, Utah
Morrison Formation 4440
Petrology 4440
Diorites
New Mexico
--Sacramento Mountains 3195
Dipmeter Surveys 2055
Diptera
Connecticut 649, 653
Disaster Response
California
--San Fernando Valley 433
Disease 3076
Disposal Sites
California
--Southern 476
Disposal Wells
Alabama 59
--Baldwin County 116
--Escambia County 116
--Mobile County 116
Mississippi 2545
Ditches
Hawaii
--Hawaiian Islands 1044
--Waimea 1009
Divide County, North Dakota
Geology 3325, 3326, 3327
Glaciation 3325
Groundwater 3325, 3326, 3327,
  3492
Land Use 3492
Minerals 3492
Pollution 3492
Water 3326, 3327
Dixidae
Connecticut 649
Dixie County, Florida
Geology 743
Dockum Group
Texas
--Panhandle 4332
Dodge County, Wisconsin
Soils 4759
Dollar Bay Formation
Florida 882
Dolly Ridge Formation
West Virginia
--Eastern 4698
--Pendleton County 4698
Dolomi Area, Alaska
Geology 262

Dolomi Area, Alaska (Cont.)
  Mineral Deposits 262
Dolomites 1520, 4686, 4740
  California 424, 427
  --Coast Ranges 435
  Florida 875
  Illinois 1193, 1201, 1271
  --Boone County 1235
  --Dekalb County 1235
  --Jersey County 1304
  Indiana 1487, 1503, 1570
  --Northeastern 1532
  --Southeastern 1500
  Kentucky 1500
  Montana 2694
  Ohio 3574
  --Allen County 3521
  --Marion 3611
  Pennsylvania 3864, 3865
  --Central 3797
  --Dauphin County 3794
  Tennessee 4109
  Texas 4172, 4291
Dolomitic Oil Production
  Kentucky
  --Adair County 2101
Dona Ana County, New Mexico
  Geology 3108, 3113, 3187,
    3231
  Groundwater 3231
  Tertiary 3187
Dona Ana Mountains, New Mexico
  Geology 3201
Dorchester County, Maryland
  Water 2295
Douglas County, Nevada
  Geology 2876
  Minerals 2876
Douglas County, Oregon
  Geology 3717, 3729
  Minerals 3717
Douglas County, South Dakota
  Geology 4015, 4016
  Gravel 4036
  Sand 4036
  Water 4015, 4017
Dover, Delaware
  Aquifers 729
Dover, Vermont
  Magnetic Data 4485
  Ultramafic Bodies 4485
Drainage
  Alabama 40, 41, 133
  --Black Warrior River Basin
    12
  --Cane Creek Basin 122
  --Choctawhatchee River Ba-
    sin 13
  --Jefferson County 11

--Oakman 122
--Walker County 122
Hawaii 1031
Idaho 1167
--Snake River Plain 1168
Indiana
--Upper East Fork Drainage
  Basin 1594
--Upper Patoka Drainage Basin
  1593
Kansas
--Upper Tuttle Creek Reservoir
  1774
Maryland
--Western 2298
Montana
--Gallatin River 2820
--Southwestern 2820
South Carolina
--Enoree River 3991
--Pacolet River 3991
--Saluda River 3992
--Savannah River 3992
--Tyger River 3991
Drawdowns
Wisconsin
--Dane County 4816
Drift
Illinois 1272
--Central 1234
--Danville 1313
--Northern 1210
--Shelbyville 1315
Iowa
--Northeast 1639, 1644
Minnesota
--Red River Valley 3462
Missouri
--Andrew County 2659
--Buchanan County 2658
--Carroll County 2657
--Chariton County 2656
--Davies County 2653
--Dekalb County 2652
--Gentry County 2651
--Grundy County 2645
--Harrison County 2647
--Linn County 2655
--Livingston County 2650
--Mercer County 2646
--North 2580
--Putnam County 2609, 2648
--Sullivan County 2654
--Worth County 2649
North Dakota
--Logan County 3425
--Red River Valley 3462
--Renville County 3342
--Traill County 3339

--Ward County 3342
--Williams County 3336
Drill Cores
Illinois
--Illinois Basin 1267
Michigan
--Chippewa County 2359
--Mackinac County 2359
--Schoolcraft County 2359
Nevada 2936
Drill Cuttings 3691
Drill Holes
Iowa 1652
Missouri 2636
Vermont
--Monkton Kaolin Deposit
4486
Drill-Stem Tests 2081
Drilling
Alabama
--Piedmont Area 123
Illinois 1377
--Boone County 1453
--Cook County 1385, 1386
--Dekalb County 1382, 1453
--Du Page County 1378, 1429
--Kane County 1382, 1429,
1451
--Kendall County 1382, 1429
--Lake County 1385
--Lee County 1447
--McHenry County 1383
--Ogle County 1447
--Will County 1386
Kentucky
--Eastern 2043, 2066, 2089
Minnesota 2408
West Virginia 2082, 4712
Drilling Contractors 2019
Drilling Logs
Hawaii
--Alaeloa 1026
--Haena 1032
--Hamakua 1020
--Hamakuapoko 1025
--Hanalei 998
--Hanapepe 996, 1033
--Hilo 1055
--Kalaheo 1016
--Kamalo 1002
--Kamiloloa 1013
--Kauai 999, 1016, 1032,
1033, 1035
--Kawaihae 1015
--Keahole 1039
--Keei 1014
--Kepaniwai 1056
--Kohala 994, 997, 1015,
1027

--Kona 1014, 1039, 1054
--Lahaina 1000, 1028
--Lihue 999
--Maalaea 1034
--Maka Ridge 998
--Makena 1019
--Malama-Ki 1003
--Maui 995, 1000, 1019, 1025,
1026, 1028, 1034, 1056, 1057
--Molokai 1002, 1013
--Pohakuloa 1020
--Pulama 1017
--Puna 1003, 1017
--Waikapu 995, 1057
--Waimea 1027, 1035
Tennessee
--Morgan County 4141
--Scott County 4142
West Virginia
--Little Kanawha River Basin
4750
--Monongahela River Basin 4749
Droughts
Georgia 892
Drummond, Montana
Geology 2845
Dry Creek, California
Erosion 575
Dry Holes
California
--Gobernadora Quadrangle 449
--Orange County 449
Dry Limestone Injection 3586
Dry Mountain Quadrangle, California
Geology 535
Dry Pass Area, Alaska
Geology 242
Dry Valley, Idaho
Precipitation 1140
Du Bois, Pennsylvania
Groundwater 3956
Dubuque Formation
Iowa
--Northern 2439
Minnesota 2439
Duchesne River Formation
Utah
--Uinta Basin 4370
Dugway Valley, Utah
Water 4458
Duluth, Minnesota
Fond Du Lac Formation 2418
Gabbro Complex 2383
Geology 2383
Petrology 2418
Stratigraphy 2418
Duluth Complex
Minnesota 2420
--Copper 2407

Duluth Complex (Cont.)
 Minnesota (Cont.)
  --Nickel 2407
Dumplin Valley, Tennessee
 Fault Zone 4150
Dundee Limestone
 Ohio
  --Northwestern 3587
Dunes
 Michigan 2339, 2341, 2361
  --Lake Michigan 2364
 Texas
  --Mustang Island 4327
  --Padre Island 4327
Dunkard Group
 Appalachian Basin 4704
 Kansas 4704
Dunn Center, North Dakota
 Geology 3491
 Groundwater 3491
 Hydrogeochemistry 3491
Dunn County, North Dakota
 Groundwater 3386, 3387
 Water 3386
Dunn County, Wisconsin
 Irrigation 4827
Du Page County, Illinois
 Drilling 1378, 1429
Duval County, Florida
 Aquifers 846
 Groundwater 830
Dye
 Maryland
  --Antietam Creek 2259
  --Conococheaque Creek 2259
  --Monocaly River 2256

Eagle A-1 Quadrangle, Alaska
 Geochemistry 231
 Geology 231
Eagle County, Colorado
 Geology 604
 Groundwater 604
 Mineral Resources 604
Eagle Nest, New Mexico
 Geology 3105, 3272
 Ore 3105
Eagle Nest Quadrangle, New Mexico
 Biogeóchemistry 3148
 Geochemistry 3148
Eagle Rock Quadrangle, Virginia
 Geology 4578
Early County, Georgia
 Paleozoic 946
 Sediments 946
Early Cretaceous
 Virginia
  --Coastal Plain 4509

Early Devonian
 Indiana
  --Southwestern 1579
 Oklahoma 3615, 3616, 3617, 3618
  --Anadarko Basin 3648
  --Arkoma Basin 3656
Early Miocene
 South Dakota
  --Wounded Knee 4058
Early Ordovician
 Missouri 2569
 Texas
  --Central 4297
 Wisconsin
  --Madison 4788
Early Pennsylvanian
 Illinois
  --West Central 1356
 Indiana
  --Greene County 1489
Early Permian
 Oklahoma 3654, 3675
 Texas 3654
Early Proterozoic
 Wyoming
  --Medicine Bow Mountain 4883
Early Silurian
 Illinois
  --Southwestern 3646
 Missouri
  --Eastern 3646
 Oklahoma 3646
Early Tertiary
 Wyoming
  --Carbon County 4863
Earth Materials 1388, 1399, 1462
Earth Movements 430
Earth Resistivity
 Illinois 1232
Earth Tides 1577
Earthquakes
 Alaska 323
 California 471, 477, 478, 480,
  484, 489
  --Chittenden 515
  --Coyote Lake 464
  --Inglewood 555
  --Newport 555
  --Orange County 454
  --Oroville 565
  --San Fernando 429, 430, 431,
   432, 433, 434, 450
  --Santa Barbara 461, 462, 585
  --Santa Cruz County 515
  --Southern 470, 555
 Illinois
  --Northern 1435
  --Southern 1400
 Indiana 1571

Kansas 1809
Minnesota 2436
Mississippi Valley 1200
Missouri 2593
New Mexico 3132, 3156, 3180
Ohio 3514
Tennessee 4128, 4132
Washington
--Western 4647, 4649, 4650,
4662, 4663, 4664
West Virginia 4730
East Glades, Florida
Salinity 853
East Grand Forks, North Dakota
Floods 3409, 3445
East Liberty Quadrangle, Ohio
Glacial Geology 3567
East Shore Area, Utah
Groundwater 4448
East Thunder Mountain Mining Dis-
trict, Idaho
Geology 1139
Eastern Piedmont
North Carolina 3301
Eastern Shore, Maryland
Pleistocene 2280
Stratigraphy 2280
Easton, Maryland
Groundwater 2295
Eau Claire, Wisconsin
Groundwater 4819
Irrigation 4819
Echols County, Georgia
Clay 964
Heavy Minerals 964
Phosphate 971
Phosphorite 966
Sand 964
Ecology
Wisconsin
--Baraboo District 4801
--Menominee County 4761
Econfina Creek Basin, Florida
Water Resources 773, 828
Economic Analysis 2079
Economic Development
Kansas 1840
Economic Geology
California
--French Gulch Quadrangle
502
--Inyo County 490
--Modoc District 490
--Panamint Butte Quadrangle
490
--Shasta County 502
--Trinity County 502
Georgia
--Chatham County 901

--Cochran Quadrangle 900
--Perry Quadrangle 900
Idaho
--Lewiston 1146
--Snake River 1146
Kentucky
--Allen County 1931
--Callaway County 1932
--Hancock County 1934
--Lincoln County 2131
--McCracken County 1937
--Marshall County 1935
--Simpson County 1933
--Warren County 1936
Maine
--Knox County 2214
Montana
--Broadwater County 2691
--Central 2710
--Great Falls-Lewistown Coal
Field 2710
New Mexico 3125
Pennsylvania
--Greene County 3895
--Westmoreland County 3894
Virginia
--Daleville Quadrangle 4597
Economic Growth
Kentucky 2050
Economics
Colorado 589
Connecticut
--Coast 695
Kentucky
--Adair County 2101
Mississippi 2484
West Virginia 4758
Ecosystems 3005
Ectoprocta
Alabama 53
Eddleman Coal
Texas
--North-Central 4299
Eddy County, North Dakota
Geology 3322, 3323, 3324, 3441
Glaciation 3322
Groundwater 3322, 3323, 3324
Water 3323, 3324
Eddy Flux Techniques
Utah
--Great Salt Lake 4455, 4464
Edestidae
Mid-Continent 2613
Edgar County, Illinois
Coal 1236
Geology 1236
Pennsylvanian 1236
Edgar Creek, Alaska
Mineral Occurrences 340

Edinburg Quadrangle, Virginia
4589
Edmonson County, Kentucky
Brownsville Paleovalley 1967
Coal 1938
Kyrock Sandstone 1967
Pennsylvanian 1967
Sedimentology 1967
Edmunds County, South Dakota
Geology 4024, 4025
Water 4024
Edwards Formation
Texas
--Central 4309
Edwards Plateau
Texas
--Southern 4215
Edwards Reef
Texas
--Central 4249
Effingham, Illinois
Limestone 1310
Effingham County, Georgia
Heavy Minerals 968
Phosphate 965
Phosphate Sand 968
Sand 965, 968
Effluents
Hawaii
--Oahu 1045
--Waimanalo 1045
Idaho
--Coeur D'Alene River Basin
1169
Wisconsin 4808
El Cerrito, California
Geology 458
Geophysics 458
Eldorado Mountains, Nevada
Geology 2881
Granite Complex 2881
Electrical Energy
Washington 4644
Electrical Logs 2251
Mississippi 2541
Electrical Resistivity
Alabama
--Mobile 61
--Piedmont Area 115
Illinois 1232
Electronic Data Processing
Missouri 2597
Elgin Sandstone
Kansas
--South-Central 1727
Elizabeth Mine, Vermont
Geology 4489
Elk Run Gas Pool, Pennsylvania
Geology 3878

Elkhorn, Montana
Geology 2845
Elko County, Nevada
Geology 2866, 2867
Ellensburg Highway, Washington
Geology 4648
Ellesmeroceratida 3236
Elliott Knob Quadrangle, Virginia
Geology 4575
Ellis County, Oklahoma
Cenozoic Rocks 3668
Geology 3668
Ellsworth County, Kansas
Clay 1685
Geology 1752
Groundwater 1752
El Paso Group
New Mexico 3250, 3253
Elsey Formation
Missouri
--Southwestern 2618, 2619
Elsinore Fault Zone, California 572
El Toro Quadrangle, California
Geology 551
Elysian Park, California
Geology 542
Embayments
Georgia
--Southwestern 947
Idaho
--Clearwater 1148
Maryland
--Salisbury Embayment 2307
Emergency Operations
California
--Los Angeles County 43
Emery Coal Field, Utah 4392
Emery County, Utah
Geology 4386
Uranium 4386
Vanadium 4386
Emmons County, North Dakota
Groundwater 3380, 3381
Water 3380
Wells 3380
Endangered Species
Connecticut 700
Endoceroid 3261
Alaska 3248
Quebec 3247
Endothyrid Foraminifera
Kansas
--Cherokee County 1776
Energy 1416, 1478, 1440, 3033,
3081, 3763, 4727
Alabama
--Wilcox County 176
--Winston County 181
Alaska 302

California 424
Colorado 589, 617, 623
--Cheyenne Basin 608
--Denver Basin 608
New Mexico 3118, 3134, 3194,
3202, 3221, 3226
Texas 4216, 4218
Utah 4044
Washington 4638, 4639, 4640,
4641, 4642, 4643, 4644
Energy Conversion 3071
Energy Distribution Systems
Alabama 156
Engineering Characteristics
Illinois
--McHenry County 1395
Engineering Economy
New Jersey
--Passaic River Basin 2974
Engineering Geology
Alabama
--Jefferson County 14
California
--Lake County 563
--Mendocino County 563
--Sonoma County 563
Colorado
--Crystal Valley 604
--Eagle County 604
--Garfield County 604
--Gunnison County 604
--Larimer County 602
--Pitkin County 604
--Planning District 1 598
--Planning District 7 599
--Planning District 9 600
--Planning District 13 599
--Roaring Fork Valley 604
--Weld County 602
--Windsor Study Area 602
Kentucky
--Calvert City Quadrangle
1943
--Livingston County 1943
--Marshall County 1943
Missouri 2554
--Creve Coeur Quadrangle
2553
--Jefferson County 2552
--Maxville Quadrangle 2552
--St. Louis County 2552,
2553, 2555
Oregon
--Tualatin Valley 3702
Washington 4652, 4653, 4654,
4655, 4656
Englewood Formation
North Dakota 3311
South Dakota 3311

Enoree River, South Carolina
Drainage Basin 3991
Fluvial Monazite 3991
Enosburg, Vermont
Geology 4476
Environment 1394, 1416, 1418, 1425,
3027, 3546
Alabama 187
Georgia 960
Illinois
--Illinois Waterway 1334
Indiana
--Marion County 1610
Maryland 2245
Michigan
--Monroe County 2354
Minnesota
--Minneapolis 2389
--St. Paul 2389
Missouri 2549B
North Dakota 3402, 3458
Virginia
--Coastal Plain 4510
Vermont
--Barre 4493
--Brandon 4494
--Brattleboro 4498
--Burlington 4495
--Hardwick 4496
--Johnson 4496
--Middlebury 4495
--Milton 4497
--Montpelier 4493
--Rutland 4494
--St. Albans 4497
--Windsor 4498
Environmental Geology 2549A, 4316
Alabama
--Colbert County 6
--Franklin County 6
--Huntsville 8
--Lauderdale County 6
--Madison County 1, 3, 4, 5,
8
--Maysville Area 3
--Meridianville Quadrangle 1
--Triana Area 4
--West-Central 7
California
--Santa Clara County 456
Colorado 586, 591, 635, 643
--Crystal Valley 604
--Eagle County 604
--Garfield County 604
--Gunnison County 604
--Larimer County 602
--Pitkin County 604
--Roaring Fork Valley 604
--Weld County 602

Environmental Geology (Cont.)
Colorado (Cont.)
--Windsor Study Area 602
Florida
--Tallahassee 883
--Tampa 886
Illinois
--Northeastern 1379
Indiana
--Allen County 1604
--Evansville 1603
--Hendricks County 1567
--Johnson County 1573
--Lake County 1602
--Marion County 1610
--Morgan County 1572
--Porter County 1602
--Southwestern 1603
Mississippi
--Adams County 2540
--Hinds County 2537, 2538
Oregon
--Clatsop County 3716, 3721
--Coos County 3729
--Douglas County 3729
--Lane County 3727
--Lincoln County 3723
--Linn County 3726
--Tillamook County 3716,
3721
Pennsylvania 3782
--Harrisburg 3784
--York 3786
--York County 3786
Tennessee
--Bellevue Quadrangle 4131
--Kingston Springs Quadrangle
4126
Texas
--Austin 4321
--Coast 4178
--East 4200, 4333
Utah
--Bear Lake 4369
--Rich County 4369
West Virginia 4726
Environmental Health
Developing Countries 3069
Environmental Pollution 1407
Environmental Protection
Alabama 180
Eocene
California
--San Diego County 539
Florida 739
Georgia 953
Gulf Coast 196
Kentucky
--Mayfield 1942

Mississippi 2519
--Hinds County 2489
Oregon
--Southwestern 3725
Tennessee
--Western 4119
Texas 4217, 4282
--East 4226
--Gulf Coast Basin 4212
Utah
--Uinta Basin 4370, 4373
Wyoming
--Southwestern 4890
--Washakie County 4848
Eostaffella
New Mexico 3258
Eostaffella Fusulinid
Texas 3258
Equus-Beds
Kansas 1726
Erie County, Ohio
Devonian 3571
Geology 3561
Pre-Olentangy 3571
Rocks 3571
Silurian 3571
Stratigraphy 3571
Erie County, Pennsylvania
Geography 3828
Geology 3828
Isotopic Dates 3851
Medina Production 3879
Oriskany Production 3879
Precambrian 3851
Erionite
Nevada 2880
Erosion 2239
California
--Dry Creek 575
--Sonoma County 575
Connecticut
--Coast 695
Illinois
--Lake Michigan 1452
Maryland
--South Mountain Anticlinorium
2266
Ohio
--Lake County 3600
--Lucas County 3608
Texas
--Panhandle 4234
Wisconsin 4831
Escalante, Utah
Water 4460
Escalante-Upper Valley Coal Area,
Utah 4410
Escambia County, Alabama
Disposal Wells 116

Geology 21
Groundwater 21
Escambia County, Florida
  Geology 740
  Water Resources 766, 827
Escarpment
  Texas
  --Central 4194
  --Panhandle 4234
Esker
  North Dakota
  --Grand Forks County 3442
  --Walsh County 3442
Esmeralda County, Nevada
  Geology 2878
  Minerals 2878
Essex County, Vermont
  Geochemistry 4491
Essex Quadrangle, Connecticut
  Bedrock Geology 671
  Surficial Geology 687
Estero Americano, California
  Geology 459
Estill County, Kentucky
  Geology 2029
  Sandstone 2029
Estuaries
  Alabama 127, 187
  Delaware
  --Delaware Estuary 704
  New Jersey
  --Sandy Hook Bay 3032
Eufaula Bauxite District, Alabama
  Bauxite 55
  Ceramic Investigations 152
  Kaolin 55
Eureka, South Dakota
  Groundwater 4105
Eureka County, Nevada
  Geology 2863
  Minerals 2863
Eureka Creek, Alaska
  Geochemistry 255
  Geology 255, 281
Evangeline Aquifer, Louisiana
  Geohydrology 2180
Evansville, Indiana
  Geology 1603
Evansville Valley Sedimentation
  Illinois
  --Illinois Basin 1240
Evaporation
  Hawaii 1077, 1112
  Utah
  --Great Salt Lake 4453,
    4454, 4455, 4464
Evaporites 1161
  Kansas
  --Western 1781

Nevada 2888
Oklahoma
  --Anadarko Basin 3674
  --Western 3624
Texas
  --Anadarko Basin 3624
  --Panhandle 3674
West Virginia 4744
Everona Formation
  Virginia 4517
Exchange, Illinois
  Limestone 1292
  Oil Production 1292
Exploration 736, 2079
  Alabama 128
  Florida 765
  Georgia 944, 958
  Kentucky 2067
  --Butler County 2110
  --Logan County 2103
  --South Central 2103
  --Western 2080
  Missouri 2583
  North Carolina 3300
Explosives 2068

Fabius Area, Alabama
  Coal 51
Facies
  California
  --Coast Ranges 419
  Illinois 2130
  Indiana 1570, 2130
  Iowa
  --Eastern 1643
  Kansas 1743, 1865
  --Eastern 1821
  --Southwest 1766
  Kentucky 2130
  Maryland
  --Coastal Plain 2299
  Missouri 2611
  --St. Francois Mountains 2621
  New Jersey
  --Brigantine Inlet 3045
  New Mexico
  --North-Central 3245
  North Dakota 3496
  --Central 3497
  Ohio
  --Northwestern 3587
  Texas
  --Coast 4302, 4334
  --North 4315
  --North-Central 4199, 4248
  --Palo Duro Basin 4337
  --Paluxy River Valley 4171
  --Panhandle 4232

Facies (Cont.)
 Texas (Cont.)
  --Somervell County 4171
 Utah
  --Uinta Basin 4373
 Wyoming
  --Washakie County 4848
Fairburn Quadrangle, South Dakota
 Geology 4054
Fairfield Basin, Illinois
 Coal 1345
Fall River County, South Dakota
 Groundwater 4062
Fall Zone
 Delaware 722
Falls
 Indiana
  --Ohio River 1534
Farewell, Alaska
 Geochemistry 261
 Geology 261
Fargo, North Dakota
 Geology 3418
Farm Ponds
 Kentucky
  --Inner Blue Grass Region
   2127
Farmdalian Lake Deposits
 Illinois
  --Northern 1323
Farmington, Maine
 Aggregate 2205
 Glacial Lake 2205
 Glacial Marine Clays 2205
Faulk County, South Dakota
 Geology 4024, 4025
 Water 4024
Faults
 California 452, 474, 479, 490
  --Azusa Quadrangle 456
  --Big Pine 498
  --Calaveras 474
  --Elsinore Fault Zone 572
  --Hayward Fault 474
  --Hosgri 578
  --Los Angeles County 546
  --Mount Wilson Quadrangle
   546
  --Orange County 454
  --Riverside County 572
  --San Andreas Fault 421,
   423, 474, 498, 559
  --San Diego 564
  --San Fernando 450
  --San Gabriel Mountains
   546
  --San Gregorio 578
  --San Jacinto 474
  --Transverse Ranges 441, 498

Connecticut
 --Southern 696
Illinois 2121
 --Northern 1361
 --Northwestern 1347
 --Plum River 1347
 --Wabash Valley 1365
Indiana
 --Southern 1537
Iowa
 --East-Central 1649
 --Plum River Fault Zone 1649
Kentucky 2121
Maryland
 --Charles County 2260
 --Prince George's County 2260
Missouri 2629
 --Interstate 55 2642
Nebraska
 --Humboldt Fault Zone 2858
Tennessee
 --Dumplin Valley 4150
 --East 4150
 --Houston County 4137
 --Stewart County 4137
Texas 4219
 --Coast 4320
 --East 4224
West Virginia
 --Eastern 4748
Wyoming
 --Lincoln County 4888
 --Sublette County 4888
Fauna 1342, 4182, 4846
 Alabama 73
 Alaska 3249
 Connecticut 700
 Illinois 1317
  --Northern 1323
 Indiana
  --Eastern 1493
 Minnesota 2440
 Mississippi 2534
 Missouri 2673
 New Mexico
  --East-Central 3263
  --Northeastern 3214
 North Dakota 3368
  --Red Lake County 3430
  --Southeastern 3423
 Ohio
  --Western 1493
 Southwest 3265
 Texas
  --Central 4297
  --San Patricio County 4312
 Western U. S. 3270
Fayette County, Illinois
 Carper Sand Oil 1218

Fayette County, Iowa
  Gravity Surveys 1648
  Magnetic Anomalies 1648
Fayette County, Ohio
  Sub-Trenton Rocks 3553
Fayetteville Formation
  Missouri
  --Southwestern 2630
Feldspar
  California 424
  --Atascadero 569
  --San Luis Obispo County
    569
  --Santa Lucia Range 569
  Illinois 1247
  North Carolina 3298
Fern
  Illinois 1348
Ferrelview Formation
  Missouri 2622
Ferruginous Bauxite
  Hawaii
  --Kauai 1006
  --Maui 1006
Fertilizer
  Idaho 1163
Fiborn Limestone
  Michigan
  --Chippewa County 2359
  --Mackinac County 2359
  --Schoolcraft County 2359
Field Sampling
  California
  --Lakeview Mountains 519
  --Riverside County 519
Field Tests
  Mineral Elements 346
Fill
  California
  --San Francisco Bay 527,
    531, 532, 533
  Maryland
  --Salisbury 2237
Filters 735
Finney County, Kansas
  Irrigation Water 1786
Fish 3076, 4846
  Alabama
  --Birmingham County 70
  --Jefferson County 70
  Connecticut 657, 661
  New Mexico
  --Manzanita Mountains 3189
  North Dakota
  --Sheridan County 3431
Fisk Quadrangle, Alabama
  Geology 206
  Mineral Resources 206
Fitzgerald, Georgia

Aquifers 942
Flagler County, Florida
  Geology 819
  Groundwater 753
  Surface Water 755
Flagstones
  Pennsylvania
  --Endless Mountains 3838
  --Northern 3838
Flathead County, Montana
  Geochemistry 2700, 2717
  Geology 2684, 2692, 2739
  Minerals 2739
  Stream Sediments 2700
Flatwoods, Louisiana
  Water Resources 2199
Flatwoods Region, Indiana 1583
Flies
  Connecticut 649, 653
Flint
  Ohio 3511
Flint, Idaho
  Geology 1181
  Ore Deposits 1181
Flint Creek Range, Montana
  Mines and Minerals 2745
Flint Hill Quadrangle, Virginia
  Geology 4600
  Hydrogeology 4601
Flint Ridge, Ohio
  Bedrock Geology 3585
Floods 1632, 1633, 2957
  Alabama 155
  --Birmingham 2
  --Jefferson County 2, 191
  --Shades Creek Basin 191
  Colorado
  --Larimer County 606
  Florida
  --Dade County 834
  --Gadsden County 796
  Hawaii 990, 1023
  --Hawaiian Islands 1022, 1043,
    1048, 1087, 1097, 1098
  --Kauai 1038, 1046, 1054, 1110
  --Maui 1022, 1041, 1059, 1091,
    1100
  --Oahu 1022, 1038, 1040, 1047,
    1059
  --Olowalu 1091
  Illinois
  --Greater Chicago Area 1428
  Missouri 2667, 2672
  --Jefferson County 2663
  --Joplin 2668
  --Northwestern 2665
  --St. Francois County 2663
  --Ste. Genevieve County 2663
  --Springfield 2678

Floods (Cont.)
  North Dakota
    --East Grand Forks 3409,
      3445
    --Grand Forks 3409, 3445
  Ohio
    --Lucas County 3608
  Texas
    --Central 4194
Flora
  Connecticut 702
  Illinois 1319
    --West-Central 1356
  Indiana
    --Greene County 1489
  Missouri 2673
  North Carolina
    --Deep River Basin 3305
  Wyoming 4843
Florida Panhandle
  Limestone Outcrop Belt 873
Florida Peninsula
  Geomorphology 745
Florida Plateau
  Geothermal Nature 888
Flotation 3119, 3127, 3144,
  3160
  Arizona 380
Flow Measurement 3928
Flowerpot Shale
  Oklahoma
    --Jackson County 3663
Flowing Wells
  Michigan 2368
Floyd County, Georgia
  Geology 945
  Groundwater 945
Floyd River Basin, Iowa
  Groundwater 1676
Flue Gases 1397, 1414, 1431
Fluoride
  Florida
    --Alafia Basin 833
    --Peace River Basin 833
Fluorine 2114
Fluorine Chemistry
  Illinois 1357
Fluorite
  Mississippi Valley 2119
  Utah 4383
Fluoroaromatic Compounds 1364
Fluorspar 2113, 2115
  Arizona 378
  California 424
  Illinois 1276, 2120, 2121
    --Hardin County 1473
    --Illinois Fluorspar District
      1198, 1241, 1269
    --Pope County 1473

Kentucky 2120, 2121
  --Crittenden County 2123
  --Livingston County 2122
Mexico 2117
Nevada 2894
New Mexico 3267
Texas 4257
  --Brewster County 4168
United Kingdom 2116
Western United States 2118
Flushing Procedures
  Florida
    --Ft. Myers 778
Fluvanna County, Virginia
  Geology 4503
  Gravity Surveys 4574
  Magnetic Data 4574
  Minerals 4503
Fluvial Deposits
  Tennessee
    --Western 4119
  Texas
    --North-Central 4175
Fluvial Monazite
  South Carolina
    --Enoree River 3991
    --Pacolet River 3991
    --Saluda River 3992
    --Savannah River 3992
    --Tyger River 3991
Fluvial Sandstone
  Utah
    --Uinta Basin 4425, 4428
Fluvial Sediments
  Iowa 1656
Fly Ash
  Montana 2753
Folding
  West Virginia
    --Eastern 4748
Folds
  Missouri 2629
Fond Du Lac Formation
  Minnesota
    --Duluth 2418
Foraminifers
  Alabama
    --Gipsy Limestone County 58
    --Mobile Bay 140
  California 516
  Indiana
    --Northern 1498
  Kansas
    --Cherokee County 1776
  Louisiana 2158
  North Carolina
    --Grimesland 3306
    --Pitt County 3306
  Virginia

--Southeastern 4563
Washington
--Coast 4614
--Olympic Peninsula 4605
--Point Grenville 4614
--Raft River 4614
--Satsop River 4605
Ford County, Kansas
Dakota Formation 1827
Water 1827
Forest County, Pennsylvania
Geology 3876
Red Valley Sandstone 3876
Forests 2952
Alabama
--Birmingham 192
--Jefferson County 192
Alaska
--Unga Island 329
Idaho
--St. Joe National Forest
1133
Kentucky 2047
Mississippi
--West Central 2516
Tennessee
--Wilson County 4161
Wisconsin
--Menominee County 4761
Formations 4255, 4846
Alabama 87
--Gipsy Limestone County
58
Alaska
--Alaska Peninsula 297
--Alaska Range 282
Arkansas
--Backbone Anticline 390
--Sebastian County 390
California
--Los Angeles 520
--Montara Mountain Quad-
rangle 495
--San Mateo County 495
Colorado
--Cheyenne Basin 628
--Denver Basin 628
Delaware 715
Florida 751, 882
--Chipola Formation 747
--Jackson Bluff Formation
747
Georgia
--Broad Quadrangle 897
--Northern 959
Illinois 1244
Illinois Basin 1495
Indiana 1485, 1560, 1589
--Clark County 1607

--Eastern 1493
--Greene County 1575
--Latta 1575
--Northeastern 1532
--Northern 1492, 1527
--Sullivan County 1600
Iowa
--North-Central 1647
--Northern 2429, 2439
Kansas
--Southern 3620
Kentucky
--Jefferson County 1607
Maryland 2290
--Coastal Plains 2299
--Southern 2282
Minnesota 2429, 2439
--Duluth 2418
--East-Central 2424
--Northeastern 2443
--Southeastern 2421
Mississippi 2534
--Hinds County 2489
Missouri 2569, 2622
--Central 2612
--Highway 65 2617
--Southwestern 2618, 2619, 2630
Nebraska
--Western 2854
New Mexico
--Carlsbad Potash District 3104
--Center-Eastern 3198
--Fort Ann 3251, 3252
--Manzanita Mountains 3189
--North-Central 3245
--Northeastern 3215
--San Juan Basin 3183
North Carolina
--Beaufort County 3285
North Dakota
--Bakken 3311
--Bullion Creek 3489
--Cannonball 3421, 3487
--Englewood 3311
--Fort Union Group 3488
--Fox Hills 3368, 3466, 3485
--Hell Creek 3350, 3477
--Ludlow 3486
--Madison 3473, 3482
--Newcastle 3476, 3477
--Niobrara 3471
--Northern 3346
--Oahe 3464
--Pierce 3494
--Red River 3496
--Scallion Subinterval 3497
--Sentinel Butte 3453, 3475,
3483, 3488
--Slope 3489

Formations (Cont.)
North Dakota (Cont.)
--Tongue River 3453, 3475,
3483, 3488, 3495
--Williston Basin 3346
Ohio
--Newburg 3537
--Northwestern 3579
--Southwestern 3609
--Sub-Trenton 3550, 3553
--Western 1493
Oklahoma 3616, 3617, 3626,
3630, 3631, 3642, 3643,
3669, 3672, 3673
--Arbuckle Mountains 3644,
3650, 3665
--Bryan County 3636
--Eastern 3652
--Northwestern 3620
--Nowata County 3629
--Southern 3657, 3679
--Washington County 3629
Oregon
--Central 3767
--Eastern 3715
--Western 3712
Pennsylvania
--Adams County 3932
--Berks County 3933
--Central 3788
--Dauphin County 3935
--Lancaster County 3934
--Lehigh County 3795, 3941
--Montgomery County 3933
--North-Central 3855
--Northampton County 3795,
3941
--South-Central 3798
--Southeastern 3925, 3960
--York County 3932
South Dakota 4050, 4053, 4057
--Bakken 3311
--Big Badlands 4052
--Englewood 3311
--Fox Hills 3466
--Southwestern 4049
Tennessee
--East 4144
--Middle 4133
--West 4157
Texas 4172, 4184, 4340
--Burleson County 4290
--Central 4297, 4301, 4309
--Coastal Plains 4322, 4335
--East 4227
--Gulf Coast 4197, 4201,
4210, 4212, 4223
--Lee County 4290
--Marathon Basin 4292

--Panhandle 4232
--South 4190, 4298
--Southern High Plains 4286
Utah 4444
--Dinosaur Quarry Quadrangle
4440
--Southeastern 4377
--Uinta Basin 4370, 4373, 4418,
4419
Virginia 2290, 4517
--Danville Basin 4560
Washington
--Coast 4614
--Point Grenville 4614
--Raft River 4614
West Virginia 4741, 4744
--Eastern 4698
--Pendleton County 4698
--Wayne County 4699
Wyoming 4859
--Washakie County 4848
Forsyth County, Georgia
Geohydrology 957
Fort Ann, New Mexico
Fossils 3251
Fort Ann Formation
New Mexico 3252
Fort Apache Indian Reservation,
Arizona
Mineral Deposits 348
Fort Defiance Quadrangle, Virginia
Geology 4549
Fort Myers, Florida
Salt Water Intrusion 778
Fort Pillow Test Well, Tennessee
Stratigraphy 4147
Fort Sumner, New Mexico
Geology 3109
Fort Union Coal Region, Montana
Groundwater 2843
Fort Union Group
North Dakota
--Renville County 3342
--Ward County 3342
--Williams County 3336
--Williston Basin 3488
Fort Wayne, Indiana
Stone 1559
Fortymile District, Alaska
Geochemistry 224, 231
Geology 231
Fossils 1369, 3746
Alabama 73
--Black Warrior Basin 153
--Gulf Coast 196
Arkansas 387
California
--Gubernadora Quadrangle 449
--Orange County 449

Connecticut
--Connecticut Valley 652
Delaware 724
Florida 879, 891
Georgia
--Coastal Plain 975
Idaho 1124
--St. Maries River 1143
Illinois 1371
--Illinois Basin 1376
Kansas 1699, 1790
Maryland 2242
Michigan 2337, 2340
Minnesota 2385
Mississippi 2519, 2543
Nevada 2891
New Mexico 3250, 3251, 3252, 3253
North Dakota 3421
--Lake Agassiz 3430
--Logan County 3422
--Red Lake County 3430
--Sheridan County 3431
--Western 3495
Oklahoma
--Southern 3657
Oregon 3734
Pennsylvania 3790
South Dakota
--Wounded Knee 4058
Tennessee 4121
--Western 4157
Texas
--San Patricio County 4312
Virginia 4514
West Virginia 4708, 4709, 4710
Wisconsin 4768
Wyoming 4836
Foster County, North Dakota
Geology 3322, 3323, 3324, 3441
Glaciation 3322
Groundwater 3322, 3323, 3324
Water 3323, 3324
Foster Creek, Montana
Coal 2732
Geology 2732
Fourier Series 1862
Fox Hills Formation
North Dakota 3368, 3466
--Williston Basin 3485
South Dakota 3466
Foxon Fault, Connecticut 696
Fractionation 1249
Fractures
Kentucky
--Bryantsville Quadrangle 2129

--Central 2129
Texas
--Southern Edwards Plateau 4215
Fracturing 2090
Kentucky
--Oil Springs Pool 2028
Franciscan Formation
California
--Geysers Geothermal Area 583
--Montara Mountain Quadrangle 495
--San Mateo County 495
Franciscan Melanges
California 467
Franciscan Rocks
California
--Western 406
Franklin County, Alabama
Environmental Geology 6
Geology 19
Groundwater 19
Franklin County, Kansas
Geology 1680
Franklin County, Washington
Geology 4623
Frederick County, Virginia
Geology 4504
Minerals 4504
Silica 4534
Frederick Valley, Maryland
Cambro-Ordovician 2301
Carbonate Rocks 2297
Hydrogeology 2297
Paleogeography 2301
Sedimentology 2301
Stratigraphy 2301
Fredericksburg Division, Texas
Stratigraphy 4287
Freemont Peak Quadrangle, California
Geology 411
Freeport Quadrangle, Illinois
Geology 1251
French Creek State Park, Pennsylvania
Guide 3903
French Gulch Quadrangle, California
Economic Geology 502
Fresh Water
Delaware 718
Florida
--Lee County 856
--Sanibel Island 856
Kentucky
--Muhlenberg County 1939
--New Cypress Pool 1939
Louisiana
--Lake Charles 2184
--Lake Pontchartrain 2193
--Southwestern 2189
Fresh Water Fish

Fresh Water Fish (Cont.)
Connecticut 657
Friday Creek, Alaska
Mineral Preparation 333
Ore 333
Frio Formation
Louisiana 2149
Texas
--Gulf Coast 4197, 4201,
4210, 4326
--South 4190
Frisco Formation
Oklahoma 3616
--Arbuckle Mountains 3650
Frisco Quadrangle, Utah
Hydrothermal Alteration 4394
Front Range Urban Corridor,
Colorado
Aggregate 639, 640
Geology 602
Gravel 639, 640
Rock 603
Sand 639, 640
Soil 603
Front Royal Quadrangle, Virginia
Geology 4594
Hydrogeology 4594
Froth Flotation
Arizona 380
Fructifications
Illinois 1319
Fry in Situ Project, Kentucky
2062
Fuels 4727
Illinois 1266
Oregon
--Coos County 3722
Fuller's Earth
Kentucky
--Jackson Purchase 1928
Nevada 2877
Fulton County, Illinois
Limestone 1226
Mississippian 1226
Fusulinids
Missouri 2608
New Mexico 3258
--Joyita Hills 3255
--Socorro County 3255
Oklahoma
--Ardmore Basin 3637
--Carter County 3637
--Love County 3637
Texas 3258

Gabbro Complex
Minnesota
--Duluth 2383

Gabbro Lake Quadrangle, Minnesota
Duluth Complex 2420
Lower Precambrian Rocks 2449
Gadsden County, Florida
Floods 796
Gaillard Graben
Connecticut
--South-Central 667
--Southern 696
Galena Group
Illinois
--Northern 1358
Galesburg, Illinois
Geology 1484
Groundwater 1484
Galesville Sandstone
Illinois 1259
Gallatin County, Montana
Geology 2693
Gallatin River, Montana 2820
Gallinas Mountains, New Mexico
Geology 3106
Minerals 3106
Gallium
California 424
Gallup Sandstone
New Mexico
--Northwestern 3218
Galveston, Texas
Aquifers 4209
Galveston Island, Texas
Sediments 4244
Shorelines 4187
Gamma Radiation
Alabama
--Gilbertown 167
Gamma Rays
Louisiana 2157
Garfield County, California
Geology 604
Groundwater 604
Mineral Resources 604
Garfield County, Utah
Coal 4408, 4410, 4417
Geology 4380, 4408
Minerals 4380
Palynology 4362
Sandstone 4362
Upper Cretaceous 4362
Uranium 4412
Water 4460
Garnet
California 424
Garnet, Montana
Geology 2783
Metallic Resources 2784
Garnet Mountain Quadrangle, Montana
Geology 2693
Garrettsville Quadrangle, Ohio

Bedrock Geology 3555
Gas 1397, 1414, 1431, 1882,
  2041, 2064, 4875
  Appalachians
  --Central 4742
  East-Central States 1984,
    1986, 1990, 1992, 1997,
    2000, 2002, 2005, 2008,
    2009, 2011, 2013, 2015,
    2134, 2136
  Kansas 1690, 1703, 1722, 1732
  Kentucky 1906, 1908, 1911,
    1913, 1914, 1915, 1918,
    1920, 1925, 1969, 1973,
    1975, 1977, 1979, 1981
  --Butler County 1962
  --Eastern 2020
  --Muhlenberg County 1926
  --Southeastern 1960
  --Western 2059
  Louisiana 2159
  Michigan 2332, 2343
  Mississippi 2460, 2542
  --Jasper County 2451
  Missouri 2598
  --Belton 2567
  --Northwest 2557
  --Putnam County 2609
  Montana 2793
  Nebraska 2846, 2848, 2850
  Nevada 2915, 2935
  New Mexico 3142, 3163
  North Dakota 3410
  --Williston Basin 3467
  Ohio 3537, 3548, 3551, 3557
  Oklahoma
  --Craig County 3621
  Oregon 3747, 3748, 3750,
    3753
  Pennsylvania 3779, 3860
  --Amity Quadrangle 3873
  --Claysville Quadrangle 3873
  --McKean County 3881
  --Pittsburgh 3852
  --Warren County 3881
  --Warren Quadrangle 3871
  --Westmoreland County 3894
  --Youngsville Quadrangle
    3872
  South Dakota 4028, 4029
  Texas 4170, 4179, 4184
  --Northeast 4213
  Utah 4342, 4347, 4367
  --East 4398
  --South-Central 4358
  Washington 4642, 4665
  West Virginia 4674, 4687,
    4690, 4711, 4742
  --Barbour County 4679

  --Braxton County 4677
  --Brooke County 4681
  --Clay County 4677
  --Eastern 4693
  --Hancock County 4681
  --Jackson County 2075
  --Kanawha County 2075
  --Ohio County 4681
  --Pocatalico River Basin 4696
  --Upshur County 4679
  --Wayne County 4741
  Gas Accumulations 2108
  Gas Exploration 2056
  Kentucky 2036
  --Butler County 2018
  Gas Fields
  Kansas 1859
  Gas Injection
  Kentucky
  --Union County 2034
  Gas-Oil Ratios 2321
  Gas Pools
  Kentucky
  --Daviess County 2025
  Michigan 2343
  Pennsylvania
  --Jefferson County 3878
  Gas Production
  Central Appalachian Area 1909
  Kansas 1796, 1798, 1802, 1804,
    1807, 1859, 1879, 1888, 1892,
    1897, 1902
  Kentucky
  --Eastern 2020, 2105
  --Western 2088
  Gas Reservoirs
  Louisiana
  --South 2154
  Gas Storage
  Kentucky
  --Western 2035
  Gas Tests
  Iowa 1626
  Gasification 1286
  Gastropoda
  Mississippi 2515
  Texas 4282
  Geismar, Louisiana
  Groundwater 2167
  Gemstones
  California 424
  South Carolina 3979
  General Electric Test Reactor Site,
  California
  Geology 488
  General Land Offices 1880
  Geneva County, Alabama
  Mineral Resources 175
  Water Resources 175

Genevievian
Illinois 1479
Gentry County, Missouri
Water 2651
Geobotany
Alaska
--Bokan Mountain 276
--South-Eastern 276
Geochemistry 2108
Alaska
--Admiralty Island 216
--Ambler River Quadrangle 274
--Anton Larsen Bay 266
--Belt Creek 230
--Bettles Quadrangle 270
--Big Delta Quadrangle 211
--Brooks Range 277
--Central 218
--Chandalar 277
--Chatanika 213
--Chugach Mountains 260
--Circle Hot Springs 213
--Copper River Basin 248
--Cosmos Hills 272, 274
--Craig A-2 Quadrangle 283
--Diana Lakes 269
--Eagle A-1 Quadrangle 231
--East-Central 217, 224
--Elliott Highway 219
--Eureka Creek 255
--Farewell 261
--Fortymile District 224, 231
--Hollis 252
--Inmachuk River 258
--Iron Creek 226
--Kenai Lake 222
--Kenai Peninsula 244
--Kodiak Island 266
--Kontrashibuna Lake 228
--Lake Clark 228
--Libby River 230
--Little Falls Creek 227
--Medfra Quadrangle 257
--Metal Creek 260
--Minook Creek 220
--Mount Hayes Quadrangle 255
--Nellie Juan River 244
--Nenana Highway 218
--Nixon Fork 257
--Paxson 248
--Prince of Wales Island 252, 283
--Rainy Creek 255
--Rampart District 220
--Richardson 211
--Seward Peninsula 226, 230,
258, 259, 264, 268, 271
--Shungnak D-2 Quadrangle 272, 274
--Sinuk Area 264, 271
--Sinuk River 259
--Sithylemenkat Lake 270
--Slana District 210
--Solomon C-5 Quadrangle 268
--Solomon D-6 Quadrangle 226
--South-Central 210, 212, 215, 222, 229
--South-Eastern 209, 214, 283
--Talkeetna Mountains 227, 269
--Tanana 221
--Taylor Highway 217
--Tikchik Lake 225
--Twelve-Mile Creek 252
--Valdez to Chitina 223
--White River 229
--Willow Creek 222
--Wood River Lake 225
--Yukon 221
Idaho 1136
--Bighorn Crags 1135
--Blackbird Mountain 1187
--Blaine County 1166
--Juniper Mountain 1186
--Lamas County 1166
--Lemhi County 1187
--Mt. Pend Oreille Quadrangle 1183
--Owyhee County 1182, 1183
--Panther Creek 1187
--Silver City 1182
--South Mountain 1182, 1186
--Vienna District 1166
Illinois
--Illinois Basin 1277
Mexico 4331
Minnesota
--Birchdale 2428
--Indus 2428
--Koochiching County 2428
Montana
--Broadwater County 2691
--Flathead County 2700, 2717
--Lewis and Clark County 2827
--Lincoln County 2700, 2717
--Stemple Pass 2827
--Winston Mining District 2691
New Mexico
--Central 3268
--Eagle Nest Quadrangle 3148
--Lincoln County 3121
--Magdalena Mining District 3139
--Nogal Peak 3121
--Philmont Ranch Region 3146
--Rio Arriba County 3175

--South-Central 3268
North Dakota
  --West-Central 3493
Oregon
  --Mt. Hood 3777
Tennessee
  --East 4153
  --Northwest 4153
Texas
  --Coastal Plain 4335
  Matagorda Bay 4221
  --Trans-Pecos 4331
Utah
  --Mercur 4435
  --Uinta Basin 4373
Vermont
  --Caledonia County 4491
  --Essex County 4491
  --Pomfret Dome 4492
Virginia
  --Staunton Quadrangle 4585
Wisconsin
  --Southwest 4793
Geochronology
Connecticut 699
Kansas
  --Riley County 1747
Geodetic Control Network 3047
Geodetic Control Surveys
New Jersey 2944
Geodetic Measurements 430, 3053
Geodetic Surveys
New Jersey 3049
Geoenvironment
California
  --Orange County 454
Geography
Mississippi
  --Adams County 2540
Oregon
  --Coos County 3722
Pennsylvania
  --Erie County 3828
Virginia 4527
Geohydrology
Georgia
  --Bartow County 957
  --Ben Hill County 925
  --Cherokee County 957
  --Forsyth County 957
  --Irwin County 925
  --Tift County 925
  --Turner County 925
  --Worth County 925
Hawaii
  --Oahu 1045
  --Waimanalo 1045
Iowa
  --Cedar Rapids 1677

--Muscatine County 1675
--Muscatine Island 1675
Kansas
  --Grant County 1692
  --Great Bend Prairie 1826
  --Jefferson County 1757
  --Nemaha County 1819
  --Northeastern 1819
  --Sedgwick County 1711
  --Stanton County 1092
Louisiana
  --Southwestern 2180
Maryland 2296
  --Salisbury 2237
North Dakota
  --Knife River Basin 3494
Texas
  --Palo Duro Basin 4220, 4230
Geologic Evaluation 1177
Geologic Features
California
  --Death Valley 547
Nevada
  --Death Valley 547
Geologic Hazards
California
  --Marin County 460
  --Novato 460
  --San Bernardino County 554
Colorado 641, 643
  --Crested Butte 613
  --Gunnison County 613
  --Larimer County 606
Mississippi
  --Adams County 2540
Tennessee 4129
Geologic Investigations 44
Geologic Logs 2406
Geologic Maps 2056
California 414, 571
Kentucky 2050, 2085
Geologic Names
Idaho 1129
Mississippi 2532
Geological Surveys
Minnesota 2437
Mississippi 2466, 2468, 2473,
  2474, 2490
Geologists
Mississippi 2475
Geology 1389, 1390, 1394, 1405,
  1851, 3085, 3544, 4316, 4694,
  4695, 4719, 4728, 4785
Alabama 177, 195
  --Athens Area 18
  --Birmingham Red Iron District
  79
  --Cherokee County 26
  --Coastal Plains 105

324 / SUBJECT INDEX

Geology (Cont.)
Alabama (Cont.)
--Colbert County 6, 163,
 205, 207, 208
--Escambia County 21
--Fisk Quadrangle 206
--Franklin County 6, 19
--Gilbertown Oil Field 93
--Greene County 33
--Huntsville 8
--Jefferson County 10, 14,
 15
--Lauderdale County 6, 162
--Lawrence County 25
--Leighton Quadrangle 208
--Limestone County 104, 164
--Lineville East Quadrangle
 64
--Madison County 1, 3, 4,
 5, 8, 202, 203, 204, 206
--Maysville Area 3
--Mellow Valley Quadrangle
 64
--Meridianville Quadrangle
 1
--Montgomery Area 17
--Moontown Quadrangle 202
--Morgan County 23
--Newhope Quadrangle 204
--Northwestern 50
--Ofelia Quadrangle 64
--Pride Quadrangle 205
--Russellville Area 24
--St. Clair County 20
--Southeastern 35
--Southwestern 54
--Toney Quadrangle 203
--Triana Area 4
--Tuscumbia Quadrangle 207
--Valley and Ridge Province
 10
--Wadley North Quadrangle
 64
--Warrior Basin 88
--West-Central 7
--Winston County 52
Alaska 286, 290, 296
--Admiralty Island 216
--Ahtell Creek 241
--Alaska Range 237, 280,
 281
--Ambler River Quadrangle
 274
--Amphitheatre Mountains 254
--Anchorage 240, 253
--Anton Larsen Bay 266
--Bear Creek 247
--Belt Creek 230
--Bendeleben Quadrangle
 246

--Bettles Quadrangle 270
--Bluff Area 245
--Brooks Range 277, 293
--Candle Quadrangle 247
--Chandalar 277
--Chugach Mountains 260, 326
--Coast Range 236
--Copper River Basin 248
--Cosmos Hills 272, 274
--Craig A-2 Quadrangle 283
--Denali 239
--Diana Lakes 269
--Dolomi 262
--Dry Pass 242
--Eagle A-1 Quadrangle 231
--Eklutna 253
--Eureka Creek 255, 281
--Farewell 261
--Fortymile District 231
--Gulkana Glacier Area 280
--Hollis 252
--Iliamna Quadrangle 243
--Inmachuk River 258
--Iron Creek 226
--Kenai Peninsula 244
--Knight Island 251
--Kodiak Island 266
--Kontrashibuna Lake 228
--Lake Clark Region 228
--Libby River 230
--Little Falls Creek 227
--Maclaren River 239
--Medera Quadrangle 257
--Mentasta Pass 265
--Metal Creek 260
--Midas Mine 250
--Mt. Hayes Quadrangle 249,
 254, 255, 263
--Nellie Juan River 244
--Niblack 240
--Nixon Fork 257
--Omilak 246
--Otter Creek 246
--Paint River 243
--Paxson 248
--Portage Creek 238
--Prince of Wales Island 252,
 262, 283
--Prince William Sound 251
--Prudhoe Bay Oil Field 342
--Rainbow Mountain 237, 280
--Rainy Creek 249, 255
--Ruby Ridge 293
--Seward Peninsula 226, 230,
 245, 246, 247, 258, 259,
 264, 268, 271
--Shungnak D-2 Quadrangle 272,
 274
--Sinuk Area 264, 271
--Sinuk River 259

--Sithylemenkat Lake 270
--Slana District 241, 256, 265
--Solomon C-5 Quadrangle 268
--Solomon D-6 Quadrangle 226, 245
--South-Central 229, 241, 256
--Southeastern 240, 242, 252, 283
--Southwestern 228
--Spirit Mountain 275
--Sulphide Gulch 250
--Susitna River 238
--Talkeetna Mountains 227, 267, 269
--Talkeetna River 267
--Twelve-Mile Creek 252
--Valdez 250
--White River 229
Appalachian Basin 2088
Appalachians 2310
--North Carolina 4158
--Tennessee 4158
Arizona 344, 361
--Arizona Highway 85, 86, 386 354
--Arizona Highway 87, 88, 188 355
--Beaverdam Mountains 357
--Central 385
--Harquahala Mountains 383
--Highway 77 347
--Highway 177 347
--Socorro Peak 383
--U.S. Highway 666 345
--Virgin Mountains 357
--White Mountains 384
Arkansas
--Backbone Anticline 390
--Barber Quadrangle 389
--Clarksville Quadrangle 394
--Coal Hill Quadrangle 394
--Delaware Quadrangle 396
--Greenwood Quadrangle 392
--Hartman Quadrangle 394
--Johnson County 39, 393, 394, 396
--Knoxville Quadrangle 391, 396
--Lavaca Quadrangle 395
--Logan County 388, 393, 396
--New Blaine Quadrangle 393
--North-Central 399
--Northeastern 405

--Paris Quadrangle 388
--Pope County 391
--Scranton Quadrangle 393
--Sebastian County 389, 390
--Van Buren Quadrangle 395
California 424, 436, 452, 491, 503, 511, 517, 536, 556, 570
--Alameda County 408, 488
--Azusa Quadrangle 546
--Bay Area 423
--Butte County 407
--Calaveras County 428
--Canada Gobernadora Quadrangle 552
--Cascade Range 416, 424
--Central 422
--Church Creek 505
--Coachella Valley 524
--Coast Ranges 419, 421, 423, 424
--Contra Costa County 458, 497
--Dana Point Quadrangle 550
--Death Valley 508, 535
--Del Mar Quadrangle 438
--Desert Hot Springs 524
--Dry Mountain 535
--El Cerrito 458
--El Toro Quadrangle 551
--Elysian Park 542
--Escondido Quadrangle 438
--Franciscan Melanges 467
--Freemont Peak Quadrangle 411
--French Gulch Quadrangle 502
--Geysers Geothermal Area 583
--Gobernadora Quadrangle 449
--Great Basin 416, 424
--Great Valley 418, 424, 425
--Helena Quadrangle 522
--Hollister 423
--Inglewood 555
--Inyo County 490, 508, 526, 535
--Kern County 498
--Klamath Mountains 415, 424
--Laguna Beach Quadrangle 568
--La Jolla Quadrangle 438
--Lake County 563
--Lakeview Fault Segment 430
--La Mesa Quadrangle 438
--Laytonville Quarry 423
--Little Tujunga Canyon 430
--Lockwood Valley 498
--Los Angeles 542, 546
--Marin County 440, 460
--Mendocino County 423, 563
--Modoc District 490
--Modoc Plateau 416, 424
--Mojave Desert 411, 424, 500

Geology (Cont.)
  California (Cont.)
    --Monte Bello Ridge Mountain
      Study Area 456
    --Mont Diablo 497
    --Monterrey County 505
    --Morro Bay 467
    --Mt. Hamilton Quadrangle
      408
    --Mount Wilson Quadrangle
      546
    --Newhall 430
    --Newport 555
    --Northeastern 424
    --Northern 413, 422
    --Novato 460
    --Offshore Area 420
    --Ono Quadrangle 425
    --Opal Mountain Quadrangle
      411
    --Orange County 449, 454,
      534, 550, 551, 552, 553,
      567, 568
    --Ord Mountain District 494
    --Oroville Quadrangle 407
    --Panamint Butte Quadrangle
      490
    --Peninsular Ranges 424
    --Point Loma Quadrangle
      438
    --Poway Quadrangle 438
    --Ramona Pegmatite District
      504
    --Repetto Hills 542
    --Richmond 458
    --Riverside County 524
    --Sacramento Valley 423
    --Salton Trough 424
    --San Andreas Fault 421,
      423
    --San Bernardino County
      411, 451, 494, 500, 501,
      526
    --San Clemente 534
    --San Diego 438, 504, 534
    --San Fernando Valley 430
    --San Francisco 423
    --San Francisco Bay 527,
      528
    --San Gabriel Mountains
      430, 546
    --San Juan Capistrano Quad-
      rangle 553
    --San Luis Obispo County
      437
    --San Pablo 458
    --San Simeon 467
    --Santa Barbara County
      409
    --Santa Clara County 408, 456,
      457
    --Santa Clara Valley 423
    --Santa Margarita 437
    --Santa Susana Mountains 430
    --Santa Ynez Mountains 409
    --Saugus 430
    --Shadow Mountains 451
    --Shasta County 425, 502
    --Sierra Nevada 417, 423, 424
    --Slate Range 526
    --Solemint 430
    --Sonoma County 455, 459, 561,
      563
    --South County Study Area 457
    --Sylmer 430
    --Taylorsville 417
    --Tehama County 425
    --Transverse Ranges 424
    --Trinity County 502, 522
    --Valencia 430
    --Ventura County 453, 498
    --Victorville 501
    --Western 406
    --Wilson Canyon 430
    --Yosemite Valley 423
  Colorado 586, 589, 591, 593, 624,
    627, 629, 631, 633, 635, 637,
    638, 641, 643, 647
    --Boulder County 596
    --Cheyenne Basin 628
    --Crested Butte 613
    --Crystal Valley 604
    --Denver 597
    --Denver Basin 628
    --Eagle County 604
    --Garfield County 604
    --Gunnison County 604, 613
    --Jefferson County 590
    --Larimer County 602, 606
    --Mesa County 625
    --Pitkin County 604
    --Planning District 1 598
    --Planning District 7 599
    --Planning District 9 600
    --Planning District 13 599
    --Roaring Fork Valley 604
    --Weld County 602
    --Windsor Study Area 602
  Connecticut
    --Ansonia Quadrangle 679
    --Branford Quadrangle 670
    --Bridgeport Quadrangle 680
    --Central 697
    --Clinton Quadrangle 684, 685
    --Coast 695
    --Colchester Quadrangle 683
    --Collinsville Quadrangle 672
    --Deep River Quadrangle 669

--Essex Quadrangle 671, 687
--Guilford Quadrangle 684
--Haddam Quadrangle 692, 693
--Hamburg Quadrangle 675
--Hartford Quadrangle 676
--Long Hill Quadrangle 680
--Milford Quadrangle 679
--Moodus Quadrangle 683
--Naugatuck Quadrangle 691
--New Haven Quadrangle 674, 698
--New Town Quadrangle 689
--Norwalk North Quadrangle 690
--Norwalk South Quadrangle 690
--Old Lyme Quadrangle 677, 687
--Sharon Quadrangle 694
--South Canaan Quadrangle 688
--Southbury Quadrangle 686
--Spring Hill Quadrangle 682
--Torrington Quadrangle 681
--Waterbury Quadrangle 678
--West Torrington Quadrangle 673
--Woodmont Quadrangle 674
Czechoslovakia
--East 2963
Delaware
--Atlantic Seashore 720
--Fall Zone 722
--Middletown 707
--Newark 721
--Odessa 707
Florida
--Alachua County 872
--Bay County 752
--Dixie County 743
--Escambia County 740
--Flagler County 819
--Gilchrist County 743
--Glades County 824
--Hendry County 824
--Jackson Bluff 881
--Jefferson County 742
--Leon County 741
--Northern 776
--Panhandle 740
--Peninsular 776
--Putnam County 819
--St. Johns County 819
--Santa Rosa County 740
--South 887

--Tallahassee 883
--Tampa 886
Georgia 898, 903, 908, 909, 912, 951, 956
--Atlanta 896
--Bethesda Church 941
--Brevard Lineament 896
--Broad Quadrangle 897
--Catoosa County 934
--Chatham County 901
--Chattooga County 933
--Coastal Plain 906, 931
--Cochran Quadrangle 900
--Dade County 932
--Dawson County 936
--Floyd County 945
--Gordon County 954
--Greene County 941
--Habersham County 902
--Mitchell County 930
--Murray County 954
--Panola Mountain State Park 918, 919
--Perry Quadrangle 900
--Polk County 945
--Rabun County 902
--Rockdale County 939
--Southwestern 947
--Stone Mountain Park 920
--Sweetwater Creek State Park 917
--Thomas County 940
--Walker County 935
--Whitfield County 954
Gulf Coast
--East 196
Hawaii
--Hawaiian Islands 983, 984, 985, 1010
--Kahoolawe Island 981
--Kauai Island 987, 1083
--Lanai Island 981
--Maui Island 982
--Molokai Island 986
--Niihau Island 987
--Oahu Island 976, 977, 978, 980, 1058, 1092
--Waimanalo 1092
Idaho 1124, 1129, 1142
--Blackbird Mountain 1187
--Blaine County 1162, 1166
--Bonner County 1123
--Camas County 1166
--Central 1162
--Clearwater 1148
--Custer County 1162
--Delamar 1181
--East Thunder Mountain Mining District 1139

Geology (Cont.)
    Idaho (Cont.)
        --Flint 1181
        --Homestead 1156
        --Idaho Batholith 1144
        --Juniper Mountain 1186
        --Lakeview Mining District
            1176
        --Lemhi County 1187
        --Lewiston 1146
        --Mt. Pend Oreille Quad-
            rangle 1183
        --Owyhee County 1158,
            1171, 1181, 1182, 1186
        --Oxbow 1156
        --Panther Creek 1187
        --Pend Oreille Lake Region
            1154
        --Pioneer Mountains 1162
        --Reynolds Creek Experi-
            mental Watershed 1171
        --Sawtooth Range 1149
        --Selway-Bitterroot Wilder-
            ness Area 1127, 1174
        --Silver City 1158, 1181,
            1182
        --Snake River 1138, 1146,
            1156
        --Snake River Plain 1180
        --South Mountain 1182, 1186
        --U.S. Highway 93 1150
        --Valley County 1139
        --Vienna District 1166
    Illinois 1194, 1449
        --Alexander County 1461
        --Central North 1464
        --Champaign County 1362
        --Chicago 1316, 1384
        --Clark County 1236
        --Crescent City 1412
        --Decatur 1353
        --De Kalb County 1409
        --De Witt County 1225,
            1329, 1459
        --Edgar County 1236
        --Freeport Quadrangle 1251
        --Galesburg 1484
        --Hardin County 1473
        --Hennepin 1278
        --Illinois Basin 1477, 1478
        --Illinois Fluorspar District
            1198, 1241, 1269
        --Illinois Waterway 1334
        --Kewanee 1484
        --Lake County 1337, 1433
        --Lake Saline 1327
        --McClean County 1225, 1329
        --McHenry County 1294
        --Macomb 1380

        --Monmouth 1484
        --Naperville 1398
        --Northeastern 1379
        --Paducah Quadrangle 1216
        --Piatt County 1329
        --Pope County 1473
        --Rock Island 1484
        --St. Clair County 1321
        --Smithland Quadrangle 1216
        --Southeastern 1327
        --Southern 1207, 1366
        --Springfield 1353
        --Tuscola 1280
        --Union County 1461
        --Vandalia 1298
        --Western 1468
    Indiana 1536, 1596
        --Allen County 1604
        --Catlin 1528
        --Clifty Falls State Park 1612
        --Evansville 1603
        --Hendricks County 1567
        --Indiana Dunes State Park 1618
        --Lake County 1602
        --Laporte County 1605
        --McCormicks Creek State Park
            1613
        --Madison County 1601
        --Mansfield 1528
        --Marion County 1488, 1533,
            1610
        --Morgan County 1572
        --New Harmony 1582
        --Ohio River 1534
        --Parke County 1528
        --Pokagon State Park 1611
        --Porter County 1602
        --Putnam County 1528
        --Shades State Park 1614
        --Southwestern 1603
        --Spring Mill State Park 1617
        --Turkey Run State Park 1615
        --Upper East Fork Drainage
            Basin 1594
        --Upper Patoka Drainage Basin
            1593
        --Versailles State Park 1616
        --Wabash Valley 1582
        --Washington County 1499
    Iowa 1620, 1621, 1622, 1623,
        1625
        --Cerro Gordo County 1673
        --Linn County 1674
    Kansas 1679, 1779, 1794, 1849
        --Allen County 1741
        --Barton County 1775
        --Brown County 1723
        --Central 1752
        --Cheyenne Bottoms 1775

--Decatur County 1742
--Ellsworth County 1752
--Franklin County 1680
--Johnson County 1758
--Linn County 1736
--Manhattan 1763
--Miami County 1718
--Montgomery County 1818
--Neosho County 1720
--Pratt County 1764
--Rice County 1768
--Rush County 1769
--Southeastern 1818
--Trego County 1705
--Wallace County 1678
Kentucky 2138
--Adair County 2101
--Allen County 1931, 1987
--Bernheim Forest 2047
--Bluegrass Region 2127
--Calloway County 1932
--Calvert City Quadrangle 1943
--Camp Breckinridge 2052
--Carter Cave 2046
--Cascade Cave 2046
--Clark County 2029
--Cumberland Falls State Park 2045
--Estill County 2029
--Hancock County 1934
--Jackson Purchase Region 1956
--Lincoln County 2131
--Livingston County 1943, 2122
--McCracken County 1937
--Marshall County 1935, 1943
--Monroe County 2065
--Muhlenberg County 1926
--Olive Hill District 1972
--Pine Mountain 2125
--Pulaski County 1966
--Simpson County 1933
--Sloans Valley 1966
--Sulphur Lick 2065
--Tompkinsville 2065
--Warren County 1936
--Western Coal Field Region 1961
Louisiana
--Claiborne Parish 2147
Maine 2213, 2218
--Baxter State Park 2207
--Bryant Pond Quadrangle 2211
--Knox County 2214
--Moose River 2216

--Mount Blue State Park 2212, 2222
--Mount Katahdin 2207
--Northwestern 2216
--Presque Isle Quadrangle 2210
--Roach River 2216
--Sebago Lake State Park 2206
--Vassalboro 2215
--Waterville 2215
--Western 2219
--York County 2209
Maryland 2292
--Aberdeen 2286
--Baltimore 2286, 2305
--Coastal Plain 2243, 2286
--Piedmont 2244, 2286, 2305
--Southern 2243, 2293
Mexico 2117, 4331
Michigan 2318, 2319, 2322, 2323, 2324, 2325, 2339, 2362
--Baraga County 2382
--Barry County 2356
--Crystal Falls 2348, 2349, 2350, 2352
--Iron River 2345, 2346, 2347, 2348
--Keweenaw Peninsula 2381
--Michigan Basin 2360
--Monroe County 2354
--Northern Peninsula 2365
--Panola Plains 2351
--Rule River 2351
--Upper Peninsula 2353
--Wayne County 2344
Minnesota 2404, 2405, 2426, 2446
--Birchdale 2428
--Cook County 2395
--Cuyuna Range 2409, 2410
--Duluth 2383
--Goodhue County 2413
--Granite Falls 2441
--Indus 2428
--Isaac Lake Quadrangle 2444
--Itasca County 2431, 2432
--Koochiching County 2428
--McLeod County 2412
--Minneapolis 2389
--Montevideo 2441
--Northeastern 2443
--Rochester 2390
--St. Louis County 2444
--St. Paul 2389
--Southwestern 2441
--Tower-Soudan State Park 2387
--Vermilion District 2438
--Wabasha County 2413
Mississippi 2454, 2460, 2478, 2483, 2486, 2545

Geology (Cont.)
  Mississippi (Cont.)
    --Adams County 2540
    --Attala County 2464
    --Bolivar County 2467
    --Claibore County 2499,
      2501
    --Coahoma County 2467
    --Copiah County 2509
    --George County 2503
    --Highway 25 2461
    --Hinds County 2494, 2537
    --Holmes County 2467
    --Issaquena County 2467
    --Jasper County 2453
    --Rankin County 2520
    --Smith County 2525
    --Tallahatchie County 2467
    --Washington County 2467
    --Wayne County 2528
    --West Point Quadrangle
      2481
    --Yazoo County 2467
  Missouri 2549A, 2549B, 2549C,
      2558, 2559, 2561, 2562,
      2563, 2564, 2565, 2572,
      2583, 2593, 2594, 2606,
      2610, 2611, 2629, 2675
    --Bates County 2639
    --Creve Coeur Quadrangle
      2553
    --Highway 65 2617
    --Interstate 44 2602
    --Interstate 55 2642
    --Jefferson County 2552
    --Maxville Quadrangle 2552
    --Rolla 2584
    --St. Charles County 2595,
      2614
    --St. Francois Mountains
      2641
    --St. Louis County 2552,
      2553, 2555, 2614
    --Southeast 2624, 2638
    --Springfield 2678
  Montana 2781, 2792, 2797,
      2809, 2813, 2816, 2819,
      2821, 2822, 2826, 2831,
      2835, 2839, 2841, 2844
    --Bearmouth 2783
    --Broadwater County 2691
    --Carbon County 2832
    --Cascade 2706
    --Castle Mountains Mining
      District 2720
    --Central 2310
    --Clinton Mining District
      2690
    --Dawson County 2824

    --Drummond 2445
    --Eastern 2732
    --Elkhorn 2845
    --Flathead County 2684, 2692,
      2739
    --Foster Creek 2732
    --Gallatin County 2693
    --Garnet 2783
    --Garnet Mountain Quadrangle
      2693
    --Great Falls-Lewistown Coal
      Field 2710
    --Judith Basin 2704, 2749, 2767
    --Kalispell Valley 2726
    --Kootenai-Flathead 2692
    --Lake County 2692
    --Lincoln County 2684, 2739
    --Madison County 2786
    --Meagher County 2720, 2798
    --Missoula Basin 2699
    --Missoula County 2690
    --Northwestern 2726
    --Pryor Mountain Quadrangle
      2832
    --Ravalli County 2789
    --Sheridan District 2786
    --Stillwater River Valley 2741
    --Tobacco Valley 2741
    --Ulm 2706
    --Western 2783
    --Winston Mining District 2691
  Nebraska
    --Omaha 2856
    --Southeastern 2855
    --Western 2854
  Nevada 2902, 2926, 2937
    --Bunkerville Mining District
      2862
    --Central 2916
    --Churchill Butte Quadrangle
      2871
    --Churchill County 2884
    --Clark County 2861, 2862, 2881
    --Dead Mountains 2881
    --Douglas County 2876
    --Eldorado Mountains 2881
    --Elko County 2866, 2867
    --Esmeralda County 2879
    --Eureka County 2863
    --Humboldt County 2859
    --Klondike Quadrangle 2893
    --Lander County 2889
    --Lincoln County 2874
    --Lone Mountain Quadrangle
      2893
    --Lyon County 2876
    --Majura Hill 2887
    --Mineral County 2875
    --Mount Velma Quadrangle 2867

--Mud Lake Quadrangle 2893
--Newberry Mountains 2881
--Nye County 2878
--Ormsby County 2876
--Pamlico Mining District
2875
--Pershing County 2887,
2890
--Rowland Quadrangle 2866
--Storey County 2869
--Tonopah Quadrangle 2893
--Truckee Meadows 2924
--Wadsworth Quadrangle
2871
--Washoe County 2869, 2924
--White Pine County 2886
New Jersey
--Mercer County 3084
--Sussex County 3014, 3083
--Tocks Island 3014
--Warren County 3014
New Mexico 3101, 3110, 3117,
3125
--Albuquerque Basin 3266
--Animas Mountains 3179
--Apache Warm Springs 3157
--Big Hatchett Peak Quad-
rangle 3200
--Bottomless Lakes State
Park 3273
--Capitan 3273
--Capulin Mountain 3277
--Cebolla Quadrangle 3103
--Cedar Hills 3187
--Central 3268
--Cerro de Cristo Rey Uplift
3264
--Chama Quadrangle 3100
--Clayton 3277
--Columbus 3280
--Delaware Basin 3213
--Deming 3280
--Dona Ana County 3108,
3113, 3187, 3231
--Dona Ana Mountains 3201
--Eagle Nest 3105, 3272
--Fort Sumner 3109
--Gallinas Mountains 3106
--Good Sight Mountains 3223
--Grant County 3107, 3232
--Grants Uranium Region
3239
--Hidalgo County 3095, 3107,
3179, 3200, 3206, 3212
--High Plains 3277
--Hurley 3275
--Jarilla Mountains 3093
--Jornada Del Muerto Coal
Field 3222

--La Madera Quadrangle 3091
--Las Cruces 3280
--Lincoln County 3106, 3121,
3129
--Little Black Peak Quadrangle
3129
--Little Hatchet Mountains 3107
--Lordsburg 3280
--Luna County 3223
--Nogal Peak 3121
--Pecos Country 3256
--Peloncillo Mountains 3212
--Potrillo Basalt Field 3203
--Quay County 3229
--Raton 3277
--Red River 3272
--Rincon Quadrangle 3112
--Rio Arriba County 3091, 3102,
3103
--Roswell 3273
--Ruidoso 3273
--San Diego Mountains 3108
--San Juan Basin 3209
--Sandia Mountains 3262
--Sangre de Cristo Mountains
3235
--Santa Fe 3271
--Santa Rita 3275
--Selden 3187
--Sierra Alta Quadrangle 3113
--Silver City 3275, 3280
--Socorro County 3129, 3157,
3222
--South-Central 3203, 3268
--Southeast 3241, 3256
--Taos 3272
--Torrance County 3106
--Valencia County 3227
--Walnut Springs Quadrangle
3095
--Zuni Mountains 3274
North Carolina 3288, 3294, 3303
--Chapel Hill Quadrangle 3304
--Coast 3289
--Continental Shelf 3289
--Orange County 3287
--Wake County 3292
North Dakota 3408, 3420, 3450,
3458, 3459, 3465, 3499
--Adams County 3378
--Barnes County 3319. 3320,
3321
--Benson County 3362
--Bismarck 3414, 3500
--Bowman County 3378
--Burke County 3351, 3352
--Burleigh County 3316, 3317,
3318, 3452
--Cass County 3331, 3332, 3333

Geology (Cont.)
Mississippi (Cont.)
--Cavalier County 3369
--Dickey County 3390
--Divide County 3325, 3326,
3327
--Dunn Center 3491
--Eddy County 3322, 3323,
3324, 3441
--Foster County 3322, 3323,
3324, 3441
--Grand Forks 3419, 3454
--Grand Forks County 3347,
3348, 3442, 3349
--Griggs County 3373
--Highway 10 3418
--Interstate 94 3398
--Jamestown 3417
--Knife River Basin 3494
--Lake Agassiz Plain 3429
--Langdon 3480
--Lamoure County 3390
--McKenzie County 3416
--McLean County 3365
--Mandan 3414, 3500
--Mercer County 3355
--Minot 3412
--Mountrail County 3353, 3354
--Nelson County 3358
--North-Central 3404
--Northeastern 3399, 3449
--Northwest 3405
--Oliver County 3355
--Park River 3419
--Pembina County 3369
--Prairie 3407
--Ransom County 3388
--Renville County 3341, 3342
--Richland County 3328,
3329, 3330
--Rolette County 3361
--Sargent County 3388
--South-Central 3403, 3440
--Southwest 3406
--Steele County 3373
--Stutsman County 3313,
3314, 3315
--Theodore Roosevelt National
Memorial Park 3401
--Traill County 3337, 3338
--Valley City 3411
--Walsh County 3358, 3442
--Ward County 3341, 3342
--Wells County 3343, 3344,
3345
--West-Central 3493
--Williams County 3334,
3335, 3336, 3416
--Williston 3416

--Williston Basin 3438, 3485,
3487, 3488
Ohio 3508, 3515, 3527, 3531,
3532, 3543
--Adams County 3596
--Ashland County 3602
--Ashtabula County 3613
--Berlin Heights Quadrangle
3561
--Caldwell North Quadrangle
3573
--Carroll County 3566, 3569,
3572, 3588
--Carrollton Quadrangle 3572
--Champaign County 3612
--Clermont County 3578
--Clinton County 3568
--Columbiana County 3566
--Columbus 3516
--Dellroy Quadrangle 3569
--East Liberty Quadrangle 3567
--Erie County 3561
--Garrettsville Quadrangle 3555
--Guernsey County 3573
--Hamilton County 3578, 3584,
3594, 3595
--Highland County 3593, 3596
--Highway Route 23 3516
--Hocking County 3564
--Hocking Hills State Park
3526
--Holmes County 3592
--Huron County 3561
--Jefferson County 3588
--Kensington Quadrangle 3556
--Licking County 3585
--Logan County 3567, 3570
--Madeira Quadrangle 3578
--Malvern Quadrangle 3558
--Minerva Quadrangle 3566
--Moreland Oil Pool 3547
--Muskingum County 3585
--Noble County 3573
--Pike County 3596
--Portsmouth 3516
--Richland County 3589
--South Bloomington Quadrangle
3564
--Stark County 3501, 3566
--Summit County 3599
--Trumbull County 3581
--Union County 3567, 3570
--Vermilion West Quadrangle
3561
--Vinton County 3564
--Washington County 3509
--Wayne County 3547, 3562,
3563
--West Mansfield Quadrangle 3570

Oklahoma 3685, 3686, 3687
--Adair County 3667
--Arbuckle Mountains 3694
--Bryan County 3653
--Choctaw County 3647
--Craig County 3621
--Custer County 3638, 3639,
   3640
--Ellis County 3668
--Greenwood Quadrangle 392
--LeFlore County 3625,
   3664, 3676
--Latimer County 3664
--LaVaca Quadrangle 395
--Love County 3660, 3661,
   3662
--McIntosh County 3633,
   3634, 3635
--Muskogee County 3649
--Noble County 3655
--Pawnee County 3659
--Van Buren Quadrangle 395
--Winding Stair Range 3625,
   3664
--Woods County 3628
Oregon 3699, 3709, 3720, 3739,
   3752, 3757
--Bend Area 3713
--Benton County 3740
--Broken Top Quadrangle
   3771
--Clackamas County 3724,
   3741
--Clatsop County 3716, 3721
--Coast 3711, 3762
--Coos County 3722, 3729
--Crook County 3700
--Curry County 3732, 3735
--Douglas County 3717, 3729
--Eastern 3715
--Grant County 3700
--Harney County 3700
--Hood River 3733
--Jackson County 3736
--Josephine County 3742
--Klamath County 3708
--La Grande 3775
--Lake County 3708
--Lane County 3727
--Lincoln County 3723
--Linn County 3726
--Mount Hood 3777
--Multnomah County 3724
--Northern 3719
--Sherman County 3733
--Suplee 3700
--Tillamook County 3716,
   3721
--Tuatatin Valley 3702

--Upper Chetco Drainage 3730
--Wasco County 3733
--Western 3712, 3743, 4670
--Wheeler County 3714
Pennsylvania 3779, 3780, 3782,
   3787, 3792, 3812, 3856, 3883
--Amity Quadrangle 3873
--Berks County 3955
--Butler County 3927
--Claysville Quadrangle 3873
--Cornwall 3875
--Crawford County 3957
--Dauphin County 3935
--Delaware Valley 3811
--Endless Mountains 3838
--Erie County 3828
--Forest County 3876
--Greene County 3895
--Harrisburg 3784
--Horseshoe Curve 3800
--Jefferson County 3878
--Lawrence County 3926, 3927
--Lawrence Quadrangle 3927
--McKean County 3881
--Mercer County 3926, 3944
--Monroe County 3958
--Neshannock Quadrangle 3926
--Northern 3838
--Ohio Pyle State Park 3904
--Pennsylvanian Piedmont 3796
--Philadelphia 3791
--Pittsburgh 3810
--Southeastern 3925
--Stroudsburg Quadrangle 3808
--Venango County 3876
--Warren County 3881
--Washington County 3806, 3807
--York 3786
--York County 3786
--Youngsville Quadrangle 3872
South Carolina 3988
--Carolina Barite Belt 3969
--Coastal Plain 3973
--Little Mountain 3986
--Oconee County 3983
--York County 3982
South Dakota 4027, 4057
--Beadle County 4001, 4002
--Bridger 4046
--Bon Homme County 4012,
   4013
--Brown County 4021
--Campbell County 4008, 4009
--Charles Mix County 4015,
   4016
--Clay County 4004, 4005
--Day County 4020
--Douglas County 4015, 4016
--Edmunds County 4024, 4025

Geology (Cont.)
  South Dakota (Cont.)
    --Fairburn Quadrangle 4054
    --Faulk County 4024, 4025
    --McPherson County 4024,
       4025
    --Marshall County 4018,
       4019
    --Sanborn County 3998,
       3999
    --West-Central 4046
  Tennessee 4111, 4112, 4115,
     4118, 4125, 4129
    --Bellevue Quadrangle 4131
    --Boone Lake 4156
    --Carter County 4157
    --Cedars of Lebanon State
       Park 4161
    --East 4144
    --Hamilton County 4123
    --Interstate Highways 4160
    --Kingston Springs Quadrangle
       4126
    --Knox County 4114
    --Memphis 4157
    --Sullivan County 4156
    --West 4157
    --Wilson County 4161
  Texas 4166, 4177
    --Austin 4238, 4247, 4250,
       4321
    --Big Bend 4240, 4283
    --Brewster County 4283
    --Coast 4178, 4183
    --East 4200, 4333
    --Gulf 4245
    --Llano 4238, 4247
    --Longhorn Cavern 4237
    --Padre Island National Sea-
       shore 4251
    --Palo Duro Basin 4220,
       4230
    --Palo Duro Canyon 4241
    --Red River Basin 4284
    --South 4246, 4252
    --State Parks 4243
    --Trans-Pecos 4253, 4331
    --University of Texas 4169
    --West 4173
  United Kingdom 2116
  Utah 4359, 4376, 4424,
     4445
    --Bear Lake 4369
    --Beaver County 4396, 4438
    --Black Mountain 4396
    --Box Elder County 4388
    --Church Hills 4439
    --Desert Mountain 4434
    --Emery County 4386

    --Garfield County 4380, 4408
    --Iron County 4396
    --Jordan Valley 4456
    --Juab County 4350
    --Milford Flat Quadrangle 4438
    --Millard County 4439
    --Morgan County 4366
    --Northern 4355
    --Rich County 4369
    --Rush Valley 4446
    --Salt Lake City 4401, 4430
    --Salt Lake County 4344, 4387
    --Shauntie Hills 4396
    --Star District 4438
    --Tooele County 4350, 4446
    --Tooele Valley 4446
    --Uintah County 4345
  Vermont 4484
    --Averill Quadrangle 4480
    --Barre 4493
    --Brandon 4494, 4501
    --Brattleboro 4498
    --Brunswick 4475
    --Burke Quadrangle 4481
    --Burlington 4495
    --East Monkton 4487
    --Elizabeth Mine 4489
    --Enosburg 4476
    --Guildhall 4475
    --Hardwick 4477, 4496
    --Island Pond 4473
    --Johnson 4496
    --Lunenburg 4475
    --Middlebury 4495
    --Milton 4497
    --Milton Quadrangle 4479
    --Montpelier 4493
    --Pawlet Quadrangle 4483
    --Randolph Quadrangle 4474
    --Rutland 4494
    --St. Albans 4497
    --Ticonderoga Quadrangle 4501
    --Windsor 4498
    --Woodstock Quadrangle 4482
  Virginia 4521, 4526, 4537, 4544
    --Abingdon Quadrangle 4553
    --Arnold Valley Quadrangle
       4567
    --Ashby Gap Quadrangle 4590
    --Augusta Springs Quadrangle
       4575
    --Bacons Castle Quadrangle
       4572
    --Bassett Quadrangle 4580
    --Benns Church Quadrangle
       4571
    --Berryville Quadrangle 4588
    --Blacksburg Quadrangle 4551
    --Blairs Quadrangle 4538

--Bluemont Quadrangle 4568
--Bon Air Quadrangle 4555
--Bowers Hill Quadrangle 4582
--Bristol Quadrangle 4579
--Buchanan Quadrangle 4567
--Buckingham Quadrangle 4569
--Chuckatuck Quadrangle 4571
--Churchville Quadrangle 4566
--Clinchport Quadrangle 4559
--Cluster Springs Quadrangle 4542
--Coastal Plain 4507, 4508, 4509, 4510, 4516
--Conicville Quadrangle 4589
--Craigsville Quadrangle 4575
--Crimora Quadrangle 4550
--Daleville Quadrangle 4596, 4597, 4598
--Danville Quadrangle 4538
--Deerfield Quadrangle 4575
--Dillwyn Quadrangle 4564
--Eagle Rock Quadrangle 4578
--Edinburg Quadrangle 4589
--Elliott Knob Quadrangle 4575
--Flint Hill Quadrangle 4600, 4601
--Fluvanna County 4503
--Fort Defiance Quadrangle 4549
--Frederick County 4504
--Front Royal Quadrangle 4594
--Greene County 4502
--Greenfield Quadrangle 4541
--Greenville Quadrangle 4566
--Grottoes Quadrangle 4547
--Hampton Quadrangle 4595
--Hog Island Quadrangle 4572
--Holston Valley Quadrangle 4553
--Hylas Quadrangle 4577
--Lee County 2054
--Leesburg Quadrangle 4565
--Lincoln Quadrangle 4568
--Linden Quadrangle 4600
--Madison County 4502
--Martinsville Quadrangle 4570
--Massies Corner Quadrangle 4554

--Midlothian Quadrangle 4577
--Millboro Quadrangle 4562
--Mount Hermon Quadrangle 4538
--Mount Sidney Quadrangle 4548
--Mulberry Island Quadrangle 4595
--Natural Bridge Quadrangle 4567
--Newport News Quadrangle 4582, 4595
--Norfolk Quadrangle 4545, 4546
--Oak Grove Core 4557
--Omega Quadrangle 4542
--Oriskany Quadrangle 4578
--Page County 4505
--Parnassus Quadrangle 4573
--Philpott Reservoir Quadrangle 4576
--Poquoson Quadrangle 4584
--Price Quadrangle 4587
--Richmond Quadrangle 4592
--Ringgold Quadrangle 4538
--Salem Quadrangle 4591
--Salisbury Quadrangle 4578
--Seven Pines Quadrangle 4592
--Shenandoah National Park 4513
--Sherando Quadrangle 4541
--Snow Creek Quadrangle 4587
--South Boston Quadrangle 4542
--Spray Quadrangle 4587
--Staunton Quadrangle 4566
--Stephenson Quadrangle 4588
--Stokesville Quadrangle 4573
--Strasburg Quadrangle 4602
--Strom Quadrangle 4578
--Stuarts Draft Quadrangle 4566
--Studley Quadrangle 4592
--Sugarloaf Mountain Quadrangle 4567
--Toms Brook Quadrangle 4602
--Vesuvius Quadrangle 4561
--Virgilina Quadrangle 4542
--Wallace Quadrangle 4579
--Warren County 4552
--Waynesboro Quadrangle 4540
--Williamsburg Quadrangle 4572
--Windsor Quadrangle 4571
--Wolf Gap Quadrangle 4589
--Woodstock Quadrangle 4589
--Wyndale Quadrangle 4553
--Yellow Tavern Quadrangle 4592
--Yorktown Quadrangle 4584
Washington 4611, 4633, 4652, 4653, 4654, 4655, 4656, 4666, 4668

Geology (Cont.)
Washington (Cont.)
--Adams County 4623
--Cathlamet 4606
--Chinook Highway 4648
--Coast 4618, 4624
--Cowlitz County 4606
--East-Central 4623
--Ellensburg Highway 4648
--Franklin County 4623
--Grant County 4623
--Grays Harbor County 4608
--Kelso 4606
--King County 4615
--Loomis Quadrangle 4616
--Methow Valley 4620
--Okanogan County 4616,
4620
--Puget Lowland 4635
--Puget Sound 4659
--Snoqualmie Highway 4626
--Southern 3719
--Southwestern 3743
--Stevens Highway 4626
--Swauk Highway 4626
--Wahkiakum County 4606
--Western 4670
--White Pass Highway 4648
--Wynoochee Valley Quad-
rangle 4608
West Virginia 4683, 4708,
4716, 4726, 4758
--Blackwater Falls State
Park 4757
--Canaan Valley State Park
4757
--Charleston 4682
Western United States 2118
Wisconsin 4778, 4779, 4780,
4798
--Baraboo District 4799,
4800, 4801
--Columbia County 4825
--Jefferson County 4821
--Lake Beulah 4830
--Ozaukee County 4826
--St. Croix County 4820
--Southern 4802
--Walworth County 4822,
4830
--Washington County 4826
--Western 4794, 4795, 4796
Wyoming 4835, 4837, 4839,
4845, 4870, 4880
--Albany County 4860
--Arlington 4862
--Carbon County 4860
--Laramie Basin 4858
--Laramie County 4861

--Lincoln County 4864
--Morgan Quadrangle 4860
--Powder River Coal Basin
4878
--Rex Lake Quadrangle 4858
--Salt River Range 4864
--Silver Crown Mining District
4861
--Strouss Hill Quadrangle 4860
--Wagonhound Creek 4862
--Washakie County 4848
Geometrics 1126, 1130
Geomorphology
California 424
Colorado
--Larimer County 606
Florida
--Peninsula 745
Indiana
--Crawford Upland 1535
--Mitchell Plain 1535
--Southern 1535
Kentucky 2135
Minnesota 2396
Missouri
--Ozarks 2605
New Jersey
--Inner New Jersey Shelf 3063
New Mexico
--Southern 3242
Oklahoma 3685
South Dakota
--Eastern 2396
Tennessee
--Knox County 4114
Geophysics 46, 2251, 3880, 3923
Alabama 123
--Piedmont 100, 123
California
--Amador County 558
--Calaveras County 558
--Clear Lake Region 557
--Contra Costa County 458
--El Cerrito 458
--French Gulch Quadrangle 502
--Ione 558
--Richmond 458
--Sacramento County 558
--San Fernando 430
--San Pablo 458
--Shasta County 502
--Trinity County 502
Colorado 589, 627
Delaware
--Coastal Plain 728
--Middletown 707
--Odessa 707
--Piedmont 719
Florida

--Jacksonville 781
Hawaii
  --Lanai Island 981
  --Kahoolawe Island 981
  --Oahu Island 980
Illinois
  --Macomb 1380
Kentucky
  --Eastern 2072
  --Muhlenberg County 1939
  --New Cypress Pool 1939
Maryland
  --Southern 2285
Michigan
  --Dickinson County 2397
Minnesota 2411
Montana 2781, 2809, 2813,
  2819, 2821, 2822, 2826,
  2831, 2835, 2839, 2841,
  2844
New Mexico 3173
North Dakota
  --Dickey County 3391
  --Lamoore County 3391
  --McIntosh County 3396
  --Morton County 3395
  --Ramsey County 3393
Tennessee
  --Wells Creek Structure
    4112
George County, Mississippi
  Geology 2503
  Mineral Resources 2503
Geoscience
  New Mexico 3182, 3190
Geothermal Activity
  Nevada
    --Truckee Meadows 2927
    --Washoe County 2927
Geothermal Data
  Montana 2842
Geothermal Energy
  California 424
  Colorado 589, 623
  New Mexico 3134
Geothermal Exploration
  Nevada 2923
Geothermal Maps
  Montana
    --Madison Group 2828
Geothermal Nature
  Florida
    --Florida Plateau 888
Geothermal Resources
  Colorado 593, 612, 614,
    617, 636
    --Pagosa Hot Springs 645
  Idaho 1170
  Mexico 4331

Montana 2779
Oregon 3747, 3749, 3764
South Dakota 4063
Texas 4189
  --Gulf Coast 4197, 4201, 4205,
    4223, 4326
  --South 4190
  --Trans-Pecos 4331
Utah 4404
Washington 4639, 4640
Wyoming 4865
Geothermics
  New Mexico 3152
Gering Formation
  Nebraska
    --Western 2854
Germanium
  California 424
Gettysburg, South Dakota
  Groundwater 4089
Gettysburg Formation
  Pennsylvania
    --Southeastern 3960
Geysers Geothermal Area, California
  Engineering Geology 563
  Geology 583
  Slope Stability 583
Gifford Pinchot State Park, Pennsyl-
  vania
  Guide 3907
Gila River Indian Reservation, Arizona
  Mineral Resources 350
Gilbertown, Alabama
  Gamma Radiation 167
Gilbertown Oil Field, Alabama
  Subsurface Geology 93
Gilchrist County, Florida
  Geology 743
Gilgai Structure
  Texas 4196
Gilroy, California
  Shale 507
Gipsy Limestone County, Alabama
  Foraminifera 58
  Kinderhookian 58
  Mississippian 58
Girtyoceratidae 3666
Glacial Aquifers
  Illinois 1289
Glacial Drift
  Illinois 1272, 1346
    --Batavia 1312
    --Danville 1313
    --East-Central 1265
    --Mahomet Bedrock Valley
      1265
    --Shelbyville 1315
  Iowa
    --Northeast 1639

Glacial Drift (Cont.)
  Minnesota
    --Red River Valley 3462
  Missouri
    --Andrew County 2659
    --Buchanan County 3658
    --Carroll County 2657
    --Chariton County 2656
    --Davies County 2653
    --DeKalb County 2652
    --Gentry County 2651
    --Grundy County 2645
    --Harrison County 2647
    --Linn County 2655
    --Livingston County 2650
    --Mercer County 2646
    --North 2580
    --Putnam County 2609,
      2648
    --Sullivan County 2654
    --Worth County 2649
  North Dakota
    --Logan County 3425
    --Red River Valley 3462
  Pennsylvania
    --Northeastern 3822
Glacial Geology
  Illinois
    --Champaign County 1362
    --Vandalia 1298
  Michigan
    --Monroe County 2354
    --Wayne County 2344
  Wisconsin 4798
    --Baraboo District 4800
Glacial History
  Minnesota
    --Northeastern 2447
Glacial Lake Deposits
  Indiana
    --Southern 1590
Glacial Lake Sediments
  Indiana
    --Bedford 1556
    --Salt Creek Valley 1556
Glacial Lakes
  Illinois
    --Lake Douglas 1322
  Maine
    --Farmington 2205
  Michigan 2315
Glacial Marine Clays
  Maine
    --Farmington 2205
Glacial Outwash
  Illinois
    --Northwestern 1296
Glacial Pollen
  Iowa

    --Sumner Bog 1653
Glacial Stratigraphy
  Indiana
    --West-Central 1566
Glacial Tills
  Illinois 1203, 1256
    --Lake Michigan 1445
    --Northwestern 1293
Glaciation
  Montana
    --Bitterroot Range 2787
    --Lake Missoula 2787
  North Dakota 3420, 3474
    --Barnes County 3719
    --Benson County 3362, 3364
    --Burke County 3351, 3353
    --Burleigh County 3316
    --Cavalier County 3369
    --Dickey County 3390
    --Divide County 3325
    --Eddy County 3322
    --Foster County 3322
    --Grand Forks County 3347,
      3349
    --Griggs County 3373
    --Lamoure County 3390
    --McLean County 3365, 3367
    --Mercer County 3355, 3357
    --Montrail County 3353
    --Nelson County 3358, 3360
    --Oliver County 3355, 3357
    --Pembina County 3369
    --Pierce County 3362, 3364
    --Ransom County 3388
    --Rolette County 3361
    --Sargent County 3388
    --South-Central 3440
    --Steele County 3373
    --Stutsman County 3313
    --Traill County 3237, 3339
    --Walsh County 3358, 3360
    --Ward County 3342
    --Wells County 3343, 3345
    --Williams County 3334, 3336
  Ohio
    --Ashland County 3602
    --Ashtabula County 3613
    --Carroll County 3569
    --Champaign County 3612
    --Clinton County 3568
    --Dellroy Quadrangle 3569
    --East Liberty Quadrangle 3567
    --Highland County 3593
    --Holmes County 3592
    --Licking County 3560
    --Logan County 3567, 3570
    --Richland County 3589
    --Stark County 3501
    --Trumbull County 3581

--Union County 3567, 3570
--Wayne County 3563
--West Mansfield Quadrangle
  3570
West Virginia
--Morgantown 4688
Glacier County, Montana
Water 2716
Glaciers
Alaska
--Gulkana Glacier 280
--Nelchina Glacier 326
--West Fork Glacier 340
Illinois
--Western 1220
Ohio 3512
Glades County, Florida
Geology 824
Groundwater 824
Glass Manufacture
South Carolina 3972
Glauconite
Texas
--East 4242
Glen Allen, Missouri
Kaolin 2585
Glen Dean Limestone
Indiana
--Southern 1486, 1494
Kentucky 1486, 1494
Glen Rose Formation
Texas
--Paluxy River Valley 4171
--Somervell County 4171
Glenarm Series
Appalachian Piedmont 2284
Glenwood Springs, Colorado
Debris-Flow Hazards 616
Glorieta Sandstone
New Mexico
--Chaves County 3176
--Eastern 3219
Glossaries
Building Stone 1561
Masonry 1561
Glynn County, Georgia
Aquifers 926
Contamination 926
Test Drilling 929
Water Sampling 929
Gneiss
Minnesota 2419
Gobernadora Quadrangle, California
Fossils 449
Geology 449
Landslides 449
Test Holes 449
Gogebil County, Michigan

Groundwater 2379
Gold 961, 2715
Alabama
--Devils Backgone Area 125
--Tallapoosa County 125
Alaska 307, 315
--Chugach Mountains 326
--Nelchina Glacier 326
--Talkeetna Mountains 324
California 424, 426
--Calaveras County 428
--San Andreas Quadrangle 428
North Carolina 3299
Oregon 3703, 3750
South Carolina 3981
Tennessee
--Coker Creek District 4116
--Monroe County 4116
Texas 4270
Virginia 4556
Washington 4651
Wyoming
--Albany County 4850, 4854,
  4893
Gold Butte Vermiculite
Nevada
--Clark County 2913
Gold Camps 1121
Gold Hill Mining District, Utah
Mineralization 4356
Goldfield Mining District, Nevada
  2920
Goleta Mountain, California
Geology 409
Gonzales, Louisiana
Groundwater 2167
Good Sight Mountain, New Mexico
Geology 3223
Goodhue County, Minnesota
Clay 2413
Geology 2413
Goose Creek Coal Field, Utah 4391
Goose Egg Formation
Wyoming 4859
Gordon County, Georgia
Geology 954
Groundwater 954
Goshen Bay, Utah
Hydrogeology 4472
Goshen County, Wyoming 4852
Government Creek Valley, Utah
Water 4458
Graben
Connecticut
--South-Central 667
--Southern 696
Grace Mine, Pennsylvania
Magnetite 3862

Gradyville East Field, Kentucky
  Economics 2101
  Geology 2101
Grand County, Utah
  Oil 4423, 4429
  Sandstone 4423, 4429
Grand Falls Chert
  Missouri
    --Southwestern 2618
Grand Forks, North Dakota
  Floods 3409, 3445
  Geology 3419, 3454
Grand Forks County, North Dakota
  Aquifers 3349
  Cement Rock 3478
  Cretaceous 3349
  Esker 3442
  Geology 3347, 3348, 3349, 3442
  Glaciation 3347, 3349
  Groundwater 3347, 3348, 3349
  Ordovician 3349
  Springs 3348
  Water 3348
  Wells 3348
Grand Tower Limestone
  Illinois
    --Southern 1245
Graneros Shale
  Kansas
    --Central 1713
Granite
  California
    --Coast Ranges 419
    --Klamath Mountains 415
    --Mariposa County 512
    --Yosemite Valley 512
  Minnesota 2419
  Nevada 2900
  South Carolina 3995
Granite Complex
  Nevada
    --Clark County 2881
    --Dead Mountains 2881
    --Eldorado Mountains 2881
    --Newberry Mountains 2881
Granite County, Montana
  Limestone 2807
Granite Falls, Minnesota
  Geology 2441
  Precambrian Rock 2441
Grant County, Kansas
  Geohydrology 1692
  Water Levels 1856, 1875,
    1848
Grant County, New Mexico
  Geology 3107, 3232
  Minerals 3094
  Pennsylvanian 3185
  Water 3232

Grant County, North Dakota
  Groundwater 3383, 3384
  Water 3383
  Wells 3383
Grant County, Oregon
  Geology 3700
Grant County, Washington
  Geology 4623
Grantham Talc Mine, Nevada 2932
Grants, New Mexico
  Uranium 3116, 3239
Graphite
  California 424
Gravel
  Alabama 102
  California 424
    --Sacramento 562
  Colorado
    --Front Range 639, 640
  Florida 877
  Illinois
    --Boone County 1273
    --DeKalb County 1223
    --Macon County 1302
    --Mason County 1320
    --Northeastern 1215, 1359
    --Peoria County 1237
    --Rock River 1270
    --Sangamon County 1308
    --Tazewell County 1255
  Indiana 1505
    --Johnson County 1595
    --Shelby County 1595
  Mississippi 2542
  Missouri
    --Putnam County 2609
  Ohio
    --Madison County 3518
    --Maumee River 3577
  Pennsylvania
    --Northwestern 3858
    --Pittsburgh 3886
  South Dakota
    --Bon Homme County 4033
    --Charles Mix County 4036
    --Douglas County 4036
Gravimeter Observations 1578
Gravity 1577
  Alabama
    --Baldwin County 118
    --Clarke County 112
    --Lamar County 145
    --Limestone County 146
    --Marengo County 117
    --Marion County 119
    --Mobile Delta 132
    --Monroe County 109
    --Wilcox County 120
    --Winston County 121

Alaska
--Beluga Basin 284
--Cook Inlet 284
--South-Central 284
California 430, 510
--Northern 422
Georgia
--Piedmont 948
Illinois 1350
--DeWitt County 1225
--McLean County 1225
--Southeastern 1306
Iowa
--Fayette County 1648
Kansas
--Linn County 1761
Michigan
--Southern Peninsula 2342,
2355
Minnesota
--Seismic Studies 2422
Montana 2825
Nevada
--Warm Springs Valley 2912
--Washoe County 2912
New Jersey 3086
New Mexico
--Socorro County 3145
North Carolina 3296
North Dakota
--Central 3469, 3470
South Dakota
--Southeastern 4047
Tennessee
--Knox County 4114
Texas
--Presidio 4235
Utah 4365
--Desert Mountain 4368
--Juab County 4368
--San Francisco Mountains
4371
--Southwestern 4371
Virginia 4581, 4583, 4599
--Albemarle County 4574
--Eastern Shore 4586
--Fluvanna County 4574
--Northeastern 4586
--Southeastern 4593
--Southwestern 4543
West Virginia
--Allegheny Plateau 4743
--Western Valley and Ridge
4743
Wisconsin
--Seismic Studies 2422
Wyoming
--Carbon County 4853
Gravity Anomalies

Idaho 1152
Illinois
--Northern 1210
Kentucky
--Western 2038
Gravity Base Station Network
Georgia 905
Illinois 1254
Iowa 1645
Utah 4365
Virginia 4524
Washington 4657
Gray County, Kansas
Irrigation Water 1786
Grays Harbor County, Washington
Black Sands 4671
Geology 4608
Grayson County, Kentucky
Coal 1938
Great Basin
Minerals 2938
Great Basin, California
Geology 416, 424
Minerals 416
Great Bend Prairie, Kansas
Geohydrology 1826
Hydrogeology 1816
Great Falls-Lewistown Coal Field,
Montana
Economic Geology 2710
Stratigraphy 2710
Great Plains
Quaternary 2903
Uranium 1811
Great Salt Lake, Utah 4389
Brine 4432, 4449, 4450, 4461,
4463, 4466, 4467, 4469, 4471
Brine-Sediment Interface 4427
Clay Mineralogy 4427
Dikes 4470
Evaporation 4453, 4454, 4455,
4464
Minerals 4449, 4450
Salt 4467, 4470
Water 4470
Great Valley, California
Geology 418, 424, 425
Hydrogeology 418
Minerals 418
Subsidence 418
Great Valley, Pennsylvania
Caves 3820
Greeley County, Kansas
Groundwater 1814, 1824
Green County, Kentucky
Oil 2003
Green Pond Outlier 3088

Green River Formation 4846
Utah
--Uinta Basin 4373, 4418
Green Swamp, Florida
Hydrology 829
Greenbrier County, West Virginia
Hydrology 4684
Limestone Karst 4684, 4735
Greenbrier Limestone
West Virginia
--Barbour County 4679
--Upshur County 4679
Greene County, Alabama
Geology 33
Groundwater 33
Greene County, Georgia
Geology 941
Mineral Resources 941
Greene County, Indiana
Acid Mine Conditions 1575
Coxville Sandstone 1575
Early Pennsylvanian 1489
Linton Formation 1575
Plummer Field 1608
Pyrite 1575
Greene County, Pennsylvania
Clay 3839
Coal 3895
Shale 3839
Greene County, Virginia
Geology . 4502
Minerals 4502
Greenfield Quadrangle, Virginia
Geology 4541
Greenhorn Limestone
Kansas 1755, 1771
Greensand Filters 735
Greensburg Consolidated Oil Pool,
Kentucky 2003
Greenville County, South Carolina
Groundwater 3987
Greenville Quadrangle, Virginia
Geology 4566
Greenwood, Alabama
Sinkholes 141
Greenwood, Delaware
Cretaceous 726
Tertiary 726
Test Wells 726
Greenwood Quadrangle, Arkansas
Geology 392
Greenwood Quadrangle, Oklahoma
Geology 392
Gregory, South Dakota
Groundwater 4094
Grid and Contour Random Data
4522
Griggs County, North Dakota
Geology 3373

Glaciation 3373
Groundwater 3374, 3375
Water 3374
Wells 3374
Grimesland, North Carolina
Upper Miocene Foraminifera 3306
Groton, South Dakota
Groundwater 4103
Grottoes Quadrangle, Virginia
Geology 4547
Ground Displacement
California
--San Fernando Valley 430
Ground Motion
California
--San Fernando Valley 431
Ground Subsidence
Colorado
--Boulder-Weld Coal Field 605
Groundwater 1109, 1161, 1164, 1390,
1726, 2251, 2311, 2574, 3066,
3074, 3930, 4318, 4654
Alabama 81, 95, 154
--Athens Area 18
--Birmingham Valley 161
--Bryce State Hospital 83
--Cahaba Valley 161
--Cherokee County 26
--Colbert County 163
--Escambia County 21
--Franklin County 19
--Greene County 33
--Huntsville Area 82
--Jefferson County 161
--Lauderdale County 162
--Lawrence County 25
--Limestone County 164
--Marshall County 32
--Montgomery Area 17
--Morgan County 23
--Negro Colony 83
--Pickens County 30
--Piedmont Area 100, 123
--Russell County 22
--Russellville Area 24
--St. Clair County 20
--Stevenson Area 134
--Talladega County 28
--Tuscaloosa County 83
Arkansas
--Bradley County 403
--Calhoun County 403
--Coastal Plain 404
--Northeastern 405
--Ouachita County 403
California
--San Fernando Valley 430
Colorado 587, 638
--Boulder County 596

--Crystal Valley 604
--Eagle County 604
--Garfield County 604
--Gunnison County 604
--Jefferson County 590
--Pitkin County 604
--Roaring Fork Valley 604
Delaware 712, 716, 733
--Atlantic Seashore 720
--Coastal Plain 726
--New Castle County 705
--Newark 721, 730
--Piedmont 730
Florida 813
--Alafia River Basin 836
--Baker County 839
--Charlotte County 769
--Collier County 767
--Desoto County 769, 870
--Duval County 830
--Flagler County 753, 819
--Glades County 824
--Hallandale 794
--Hardee County 769, 870
--Hendry County 824
--Hillsboro Canal 842
--Hollywood 864
--Immokalee 767
--Leon County 741
--Nassau County 830
--Polk County 831
--Putnam County 753, 819
--St. Johns County 753, 819
--Southeastern 842
--Tampa 861
Georgia 922, 955
--Bainbridge 938
--Broad Quadrangle 897
--Catoosa County 934
--Chattooga County 933
--Dade County 932
--Dawson County 936
--Floyd County 945
--Gordon County 954
--Liberty County 952
--McIntosh County 952
--Mitchell County 930
--Murray County 954
--Polk County 945
--Rockdale County 939
--Thomas County 940
--Walker County 935
--Whitfield County 955
Hawaii
--Coastal Plain 1114
--Haiku 1012
--Hawaiian Islands 984,
985, 992, 993
--Kahaluu 1012

--Kauai Island 988, 1114
--Lanai Island 981
--Kahoolawe Island 981
--Maui Island 982
--Molokai Island 986, 1011,
1050
--Niihau Island 987
--Oahu Island 976, 980, 1007,
1012
--Waianae 1007
Idaho 1163, 1178
--Caribou County 1179
--Little Long Valley 1179
--Moscow Basin 1173
--Snake River Plain 1168
--Upper Dry Valley 1179
Illinois 1387
--Alexander 1461
--Galesburg 1484
--Kewanee 1484
--Macomb 1380
--Monmouth 1484
--Rock Island 1484
--Union County 1461
Indiana
--Crawford Upland 1535
--Mitchell Plain 1535
--Southern 1535
Iowa
--Cerro Gordo County 1673
--Decatur County 1667
--Floyd River Basin 1676
--Linn County 1674
--Wayne County 1668
Kansas 1691, 1704, 1712, 1719,
1721, 1787, 1814, 1817
--Allen County 1741
--Brown County 1723
--Central 1752
--Cloud County 1730
--Decatur County 1742
--Ellsworth County 1752
--Greeley County 1824
--Jewell County 1730
--Johnson County 1758
--Kansas River Valley 1767,
1867
--Lane County 1823
--Linn County 1737
--Miami County 1718
--Montgomery County 1818
--Neosho County 1720
--Pratt County 1764
--Republic County 1730
--Republican River 1730
--Rush County 1769
--Scott County 1823
--Southeastern 1818
--Topeka 1867

Groundwater (Cont.)
  Kansas (Cont.)
    --Trego County 1705
    --Wallace County 1678
    --Wichita County 1824
  Kentucky
    --Bryantsville Quadrangle
      2129
    --Central 2129
  Louisiana 2161
    --Ascension Parish 2167
    --Avoyelles Parish 2175
    --Baton Rouge 2162
    --Calcasieu Parish 2187
    --Geismar 2167
    --Gonzales 2167
    --Iberville Parish 2176
    --Monroe 2188
    --Morehouse Parish 2179
    --New Orleans 2169
    --Norco 2178
    --Plaquemine 2176
    --Southwestern 2170, 2181,
      2183, 2202
    --White Castle 2176
  Maryland
    --Anne Arundel County 2232
    --Cambridge 2295
    --Easton 2295
    --Harford County 2231,
      2238
    --Ocean City 2302
    --Piedmont 2288
    --Prince George's County
      2235
    --Worcester County 2230,
      2302
  Michigan
    --Alger County 2372
    --Baraga County 2382
    --Battle Creek 2375
    --Dickinson County 2376
    --Gogebic County 2379
    --Iron County 2378
    --Keweenaw Peninsula 2381
    --Menominee County 2373
    --Monroe County 2354
    --Ontonagon County 2380
    --Van Buren County 2374
    --Wayne County 2344
  Minnesota 2417, 2435
  Mississippi
    --Jasper County 2453
  Missouri 2550, 2556, 2607
    --Atchison County 2662
    --Holt County 2661
    --Joplin 2668
    --Macon County 2620
    --Nodaway County 2660
    --Putnam County 2609
    --Randolph County 2620
    --Saline County 2670
  Montana 2800
    --Cascade 2706
    --Cedar Creek Anticline 2785
    --Eastern 2830, 2843
    --Fort Union Coal Region 2843
    --Judith Basin 2704
    --Kalispell Valley 2726
    --Libby 2773
    --Meagher County 2798
    --Missoula Basin 2699
    --Northwestern 2726, 2773
    --Powder River Valley 2722
    --Southeastern 2722
    --Ulm 2706
  Nevada
    --Southern 2922
  New Jersey 2994, 3073
    --Mercer County 3084
    --Sussex County 3014
    --Tocks Island 3014
    --Warren County 3014
  New Mexico
    --Dona Ana County 3231
    --Magdalena Mountains 3178
    --Quay County 3229
    --Socorro County 3178
    --Union County 3228
    --Valencia County 3227
  New York
    --Long Island 2964
  North Dakota 3426
    --Adams County 3377, 3378
    --Barnes County 3319, 3320,
      3321
    --Benson County 3363, 3364
    --Bowman County 3377, 3378
    --Burke County 3352, 3353
    --Burleigh County 3316, 3317,
      3318
    --Cass County 3331, 3332, 3333
    --Cavalier County 3370, 3371
    --Dickey County 3391
    --Divide County 3325, 3326,
      3327, 3492
    --Dunn Center 3491
    --Eddy County 3322, 3323, 3324
    --Emmons County 3380, 3381
    --Foster County 3322, 3323,
      3324
    --Grand Forks County 3347,
      3348, 3349
    --Grant County 3383, 3384
    --Griggs County 3374, 3375
    --Knife River Basin 3494
    --Lamoure County 3391
    --McIntosh County 3397

--McKenzie County 3492
--McLean County 3366, 3367
--Mercer County 3356, 3357
--Morton County 3395
--Mountrail County 3352,
   3353
--Nelson County 3359, 3360
--Oliver County 3356, 3357
--Pembina County 3370,
   3371
--Pierce County 3363, 3364
--Ramsey County 3393
--Ransom County 3389
--Renville County 3341,
   3342
--Richland County 3328,
   3329, 3330
--Sargent County 3389
--Sioux County 3383, 3384
--Steele County 3374, 3375
--Stutsman County 3313,
   3314, 3315
--Traill County 3337, 3338
--Walsh County 3359, 3360
--Ward County 3341, 3342
--Wells County 3343, 3344,
   3345
--West-Central 3493
--Williams County 3334,
   3335, 3336, 3492
Ohio
--Pickaway County 3597
Oklahoma
--Caddo County 3658
--Cleveland County 3670
--Custer County 3641
--Oklahoma County 3670
Pennsylvania 3918, 3931
--Allegheny County 3919,
   3946
--Armstrong County 3945
--Beaver County 3920,
   3950
--Berks County 3933, 3955
--Bucks County 3922
--Butler County 3947
--Centre County 3959
--Clearfield County 3956
--Coastal Plains 3924
--Crawford County 3957
--Dubois 3956
--Jefferson County 3956
--Lackawanna County 3952
--Lancaster County 3954
--Luzerne County 3951
--Mercer County 3944
--Mifflintown Quadrangle
   3938
--Monroe County 3958

--Montgomery County 3933,
   3940
--Northcentral 3917
--Northeastern 3915
--Northwestern 3914
--Pittsburgh 3921
--South-Central 3916, 3938
--Southeastern 3913, 3924
--Southwestern 3912
--Washington County 3949
--Westmoreland County 3948
--York County 3953
South Carolina 3963, 3994
--Greenville County 3987
--Orangeburg County 3985
South Dakota
--Baltic 4101
--Bowdle 4075
--Britton 4078
--Brookings 4090
--Bryant 4080
--Canton 4076
--Colome 4093
--Columbia 4095
--Dell Rapids 4084
--Eureka 4105
--Fall River County 4062
--Gettysburg 4089
--Gregory 4094
--Groton 4103
--Harrisburg 4071
--Harrold 4107
--Hazel 4098
--Howard 4092
--Ipswich 4077
--Lake Andes 4082
--Lake Norden 4079
--Langford 4074
--Lennox 4091
--Marion 4072
--Mission 4086
--Mitchell 4087
--Orient 4106
--Parkston 4100
--Peever 4104
--Pierre 4097
--Sanborn County 3998, 4000
--Spencer 4099
--Viborg 4088
--Volga 4096
--Watertown 4073
--Waubay 4085
--Webster 4102
--Wessington Springs 4083
--Winner 4081
--Wolsey 4108
Tennessee
--Hamilton County 4123

Groundwater (Cont.)
  Texas
    --Coastal Plain 4322
    --Northeast 4213
  Utah
    --East Shore 4448
    --Great Salt Lake 4454
    --Jordan Valley 4456
    --San Juan County 4415
  Virginia
    --Accomack County 4532
    --Northampton County 4532
    --Shenandoah National Park
      4533
  Washington
    --Puget Sound 4659
  West Virginia
    --Berkeley County 4731
    --Mason County 4680
    --Monongahela River Basin
      4749, 4754
    --Ohio County 4675
    --Pocatalico River Basin
      4696
    --Putnam County 4680
  Wisconsin 4773, 4792, 4809,
    4810, 4828
    --Columbia County 4825
    --Eau Claire 4819
    --Jefferson County 4821
    --Lake Beulah 4830
    --Ozaukee County 4826
    --Rice Lake 4819
    --St. Croix County 4820
    --Walworth County 4822,
      4830
    --Washington County 4826
    --Waukesha County 4817
Groundwater Contamination 1457
  Alabama
    --Stevenson Area 134
Grundel Mastodon
  Missouri
    --Holt County 2615
Grundy County, Illinois
  Coal 1275
Grundy County, Missouri
  Water 2645
Guadalupe County, New Mexico
  Water 3233
Gualala River, California
  Geology 455
Guantanamo Naval Base, Cuba
  Desalination 2953
Guernsey County, Ohio
  Bedrock Geology 3573
Gueydan Formation
  Texas
    --South 4298

Guildhall, Vermont
  Geology 4475
Guilford Quadrangle, Connecticut
  Surficial Geology 684
Gulf Basin, Louisiana
  Mesozoic 2148
Gulf Coast
  Miocene 2149
  Oligocene 2149
Gulf of Mexico
  Florida 843
Gulf Trough
  Georgia
    --Southwestern 947
Gulkana Glacier, Alaska
  Geology 280
  Paleozoic 280
  Petrology 280
  Rocks 280
  Sedimentology 280
  Tertiary 280
Gum Hollow Fan Delta, Texas 4304
Gunnison County, Colorado
  Geologic Hazards 613
  Geology 604
  Groundwater 604
  Mineral Resources 604
Gwinnett County, Georgia
  Surface Water 928
Gypsum 368
  California 424
  Florida 885
  Indiana 1502
  Kentucky
    --Northwestern 1940
  Missouri
    --Putnam County 2609
  Wyoming
    --Cody 4856
    --Park County 4856

Habersham County, Georgia
  Geology 902
Hackensack Meadowlands, New Jersey
  Industrial Pollutants 3031
  Phytoplankton Bioassays 3031
Hackensack River, New Jersey
  Sewage 3013
  Wastes 3013
  Water 3013
Haddam Quadrangle, Connecticut
  Bedrock Geology 693
  Surficial Geology 692
Haena, Hawaii
  Drilling Logs 1032
  Pumping Tests 1032
Hafnium
  California 424

Hagerstown Valley, Maryland
  Carbonate Rocks 2297
  Hydrogeology 2297
Haiku, Hawaii
  Groundwater 1012
  Streamflow 1012
  Water Development Tunnels
    1012
Hallandale, Florida
  Groundwater 794
Hamakua, Hawaii
  Drilling Logs 1020
  Water Supply 1004, 1008,
    1086, 1090
Hamakuapoko, Hawaii
  Drilling Logs 1025
  Pumping Tests 1025
Hamburg Quadrangle, Connecticut
  Bedrock Geology 675
Hamilton County, Kansas
  Irrigation Water 1786
Hamilton County, Ohio
  Bedrock Geology 3578, 3584,
    3594, 3595
Hamilton County, Tennessee
  Geology 4123
Hammer Creek Formation
  Pennsylvania
  --Southeastern 3560
Hampden Basalt
  Connecticut
  --Central 697
Hanalei, Hawaii
  Drilling Logs 998
  Pumping Tests 998
Hanapepe, Hawaii
  Drilling Logs 996, 1033
  Pumping Tests 996, 1033
Hancock County, Illinois
  Coal 1230
Hancock County, Kentucky
  Economic Geology 1934
Hancock County, West Virginia
  Oil and Gas 4681.
Hand Placer Mining 300
Haney Limestone
  Illinois 2130
  Indiana 1557, 2130
  Kentucky 2130
Hanna Basin, Wyoming
  Early Tertiary 4863
  Late Cretaceous 4863
  Sediments 4863
Hanna Coal Field, Wyoming
  Coal 4887, 4892
Hanson Pool, Kentucky 2026
Haragan Formation
  Oklahoma 3618, 3643
  --Arbuckle Mountains 3650

Hardee County, Florida
  Groundwater 769, 870
Hardin County, Illinois
  Aeromagnetic Surveys 1219
  Fluorspar 1473
  Geology 1473
Harding County, South Dakota
  Newcastle Sandstone 4059
  Oil 4059
Hardwick, Vermont
  Geology 4477, 4496
Harford County, Maryland
  Groundwater 2231, 2238
  Wells 2231
Harmar Mine, Pennsylvania
  Petrography 3861
Harney County, Oregon
  Geology 3700
Harquahala Mountains, Arizona
  Geology 383
Harrisburg, Pennsylvania
  Environmental Geology 3784
Harrisburg, South Dakota
  Groundwater 4071
Harrisburg (No. 5) Coal
  Illinois
  --Southeastern 1287
Harrison County, Missouri
  Water 2647
Harrold, South Dakota
  Groundwater 4107
Hart County, Kentucky
  Brownsville Paleovalley 1967
  Kyrock Sandstone 1967
  Pennsylvanian 1967
  Sedimentology 1967
Hartford South Quadrangle, Connecticut
  Surficial Geology 676
Hartman Quadrangle, Arkansas
  Geology 394
Hartselle Formation
  Tennessee
  --Middle 4133
Hawaii Island, Hawaii
  Floods 1022, 1043, 1098
  Geology 984, 985
  Groundwater 984, 985
  Irrigation 1070
  Population Growth 1066
  Water Resources 1095, 1108
  Water Supply 1074, 1086, 1090,
    1113
Hawaiian Islands
  Floods 1048
  Geology 983
  Groundwater 992
Haymond Formation
  Texas
  --Marathon Basin 4292

Haystack Range, Wyoming 4852
Hayward Fault, California 474
Hazards
　California
　　--San Bernardino County
　　　554
　Colorado 592, 622, 641, 642
　　--Crested Butte 613
　　--Glenwood Springs 616
　　--Gunnison County 613
　　--Larimer County 606
　Mississippi
　　--Adams County 2540
　Oregon
　　--Benton County 3740
　　--Bull Run Watershed 3724
　　--Clackamas County 3724,
　　　3741
　　--Coast 3762
　　--Hood River 3733
　　--Multnomah County 3724
　　--Sherman County 3733
　　--Wasco County 3733
　Pennsylvania 3780
　Tennessee 4129
　Utah 4424
　　--Morgan County 4366
Hazel, South Dakota
　Groundwater 4098
Heat 2068
Heat Flow 1584
　Oregon 3773
　Washington 4640
　　--Lemei Rock 4660
　　--Skamania County 4660
　　--Steamboat Mountain 4660
Heat Flow Contours
　Colorado 612
Heavy Metals 1455
Heavy Minerals
　Georgia
　　--Coastal Region 970
　　--Echols County 964
　　--Effingham County 968
　Illinois
　　--Southern 1284
　New Mexico
　　--San Juan Basin 3122
　South Carolina 3984
　　--Hilton Head Island 3975
　Tennessee
　　--West 4152
　Texas
　　--Burleson County 4290
　　--Lee County 4290
　Wabash River 1258
Helena Quadrangle, California
　Geology 522
Helium

Arizona 353
Hell Creek Formation
　North Dakota 3350
　　--Southwestern 3486
Henderson County, Illinois
　Coal 1230, 1295
Henderson County, Kentucky
　Pennsylvanian Channel 2132
Hendricks County, Indiana
　Geology 1567
Hendry County, Florida
　Geology 824
　Groundwater 824
Hennepin, Illinois
　Geology 1278
　Land Use 1278
Henry County, Illinois
　Coal 1295
Henry County, Tennessee
　Anomalous Zinc 4148
Henry Mountains Coal Field, Utah
　4391
Henryhouse Formation
　Oklahoma 3630, 3631, 3642
Henrys Fork Coal Field, Utah 4391
Herendeen Bay, Alaska
　Coal 334
Hernando County, Florida
　Hydrology 821
　Suwannee Limestone 748, 749
Hetta Inlet, Alaska
　Rock 232
　Stream Sediments 232
Hickory Run State Park, Pennsylvania
　Guide 3899
Hidalgo County, New Mexico
　Geology 3095, 3107, 3179, 3200,
　　3206, 3212
　Peloncillo Mountains 3212
　Petroleum Source Rocks 3206
　Thermal Metamorphism 3206
Higginsville Formation
　Missouri
　　--Central 2612
High Bridge Group
　Kentucky
　　--Boone County 1949
　　--Central 1964
High Plains
　New Mexico 3277
High Points
　Wisconsin 4774
High Water 3079
High Water Table
　Colorado
　　--Crested Butte 613
　　--Gunnison County 613
Highland County, Ohio
　Bedrock Geology 3596

Glacial Geology 3593
Highland County, Virginia
Igneous Rocks 4523
Highland Heavy-Mineral Sand Deposits
Florida
--Northern Peninsula 871
Highways 4694, 4695
Arizona
--Highway 77 347
--Highway 177 347
--Highways 85, 86, 386
354
--Highways 87, 88, 188
355
--U. S. Highway 666 345
California
--San Fernando Valley 432
Delaware
--Piedmont 719
Idaho
--U. S. Highway 93 1150
Illinois 1387
Iowa 1657
--Southwestern 1624
Mississippi
--Highway 25 2461
Missouri
--Interstate 44 2602
--Interstate 55 2542
Nebraska
--Interstate 480 2853
North Dakota
--Highway 10 3418
--Interstate 94 3398
Ohio
--U. S. Route 23 3516
Tennessee 4160
--Interstate 75 4159
Washington
--Chinook Highway 4648
--Ellensburg Highway 4648
--Snoqualmie Highway 4626
--Stevens Highway 4626
--Swauk Highway 4626
--White Pass Highway 4648
Hillsboro Canal, Florida
Groundwater 842
Hillsboro County, Florida
Aquifers 803
Hydrogeology 803
Hillsborough County, Florida
Water Resources 760
Hilo, Hawaii
Drilling Logs 1055
Pumping Tests 1055
Water 1021, 1036
Hilton Head Island, South Carolina
Heavy Minerals 3975

Hinds County, Mississippi
Clay 2496
Cockfield Formation 3489
Eocene 2489
Geology 2537, 2538
Land Use 2537, 2538
Mineral Industries 2497
Mineral Resources 2492
Plant Microfossils 2489
Public Utilities 2537
Recreation 2537, 2538
Sanitary Landfill 2538
Topography 2537, 2538
Transportation 2537, 2538
Water Resources 2495, 2537, 2538
Hindsville Limestone
Missouri
--Southwestern 2630
History
New Mexico 3278
Hocking County, Ohio
Bedrock Geology 3564
Hocking Hills State Park, Ohio
Geology 3526
Hodgeman County, Kansas
Dakota Formation 1827
Water 1829
Hog Island Quadrangle, Virginia
Geology 4572
Hog Shooter Formation
Oklahoma
--Nowata County 3629
--Washington County 3629
Hollandale, Minnesota
Test Wells 2423
Hollis Area, Alaska
Geochemistry 252
Geology 252
Hollister, California
Geology 423
Hollywood, Florida
Groundwater 864
Holmes County, Florida
Mineral Resources 744
Holmes County, Mississippi
Geology 2467
Holmes County, Ohio
Glacial Geology 3592
Holocene
Alabama
--Mobile Bay 140
Illinois
--Lake Michigan 1352
Virginia
--Coastal Plain 4510
Wyoming 4880
Holston Valley Quadrangle, Virginia
Geology 4553

Holt County, Missouri
  Groundwater 2661
  Mastodons 2615
Home Construction
  Colorado 646, 647
Homestead, Idaho
  Geology 1156
Hood River, Oregon
  Geologic Hazards 3733
Hoolehua, Hawaii
  Irrigation 1061
  Water 1061
Hoover Dike, Florida
  Seepage 845
Hopkins County, Kentucky
  Oil 2026
Horn Mountains, Alaska
  Mordenite 335, 337
  Zeolite 335
Horn Silver Mine, Utah
  Hydrothermal Alteration 4399,
    4406
Hornblende
  Minnesota 2419
Horseshoe Curve, Pennsylvania
  Guide 3800
Hosgri-San Gregorio Fault Zone,
  California 578
Houghton, Douglass 2335
Housing
  Alabama 193
Houston, Texas
  Aquifers 4209
Houston County, Alabama
  Mineral Resources 174
  Water Resources 174
Houston County, Tennessee
  Faulting 4137
Houtzdale Quadrangle, Pennsylvania
  Coal 3874
  Petrography 3874
Howard, South Dakota
  Groundwater 4092
Howell County, Missouri
  Brown Iron Ore 2570
Hudspeth County, Texas
  Asbestos 4185
Humboldt County, Nevada
  Geology 2859
  Minerals 2859
Humboldt Fault Zone, Nebraska
  Tarkio Limestone 2858
Huntersville Chert
  West Virginia
    --Eastern 4693
Huntley Mountain Formation
  Pennsylvania
    --North-Central 3855
Hunton Group

Oklahoma 3622
  --Anadarko Basin 3648
  --Arkoma Basin 3656
Huntsville, Alabama
  Environmental Geology 8
  Groundwater 82
  Hydrology 8
Hurley, New Mexico 3275
Huron County, Ohio
  Geology 3561
Hurricanes
  Alabama 182
  Texas 4178, 4296, 4302
Hydraulics
  Delaware
    --Dover 729
    --Piney Point Aquifer 729
  Florida
    --Miami 862
  Ohio
    --Marion 3611
Hydrocarbons
  Delaware
    --Baltimore Canyon Trough 734
  New Mexico
    --Roosevelt County 3184
  Ohio
    --Lake Erie 3517
  Texas
    --South 4313
  Utah
    --Uinta Basin 4433
Hydrochloric Acid Treatment
  Kansas 1701
Hydrochlorite 3120
Hydrodynamics 3043
  Illinois
    --Illinois Basin 1326
Hydroelectric Energy 3081
Hydrogeochemistry 1164
  North Dakota
    --Dunn Center 3491
Hydrogeology 1427, 2991
  Alabama 129
  California
    --Bay Area 423
    --Great Valley Province 418
    --Santa Clara Valley 423
  Colorado 587, 614, 629
    --Cheyenne Basin 620
    --Pagosa Hot Springs 645
  Florida
    --Hernando County 821
    --Hillsboro County 803
    --Old Tampa Bay 855
    --Pasco County 821
    --Southeastern 847
  Idaho 1165
  Illinois

--East-Central 1265
--Lake County 1433
--Mahomet Bedrock Valley
  1265
--Northeastern 1402, 1456
--Shelbyville 1391
Iowa 1630
--Northwest 1661
Kansas
--Great Bend Prairie 1816
--Greeley County 1815
--Lane County 1815
--Scott County 1815
--South-Central 1816
--Wichita County 1815
Maryland 2292, 2304
--Frederick Valley 2297
--Hagerstown Valley 2297
--Somerset 2226
--Southern 2228
--Western 2298
Missouri 2675
--St. Louis 2674
Montana 2812
--Colstrip 2768
--Decker 2762
--Gallatin River 2820
--Silver Bow Creek Drainage
  Area 2734
--Southeastern 2768
--Southwestern 2820
New Jersey 3073
New Mexico
--Chaves County 3147
--Roswell Artesian Basin
  3147
North Dakota
--Bismarck 3500
--Mandan 3500
Ohio 3536
Pennsylvania
--Lebanon Valley 3929
Texas
--Galveston 4209
--Houston 4209
--North-Central 4199
Utah
--Bonneville Salt Flats 4468
--Goshen Bay 4472
--Price River 4451
--Utah Lake 4472
Virginia
--Flint Hill Quadrangle
  4601
--Front Royal Quadrangle
  4594
--Linden Quadrangle 4601
Wisconsin
--Southeastern 4805

Hydrology 1726, 2954, 3004, 3067,
  3930
Alabama 43, 44, 45, 46, 47, 48,
  49, 177
--Huntsville 8
--Madison County 1, 3, 4, 5,
  8
--Maysville Area 3
--Meridianville Quadrangle 1
--Mobile 61
--Triana Area 4
Alaska
--Prudhoe Bay Oil Field 342
Australia 2993
Delaware 708
--Middletown 707
--Newark 725
--Odessa 707
Florida
--Alafia River Basin 836
--Brooklyn Lake 820
--Collier County 850
--Dade County 834
--Green Swamp 829
--Keystone Heights 820
--Lakeland Ridge 851
--Middle Gulf 843
--Old Tampa Bay 793
--Orange County 859
--Peace River Basin 836
--Polk County 851
--Seminole County 868
--Tallahassee 883
--Tampa 861, 866
--Tampa Bypass Canal System
  869
--Walt Disney World Develop-
  ment 866
Georgia
--Glynn County 926
Hawaii 1060
--Molokai 1050
--Oahu 1092
--Waimanalo 1092
Indiana
--Crawford Upland 1535
--Mitchell Plain 1535
--Southern 1535
Kansas
--Kansas River Valley 1867
--Rice County 1768
Kentucky
--Western Coal Field Region
  1961
Michigan
--Upper Peninsula 2370
Minnesota
--Minneapolis 2397
--St. Paul 2397

Hydrology (Cont.)
  Missouri 2594
    --St. Charles County 2595
  New Jersey 2994
  New York
    --Long Island 3019
  North Dakota
    --Dunn Center 3491
    --West-Central 3493
  Pennsylvania
    --Adams County 3932
    --Butler County 3927
    --Chester County 3936
    --Dauphin County 3935
    --Lancaster County 3934
    --Lawrence Quadrangle 3927
    --Lehigh County 3941
    --Luzerne County 3939
    --Mercer County 3926
    --Neshannock Quadrangle
      3926
    --Northampton County 3941
    --South-Central 3937
    --Southeastern 3925
    --Wyoming Valley 3939
    --York County 3932
  South Dakota 4057
    --Clay County 4004, 4006
    --Lake Kampeska 4056
    --Lake Poinsett 4055
  Texas
    --East 4226
  Tropics 2995
  Utah
    --Great Salt Lake 4449
  West Virginia
    --Berkeley County 4731
    --Coal River Basin 4753
    --Greenbrier County 4684
    --Mason County 4680
    --Monongahela River Basin
      4754
    --Pocatalico River Basin
      4696
    --Putnam County 4680
  Wisconsin
    --Dane County 4818
    --Lake Beulah 4830
    --Walworth County 4830
Hydrolysis 3127
Hydrometallurgy 2898
  New Mexico
    --Questa Mine 3135
Hydropower
  Hawaii
    --Waialeale 1120
Hydrostatic Pressure 2990
Hydrothermal Alteration
  Utah

    --Beaver County 4394, 4396,
      4402, 4406
    --Black Mountains 4396
    --Frisco Quadrangle 4394
    --Horn Silver Mine 4406
    --Iron County 4396
    --Shauntie Hills 4396
    --Wah Wah Range 4402
Hydrothermal Deposits
  Nevada
    --Central 2921
    --Western 2921
Hydrothermal Resources
  Colorado 629
Hylas Quadrangle, Virginia
  Geology 4577
Hyperoscelidae
  Connecticut 649

Iberville Parish, Louisiana
  Groundwater 2176
Ice Age
  Indiana
    --Pokagon State Park 1611
Ice-Contact Deposits
  North Dakota
    --Logan County 3422
Ice Disintegration
  Indiana
    --West-Central 1563
Ice Margins
  Ohio
    --Northeastern 3522
Ice Mines
  Pennsylvania
    --Trough Creek State Park 3898
Ice-Push Deformation
  Kansas
    --Northeastern 1708
Ice-Thrust Bedrock
  North Dakota
    --Lavalier County 3443
Idaho Batholith
  Geology 1144
  Petrology 2788
  Road Log 1144
Igneous Complex
  Michigan
    --Northern Peninsula 2365
  Utah
    --Beaver County 4407
    --Wah Wah Pass 4407
Igneous Rocks 2399, 2899, 2900
  Florida 750
  Missouri
    --St. Francois Mountains 2631
  Pennsylvania
    --Chester County 3936

Virginia
--Augusta County 4523
--Bath County 4523
--Highland County 4523
--Rockingham County 4523
Western United States 2868
Iliamna Quadrangle, Alaska
Geology 243
Illinoian
Illinois 1317
--Central 1475
Illinoian Drift
Illinois
--Central 1234
Illinoian Glacial Boundary
Illinois
--Western 1220
Illinois Basin 1993, 2051
Aquifers 1326
Carbondale Formation 1195
Chesterian 1277, 1314, 1495,
1924
Coal 1195, 1354, 1376
Cross-Bedding 1924
Cyclothems 1192
Fossils 1376
Geochemistry 1277
Hydrodynamics 1326
Late Paleozoic 1480
Late Pennsylvanian 1192, 1267
Menard Formation 1495
Mineral Matter 1354
Mississippian 1522
Oil 1224
Oolite Bodies 1522
Paleobotany 1335
Paleogeology 1314
Paleozoic 1480
Peat 1376
Pennsylvanian 1195, 1240, 1314,
1376, 1477, 1478
Ste. Genevieve Limestone 1522
Sandstone 1480, 1924
Sedimentation 1240
Spoon Formation 1195
Spores 1192
Strata 1192, 1267
Sub-Pennsylvanian 1314
Upper Mississippian 1277,
1314
Waltersburg Crudes 1277
Illinois Fluorspar District, Illinois
Geology 1198, 1241, 1269
Illinois River, Illinois
Sediments 1432
Trace Elements 1432
Illinois River Valley, Illinois
Landslides 1424
Illinois Valley, Illinois

Seismic Refraction 1363
Illinois Waterway, Illinois
Geology 1334
Immokalee, Florida
Groundwater 767
Imperial County, California
Mineral Resources 446
Mines 446
Independent Oil Producers 2109
Index Maps
Coal 2834
Indexes
Areal Geology
--Missouri 2562
Coal
--Colorado 595
Core Holes
--Georgia 962
Geology
--California 491, 556, 571
--Colorado 591
--Hawaii 978
--Illinois 1194
--Kansas 1779, 1793
--Louisiana 2160
--Michigan 2324
--Montana 2797, 2816, 2840
--Oregon 3758
--Washington 4611, 4666
Geophysics
--Minnesota 2411
Groundwater
--Kansas 1719
--Montana 2800
Mapping
--Washington 4666
Maps
--California 543
--Michigan 2325
Metallic Mineral Resources
--Montana 2833
Minerals
--Washington 4611
Mining Properties
--Arizona 358, 360, 362, 363
Quaternary
--Minnesota 2426
Salt Lake Mining Review 4364
Stratigraphy
--Great Basin 2873
Test Samples
--New Mexico 3142, 3163
Water
--Florida 878
--Hawaii 978
Indian Creek District, Montana
Ore 2683
Indian River County, Florida
Water Resources 867

Indian Village Sites
  Louisiana 2140
  --Sicily Island 2139
  Mississippi 2140
Indiana Dunes State Park, Indiana
  Geology 1599, 1618
Indians
  Louisiana 2142
Indium
  California 424
Induced Infiltration
  Alabama
  --Alabama River 142
Indus, Minnesota
  Geochemistry 2428
  Geology 2428
  Sulfide Mineralization 2428
Industrial Growth
  New Jersey
  --Passaic River Basin 2974
Industrial Metals
  Illinois 1373
Industrial Minerals 884, 956,
  1392, 1631, 1872, 2909, 2940,
  3544, 3680, 3883
  Colorado 631
  Illinois 1373
  Kansas 1857
  Kentucky 2124
  Montana 2718, 2837
  Nevada
  --Nuneral County 2911
Industrial Monitoring 3077
Industrial Pollutants
  New Jersey
  --Hackensack Meadowlands
    3031
Industrial Rock
  Nevada
  --Storey County 2870
  --Washoe County 2870
Industrial Sand
  Kentucky
  --Pike County 1953
  Michigan 2328
Industrial Wastes 2071
  Illinois 1282
  Oklahoma 3681, 3682, 3683,
    3684
Industrial Water
  Kentucky 1947
Industry
  Montana 2805
Inert Gas Injection
  Kentucky
  --Union County 2034
Infiltration
  Alabama
  --Alabama River 142

Infinite Well Fields 1871
Ingleside Fauna
  Texas
  --San Patricio County 4312
Inglewood-Newport Structural Zone,
  California
  Earthquakes 555
  Geology 555
Ingomar, Montana
  Bentonite 2814
Injection
  Kentucky
  --Union County 2034
Inlets
  New Jersey 3023
Inliers
  Wisconsin
  --South-Central 4787
Inmachuk River, Alaska
  Geochemistry 258
  Geology 258
Insects
  Connecticut 649, 653, 663
Interstate Highway System
  Illinois 1387
Intracoastal Waterway
  Florida
  --Venice 825
Intrusive Rocks
  Missouri
  --Southeast 2627
Intrusives
  Minnesota 2419
Invertebrates
  Delaware 710
  Florida 891
  Mississippi
  --Clarke County 2536
Inyo County, California
  Cambrian 521
  Economic Geology 490
  Geology 508, 526
  Petrogeology 540
  Precambrian 521
  Sedimentary Rock 521
  Stratigraphy 540
  Talc 525
Iodine
  California 424
Ione Area, California
  Geophysics 558
Iowan Drift
  Iowa
  --Northeastern 1644
Ipswich, South Dakota
  Groundwater 4077
Iridium
  California 424
Iron 2358, 3140

California 424
Maryland
--Piedmont Upland 2234
Missouri 2583
New Mexico
--Rio Arriba County 3175
Tennessee 3134
Utah 4361
Wyoming
--Carbon County 4853
Iron Bacteria
Alabama 154
Iron County, Michigan
Groundwater 2378
Iron County, Utah
Geology 4396
Hydrothermal Alteration 4396
Iron Creek, Alaska
Geochemistry 226
Geology 226
Iron Formation
Minnesota
--Mesabi Range 2384
Iron Mountain Mine, California
Geophysics 502
Iron Ore 2329
Alabama
--Barbour County 166
--Birmingham Red Iron District 79
--Butler County 170
--Crenshaw County 92, 170
--Pike County 170
Indiana 1506
Louisiana
--Central North 2151
Minnesota
--Vermilion District 2438
--Zenith Mine 2438
Mississippi 2477
Missouri
--Howell County 2570
--Oregon County 2571
--Wayne County 2578
Tennessee
--East 4140
Texas
--Llano Region 4288
Iron River, Michigan
Geology 2345, 2346, 2347, 2348
Magnetic Data 2345, 2346, 2347, 2348
Ironstone
Texas
--East 4242
Ironton Sandstone
Illinois 1259
Ironworks

Indiana 1506
Irrigation
Hawaii 1062
--Coastal Plain 1114
--Hawaii Island 1070
--Hoolehua 1961
--Kauai 1063, 1114
--Maui 1064
--Molokai 1011, 1061
--Oahu 1065
--Waimanalo 1067
Idaho 1163
Kansas
--Cedar Bluff Irrigation District 1783
--Finney County 1786
--Ford County 1789
--Gray County 1786
--Great Bend Prairie 1789
--Hamilton County 1786
--Kearny County 1786
--Southwestern 1788
--West-Central 1784
Wisconsin
--Central 4824
--Dunn County 4827
--Eau Claire 4819
--Rice Lake 4819
Irwin County, Georgia
Geohydrology 925
Phosphate 967
Isaac Lake Quadrangle, Minnesota
Geology 2444
Island Mountain Mine, California
Magnetic Study 506
Island Pond, Vermont
Geology 4473
Isoelectricity 3133
Isotopic Age
Pennsylvania 3842
--Erie County 3851
Utah 4354
Isotopic Fractionation 1249
Isotopic Studies
Utah
--Uinta Basin 4433
Issaquena County, Mississippi
Geology 2467
Itasca County, Minnesota
Cumulus Mineralogy 2432
Deer Lake Complex 2431, 2432
Geology 2431
Petrology 2432
Sulfide Minerals 2433
Izee, Oregon
Geology 3700

Jackson Bluff, Florida

Jackson Bluff, Florida (Cont.)
  Geology 881
Jackson Bluff Formation
  Florida 747
Jackson County, Alabama
  Coal 51
Jackson County, Oklahoma
  Copper 3663
  Flowerpot Shale 3663
  Permian 3663
Jackson County, Oregon
  Land Use Geology 3736
Jackson County, South Dakota
  Miocene Channels 4064
Jackson County, West Virginia
  Newburg Gas 2075
Jackson Eocene
  Mississippi 2519
Jackson Formation
  Mississippi 2534
Jackson Group
  Texas 4179
Jackson Purchase, Kentucky
  Fuller's Earth 1928
  Geology 1956
  Paleozoic 1956
  Petroleum 1956
  Water 2097
Jackson Quadrangle, Mississippi
  Environmental Geology 2538
Jacksonburg Formation
  Pennsylvania
    --Lehigh County 3795
    --Northampton County 3795
Jacksonville, Florida
  Geophysics 781
  Water 774, 781, 815
Jacobs Chapel Bed
  Indiana
    --Southern 1501
Jade
  California 481
James Lake Quadrangle, Wyoming
  Structural Geology 4860
Jamestown, North Dakota
  Geology 3417
Janes Island State Park Test Well,
  Maryland
  Hydrogeology 2226
Jarilla Mountains, New Mexico
  Geology 3093
Jasper Aquifer
  Louisiana 2180
Jasper County, Mississippi
  Gas 2451
  Groundwater 2453
  Mineral Resources 2450
  Oil 2451
  Subsurface Stratigraphy 2452
  Subsurface Structure 2453
Jeff Davis County, Georgia
  Phosphate 971
Jefferson County, Alabama
  Areal Geology 10, 15
  Drainage 11
  Engineering Geology 14
  Fish 70
  Flooding 191
  Floods 2
  Forest Resources 192
  Groundwater 161
  Streamflow 16
Jefferson County, Colorado
  Geology 590
  Soil 590
  Water 590
Jefferson County, Florida
  Geology 742
Jefferson County, Kansas
  Geohydrology 1757
Jefferson County, Kentucky
  Conodonts 1607
  Louisville Limestone 1607
  Silurian 1607
  Wabash Formation 1607
Jefferson County, Missouri
  Engineering Geology 2552
  Floods 2663
  Paleozoic 2577
  Water 2674
Jefferson County, Ohio
  Bedrock Geology 3588
Jefferson County, Pennsylvania
  Geology 3878
  Groundwater 3956
Jefferson County, Wisconsin
  Geology 4821
  Groundwater 4821
  Soils 4762
Jeffery Quarry, Arkansas
  Cookeite 386
  Quartz 386
  Rectorite 386
Jellico, Tennessee
  Mississippian 4159
  Pennsylvanian 4159
Jemez Mountains, New Mexico
  Subsurface Temperatures 3205
Jeptha Knob Structure, Kentucky 1995
Jerome, Arizona
  United Verde Open Pit 349
Jersey County, Illinois
  Dolomite 1304
  Limestone 1304
Jewell County, Kansas
  Groundwater 1730
Joe's Pond, Vermont
  Morphometry 4500

Sedimentation 4500
Johns Peak, Utah
  Coal 4385
Johnson, Vermont
  Geology 4456
Johnson County, Arkansas
  Geology 391, 393, 394, 396
Johnson County, Indiana
  Geology 1573
  Gravel 1595
  Land Use 1573
  Sand 1595
Johnson County, Kansas
  Geology 1758
  Groundwater 1758
Johnson City Lake, Illinois
  Sediments 1458
  Trace Elements 1458
Joplin, Missouri
  Water 2668
Jordan Valley, Utah
  Geology 4456
  Groundwater 4456
Jornada Del Muerto Coal Field,
  New Mexico
  Geology 3222
Josephine County, Oregon
  Geology 3742
  Minerals 3742
  Mines 3766
Joyita Hills, New Mexico
  Fusulinids 3255
Joyita Uplift
  New Mexico
  --Central 3254
Juab County, Utah
  Geology 4350
  Gravity Data 4368
  Magnetic Data 4368
  Minerals 4372
Judith Basin, Montana
  Geology 2704, 2749, 2767
  Groundwater 2704
  Water 2749, 2767
Juniata County, Pennsylvania
  Phosphate 3836
Juniper Mountain, Idaho
  Geochemistry 1186
  Geology 1186
Jurassic
  Alabama 2508
  Mississippi 2506, 2508
  --Central 2507
  --Southern 2507

K/AR Age
  Minnesota 2419
K-Feldspar

California
  --Atascadero 569
  --San Luis Obispo County 569
  --Santa Lucia Range 569
Kahakuloa, Hawaii
  Water 1115
Kahaluu, Hawaii
  Groundwater 1012
  Streamflow 1012
  Water Development Tunnels 1012
Kahuku, Hawaii
  Water Resources 1030
Kaiparowits Plateau, Utah
  Coal 4390, 4410
Kalaneo, Hawaii
  Drilling Logs 1016
  Pumping Tests 1016
Kalispell Valley, Montana
  Geology 2726
  Groundwater 2726
  Water 2707
Kalkar Quarry, California
  Mineralogy 518
Kamalo, Hawaii
  Drilling Logs 1002
  Pumping Tests 1002
Kamalo Exploratory Wells, Hawaii
  Drilling Logs 1002
  Pumping Tests 1002
Kamiloloa, Hawaii
  Drilling Logs 1013
  Pumping Tests 1013
Kanab, Utah 4341
Kanawha County, West Virginia
  Newburg Gas 2075
Kane County, Illinois
  Drilling 1382, 1429, 1451
Kane County, Utah
  Coal 4417
  Water 4452, 4457
Kankakee County, Illinois
  Coal 1275
Kansan Drift
  Illinois
  --Central 1234
Kansan Formation
  Illinois 1317
Kansas City Group
  Kansas
  --Eastern 1740
  Missouri 2611
Kansas River Basin, Kansas
  Water Quality 1762
Kansas River Valley, Kansas
  Groundwater 1767, 1867
  Hydrology 1867
Kantishna Mining District, Alaska
  Mineral Preparation 333
  Ore 333

Kaolin
  Alabama
    --Eufala Bauxite District 55
  Missouri
    --Bollinger County 2585
    --Glen Allen 2585
    --Lutesville 2585
  Vermont
    --East Monkton 4487
    --Monkton Kaolin Deposit
      4486
Kaolin Clay
  Minnesota 2446
    --Brown County 2414
    --Minnesota River Valley
      2414
    --Redwood County 2414
    --Renville County 2414
Kapaa, Hawaii
  Oceanography 1085
Karst
  Alabama 60
  Arid Regions 60
  Kentucky 2135, 2138
  Semiarid Regions 60
  Southeast 3000
  Southwestern States 49
  West Virginia
    --Greenbrier County 4684,
      4735
    --Monroe County 4735
Kaskaskia Valley, Illinois
  Limestone 1202
Kau District, Hawaii
  Water Resources 1018
Kauai Island, Hawaii
  Drilling Tests 996, 998, 999,
    1016, 1032, 1033, 1035
  Ferruginous Bauxite 1006
  Floods 1038, 1046, 1059, 1110
  Geology 988
  Groundwater 988, 1114
  Irrigation 1063, 1114
  Plastic Clay 1006
  Pumping Tests 996, 998, 999,
    1016, 1032, 1033, 1035
  Sewage 1085
  Soil 1102
  Water Supply 1080, 1082, 1083,
    1101
Kaunakakai, Hawaii
  Water 1078
Kawaihae, Hawaii
  Drilling Logs 1015
  Pumping Tests 1015
Kawaihee Exploratory Wells, Hawaii
  Drilling Logs 994, 997
  Pumping Tests 994, 997
Keahole, Hawaii

  Drilling Logs 1039
  Pumping Tests 1039
Kearny County, Kansas
  Irrigation Water 1786
Keefer Sandstone
  West Virginia 4691
Keei, Hawaii
  Drilling Logs 1014
  Pumping Tests 1014
Kehena Ditch, Hawaii
  Water 1044
Kekaha-Mana Coastal Plain, Hawaii
  Groundwater 1114
  Irrigation 1114
Kelso, Washington
  Geology 4606
  Minerals 4606
Kenai Group
  Alaska
    --Cook Inlet 331
Kenai Lake, Alaska
  Geochemistry 222
Kenai Peninsula, Alaska
  Geochemistry 244
  Geology 244
Kendall County, Illinois
  Drilling 1382, 1429
Kennecott Copper Company, Utah
  Tailings 4420
Kensington Quadrangle, Ohio
  Geology 3556
Kent County, Delaware
  Aquifers 732
Keokuk Formation
  Illinois 1244
Keokuk Limestone
  Kansas
    --Cherokee County 1717
Kern County, California
  Geology 498
Ketchikan District, Alaska
  Mines 327
Kewanee, Illinois
  Geology 1484
  Groundwater 1484
Keweenaw Peninsula, Michigan
  Geology 2381
  Groundwater 2381
Keweenawan
  Michigan
    --Northern Peninsula 2365
  Minnesota
    --Southeastern 2427
Keystone Gold-Copper Prospect Area,
  Wyoming 4850
Keystone Heights, Florida
  Hydrology 820
Kihei, Hawaii
  Floods 1091

Kihei-Makena, Hawaii
Water 1099
Kimberlite
Colorado 4889
--Larimer County 4882
Kansas
--Riley County 1747, 1748,
1749, 1751
Wyoming 4889
--Albany County 4882
Kindblade Formation
Oklahoma
--Arbuckle Mountains 3665
Kinderhookian
Alabama
--Gipsy Limestone County
58
Kansas
--Western 1756
Missouri
--Southwestern 2625
Nebraska
--Southwestern 1756
King County, Washington
Geology 4615, 4659
Groundwater 4659
Minerals 4615, 4659
Kingsport Formation
Tennessee
--East 4144
Kingston Range, California
Talc 525
Kinzua Quadrangle, Pennsylvania
Oil and Gas 3881
Kirkbyacean Ostracods
Indiana 1591
Kirwin Mineralized Area, Wyoming
4849
Klamath County, Oregon
Geology 3708
Minerals 3708
Klamath Mountains, California
Economic Deposits 415
Geology 415, 424
Granitic History 415
Metamorphic History 415
Klondike Quadrangle, Nevada
Geology 2893
Knife River Basin, North Dakota
Geology 3494
Geohydrology 3494
Knifley Sandstone
Kentucky
--South-Central 1959
Knight Island, Alaska
Geology 251
Minerals 251
Knightstown Moraine
Indiana 1588

Knox County, Maine
Geology 2214
Physical Resources 2221
Knox County, Tennessee
Geology 4114
Knox Dolomite Oil Production
Kentucky
--Adair County 2101
Knox Formation
Kentucky
--Northeastern 2040
Knoxville Quadrangle, Arkansas
Geology 391, 396
Kodiak Island, Alaska
Geochemistry 266
Geology 266
Stream Sediments 266
Kohakohau Dam, Hawaii 1113
Kohala, Hawaii
Drilling Logs 994, 997, 1015,
1027
Pumping Tests 994, 997, 1015
Water Supply 1004, 1008, 1027,
1086, 1096, 1113
Kohala Mountain, Hawaii
Water Resources 1005
Kokee Project, Hawaii
Geology 1010
Kolob-Harmony Coal Field, Utah 4390
Kona, Hawaii
Drilling Logs 1014, 1039, 1054
Irrigation 1070
Population Growth 1066
Pumping Tests 1014, 1039, 1054
Rainfall 1068, 1069
Water Resources 1037
Water Supply 1071
Kontrashibuna Lake, Alaska
Geochemistry 228
Geology 228
Kootenai-Flathead Area, Montana
Geology 2684, 2692
Kootenay Arc, Washington
Lead 4613
Zinc 4613
Kramer Junction, California
Colemanite 482
Kyanite
Alabama
--Cleburne County 99
--Turkey Heaven Mountain 99
California 424
South Carolina
--Little Mountain 3986
Kyrock Sandstone
Kentucky
--Edmonson County 1967
--Hart County 1967

Labarge Fault, Wyoming
  Geometry 4888
Lackawanna County, Pennsylvania
  Groundwater 3952
Lacustrine Sandstone
  Utah
    --Uinta Basin 4425, 4428
Laguna Beach Quadrangle, Cal-
  ifornia
  Geology 568
Lahaina, Hawaii
  Drilling Logs 1000, 1028
  Pumping Tests 1000, 1028
Lahaina District, Hawaii
  Water Resources 1042
  Water Supply 1081, 1094
Lairdtract Well Field, Delaware
  Aquifer Tests 731
  Well Tests 731
La Jolla Quadrangle, California
  Geology 438
Lake Agassiz, North Dakota
  Fossils 3430
  Geology 3429
  Mollusks 3430
  Sediments 3490
  Stratigraphy 3490
Lake Andes, South Dakota
  Groundwater 4082
Lake Beulah, Wisconsin
  Geology 4830
  Groundwater 4830
  Hydrology 4830
Lake Bonneville, Nevada
  Quaternary 2903
Lake Charles, Louisiana
  Brackish Water 2184
  Freshwater Aquifers 2184
  Gas 2184
Lake Chatuge, Georgia
  Rock 904
Lake Clark Region, Alaska
  Geochemistry 228
  Geology 228
Lake County, California
  Engineering Geology 563
Lake County, Illinois
  Drilling 1385
  Geology 1337, 1433
  Hydrogeology 1433
  Soil 1433
Lake County, Indiana
  Geology 1602
Lake County, Montana
  Geology 2692
Lake County, Ohio
  Shore Erosion 3600
Lake County, Oregon
  Geology 3708

  Minerals 3708
  Pluvial 3776
Lake Deposits
  Illinois
    --Northern 1323
  Indiana
    --Southern 1590
Lake Douglas, Illinois
  Sedimentation 1322
Lake Du Quoin, Illinois
  Sediments 1458
  Trace Elements 1458
Lake Erie 2976
  Medina Production 3879
  Oriskany Production 3879
Lake Erie, Ohio
  Bathythermograph Recordings 3529
  Features 3523
  Hydrocarbons 3517
  Limnology 3580
  Reefs 3583
  Sediments 3603
  Shore Erosion 3600, 3608
  Shore Structures 3554
  Storms 3534
  Water Movement 3575
  Water Properties 3559
Lake Jackson, Florida
  Water Levels 835
Lake Kampeska, South Dakota
  Hydrology 4056
Lake Kerr, Florida
  Water Balance 860
Lake Lahontan, Nevada
  Quaternary 2903
Lake Lemon, Indiana
  Sedimentation 1564
Lake Michigan, Illinois
  Erosion 1452
  Sediments 1331, 1352, 1406, 1408,
    1411, 1413, 1415, 1417, 1420,
    1423, 1430, 1434, 1436, 1445,
    1450, 1460
  Shores 1452
  Trace Elements 1436
Lake Michigan, Michigan
  Dunes 2364
Lake Missola, Montana
  Glaciation 2787
  Pleistocene 2787
Lake Monongahela Deposits
  West Virginia
    --Morgantown 4688
Lake Norden, South Dakota
  Groundwater 4079
Lake Okeechobee, Florida
  Biological Conditions 858
  Chemical Conditions 858
  Seepage 775, 845

Lake Poinsett, South Dakota
  Hydrology 4055
Lake Pontchartrain, Louisiana
  Fresh Water 2193
  Water Resources 2172
Lake Sakakawea Bluffs, North
  Dakota
  Stratigraphy 3481
Lake Saline, Illinois
  Geology 1327
  Paleontology 1327
  Sediments 1436
  Trace Elements 1436
Lake Sand, Louisiana
  Gas Reservoirs 2154
  Miocene 2153
Lake Tahoe Basin
  Bibliography 468
Lake Travis, Texas
  Land Capability 4319
Lake Washington, Minnesota
  Magnetic Anomaly 2415
Lakeland Ridge, Florida
  Hydrology 851
Lakes
  Florida 756
  --Orange 763
  --Tampa 861
  Illinois
  --Lake Douglas 1322
  Indiana
  --Bedford 1556
  --Salt Creek Valley 1556
  Michigan 2315
  Missouri 2675, 2677
  Nevada 2895
  Wisconsin 4783
Lakeview Fault Segment, Cali-
  fornia 430
Lakeview Mining District, Idaho
  Geology 1176
  Mineral Resources 1176
Lakeview Mountains, California
  519
La Madera Quadrangle, New Mex-
  ico
  Precambrian Geology 3091
Lamar County, Alabama
  Gravity Surveys 145
La Mesa Quadrangle, California
  Geology 438
Lamoure County, North Dakota
  Geology 2390
  Geophysics 3391
  Glaciation 3390
  Groundwater 3391
  Springs 3391
  Water 3391
  Wells 3391

Lanai Island, Hawaii
  Geology 981
  Geophysics 981
  Groundwater 981
  Petrography 981
  Soil 1105
Lancaster County, Pennsylvania
  Groundwater 3954
  Hydrology 3934
  Limestone 3848
  New Oxford Formation 3934
Land 1446, 4785
  Michigan
  --Wayne County 2344
  Missouri
  --St. Charles County 2595
  Texas
  --Corpus Christi 4330
  --East 4200
  --Lake Travis 4319
  --Mustang Island 4327
  --Padre Island 4327
  Utah
  --Cedar Breaks 4411
  --Sanpete County 4358
  Wisconsin 4832
Land Modification
  Virginia
  --Warren County 4552
Land Oriented Reference Data
  New Jersey 3015, 3042
Land Planning
  Colorado
  --Front Range 639
Land Reclamation
  Kansas
  --Southeast 1834
  North Dakota 3465
  --West-Central 3493
Land Redevelopment 1896, 1903
Land Snails
  Louisiana 2457
  Mississippi 2457
  --Southern 2458
Land Subsidence
  California
  --Great Valley Province 418
  Kansas
  --Central 1780
Land Surveys 3075
Land Use 1405, 1439, 3040
  Colorado 641, 643
  --Boulder-Weld Coal Field 605
  --Larimer County 606
  Florida 789
  Illinois 1422
  --Chicago 1384
  --Hennepin 1278

Land Use (Cont.)
  Indiana
    --Johnson County 1573
    --La Porte County 1605
    --Morgan County 1572
    --Southern 1590
  Iowa 1654
  Mexico 3048
  Minnesota
    --Clay County 3484
  Mississippi
    --Adams County 2540
    --Hinds County 2537, 2538
  Missouri 2592
  North Dakota
    --Bismarck 3500
    --Cass County 3484
    --Divide County 3492
    --McKenzie County 3492
    --Mandan 3500
    --Williams County 3492
  Oregon
    --Curry County 3732
    --Jackson County 3736
  Pennsylvania 3782
    --Washington County 3806,
      3807
  Texas
    --Coast 4180
  Washington
    --Puget Sound 4655
  West Virginia 4736, 4737
    --Charleston 4682
  Wisconsin 4814
Lander County, Nevada
  Geology 2889
  Minerals 2889
Landfill Leachate 735
Landfills 1427, 1454, 1455,
  4694, 4719
  Alabama 129
  Florida
    --Old Tampa Bay 855
  Illinois 1403
    --Northeastern 1402
  Mississippi
    --Hinds County 2538
  Ohio 3536
  Tennessee 4125
  Texas
    --Coast 4183
  West Virginia
    --Central-West 4723
    --Eastern Panhandle 4720
    --Northern 4725
    --Northwestern 4724
    --Southeastern 4721
    --Southwestern 4722
  Wisconsin 4823

Landscapes
  Washington
    --Cascade Range 4629
Landslides 4694
  California 560
    --Azusa Quadrangle 546
    --Gobernadora Quadrangle 449
    --Mount Wilson Quadrangle 546
    --Orange County 449
    --San Bernardino County 577
    --San Fernando Valley 430
    --San Gabriel Mountains 546
    --Wrightwood 577
  Colorado
    --Crested Butte 613
    --Gunnison County 613
    --Larimer County 606
  Illinois
    --Illinois River Valley 1424
  Mississippi
    --Acker Lake 2546
    --Monroe County 2546
  Tennessee
    --Nashville 4127
  Utah 4363
  Washington
    --Seattle 4646
  West Virginia 4733
Landstat Analyses
  Texas
    --Coastal Zone 4328
Lane County, Kansas
  Groundwater 1814, 1823
Lane County, Oregon
  Environmental Geology 3727
Langdon, North Dakota
  Geology 3480
Langford, South Dakota
  Groundwater 4074
Lanier County, Georgia
  Phosphate 965, 971
  Sand 965
Lansing Group
  Kansas
    --Southeastern 1695
Lapalli Tuffs
  Missouri
    --Washington County 2623
Lapies
  Kentucky 2135
La Porte City, Iowa
  Chert 1641
  Devonian 1641
  Subsurface Geology 1641
Laporte County, Indiana
  Geology 1605
  Land Use 1605
Laramie Basin, Wyoming
  Structural Geology 4858

Laramie County, Wyoming
  Geology 4861
  Minerals 4861
Laramie Formation
  Colorado
    --Cheyenne Basin 628
    --Denver Basin 628
Larimer County, Colorado
  Diamond 4882
  Floods 606
  Geologic Hazards 606
  Geology 602, 606
  Geomorphic Features 606
  Kimberlite Diatremes 4882
  Land Use 606
  Landslides 606
La Salle County, Illinois
  Coal 1275
  Conodonts 1472
  Pennsylvanian 1472
La Salle Parish, Louisiana
  Burial Mound 2141
La Salsan Juan Coal Field, Utah
  4391
Las Cruces, New Mexico
  Geology 3280
Last Chance Range, California
  Geology 535
Latah County, Idaho
  Water 1132, 1157
Late Cambrian
  Missouri 2569
  --St. Francois Mountains
    2635
  --Southern 2635
  Oklahoma
    --Arbuckle Mountains 3632
    --Delaware County 2635
    --Murray County 3632
    --Wichita Mountains 3651
  Texas
    --Central 4297
  Wisconsin
    --Madison 4788
Late Canadian
  Southwest 3265
Late Cenozoic
  New Mexico
    --Northeastern 3214
    --Southeastern 3199
  North Dakota
    --Riverdale 3481
  Utah
    --West-Central 4447
Late Cretaceous
  North Dakota
    --Williston Basin 3485
  Wyoming
    --Carbon County 4863

Late Mississippian
  Illinois 1479
Late Ordovician
  Illinois
    --Southwestern 3646
  Missouri
    --Eastern 3646
  Oklahoma 3646
    --Anadarko Basin 3648
    --Arkoma Basin 3656
Late Paleocene
  North Dakota
    --Western 3495
Late Paleozoic
  Illinois
    --Illinois Basin 1480
    --Southern 1469
  Oklahoma
    --Arbuckle Mountains 3693
    --Ouachita Mountains 3693
Late Pennsylvanian
  Illinois
    --Illinois Basin 1192, 1267
  Kansas
    --Eastern 1744
  Oklahoma 3654
  Texas 3654
Late Pleistocene
  Illinois
    --Lake Michigan 1436
    --Lake Saline 1436
  Mississippi
    --Mississippi Sound 2480
  North Dakota
    --Sheridan County 3431
    --Southeastern 3423
  Texas
    --San Patricio County 4312
Late Quaternary
  Minnesota
    --Red River Valley 3462
  North Dakota 3464
    --Red River Valley 3462
Late Silurian
  Indiana
    --Southwestern 1579
Late Wisconsinan
  Indiana
    --Eastern 1523
  North Dakota
    --Logan County 3422
  Pennsylvania
    --Northeastern 3822
Late Woodfordian Jules 1342
Later Tertiary
  Idaho
    --Weiser 1137
Latimer County, Oklahoma
  Geology 3664

Latta, Indiana
  Acid Mine Conditions 1575
  Coxville Sandstone 1575
  Linton Formation 1575
  Pyrite 1575
Lauderdale County, Alabama
  Environmental Geology 6
  Geology 162
  Groundwater 162
Lauderdale County, Tennessee
  Stratigraphy 4147
Lava Tubes
  Oregon
  --Bend Area 3713
Lavaca Quadrangle, Arkansas
  Geology 395
Lavaca Quadrangle, Oklahoma
  Geology 395
Law 3034
  Alaska 311
Lawrence County, Alabama
  Geology 25
  Groundwater 25
Lawrence County, Pennsylvania
  Geology 3926, 3927
  Hydrology 3926, 3927
Laytonville Quarry, California
  Geology 423
  Mineralogy 423
Leachate 735, 1454, 1455
  Wisconsin 4823
Leached Outcrops 2865
Leaching 3120, 3153
  Alabama
  --Crenshaw County 95
Lead
  Alaska
  --Chugach Mountains 326
  --Nelchina Glacier 326
  British Columbia 4613
  California 424
  --Death Valley 508
  --Inyo County 490, 508
  --Modoc District 490
  --Panamint Butte Quadrangle 490
  --Plumas County 544
  Georgia
  --Lincoln County 950
  --Metasville Quadrangle 950
  --Wilkes County 950
  Kansas
  --Linn County 1761
  Maryland
  --Piedmont Upland 2234
  Missouri
  --Southeast 2638
  Pennsylvania 3891
  Tennessee

  --East 4153
  --Northwest 4153
  Utah 4436
  Virginia
  --Staunton Quadrangle 4585
  Washington
  --Kootenay Arc 4613
  --Northeastern 4613
  --Stevens County 4622
Leavenworth Limestone
  Kansas 1865
Lebanon Valley, Pennsylvania
  Carbonate Rocks 3929
  Hydrogeology 3929
Lee County, Florida
  Fresh Water 856
  Saline Water 792, 856
Lee County, Illinois
  Drilling 1447
Lee County, Kentucky
  Waterflooding 2039
Lee County, Texas
  Heavy Minerals 4290
  Wellborn Formation 4290
Lee County, Virginia
  Geology 2054
Lee Creek Member
  Indiana
  --Southeastern 1500
  Kentucky 1500
Lee Formation 2107
Leesburg Quadrangle, Virginia
  Geology 4565
Le Flore County, Oklahoma
  Geology 3625, 3664, 3676
  Lynn Mountain Syncline 3676
Legal Constraints
  Idaho 1178
Legal Guides
  California 472
Legislation
  Hawaii 993
Lehigh County, Pennsylvania
  Hydrology 3941
  Jacksonburg Formation 3795
  Martinsburg Formation 3941
  Water 3942
Leighton Quadrangle, Alabama
  Geology 208
  Mineral Resources 208
Lemei Rock, Washington
  Heat Flow 4660
Lemhi County, Idaho
  Geochemistry 1187
  Geology 1187
Lemhi Pass, Idaho
  Mineralogy 1147
  Rare Earth Deposits 1147
  Thorium Deposits 1147

Lemhi Pass District, Montana
  Thorium 2804
Lennox, South Dakota
  Groundwater 4091
Leon County, Florida
  Geology 741
  Groundwater 741
Leonard Harrison State Park,
  Pennsylvania
  Guide 3902
Lesterville, South Dakota
  Water 4068
Lettsworth, Louisiana
  Water Resources 2196
Lewis and Clark County, Montana
  Geochemistry 2827
Lewiston, Idaho
  Carbonate Rocks 1146
  Economic Geology 1146
Libby, Montana
  Groundwater 2773
Libby River, Alaska
  Geochemistry 230
  Geology 230
Liberty County, Georgia
  Groundwater Pumping 952
  Phosphate 971
Lichens
  Connecticut 703
Licking County, Ohio
  Bedrock Geology 3585
  Glacial Map 3560
Lightweight Aggregate
  Indiana
    --Northwestern 1587
Lignite
  Alabama
    --Marengo County 149
    --South 56
  Colorado
    --Cheyenne Basin 628
    --Denver Basin 621, 628
  Louisiana
    --Northwest 2156
  North Dakota 3433, 3434
    --Dunn Center 3491
    --Western 3460
  Texas 4216, 4217, 4277,
    4314, 4339
    --Coastal Plain 4285
    --East 4200, 4226, 4333
    --Gulf Coast 4325
Lihue, Hawaii
  Drilling Logs 999
  Pumping Tests 999
Limberlost Dolomite
  Indiana 1570
Lime
  Indiana 1511

Lime Products
  California 424
Limestone 1260, 1299, 1520, 1689,
  4747
  Alabama 53
    --Stevenson Area 134
  California 424, 427
    --Coast Ranges 435
  Florida 875
    --Hernando County 748, 749
    --Panhandle 873
    --Pasco County 748, 749
  Illinois 1193, 1201, 1271, 1330,
    2130
    --Effingham 1310
    --Exchange 1292
    --Fulton County 1226
    --Jersey County 1304
    --Kaskaskia Valley 1202
    --McDonough County 1226
    --Madison County 1246
    --Marion County 1292
    --Schuyler County 1226
    --Scott County 1328
    --Southern 1245
  Illinois Basin 1522
  Indiana 1487, 1490, 1503, 1504,
    1557, 1585, 2130
    --Clark County 1607
    --Southeastern 1500
    --Southern 1486, 1494
  Kansas 1716, 1743, 1755, 1771,
    1831, 1853, 1865
    --Cherokee County 1717
    --Eastern 1740, 1744, 1821
    --Southeastern 1695, 1766
  Kentucky 1486, 1494, 1500, 2130
    --Appalachian Region 1929
    --Barkley Lake 1941
    --Bowling Green 1963
    --Jefferson County 1607
    --Northwestern 1940
    --Ohio River Valley 2016
    --Somerset 1954
    --South-Central 1959
    --Warren County 1963
    --Western 1930, 2006
  Michigan
    --Chippewa County 2359
    --Mackinac County 2359
    --Schoolcraft County 2359
  Missouri
    --Interstate 55 2642
    --Macon County 2620
    --Putnam County 2609
    --Randolph County 2620
    --Southwestern 2619, 2630
  Montana 2694
    --Granite County 2807

Limestone (Cont.)
Montana (Cont.)
  --Rattler Gulch 2807
Nebraska
  --Southeastern 2858
New Mexico 3251
New York 3832
North Dakota 3436
Ohio 3574
  --Allen County 3521
  --Marion 3611
  --Northwestern 3587
Oklahoma 1853
Pennsylvania 3832, 3864, 3865,
  3869
  --Central 3802
  --Dauphin County 3794
  --Lancaster County 3848
  --Mount Joy 3848
South Carolina
  --Coastal Plains 3977
Tennessee 4109
  --Lower Cumberland 1930
  --Middle 4133
Texas 4291
  --North-Central 4192
Washington
  --Western 4604
West Virginia 4744
  --Barbour County 4679
  --Greenbrier County 4684,
  4735
  --Monroe County 4735
  --Upshur County 4679
Limestone County, Alabama
  Geology 104, 164
  Gravity Surveys 146
  Groundwater 164
  Mineral Resources 104
Limestone Injection 3586
Limestone Terranes 43, 44, 45,
  46, 47, 48, 49
  Alabama
  --Columbiana 114
  Southwestern States 49
Limnology
  Ohio
  --Lake Erie 3580
Lincoln County, Georgia
  Copper 950
  Lead 950
  Stream Sediments 950
  Zinc 950
Lincoln County, Kentucky
  Economic Geology 2131
Lincoln County, Montana
  Geochemistry 2700, 2717
  Geology 2684, 2739
  Minerals 2739

Stream Sediments 2700
Lincoln County, Nevada
  Geology 2874
  Minerals 2874
Lincoln County, New Mexico
  Geochemistry 3121
  Geology 3106, 3121, 3129
  Minerals 3106
  Molybdenum 3121
Lincoln County, Oregon
  Environmental Geology 3723
Lincoln County, Wyoming
  Faults 4888
  Petrography 4864
  Rocks 4864
Lincoln Fold
  Illinois 1465
Lincoln Quadrangle, Virginia
  Geology 4568
Linden Quadrangle, Virginia
  Geology 4600
  Hydrogeology 4601
Lineations
  Texas
  --Coast 4320
Lineville East Quadrangle, Alabama
  Geology 64
Lingle Formation
  Illinois
  --Southern 1217
Linn County, Iowa
  Geology 1674
  Groundwater 1674
Linn County, Kansas
  Geology 1737
  Gravity Survey 1761
  Groundwater 1737
  Lead Mine 1761
  Magnetic Survey 1761
  Zinc Mine 1761
Linn County, Missouri
  Water 2655
Linn County, Oregon
  Environmental Geology 3726
Linton Formation
  Indiana
  --Greene County 1575
  --Latta 1575
Linton Member
  North Dakota 3466
  South Dakota 3466
Liquefaction 1286
Liquid Propane Gas
  Nebraska
  --Omaha 2856
Liquid Wastes
  Maryland 2292
  Ohio 3538
  Pennsylvania 3783

Lithification
West Virginia 4678
Lithium
California 424
Lithofacies
Delaware 715
Illinois 2130
Indiana 2130
Kentucky 2130
Lithography
Texas
--South 4298
Lithology 3829
California
--Inyo County 490
--Modoc District 490
--Panamint Butte Quadrangle
490
North Dakota 3368
Wyoming
--Carbon Basin 4886
--South-Central 4886
Lithostratigraphy 739
Indiana 1598
Minnesota
--Southeastern 2403
Nebraska 2847
Pennsylvania
--Central 3788, 3803
--Clearfield County 3833
Wisconsin
--Madison 4788
--Western 4796
Little Arkansas River Basin,
Kansas
Saline Water 1785
Little Bigelow Mountain, Maine
Geology 2219
Little Black Peak Quadrangle,
New Mexico
Geology 3129
Little Falls Creek, Alaska
Geochemistry 227
Geology 227
Little Grassy Lake, Illinois
Sediments 1458
Trace Elements 1458
Little Hatchet Mountains, New
Mexico
Geology 3107
Little Kanawha River Basin,
West Virginia
Drillers Logs 4750
Sediments 4750
Springs 4750
Streamflow 4750
Test Borings 4750
Water 4750, 4755
Wells 4750

Little Long Valley, Idaho
Groundwater 1179
Precipitation 1140
Little Mountain, South Carolina
Geology 3986
Kyanite 3986
Little Pine Mountain, California
Geology 409
Little Rock, Arkansas
Cookeite 386
Quartz 386
Rectorite 386
Little Tujunga Canyon, California
Earthquakes 430
Geology 430
Livingston County, Illinois
Coal 1275
Livingston County, Kentucky
Engineering Geology 1943
Fluorspar 2122
Geology 2122
Livingston County, Missouri
Water 2650
Llano, Texas
Cambrian 4288
Conodonts 4303
Geology 4238, 4247
Iron Ore 4288
Rocks 4222
Sandstone 4288
Lockwood Valley, California
Geology 498
Loess
Illinois 1283
--Central Western 1467
Mississippi 2457, 2513, 2514
South Dakota
--Southwestern 4051
Logan County, Arkansas
Geology 388, 393, 396
Logan County, Kentucky
Exploration 2103
Logan County, North Dakota
Glacial Drift 3425
Mollusca 3422
Logan County, Ohio
Devonian 3571
Glacial Geology 3567, 3570
Pre-Olentangy 3571
Rocks 3571
Silurian 3571
Stratigraphy 3571
Logging 2022, 2406, 3880
Delaware
--Coastal Plain 728
Kentucky
--Western 2035
Mississippi 2491, 2541
Lone Mountain Quadrangle, Nevada

Lone Mountain Quadrangle, Nevada
(Cont.)
  Geology 2893
Long County, Georgia
  Phosphate 971
Long Hill Quadrangle, Connecticut
  Bedrock Geology 680
Long Island, New York
  Groundwater 2964
  Hydrology 3019
  Sediments 2980
Longhorn Caverns, Texas
  Geology 4237
Loomis Quadrangle, Washington
  Geology 4616
  Minerals 4616
Lordsburg, New Mexico
  Geology 3280
Lorman, Mississippi
  Water Supply 2544
Los Alamos County, New Mexico
  Clay 3164
Los Angeles Basin, California
  Pliocene 523
  Sedimentary History 523
Los Angeles County, California
  Aggregates 580
  Castaic Formation 520
  Clay 537
  Faulting 546
  Geology 542, 546
  Landslides 546
  Mineral Land 584
  Miocene 520
  Pliocene 523
  Sedimentation 520, 523
Lost Burro Formation
  California 540
Lost Creek Coal Field, Utah 4391
Lost River Chert
  Kentucky 1998
Louisville Limestone
  Indiana
    --Clark County 1607
  Kentucky
    --Jefferson County 1607
Love County, Oklahoma
  Pennsylvanian Fusulinids 3637
Lowe, E. N. 2469
Lowell, Vermont
  Magnetic Data 4490
  Ultramafic Bodies 4490
Lower Cabaniss
  Missouri
    --Western 2616
Lower Carbondale
  Kentucky
    --Western 1952
Lower Cretaceous

  Florida 882
  Kansas 1787
  Mexico
    --Coahuila 4300
  Oklahoma
    --Southern 3657
  Texas 4172, 4291, 4294
    --Central 4306
    --North-Central 4199, 4329
    --Northeast 4213
    --South 4313
  Virginia
    --Coastal Plain 4508
Lower Dockum Group
  Texas
    --Panhandle 4332
Lower Eocene
  Tennessee
    --Western 4119
  Wyoming
    --Washakie County 4848
Lower Jackson
  Mississippi 2534
Lower Keweenawan
  Michigan
    --Northern Peninsula 2365
Lower Mississippian
  Kansas
    --Western 1756
  Missouri
    --Southwestern 2619, 2625
  Nebraska
    --Southwestern 1756
Lower Ordovician
  Ohio 3507
  Oklahoma
    --Arbuckle Mountains 3665
  Pennsylvania
    --Central 3797, 3802, 3809
  Tennessee
    --East 4144
Lower Osagean
  Missouri
    --Southeastern 2619
Lower Paleozoic
  Appalachians 4742
  Maryland 2247
  Nebraska 2857
  West Virginia 4742
  Wisconsin
    --Western 4795
Lower Pennsylvanian
  Oklahoma 3626
  Virginia
    --Southwestern 4511
Lower Permian
  Kansas 1681, 1699, 1853
    --Southern 1782
  Oklahoma 1853

--Northern 1782
Texas
--Palo Duro Basin 4337
Lower Pliocene
South Dakota 4053
Lower Precambrian
Minnesota
--East-Central 2434
--Northeastern 2425, 2449
--Vermilion District 2391,
2425
Lower Tertiary
Texas
--Gulf Coast 4210, 4225
Lower Willwood Formation
Wyoming
--Washakie County 4848
Lower Wolfcamp Environment
Texas
--North-Central 4174
Lowndes County, Georgia
Phosphate 965, 971
Phosphorite 966
Sand 965
Loysville Quadrangle, Pennsylvania
Groundwater 3938
Lucas County, Ohio
Floods 3608
Shore Erosion 3608
Ludlow, Vermont
Magnetic Data 4485
Ultramafic Bodies 4485
Ludlow Formation
North Dakota
--Southwestern 3486
Luna County, New Mexico
Geology 3223
Petrography 3186
Trace Base Metals 3186
Tres Hermanas Stock 3186
Lunenburg, Vermont
Geology 4475
Lutesville, Missouri
Kaolin 2585
Luzerne County, Pennsylvania
Groundwater 3951
Hydrology 3939
Pleistocene Sediments 3939
Lynn Mountain Syncline, Oklahoma
Geology 3676
Lyon County, Nevada
Geology 2876
Minerals 2876

Maalaea, Hawaii
Drilling Log 1034
Pumping Tests 1034
McCartys Basalt Field, New

Mexico 3167
McCone County, Montana
Coal 2738
McConnells Mill State Park, Pennsyl-
vania
Guide 3907
McCormicks Creek State Park, Indiana
Geology 1613
McCracken County, Kentucky
Economic Geology 1937
McDonough County, Illinois
Coal 1230
Limestone 1226
Mississippian 1226
McDowell County, West Virginia
Coal 4706
McGregor Isles, Florida
Saline Water Intrusion 792
McHenry County, Illinois
Dam Sites 1401
Drilling 1383
Geology 1294, 1395, 1401
Planning 1294
Reservoir Sites 1401
Surface Materials 1395
McIntosh County, Georgia
Groundwater Pumping 952
McIntosh County, North Dakota
Geophysics 3397
Groundwater 3397
Water 3397
Wells 3397
McIntosh County, Oklahoma
Geology 3633, 3634, 3635
Minerals 3634
Petroleum 3633, 3635
McKean County, Pennsylvania
Oil and Gas 3881
McKenzie County, North Dakota
Geology 3416
Groundwater 3492
Land Use 3492
Minerals 3492
Pollution 3492
McKenzie Formation
West Virginia 4741
--Wayne County 4699
Mackinac County, Michigan
Drill Cores 2359
Fiborn Limestone 2359
Maclaren River, Alaska
Geology 239
Mineral Deposits 239
Stream Sediments 324
McLean County, Illinois
Coal 1329
Geology 1225, 1329
Gravity Field 1225
Niagaran Reefs 1233

McLean County, Illinois (Cont.)
 Oil 1233
 Pennsylvanian 1329
McLean County, North Dakota
 Aquifers 3367
 Geology 3365
 Glaciation 3365, 3367
 Groundwater 3366, 3367
 Water 3366
 Wells 3366
McLeod County, Minnesota
 Geology 2412
 Magnetic Map 2412
McNairy Sand
 Tennessee
 --West 4152
Macomb, Illinois
 Geology 1380
 Geophysics 1380
 Groundwater 1380
Macon County, Illinois
 Gravel 1302
 Sand 1302
Macon County, Missouri
 Minerals 2620
McPherson County, South Dakota
 Geology 4024, 4025
 Water 4024
Macrofauna
 Delaware 710
Macropaleontology
 Mississippi
 --Clarke County 2536
Madeira Quadrangle, Ohio
 Bedrock Geology 3578
Madison, Wisconsin
 Early Ordovician Rocks 4788
 Late Cambrian Rocks 4788
Madison Area, Alabama
 Environmental Geology 5
 Hydrology 5
Madison County, Alabama
 Caves 110
 Environmental Geology 1, 3, 4,
 5, 8
 Geology 202, 203, 204, 206
 Hydrology 1, 3, 4, 5, 8
 Mineral Resources 202, 203,
 204, 206
Madison County, Illinois
 Limestone 1246
Madison County, Indiana
 Geology 1601
Madison County, Montana
 Bedrock Geology 2786
Madison County, Ohio
 Gravel 3518
 Sand 3518
Madison County, Virginia

 Geology 4502
 Minerals 4502
Madison Formation
 North Dakota 3482
 --Northwestern 3473
Madison Group
 Montana 2828
Madison Quadrangle, Mississippi
 Environmental Geology 2538
Maestrichtian
 North Dakota 3466
 South Dakota 3466
Mafic Intrusive
 Missouri
 --Southeast 2627
Mafic Rocks
 Maryland
 --Piedmont 2303
Magdalena, New Mexico
 Springs 3169
 Wells 3169
Magdalena Mining District, New Mexico
 Geochemistry 3139
 Mercury 3139
 Stream Sediments 3139
Magdalena Mountains, New Mexico
 Groundwater 3178
Magma 3738
Magnesium 374
Magnesium Compounds
 California 424
Magnetic Anomalies
 California
 --San Fernando 430
 Iowa
 --Fayette County 1648
 Maryland
 --Chesapeake Bay 2308
 Michigan
 --Southern Peninsula 2342
 Minnesota
 --Lake Washington 2415
 --Meeker County 2415
 North Dakota
 --Pembina County 3479
Magnetic Concentration
 Alabama
 --Crenshaw County 92
Magnetic Data
 California
 --Northern 422
 Michigan
 --Crystal Falls 2348, 2349,
 2350, 2352
 --Iron River 2345, 2346, 2347,
 2348
 --Panola Plains 2351
 --Rule River 2351

Virginia
--Albemarle County 4574
--Coastal Plain 4558
--Fluvanna County 4574
--Piedmont 4519
Vermont
--Dover 4485
--Lowell 4490
--Ludlow 4485
--Windham 4485
West Virginia
--Allegheny Plateau 4743
--Western Valley and Ridge
4743
Magnetic Maps
Minnesota
--McLeod County 2412
Magnetic Studies
California
--Island Mountain Mine 506
--Trinity County 506
Utah
--Desert Mountain 4368
--Juab County 4368
Magnetic Surveys
Kansas
--Linn County 1761
Nevada
--Reno 2925
Magnetite
Pennsylvania
--Cornwall 3875
Magothy Aquifer, Maryland
Evaluation 2300
Water Levels 2306
Magothy Sediments
Middle Atlantic Coastal Plains
2289
Mahantango Formation
Pennsylvania
--South-Central 2798
Mahomet Bedrock Valley, Illinois
Glacial Deposits 1265
Hydrogeology 1265
Majuba Hill, Nevada
Geology 2887
Minerals 2887
Maka Ridge, Hawaii
Drilling Logs 998
Pumping Tests 998
Makena, Hawaii
Drilling Logs 1019
Pumping Tests 1019
Malama-Ki, Hawaii
Drilling Logs 1003
Pumping Tests 1003
Malama-Ki Well 99, Hawaii
Drilling Logs 1003
Pumping Tests 1003

Malvern Quadrangle, Ohio
Geology 3558
Mammals
North Dakota
--Western 3495
Wyoming
--Washakie County 4848
Mammut Americanus
Missouri
--Holt County 2615
Man
Mississippi
--Adams County 2540
Mancos Shale 4437
Mandan, North Dakota
Geology 3414, 3500
Hydrogeology 3500
Land Use 3500
Manganese 369
California 424
Manhattan, Kansas
Structural Geology 1763
Mansfield, Indiana
Geology 1528
Mantell
Western United States 3270
Mantle 3738
Manzanita Mountains, New Mexico
Fishes 3189
Pennsylvanian 3189
Wild Cow Formation 3189
Mapping 1177, 1407, 1672, 1673,
1837, 1841
Alaska
--Coast Range 236
Georgia 951
Illinois 1307
Mexico 3048
New Jersey 2944, 3050, 3078
Oregon 3757
Washington 4666
Maps 1126, 2056
Alabama 193, 194
Alaska 313
Appalachians 2310
California 414, 424, 474, 543,
571
Kansas 1808
Kentucky 2050, 2085
Michigan 2325
--Southern Peninsula 2342,
2355
Minnesota 2426
--McLeod County 2412
Missouri 2562
Montana 2825, 2834
--Madison Group 2828
Nebraska
--Southeastern 2855

Maps (Cont.)
  New Jersey 3086
  North Dakota 3499
    --Central 3469, 3470
  Ohio
    --Licking County 3560
  Oregon 3745
  Pennsylvania
    --Southeastern 3849
  South Carolina 3990
  South Dakota 4027
Maquoketa Formation
  Iowa
    --Northern 2439
  Minnesota 2439
Maquoketa Group
  Indiana 1598
Marathon Basin, Texas
  Haymond Formation 4292
  Pennsylvanian 4292
  Sedimentary Petrology 4292
Marengo County, Alabama
  Gravity Surveys 117
  Lignite 149
  Mineral Resources 149
Marin County, California
  Geologic Hazards 460
  Geology 440, 460
Marine Boundaries 3052
Marine Clays
  Maine
    --Farmington 2205
Marine Mining
  Alaska 319
Marine Paleocurrent
  Kansas
    --Eastern 1740
Marine Pollution 3025
Marine Sediment Environments
  Connecticut
    --Norwalk Islands 650
Marine Terraces
  California 560
Marine Transition
  Pennsylvania
    --Southern 3813
Marion, South Dakota
  Groundwater 4072
Marion County, Alabama
  Gravity Surveys 119
  Mineral Resources 136
  Radiation Surveys 89
Marion County, Illinois
  Limestone 1292
  Oil Production 1292
Marion County, Indiana
  Environment 1610
  Geology 1488, 1533, 1610
Mariposa County, California

Granitic Intrusive Units 512
Marksville Period
  Louisiana
    --Lasalle Parish 2141
Marl 1431, 1444
  Indiana 1508
Marshall County, Alabama
  Groundwater 32
Marshall County, Illinois
  Coal 1204, 1275
Marshall County, Kentucky
  Geology 1935, 1943
Marshall County, South Dakota
  Geology 4018, 4019
  Water 4018, 4019
Marthaville, Louisiana
  Water Resources 2200
Martinsburg Formation
  Pennsylvania
    --Dauphin County 3935
    --Lehigh County 3941
    --Northampton County 3941
Martinsville East Quadrangle, Virginia
  Geology 4587
Martinsville West Quadrangle, Virginia
  Geology 4570
Mascot Formation
  Tennessee
    --East 4144
Mason County, Illinois
  Gravel 1320
  Sand 1320
Mason County, West Virginia
  Groundwater 4680
Masonry 1561
Massies Corner Quadrangle, Virginia
  Geology 4554
Mastodons
  Missouri
    --Holt County 2615
Matagorda Bay, Texas
  Geochemistry 4221
  Sediments 4221
Matagorda Island, Texas
  Shoreline 4202
Matagorda Peninsula, Texas
  Shoreline 4204
Matanuska Valley, Alaska
  Zeolite 332
Materials 2975
Maui Island, Hawaii
  Drilling Logs 995, 1000, 1019,
    1025, 1026, 1028, 1034, 1056,
    1057
  Ferruginous Bauxite 1006
  Floods 1022, 1041, 1059, 1091,
    1100
  Geology 982
  Groundwater 982

Irrigation 1064
Pumping Tests 995, 1000,
    1019, 1025, 1026, 1028,
    1056, 1057
Soil 1106
Water Resources 1042, 1049,
    1051, 1052, 1053, 1081,
    1094, 1099
Maumee River, Ohio
    Gravel 3577
    Sand 3577
Mauna Kea, Hawaii
    Water Resources 1005
Maury Formation
    Alabama
    --Gipsy Limestone County
        58
Maxville Quadrangle, Missouri
    Engineering Geology 2552
Mayfield, Kentucky
    Aquifer 1942
    Eocene 1942
    Pumping Test 1942
Maysville Area, Alabama
    Environmental Geology 3
    Hydrology 3
Maysville Conodonts
    Indiana 1996
    Kentucky 1996
    Ohio 1996
Mazon Creek Compression 1319
Meade County, Kansas
    Pleistocene 1820
Meagher County, Montana
    Geology 2720, 2798
    Groundwater 2798
    Ore 2720
Mean-Dip Maps 1126
Measured Sections
    Virginia
    --Blue Ridge 4525
Measurements
    New Jersey 3054
Medfra Quadrangle, Alaska
    Geochemistry 257
    Geology 257
Medicine Bow Mountains, Wyoming
    Early Protozoic 4883
    Metasedimentary Rocks 4883
    Precambrian Rocks 4847
    Stratigraphy 4887
    Uranium 4883
Meeker County, Minnesota
    Magnetic Anomaly 2415
Megacyclothem
    Kansas
    --Southern 1782
    Oklahoma
    --Northern 1782

Megalopteris
    Illinois
    --West Central 1356
Megaphyton
    Illinois 1348
Mellow Valley Quadrangle, Alabama
    Geology 64
Memphis, Tennessee
    Geology 4157
Menard Formation
    Illinois Basin 1495
Mendocino County, California
    Engineering Geology 563
    Geology 423
    Mineralogy 423
Menominee County, Michigan
    Groundwater 2373
Menominee County, Wisconsin
    Forest Ecology 4761
    Soil 4761
Mentasta Pass Area, Alaska
    Geology 265
Meramecian Stage
    Kansas 1736
    --Western 1683
Mercer County, Illinois
    Coal 1295
Mercer County, Missouri
    Water 2646
Mercer County, New Jersey
    Geology 3084
    Groundwater 3084
Mercer County, North Dakota
    Cenozoic 3355
    Cretaceous 3357
    Geology 3355
    Glaciation 3355, 3357
    Groundwater 3356, 3357
    Springs 3356
    Tertiary 3357
    Water 3357
    Wells 3356
Mercer Quadrangle, Pennsylvania
    Geology 3926, 3927, 3944
    Groundwater 3944
    Hydrology 3926, 3927
Mercur, Utah
    Geochemistry 4435
Mercury 372
    Alabama
    --Mobile River 137
    --Tennessee River 137
    --Tombigbee River 137
    --Warrior River 137
    Arkansas
    --Southwest 400
    California 424
    Illinois 1419, 1441

Mercury (Cont.)
New Mexico 3216
--Magdalena Mining District
3139
Ohio
--Lake Erie 3535
Texas
--Trans-Pecos 4278
Mercury Sulfide Ores 2898
Meredosia Channel, Illinois
Seismic Reaction 1344
Meridianville Quadrangle, Alabama
Environmental Geology 1
Hydrology 1
Merostomes
New Mexico 3253
Mesa County, Colorado
Mining Resources 625
Mesa Rica Sandstone
New Mexico
--Tucumcari Basin 3211
Mesabi Range, Minnesota
Biwabik Iron Formation 2384
Mesozoic
Alaska
--Alaska Peninsula 297
California
--Atascadero 569
--San Luis Obispo County
569
--Santa Lucia Range 569
Connecticut
--Central 668
Minnesota
--Southeastern 2393
Mississippi
--Central 2456
--Southern 2456
New Mexico
--South-Central 3090
--Southwestern 3090
Virginia
--Gulf Basin 2148
--Northern Coast 2148
Metal Creek, Alaska
Geochemistry 260
Geology 260
Metallurgy 2909
Minnesota
--Cuyuna Range 2409
Metals 735, 4627
Alabama 113
Illinois 1373
Montana
--Bearmouth 2784
--Garnet 2784
--Powell County 2763
New Mexico
--Luna County 3186

North Carolina
--Carolina Slate Belt 3290
Pennsylvania 3868
Virginia 4530
Washington
--Cascade Range 4610
Western United States 2868
Wisconsin
--Upper Mississippi Valley
4786, 4803, 4804
Metamorphic Rocks
Alaska
--Teklanika River 285
--Toklat River 285
California
--Ben Lomond Mountain 514
--Coast Ranges 419
--Klamath Mountains 415
--Santa Cruz County 514
Florida 750
Pennsylvania
--Chester County 3936
Metamorphic Stratigraphy
Maine
--Western 2219
Metamorphism
Maine
--Vassalboro 2215
--Waterville 2215
Metasedimentary Rocks
Wyoming
--Medicine Bow Mountain 4883
Metasville Quadrangle, Georgia
Copper 950
Lead 950
Stream Sediments 950
Zinc 950
Meteorites 3756
Michigan 2316
North Dakota 3447
Methane
Colorado 632
Louisiana
--Southwestern 2189
Utah 4442
Methow Valley, Washington
Geology 4620
Miami, Florida
Aquifers 862
Water Levels 862
Miami County, Kansas
Geology 1718
Groundwater 1718
Miami River, Florida
Salt Water 832
Mica 366
California 424
New Mexico 3159
Michigan Basin

Conodonts 1519
Devonian 2363
Geology 2360
Middle Devonian 1519
Mississippian 2363
Paleozoic 2360
Potassium Salts 2366
Shale 2363
Stratigraphy 1519, 2360, 2363
Microcrystalline Silica 1300
Microfauna
Mississippi 2498
Microfossils
Kansas 1699
Mississippi
--Hinds County 2489
Oklahoma
--Southern 3657
South Dakota
--Wounded Knee 4058
Micropaleontology 4437
Missouri
--St. Charles County 2614
--St. Louis County 2614
Micropeel Technique 1688
Micropollution 3011
Microrelief Structures
Texas
--Coast 4196
Microvertebrates 4034
Middle Atlantic Coastal Plain
Cretaceous 2289
Petrology 2289
Sediments 2289
Middle Chester
Indiana
--Southern 1486, 1494
Kentucky 1486, 1494
Middle Devonian
Illinois 1301
--Southern 1297
Indiana 1558, 1609
--Northern 1527
Michigan Basin 1519
Ohio
--Northwestern 3579
Pennsylvania 3789
--Central 3821
--Western 3892
Middle Mississippian
Illinois 1330
Middle Ordovician
Illinois 1191
Kentucky
--Boone County 1949
--Central 1964
Minnesota 2440, 2442
Missouri
--St. Charles County 2614

--St. Louis County 2614
Ohio
--Southwestern 3614
Oklahoma
--Southern 3679
Pennsylvania
--Central 3788, 3799, 3803,
    3809
--Western 3799
Tennessee
--Sequatchie Valley 4151
Middle Paleozoic
Nebraska 2857
Middle Precambrian
Minnesota
--East-Central 2424, 2434
--Northeastern 2443
Middle Silurian
West Virginia 4741
--Wayne County 4699
Middlebury, Vermont
Geology 4495
Middlesboro, Kentucky
Water 1955
Middletown, Delaware
Columbia Sediments 707
Geology 707
Geophysics 707
Hydrology 707
Midlothian Quadrangle, Virginia
Geology 4577
Midway Formation
Mississippi 2479
Mifflintown Quadrangle, Pennsylvania
Groundwater 3938
Migmatitic Terranes
Minnesota
--Minnesota River Valley 2394
Milford Flat Quadrangle, Utah
Geology 4438
Mineralogy 4438
Milford Quadrangle, Connecticut
Surficial Geology 679
Millard County, Utah
Geology 4439
Mineralization 4439
Millboro Quadrangle, Virginia
Geology 4562
Millerella Fusulinid
New Mexico 3258
Texas 3258
Milton, Vermont
Geology 4497
Milton Quadrangle, Vermont
Geology 4479
Mine Drainage 1185
Alabama
--Cane Creek Basin 122
--Oakman 122

Mine Drainage (Cont.)
Alabama (Cont.)
--Walker County 122
Mine Water
Montana 2842
Mined Land 1896, 1903
Mineral Analyses
Alabama
--Warrior Basin 86
Mineral Belts
Idaho 1131
Mineral Conservation
Colorado 635
Mineral County, Nevada
Geology 2875
Minerals 2875, 2911
Mineral Economics
California 427
Mineral Elements 346
Mineral Fuels
Illinois 1266
Mineral Industry
Alabama 97, 168
Arizona 359
California 424
Colorado
--Mesa County 625
Florida 801, 805, 806, 807,
808, 809
Georgia 915
Idaho 1159
Illinois 1285, 1375
Kansas 1828, 1829, 1830, 1833,
1839, 1845, 1855, 1873, 1878,
1889, 1893, 1899
Kentucky 1919, 1921, 1922,
1968, 1971, 1974, 1976, 1978,
1980, 1983, 1985, 1989, 1991,
1994, 1999, 2001, 2004, 2007,
2010, 2012, 2014, 2133
Maryland 2248, 2249, 2252,
2253, 2254, 2255, 2257, 2261,
2262, 2264, 2265, 2271, 2272,
2278
Mississippi
--Hinds County 2497
--Rankin County 2522
Missouri 2566
Montana 2681, 2688, 2698,
2703, 2714, 2725, 2731,
2737, 2743, 2748, 2751,
2756, 2760, 2766, 2770,
2775, 2778
New Mexico 3192
North Carolina 3293
Pennsylvania 3823, 3825,
3826, 3827, 3830, 3831,
3835, 3841, 3843, 3845,
3846, 3850, 3853, 3854,
3857, 3859
Texas 4259, 4260, 4261, 4262, 4263,
4264, 4265, 4266, 4267, 4268,
4269, 4272, 4273, 4274, 4280
Virginia 4515, 4518, 4520
Wisconsin 4765, 4766
Mineral Lands
California
--Los Angeles 584
Wisconsin 4832
Mineral Laws
Alabama 190
Mineral Paragenesis
Kentucky
--Central Kentucky Mineral Dis-
trict 1982
Mineral Preparation
Alaska
--Friday Creek 333
--Kantishna Mining District 333
Mineral Producers
California 475, 485
Florida 783
Georgia 914
Kansas 1829, 1835, 1873, 1889
Maryland 2250, 2258, 2277
Minnesota 2398, 2401, 2402
Mississippi 2542
Missouri 2560, 2589
South Carolina 3989
Wisconsin 4797
Mineral Production
Illinois 1213, 1229, 1248, 1263,
1274, 1291, 1303, 1311, 1325,
1333
Mineral Properties 312
Mineral Rights
Alaska 299
New Mexico 3115
Mineral Technology
New Mexico 3101, 3110, 3117
Mineral Wells 2327
Mineralization
Alabama
--Clay County 108
--Cragford District 108
--Randolph County 108
Alaska
--Alaska Peninsula 330
--Midas Mine 250
--Stepovak Bay 330
--Sulphide Gulch 250
--Valdez 250
California
--Music Valley 496
--Riverside County 496
Minnesota
--Birchdale 2428
--Indus 2428

--Koochiching County 2428
New Mexico
--Apache Warm Springs
3157
--Socorro County 3157
Texas
--North 4315
Utah
--Beaver County 4402
--Church Hills 4439
--Millard County 4439
--Tooele County 4356
--Wah Wah Range 4402
Wyoming
--Park County 4849
Mineralogy 1397, 1414
Alaska
--Healy 287
California 514
--Kalker Quarry 518
--Laytonsville Quarry 423
--Mendocino County 423
--Santa Cruz County 518
Colorado 594
Idaho
--Lemhi Pass 1147
Illinois 1203, 1256
Michigan 2317
Minnesota 2446
--Southeastern 2421
Missouri
--Western 2616
Pennsylvania
--Eastern 3897
Utah
--Beaver County 4438
--Great Salt Lake 4427
--Milford Flat Quadrangle
4438
--Starr District 4438
Washington
--Grays Harbor 4671
West Virginia
--Morgantown 4688
Minerals 884, 956, 1381, 1425,
1576, 1631, 1709, 1869, 1872,
1901, 2320, 2883, 2906, 2907,
2909, 2940, 3124, 3143, 3151,
3155, 3191, 3544, 3680, 3883,
4627, 4694, 4697, 4811, 4812
Alabama 96, 148, 151, 183
--Barbour County 173
--Butler County 172
--Calhoun County 139
--Colbert County 205, 207,
208
--Fisk Quadrangle 206
--Geneva County 175
--Houston County 174

--Leighton Quadrangle 208
--Limestone County 104
--Madison County 202, 203,
204, 206
--Marengo County 149
--Marion County 136
--Moontown Quadrangle 202
--New Hope Quadrangle 204
--Perry County 144
--Pride Quadrangle 205
--Toney Quadrangle 203
--Tuscumbia Quadrangle 207
--Wilcox County 124, 176
--Winston County 181
Alaska 302
--Ahtell Creek 241
--Alaska Range 340
--Cantwell 339
--Coast Range 236
--Denali 239
--Dolomi 262
--Edgar Creek 340
--Horn Mountains 335, 337
--Knight Island 251
--Maclaren River 239
--Mount Hayes Quadrangle 249
--Prince of Wales Island 262
--Prince William Sound 251
--Rainy Creek 249
--Slana District 241
--South-Central 241, 335, 337,
339
--Tanana Region 328
--West Fork Glacier Area 340
--Wood River 340
--Yukon Region 328
Arizona 344, 351, 352, 361, 377
--Fort Apache Indian Reserva-
tion 348
--Gila River Indian Reservation
350
California 412, 465
--Alpine County 447
--Coast Ranges 424
--Contra Costa County 497
--Death Valley National Monu-
ment 566
--Great Valley Province 418
--Imperial County 446
--Klamath Mountains 415
--Monterey County 444
--Mount Diablo 497
--Offshore 420
--Orange County 454
--Ord Mountain District 494
--San Bernardino County 494,
501
--San Diego County 442
--Shasta County 445

Minerals (Cont.)
California (Cont.)
--Sierra Nevada 417
--Trinity County 443
--Ventura County 453
--Victorville 501
Colorado 586, 594, 624, 631,
641, 643
--Crystal Valley 604
--Eagle County 604
--Garfield County 604
--Gunnison County 604
--Mesa County 625
--Pitkin County 604
--Roaring Fork Valley 604
Florida
--Holmes County 744
--Walton County 744
--Washington County 744
Georgia 911
--Bethesda Church 941
--Echols County 964
--Greene County 941
Great Basin 2938
Idaho 1188
--Bonner County 1123
--Lakeview Mining District
1176
--Owyhee County 1158
--Salmon River Drainage
Basin 1167
--Silver City 1158
Illinois 1283, 1332, 1368,
1370, 1373, 1374
--Illinois Basin 1354
Indiana
--Washington County 1499
Kansas 1750, 1791, 1840, 1857,
1890
Kentucky 2032, 2124
Maryland
--Southern 2293
Michigan 2312, 2313, 2340
--Dickinson County 2357
Minnesota 2386
--Itasca County 2433
Missouri 2517, 2543, 2551,
2566, 2588, 2592, 2600,
2607, 2611, 2644
--Bates County 2639
--Copiah County 2509
--Hinds County 2492
--Interstate 44 2602
--Macon County 2620
--Putnam County 2609
--Randolph County 2620
--Rankin County 2520
--St. Charles County 2595
--Smith County 2525

--Wayne County 2528
Montana 2718, 2791, 2837
--Beaverhead County 2746
--Eastern 2796
--Flathead County 2739
--Flint Creek Range 2745
--Lincoln County 2739
--Powell County 2763
--Sanders County 2682
Nevada 2864, 2882, 2885
--Bunkerville Mining District
2862
--Churchill County 2884
--Clark County 2861, 2862
--Douglas County 2876
--Esmeralda County 2879
--Eureka County 2863
--Humboldt County 2859
--Lander County 2899
--Lincoln County 2874
--Lyon County 2876
--Majuba Hill 2887
--Mineral County 2875, 2911
--Nye County 2878
--Ormsby County 2876
--Pamlico Mining District 2875
--Pershing County 2887, 2890
--Storey County 2869, 2870
--Washoe County 2869, 2870
--White Pine County 2886
New Mexico 3098, 3192
--Bernalillo County 3092
--Bonito Mining District 3177
--Delaware Basin 3213
--East Central 3193
--Gallinas Mountains 3106
--Grant County 3094
--Lincoln County 3106
--Nogol Mining District 3177
--Northeastern 3214
--Rio Arriba County 3102
--Sandoval County 3092
--Santa Fe County 3092
--Torrance County 3106
North Carolina 3294, 3302, 3303
--Carolina Slate Belt 3290
--Orange County 3287
--Wake County 3292
North Dakota 3372
--Divide County 3492
--McKenzie County 3492
--Williams County 3492
Ohio
--Washington County 3509
Oklahoma 3685
--Bryan County 3653
--Choctaw County 3647
--Custer County 3638
--Love County 3660, 3661, 3662

--McIntosh County 3634
--Muskogee County 3649
--Noble County 3655
--Woods County 3628
Oregon 3706, 3709, 3720,
  3739, 3744, 3745
--Coos County 3722
--Curry County 3735
--Deschutes County 3731
--Douglas County 3717
--Josephine County 3742
--Klamath County 3708
--Lake County 3708
--Upper Chetco Drainage
  3730
Pennsylvania 3863
--Grace Mine 3862
South Carolina 3962, 3984,
  3989, 3993
--Hilton Head Island 3975
--Oconee County 3983
--York County 3982
South Dakota 3997, 4041
Southwest 3181
Tennessee 4110
--East 4135
--Hamilton County 4123
--Knox County 4114
--West 4152
Texas 4163, 4165, 4176,
  4181, 4239
--Burleson County 4290
--Chinati Mountains 4181
--Dallas 4214
--East 4289
--Lee County 4290
--Presidio County 4181
--West 4307
Utah 4346, 4349, 4374, 4378,
  4379, 4381, 4384
--Box Elder County 4388
--Garfield County 4380
--Great Salt Lake 4449,
  4450
--Juab County 4372
--Piute County 4375
--San Juan County 4414,
  4415, 4416
--Sanpete County 4358
--Tooele County 4372
Virginia 4521, 4526, 4537
--Fluvanna County 4503
--Frederick County 4504
--Greene County 4502
--Madison County 4502
--Page County 4505
Vermont 4499
Washington 4611, 4633
--Cathlamet 4606

--Cowlitz County 4606
--Kelso 4606
--King County 4615
--Loomis Quadrangle 4616
--Okanogan County 4616
--Puget Sound 4659
--Wahkiakum County 4606
--Whatcom County 4609
West Virginia 4673, 4713, 4714
--Charleston 4682
Western North America 2939
Wisconsin 4767, 4771, 4777, 4782,
  4784, 4814, 4832
Wyoming 4833, 4838, 4872
--Laramie County 4861
--Silver Crown Mining District
  4861
Miners
  California 472
  Nevada 2943
Minerva Quadrangle, Ohio
  Bedrock Geology 3566
Mines 343, 2330, 2358, 2910, 3123,
  3847, 4812
  Alaska 310
  --Ketchikan District 327
  California 485
  --Alpine County 447
  --Death Valley National Monu-
    ment 566
  --Imperial County 446
  --Monterey County 444
  --Orange County 454
  --San Diego County 442
  --Shasta County 445
  --Trinity County 443
  Colorado 609, 610
  --Mesa County 625
  Indiana 1597
  --Greene County 1575
  --Latta 1575
  Kansas
  --Linn County 1761
  Maryland
  --Western 2298
  Michigan
  --Crystal Falls 2352
  --Iron River 2345, 2346
  Missouri
  --Southeast 2638
  Montana 2680, 2687, 2697, 2702,
    2712, 2724, 2730, 2736, 2742,
    2747, 2750, 2755, 2759, 2765,
    2769, 2838
  --Beaverhead County 2746
  --Flint Creek Range 2745
  --Sanders County 2682
  Nevada 2908
  --Goldfield Mining District 2920

Mines (Cont.)
New Mexico 3131
North Dakota
--Dunn Center 3491
Oregon
--Josephine County 3766
Pennsylvania 3868
--Allegheny County 3861
--Juniata County 3836
--Westmoreland County
3861
Tennessee 4110
--East 4144
Texas
--South 4246
Washington
--Whatcom County 4609
Wisconsin 4782
Mining 343, 473, 483, 961, 1184,
2719, 2764, 2909, 2982, 3153,
4811, 4829
Alabama 183
Alaska 300, 311, 314, 319,
321
--Native Lands 316
Arizona
--Cochise County 358
--Pima County 360
--Santa Cruz County 362
--Yuma County 363
Colorado 586, 607
--Cheyenne Basin 628
--Denver Basin 628
--Mesa County 625
Florida 789
Idaho
--Coeur D'Alene 1153
--South-Central 1151
Iowa 1636
Kansas
--Southeast 1834
Michigan 2361
Minnesota
--Cuyuna Range 2409, 2410
Mississippi 2472
Montana 2679, 2686, 2689,
2696, 2701, 2712, 2723,
2730, 2736, 2742, 2747,
2750, 2755, 2759, 2769,
2765, 2774, 2777, 2780,
2782
--Colstrip 2768
--Southeastern 2768
Nevada 2901, 2926
New Mexico 3128, 3138, 3168,
3282
North Dakota 3402, 3434, 3465
--West-Central 3493
Pennsylvania 3856

South Dakota 4031
Tennessee 4139
Texas 4176
--East 4200
Utah 4352, 4357, 4364
Washington 4619, 4625, 4630,
4631, 4632, 4634, 4636, 4661,
4667
--Okanogon County 4637
--St. Helens Mining District
4658
--Washougal Mining District
4658
West Virginia 4703, 4715
Western United States 2868
Wisconsin 4781, 4814
Minneapolis, Minnesota
Artesian Basin 2397
Environment 2389
Geology 2389
Hydrogeology 2397
Minnesota River Valley, Minnesota
Kaolin Clay 2414
Precambrian 2394
Terranes 2394
Minook Creek, Alaska
Geochemistry 220
Minot, North Dakota
Geology 3412
Miocene
California
--Los Angeles 520
Louisiana 2158
--South 2153
Maryland 2245, 2290
Nebraska
--Western 2854
North Carolina
--Grimesland 3306
--Pitt County 3306
South Dakota
--Cedar Pass 4064
--Jackson County 4064
--Southwestern 4049
--Wounded Knee 4058
Texas
--Coastal Plain 4334
Virginia 2290
--Coastal Plain 4508
--Southeastern 4506, 4563
Miogeosynclines
Nevada 4351
Utah 4351
Miospores
Iowa
--Southern 1665
Mission, South Dakota
Groundwater 4086
Mississippi Sound, Mississippi

Late Pleistocene 2480
Mississippi Valley
  Barite 2119
  Cambrian 2632
  Fluorite 2119
Mississippi Valley, Wisconsin
  Base Metals 4786, 4803, 4804
Mississippian
  Alabama 195
  --Gipsy Limestone County
    58
  Arkansas 3696
  --Northwestern 3695
  Illinois 1244, 1281, 1330,
    1479, 2130
  --Fulton County 1226
  --Illinois Basin 1277, 1314
  --McDonough County 1226
  --Schuyler County 1226
  --Southern 1257
  --Southwestern 1242
  Illinois Basin 1522
  Indiana 1490, 1525, 1526,
    1557, 1589, 2130
  --South-Central 1957, 2094
  --Southern 1501, 1580
  Kansas 1736
  --Western 1683, 1756
  Kentucky 1525, 2130, 2137
  --South-Central 1959
  --Southeastern 2106
  --Western 1957, 2094
  Michigan
  --Michigan Basin 2363
  Missouri 2604
  --St. Charles County 2614
  --St. Louis County 2614
  --Southwestern 2619, 2625
  Nebraska 2847
  --Southwestern 1756
  New Mexico
  --North-Central 3245
  North Dakota 3482
  --Central 3497
  Oklahoma 3696
  --Northeastern 3695
  Tennessee
  --Campbell 4159
  --Interstate 75 4159
  --Jellico 4159
  --Middle 4133
  --Northeastern 2106
  Texas 4236
  Virginia
  --Southwestern 4511
  West Virginia
  --Greenbrier County 4684
Missoula Basin, Montana
  Geology 2699

Groundwater 2699
Missoula County, Montana
  Geology 2690
  Ore 2690
Missoula Valley, Montana
  Water 2685
Missouri River, Iowa 1650
Missouri River Valley
  Pleistocene 1891
  Stratigraphy 1891
Missouri River Valley, Missouri
  Pleistocene Stratigraphy 2550
Missouri Valley
  Fox Hills Formation 3466
  Upper Cretaceous 3466
Missourian
  Oklahoma
  --Nowata County 3629
  --Washington County 3629
Mitchell, South Dakota
  Groundwater 4087
Mitchell County, Georgia
  Geology 930
  Groundwater 930
Mitchell Plain, Indiana
  Geomorphology 1535
  Groundwater 1535
  Hydrology 1535
Mitchell Quadrangle, Oregon
  Bedrock Geology 3714
Moab, Utah
  Brines 4403
Mobile, Alabama
  Electrical Resistivity 61
  Hydrology 61
  Water 150
Mobile Bay, Alabama
  Holocene Foraminifera 140
  Salinity Changes 140
Mobile County, Alabama
  Disposal 116
Mobile Delta, Alabama
  Gravity Surveys 132
Mobile River, Alabama
  Mercury 137
Modoc District, California
  Economic Geology 490
Modoc Plateau, California
  Geology 416, 424
  Minerals 416
Moenkopi Formation
  Utah
  --Southeastern 4377
Mogollon Rim, Arizona
  Paleozoic 382
  Uranium 382
Mohawkian Series
  Quebec 3247

Mojave Desert, California
  Geology 411, 424, 500
Mollusks 1342
  Illinois 1317
  --Northeastern 1343
  Mississippi 2554
  New Mexico
    --East-Central 3263
    --Northeastern 3214
    --Southeastern 3199
  North Dakota
    --Lake Agassiz 3430
    --Logan County 3422
    --Red Lake County 3430
    --Southeastern 3423
  Ohio 3502, 3503, 3504, 3505
  Texas
    --Copano Bay 4338
Molokai Irrigation Tunnel, Hawaii
  Groundwater Storage 1011
Molokai Island, Hawaii
  Drilling Logs 1002, 1013
  Geology 986
  Groundwater 986, 1011, 1050
  Hydrology 1050
  Irrigation 1011, 1061
  Loans 1076
  Pumping Tests 1002, 1013
  Soil 1104
  Streams 1050
  Water 991, 1061, 1075, 1078,
    1093
Molybdenite 3120, 3160, 3162
Molybdenum
  California 424
    --Plumas County 544
  New Mexico 3089
    --Lincoln County 3121
    --Nogal Peak 3121
    --Questa Mine 3135
Molybdenum Sulfide
  Arizona 380
Monazite
  South Carolina
    --Enoree River 3991
    --Palolet River 3991
    --Saluda River 3992
    --Savannah River 3992
    --Tyber River 3991
Monkton, Vermont
  Geology 4487
Monkton Kaolin Deposit, Vermont
  Resistivity Survey 4486
Monmouth, Illinois
  Geology 1484
  Groundwater 1484
Monocacy River, Maryland
  Soluble Dye 2256
Monongahela River Basin, West

Virginia
  Groundwater 4749, 4754
  Hydrology 4754
  Springs 4749
  Test Borings 4749
  Wells 4749
Monroe, Louisiana
  Groundwater 2188
Monroe County, Alabama
  Gravity Surveys 109
Monroe County, Indiana 1583
  Sedimentation 1564
Monroe County, Kentucky
  Subsurface Geology 2065
Monroe County, Michigan
  Environment 2354
  Geology 2354
  Groundwater 2354
Monroe County, Mississippi
  Landslides 2546
Monroe County, Ohio
  Coal 3520
Monroe County, Pennsylvania
  Geology 3958
  Groundwater 3958
Monroe County, Tennessee
  Gold 4116
Monroe County, West Virginia
  Limestone Karst 4735
Monroe Creek, South Dakota 4058
Montara Mountain Quadrangle, Cali-
  fornia
  Franciscan Formation 495
Monte Bello Ridge Mountain Study
  Area, California
  Environmental Geology 456
Monte Carlo Method 3191
Monteagle Limestone
  Tennessee
    --Middle 4133
Monterey County, California
  Mineral Resources 444
  Mines 444
  Stratigraphic Geology 505
  Tertiary 505
Montevideo, Minnesota
  Geology 2441
  Precambrian Rock 2441
Montgomery Area, Alabama
  Geology 17
  Groundwater 17
Montgomery County, Alabama
  Geology 17
  Groundwater 17
Montgomery County, Georgia
  Phosphate 971
Montogmery County, Kansas
  Geology 1818
  Groundwater 1818

Montgomery County, Pennsylvania
  Brunswick Formation 3932
  Groundwater 3933, 3940
Montgomery County, Tennessee
  Sinknoles 4130
Montezuma, Georgia
  Aquifers 942
Montmorillonite
  Nevada 2877
Montpelier, Vermont
  Geology 4493
Montsweag Bay, Maine
  Sedimentation 2220
Monuments
  New Jersey 3080
Moodus Quadrangle, Connecticut
  Bedrock Geology 683
Moodys Branch Formation
  Mississippi 2534
Moontown Quadrangle, Alabama
  Geology 202
  Mineral Resources 202
Moore Hollow Group
  Texas
  --Central 4323
Moorhead Coal Field, Montana
  Coal 2744
Moose River, Maine
  Geology 2216
Moraine State Park, Pennsylvania
  Guide 3901
Moraines
  Illinois
  --Shelbyville 1315
  Indiana 1588
  New Jersey
  --Sussex County 3083
Mordenite
  Alaska
  --Horn Mountains 335, 337
  --South-Central 335, 337
Morehouse Parish, Louisiana
  Groundwater 2179
Moreland Oil Pool, Ohio
  Geology 3547
Morgan County, Alabama
  Geology 23
  Groundwater 23
Morgan County, Indiana
  Geology 1572
  Land Use 1572
Morgan County, Tennessee
  Drillers' Logs 4141
  Well Samples 4141
Morgan County, Utah
  Geologic Hazards 4366
Morgan Quadrangle, Wyoming
  Structural Geology 4860
Morgantown, West Virginia

Lake Monongahela Deposits 4688
Morphology 1130
  Connecticut
  --Coast 665
  Virginia
  --Southeastern 4506
Morphometry
  Vermont
  --West Danville 4500
Morris County, New Jersey
  Water 3087
Morrison Formation
  New Mexico
  --San Juan Basin 3183
  Utah
  --Dinosaur Quarry Quadrangle
  4440
Morro Bay, California
  Geology 467
Morse, C. 4471
Morse, W. C. 2470
Morton County, North Dakota
  Geophysics 3395
  Groundwater 3395
  Water 3395
  Wells 3395
Moscow Basin, Idaho
  Groundwater 1173
  Water 1157
Mosquitoes 3076
Mount Blue State Park, Maine
  Geology 2212, 2222
Mount Carmel Fault, Indiana
  Guidebook 1937
Mount Diablo, California
  Geology 497
  Mineral Resources 497
Mount Galen, Alaska
  Tertiary 294
  Volcanic Formations 294
  Volcanoes 294
Mount Hamilton Quadrangle, California
  Geology 408
Mount Hayes A-6 Quadrangle, Alaska
  Geochemistry 255
  Geology 249, 254, 263
  Mineral Deposits 249
  Rock 234
  Stream Sediments 234
Mount Hermon Quadrangle, Virginia
  Geology 4538
Mount Hood, Oregon
  Geochemistry 3777
  Geology 3777
Mount Joy, Pennsylvania
  Limestone 3848
Mount Katahdin, Maine
  Geology 2207
Mount Pend Oreille Quadrangle, Idaho

Mount Pend Oreille Quadrangle,
  Idaho (Cont.)
  Geochemistry 1183
  Geology 1183
Mount St. Helens, Washington
  Eruption 4669
Mount Sidney Quadrangle, Virginia
  Geology 4548
Mount Velma Quadrangle, Nevada
  Geology 2869
Mount Wilson Quadrangle, Califor-
  nia
  Faulting 546
  Geology 546
  Landslides 546
Mountains 2957, 3065
Mountrail County, North Dakota
  Aquifers 3353
  Geology 3352, 3354
  Groundwater 3352, 3353
  Wells 3352
Mud 1299
  California
  --San Francisco Bay 530
Mud Facies
  Missouri
  --St. Francois Mountains
    2621
Mud Flows
  California
  --San Bernardino County
    577
  --Wrightwood 577
  Colorado
  --Crested Butte 613
  --Gunnison County 613
Mud Lake Quadrangle, Nevada
  Geology 2893
Muhlenberg County, Kentucky
  Aquifers 1939
  Coal 1938
  Fresh Water 1939
  Gas 1926
  Geology 1926
  Oil 1926
Mulberry Island Quadrangle, Vir-
  ginia
  Geology 4595
Muldraugh Formation
  Indiana 1589
Multiaquifer Wells 3923
Multivariate Analysis 4747
Multnomah County, Oregon
  Geologic Hazards 3724
  Rocks 3772
Munsungun Anticlinorium, Maine
  Stratigraphy 2217
Murphy Syncline, Georgia 894
Murray County, Georgia

Geology 954
  Groundwater 954
Murray County, Oklahoma
  Arbuckle Group 3632
  Late Cambrian 3632
  Ordovician 3632
  Timbered Hills Group 3632
  Trilobites 3632
Muscatatuck Group
  Indiana 1558
Muscatine County, Iowa
  Geohydrology 1675
Muscatine Island, Iowa
  Geohydrology 1675
Muscovite
  California 424
Music Valley, California
  Mineralization 496
Muskingum County, Ohio
  Bedrock Geology 3585
Muskogee County, Oklahoma
  Geology 3649
  Minerals 3649
Muskrats 701
Mustang Island, Texas
  Dunes 4327
  Land 4327
  Shoreline 4206
  Water 4327
Myakka River Basin, Florida
  Streamflow 840
  Test Wells 772

Naperville, Illinois
  Geology 1398
Nashville, Tennessee
  Landslides 4127
Nassau County, Florida
  Groundwater 830
Natchitoches Parish, Louisiana
  Water Resources 2164
National Accelerator Laboratory Site,
  Illinois
  Glacial Deposits 1312
National Forests
  Idaho
  --St. Joe National Forest 1133
National Bridge Quadrangle, Virginia
  Geology 4567
Natural Bridges
  Tennessee 4124
Natural Environments
  Georgia 960
Natural Gas
  Alabama 101, 128, 169, 185, 197,
    198, 200
  --North 42
  Arizona 353

California 424
Colorado 611
--Mesa County 625
Florida 770, 771, 780, 782,
788, 797
Illinois
--Ashmore Gas Area 1243
--Coles County 1243
Illinois Basin 1993
Indiana 1592
Kentucky
--Eastern 1912, 1916, 1917
Ohio 3519
Natural Resources 3075
Missouri 2592, 2594, 2597,
2601
Naugatuck Quadrangle, Connecticut
Surficial Geology 691
Nautiloids 3236, 3237
New Mexico 3246
Needles Range, Utah
Tertiary Volcanic Rocks 4421
Negro Colony, Alabama
Groundwater 83
Nelchina Glacier, Alaska
Copper 326
Gold 326
Lead 326
Silver 326
Nellie Juan River, Alaska
Geochemistry 244
Geology 244
Nelson County, North Dakota
Aquifers 3360
Geology 3358
Glaciation 3358, 3360
Groundwater 3359, 3360
Water 3359, 3360
Wells 3359
Nemaha County, Kansas
Geohydrology 1819
Nenana Highway, Alaska
Geochemistry 218
Neosho County, Kansas
Geology 1720
Groundwater 1720
Neosho River Valley, Kansas
Pleistocene 1729
Sedimentation 1729
Neshannock Quadrangle, Pennsylvania
Geology 3926
Hydrology 3926
Ness County, Kansas
Water 1825
New Albany Shale
Indiana 1518
--Delphi 1562
--Northwestern 1587

--Southern 1501
New Blaine Quadrangle, Arkansas
Geology 393
New Castle County, Delaware
Channels 713
Groundwater 705
Pleistocene 713
New Cypress Pool, Kentucky
Aquifers 1939
Freshwater 1939
New Harmony, Indiana
Geology 1582
New Haven Quadrangle, Connecticut
Geology 674, 698
New Hope Quadrangle, Alabama
Geology 204
Mineral Resources 204
New Middle Devonian
Indiana 1558
New Mississippian
Tennessee
--Middle 4133
New Orleans, Louisiana
Groundwater 2169
New Oxford Formation
Pennsylvania
--Adams County 3932
--Lancaster County 3934
--York County 3932
New Permian
Oklahoma 3669
New Providence Shale
Indiana 1490
New River, West Virginia
Springs 4752
Streamflow 4752
Water 4752
Wells 4752
New Stanton, Pennsylvania
Coal 3894
Geology 3894
New York, New York
Pollution 2986
Water 2986
Newark, Delaware
Aquifer Tests 731
Geology 721
Groundwater 721, 730
Hydrogeology 725
Well Tests 731
Newark-Gettysburg Basin, Pennsylvania
Triassic Sediments 3793
Newberry Mountains, Nevada
Geology 2881
Granite Complex 2881
Newburg Formation
Ohio 3537
West Virginia 4683, 4690

Newburg Gas
  West Virginia
  --Jackson County 2075
  --Kanawha County 2075
Newburg Oil Field, North Dakota
  3457
Newburg Sand
  West Virginia 2082
Newcastle Formation
  North Dakota
  --Williston Basin 3477
Newcastle Sand
  North Dakota 3476
Newcastle Sandstone
  South Dakota
  --Butte County 4059
  --Harding County 4059
Newhall, California
  Earthquakes 430
  Geology 430
Newport-Inglewood Structural Zone,
  California
  Earthquakes 555
  Geology 555
Newport News North Quadrangle,
  Virginia
  Geology 4595
Newport News South Quadrangle,
  Virginia
  Geology 4582
Newtown Quadrangle, Connecticut
  Bedrock Geology 689
Niagaran Reefs
  Illinois
  --Chicago 1463
  --De Witt County 1233
  --McLean County 1233
  --Southwestern 1338
  --Thornton 1466
Niagaran Series
  Ohio
  --Northeastern 3552
Niblack Anchorage Area, Alaska
  Geology 240
Nickel
  Alaska
  --Spirit Mountain 276
  California 424
  Minnesota
  --Duluth Complex 2407
  Oregon 3765
Niihau Island, Hawaii
  Geology 987
  Groundwater 987
  Petrography 987
Ninole, Hawaii
  Water Supply 1021
Niobium
  California 424

New Mexico
  --San Juan Basin 3122
Niobrara Formation
  North Dakota
  --Eastern 3471
Nippewalla Group
  Kansas
  --Western 1781
Nitrate 4318
Nitrate Contamination
  Delaware 723
Nitrogen Isotopes 4318
Nittany Anticlinorium, West Virginia
  Trenton Group 4698
Nittany Dolomite
  Pennsylvania
  --Central 3797
Nixon Fork, Alaska
  Geochemistry 257
  Geology 257
Noble County, Ohio
  Bedrock Geology 3573
  Coal 3520
Noble County, Oklahoma
  Geology 3655
  Minerals 3655
Nockamixon State Park, Pennsylvania
  Guide 3911
Nodaway County, Missouri
  Groundwater 2660
Noeggerathiales
  Illinois
  --West Central 1356
Nogal Mining District, New Mexico
  Minerals 3177
Nogal Peak, New Mexico
  Geochemistry 3121
  Geology 3121
  Nogal Peak 3121
Noise 3021
Noise Pollution 3067
Nonfenestrate Ectoprocta
  Alabama 53
Nonmetals 4627
Nonnumeric Sequences 1861
Norco, Louisiana
  Groundwater 2178
Norfolk Quadrangle, Virginia
  Geology 4545, 4546
Normapolles Pollen
  Western United States 3224
North Dakota Geological Survey 3468
North Horn Mountain, Utah
  Coal 4385
Northampton County, Pennsylvania
  Hydrology 3941
  Jacksonburg Formation 3795
  Martinsburg Formation 3941
Northampton County, Virginia

Groundwater 4532
Northumberland, Nevada
  Caldera 2918
  Tuff 2918
Northern Kettle Moraine State
  Park, Wisconsin 4798
Norwalk Islands, Connecticut
  Marine Sediment Environments
  650
Norwalk North Quadrangle, Con-
  necticut
  Bedrock Geology 690
Norwalk South Quadrangle, Con-
  necticut
  Bedrock Geology 690
Novato, California
  Geologic Hazards 460
  Geology 460
Nowata County, Oklahoma
  Hogshooter Formation 3629
  Missourian 3629
  Petrology 3629
Nuclear Waste Disposal
  Texas
  --East 4228
  --Palo Duro Basin 4220,
  4230
Nueces Bay, Texas 4304
Nye County, Nevada
  Geology 2878
  Minerals 2878

Oahe Formation
  North Dakota 3464
Oahu Island, Hawaii
  Deep Well Disposal 1045, 1092
  Floods 1022, 1038, 1040, 1047,
  1059
  Geohydrology 1045
  Geologic Map 977
  Geology 976, 978, 980, 1058
  Geophysics 980
  Groundwater 976, 980, 1007,
  1012
  Guide 977
  Irrigation 1065
  Petrography 980
  Sediments 1024
  Sewage 1045
  Streamflow 1012
  Streams 1024
  Water Resources 1001, 1012,
  1030
  Water Supply 978
  Water Withdrawals 1089
  Wells 979
Oak Grove Core
  Virginia 4557

Oakman, Alabama
  Acid Mine Drainage 122
Observation Wells
  Montana 2711, 2721, 2729, 2735
  Pennsylvania 3931
Ocean City, Maryland
  Groundwater 2302
Ocean Outfall
  Hawaii
  --Kapaa 1085
  --Kauai 1085
  --Waimanalo 1084
Oceanography
  Hawaii
  --Kapaa 1085
  --Kauai 1085
Oceans 3071
Ochoan Rocks
  New Mexico
  --Delaware Basin 3213
Oconee County, South Carolina
  Geology 3983
  Minerals 3983
Odessa, Delaware
  Columbia Sediments 707
  Geology 707
  Geophysics 707
  Hydrology 707
Ofelia Quadrangle, Alabama
  Geology 64
Offshore Areas
  Alabama 127, 187
  Offshore Drilling 736
  Offshore Exploration 736
  Offshore Production 736
Ogallala Formation
  New Mexico
  --Central-Eastern 3198
  --Northeastern 3215
  Texas 4340
  --Southern High Plains 4286
Ogdensburg-Culvers Gap Moraine,
  New Jersey 3083
Ogle County, Illinois
  Drilling 1447
Ohio County, Kentucky
  Coal 1938
Ohio County, West Virginia
  Groundwater 4675
  Oil and Gas 4681
Ohio River, Indiana
  Geology 1574
Ohio River Valley, Kentucky
  Limestone 2016
  Pumping 1945
Ohio Valley, Kentucky
  Limestone 1930
Ohiopyle State Park, Pennsylvania
  Guide 3904

Oil 1882, 2108, 2591, 4875
Alabama 101, 128, 169, 185,
    197, 198
--North 42
Appalachians
--Central 4742
Arizona 353
Colorado
--Mesa County 625
East-Central States 1984,
    1986, 1990, 1992, 1997,
    2000, 2002, 2005, 2008,
    2009, 2011, 2013, 2015,
    2134, 2136
Florida 770, 771, 780, 782,
    788, 797
Illinois 1339, 1441
--Dewitt County 1204, 1233
--Illinois Basin 1224
--McLean County 1233
--Wapella East Oil Pool
    1204
Kansas 1690, 1703, 1714,
    1722, 1732, 1905
--Southeastern 1702
Kentucky 1906, 1908, 1911,
    1913, 1914, 1915, 1918,
    1920, 1925, 1969, 1973,
    1975, 1977, 1979, 1981
--Butler County 1962
--Green County 2003
--Muhlenberg County 1926
--Southeastern 1960
--Taylor County 2003
--Western 2059
Louisiana 2159
Michigan 2322, 2343
--Michigan Basin 2363
Mississippi 2460, 2542
--Jasper County 2451
--Tinsley Field 2533
Missouri 2598
--Belton 2567
--Northwest 2557
--Putnam County 2609
Nebraska 2846, 2848, 2850
Nevada 2915, 2935
New Mexico 3141, 3142, 3163,
    3220
North Carolina 3300
North Dakota 3427, 3428, 3432,
    3435, 3437, 3439, 3446,
    3448, 3451, 3455, 3456,
    3461
--Williston Basin 3467
Ohio 3537, 3548, 3551, 3557
--Moreland Oil Pool 3547
--Wayne County 3547
Oklahoma

--Craig County 3621
Oregon 3747, 3748, 3751, 3753
Pennsylvania 3779
--Amity Quadrangle 3873
--Claysville Quadrangle 3873
--McKean County 3881
--Pittsburgh 3852, 3889
--Warren County 3881
--Warren Quadrangle 3871
--Youngsville Quadrangle 3872
South Dakota 4028, 4029, 4035
--Butte County 4059
--Harding County 4059
Texas 4170, 4179, 4184
--Northeast 4213
Utah 4342, 4347, 4367, 4431, 4445
--East 4398
--Grand County 4423, 4429
--P.R. Spring Area 4423, 4425
--Rozel Point 4395
--South-Central 4398
--Uinta Basin 4418, 4425, 4428
--Uintah County 4423, 4429
Washington 4642, 4665
West Virginia 4674, 4687, 4711,
    4742
--Barbour County 4679
--Braxton County 4677
--Brooke County 4681
--Clay County 4677
--Hancock County 4681
--Ohio County 4681
--Pocatalico River Basin 4696
--Upshur County 4679
Oil Conservation 2041, 2064
Oil Exploration 2056
Alabama 128
Iowa 1628
Kentucky 2036
--Butler County 2018
--Hopkins County 2026
Oil Fields 2071, 2073
Alabama 180, 200
--Gilbertown Oil Field 93
--South 201
Alaska
--Prudhoe Bay 342
Colorado 611
Indiana 1585
Kansas 1859
Mississippi 2545
North Dakota
--Nerburg 3457
--South Westhope 3457
Oil Industry
Illinois 1214
Northeast of Mississippi River
    2084
Oil Production

Alabama 80
Illinois
--Exchange 1292
--Marion County 1292
--Tuscola 1280
Indiana 1538, 1539, 1540,
1541, 1542, 1543, 1544,
1545, 1546, 1547, 1548,
1549, 1550, 1551, 1552,
1553, 1554, 1555
Kansas 1796, 1798, 1800,
1802, 1804, 1806, 1807,
1859, 1879, 1888, 1892,
1897, 1902
Kentucky 2024, 2031, 2044,
2057, 2076, 2086, 2095,
2112
--Adair County 2101
--Eastern 2105
--Western 2038
Oil Recovery
Kansas 1797, 1799, 1801,
1803, 1805
Oil Sand
Alabama 147
Oil Seeps
Georgia 923
Utah
--Rozer Point 4395
Oil Shale
Utah
--Uinta Basin 4418
Oil Springs Pool, Kentucky
Fracturing 2028
Oil Tests
Georgia
--Coastal Plain 893
Iowa 1626
Okaloosa County, Florida
Water Resources 784
Okanogan County, Washington
Copper 4617
Geology 4616, 4620
Minerals 4616
Mining 4637
Sediments 4617
Oklahoma County, Oklahoma
Groundwater 3670
Old Lyme Quadrangle, Connecti-
cut
Bedrock Geology 677
Surficial Geology 687
Old Tampa Bay, Florida
Hydrology 793
Sanitary Landfills 855
Olentangy
Ohio
--Erie County 3571
--Logan County 3571

Oligocene
Georgia 953
Louisiana 2158
South Dakota
--Big Badlands 4052
Southeastern States 39
Utah
--Uinta Basin 4370
Olive Hill District, Kentucky
Clay 1972
Geology 1972
Oliver County, North Dakota
Cenozoic 3355
Cretaceous 3355
Geology 3355
Glaciation 3355, 3357
Groundwater 3356, 3357
Springs 3356
Tertiary 3357
Water 3357
Wells 3356
Olowalu, Hawaii
Floods 1091
Olympic Peninsula, Washington
Foraminifera 4605
Stratigraphy 4605
Omaha, Nebraska
Geology 2856
Omega Quadrangle, Virginia
Geology 4542
Omilak Area, Alaska
Geology 246
Oneida, Idaho
Perlite 1145
Ono Quadrangle, California
Geology 425
Ontonagon County, Michigan
Groundwater 2380
Oolite Bodies
Illinois Basin 1522
Opal Mountain Quadrangle, California
Geology 411
Opals
South Carolina
--Coastal Plain 3980
Operations Research 2079
Ophiolites 3737
Orange County, California
Earthquakes 454
Faults 454
Fossils 445
Geologic-Environmental Maps 454
Geology 449, 454, 534, 550, 552,
553, 567, 568
Landslides 449, 454
Mineral Resources 454
Mines 454
Pliocene 523
Sedimentary History 523

Orange County, California (Cont.)
  Slope Stability 534
  Test Holes 449
Orange County, Florida
  Aquifers 859
  Lake Levels 763
  Water Resources 757, 817,
    837
Orange County, North Carolina
  Geology 3287
  Minerals 3287
Orangeburg County, South Carolina
  Groundwater 3985
Ord Mountain District, California
  Geology 494
  Mineral Deposits 494
Ordovician
  Illinois 1191, 1339
    --Northeastern 1481
    --Southwestern 3646
  Indiana 1598, 1996
  Kansas 1905
  Kentucky 1996
    --Boone County 1949
    --Central 1964
  Minnesota 2440, 2442
  Missouri 2569
    --Eastern 3646
    --St. Charles County 2614
    --St. Louis County 2614
  Nevada 2891
  Ohio 1996, 3507
    --Southwestern 3614
  Oklahoma 3646
    --Anadarko Basin 3648
    --Arbuckle Mountains 3632,
      3644, 3665
    --Arkoma Basin 3656
    --Murray County 3632
    --Southern 3679
    --Wichita Mountains 3651
  Pennsylvania
    --Central 3788, 3797, 3802,
      3803, 3809
    --Western 3892
  Tennessee
    --East 4144
    --Sequatchie Valley 4151
  Texas
    --Central 4297
  Virginia
    --Lee County 2054
  West Virginia
    --Eastern 4698
    --Pendleton County 4698
  Wisconsin
    --Madison 4788
    --Southwest 4789
    --Western 3794, 3795, 3796

Ore 2329, 2898, 3124, 3196
  Alabama
    --Barbour County 166
    --Birmingham Red Iron District
      79
    --Butler County 170
    --Clay County 85
    --Crenshaw County 92, 170
    --Pike County 170
    --Pyriton 85
  Alaska
    --Friday Creek 333
    --Katishna Mining District 333
  Idaho
    --Delamar 1181
    --Flint 1181
    --Owyhee County 1181
    --Silver City 1181
  Indiana 1506
  Louisiana
    --Central North 2151
  Minnesota
    --Vermilion District 2438
    --Zenith Mine 2438
  Mississippi 2477
  Missouri
    --Howell County 2570
    --Oregon County 2571
    --Southeast 2638
    --Washington County 2633
    --Wayne County 2578
  Montana
    --Broadwater County 2683
    --Castle Mountains Mining Dis-
      trict 2720
    --Clinton Mining District 2690
    --Indian Creek District 2683
    --Meagher County 2720
    --Missoula County 2690
  New Mexico
    --Eagle Nest 3105
    --Otero County 3097
    --Sacramento (Hill Rolls) Mining
      District 3097
  South Carolina
    --Carolina Barite Belt 3969
  Southeast 4138
  Tennessee
    --East 4140
  Texas
    --Llano Region 4288
  Utah
    --Big Cottonwood Mining District
      4387
    --Salt Lake County 4387
  Virginia 4530
  Washington
    --Stevens County 4622
Oread Limestone

Kansas 1743
Oregon County, Missouri
  Brown Iron Ore 2571
Organic Contamination
  Florida
    --Broward County 857
    --Plantation Canal 857
  Orient, South Dakota
    Groundwater 4106
Oriskany Quadrangle, Virginia
  Geology 4578
Oriskany Sandstone
  West Virginia
    --Eastern 4693
Ormsby County, Nevada
  Geology 2876
  Minerals 2876
Oro Grande Series
  California
    --San Bernardino County
      501
    --Victorville 501
Oroville, California
  Earthquakes 565
Oroville Quadrangle, California
  Geology 407
Orthokarstenia Dietrich
  California 516
Orthophotoquads
  South Carolina 3990
Osage County, Oklahoma
  Carbonate Rocks 3677
  Shale 3677
Osage Group
  Iowa
    --Southeastern 1638
  Missouri
    --Southwestern 2619, 2625
Osmium
  California 424
Osteology 4382
  Florida
    --Reddick 738
Ostracods
  Alabama
    --Northwestern 50
  Indiana 1591
  Iowa
    --Northern 2439
  Louisiana 2149, 2150
    --Winn Parish 2150
  Minnesota 2439
  Mississippi 2519
  Oklahoma 3630, 3643
  Southeastern United States
    3284
  Virginia
    --Southeastern 4563
Otero County, New Mexico

Geology 3093, 3097
  Ore 3097
Otter Creek, Alaska
  Geology 246
Ouachita County, Arkansas
  Groundwater 403
Ouachita Mountains, Oklahoma
  Conodonts 3653
  Late Paleozoic 3693
  Mississippian 3696
  Pennsylvanian 3696
Ouachita Parish, Louisiana
  Water Resources 2174
Outcrops 2865
  Washington
    --Coast 4645
Outliers 3088
  Nebraska
    --Western 2854
Outwash
  Illinois
    --Northwestern 1296
Overthrust Belts 3875
Owen County, Indiana 1583
Owyhee County, Idaho
  Cenozoic 1171
  Geochemistry 1182, 1186
  Geology 1158, 1171, 1181, 1182,
    1186
  Mineral Resources 1158
  Ore Deposits 1181
Oxbow
  Idaho 1156
Oxides 2899
Oyster Shells
  North Carolina
    --Albemarle Sound 3291
Ozark Region
  Cambrian 2632
Ozarks 2605
  Mississippian 3696
  Pennsylvanian 3696
Ozaukee County, Wisconsin
  Geology 4826
  Groundwater 4826

P.R. Spring Area, Utah
  Oil 4423, 4425
  Sandstone 4423, 4425, 4428
Pacolet River, South Carolina
  Drainage Basin 3991
  Fluvial Monazite 3991
Padre Island, Texas
  Dunes 4327
  Land 4327
  Shoreline 4151, 4206, 4207
  Water 4327
Padre Island National Seashore, Texas

Padre Island National Seashore,
  Texas (Cont.)
  Geology 4251
Paducah Quadrangle, Illinois
  Geology 1216
Page County, Virginia
  Geology 4505
  Minerals 4505
  Silica 4534
Pagosa Hot Springs, Colorado
  Geothermal Investigations 645
  Hydrogeology 645
Paint River, Alaska
  Geology 243
Painted Rocks Lake, Montana
  2810
Paleobotany
  Idaho
    --Weiser 1137
  Illinois
    --Illinois Basin 1336
  West Virginia 4710
Paleocene
  Alaska
    --Alaska Range 282
  Florida 739
  Gulf Coast
    --East 196
  Maryland
    --Salisbury Embayment
      2307
  North Carolina
    --Coastal Plain 3308
  North Dakota 3421, 3483,
    3489
    --Southwestern 3486
    --Western 3495
    --Williston Basin 3453,
      3488
  Tennessee
    --Western 4119
Paleoecology 4437
  Kansas 1681, 1699
  Maryland 2290
  North Dakota 3368
  Virginia 2290
  Washington
    --Coast 4614
    --Point Grenville 4614
    --Raft River 4614
Paleogeography 4182
  Appalachians
    --Central 4742
  Maryland
    --Frederick Valley 2301
  Texas
    --Palo Duro Basin 4337
  West Virginia 4699, 4742
Paleogeology

Illinois
  --Illinois Basin 1314
Virginia
  --Coastal Plain 4509
Paleohydrology
  Alabama 60
    --Arid Regions 60
Paleontology 3746, 3846
  Florida
    --Reddick 738
  Georgia
    --Cochran Quadrangle 900
    --Perry Quadrangle 900
  Illinois
    --Lake Saline 1327
    --Northeastern 1343
    --Southwestern 1327
  Indiana 1525
    --Cincinnati Arch 1496
    --Northeastern 1532
    --Southeastern 1500
  Kentucky 1500, 1525
  Louisiana
    --Plaquemines Parish 2146
  Missouri 2569, 2608
  New Mexico
    --Northeastern 3215
  North Dakota 3424, 3466
  Pennsylvania
    --South-Central 3798
  South Dakota 3466
  Wyoming
    --Washakie County 4848
Paleovalley
  Kentucky
    --Edmonson County 1967
    --Hart County 1967
Paleozoic 899, 3088
  Alaska
    --Alaska Range 280
    --Gulkana Glacier 280
    --Rainbow Mountain 280
  Appalachians 4742
  Arizona
    --East-Central 382
  Georgia
    --Chattooga County 933
    --Coastal Plain 946
  Illinois
    --Illinois Basin 1480
    --Southern 1469
    --Western 1465
  Kansas
    --Northwestern 1696
  Kentucky
    --Jackson Purchase Region
      1956
  Maryland 2247
  Michigan

--Michigan Basin 2360
Minnesota
--Southern 2393, 2403
Missouri
--Jefferson County 2577
--St. Louis 2577
Nebraska 2857
New Mexico
--South-Central 3090
--Southwestern 3090
North Dakota 3436
--Eastern 3312, 3478
--Western 3444
Oklahoma
--Arbuckle Mountains 3693
--Ouachita Mountains 3693
West Virginia 4742
Wisconsin 4791, 4807
--Baraboo District 4799
--Western 4795
Palladium
California 424
Palo Duro Basin, Texas
Geohydrology 4220, 4230
Geology 4220, 4230
Lower Permian Facies 4337
Pennsylvanian 4231
Petroleum Rocks 4233
Palo Duro Canyon, Texas
Geology 4241
Paluxy Formation
Texas
--Northeast 4213
Paluxy River Valley, Texas
Glen Rose 4171
Palynology
Georgia
--Coastal Plains 946
Illinois 1301
--Lake Michigan 1352
Iowa 1664
Kansas 4704
Oklahoma
--Bryan County 3636
Texas
--North-Central 4299
Utah
--Garfield County 4362
West Virginia
--Morgantown 4688
Pamlico Mining District, Nevada
Geology 2875
Minerals 2875
Pan Evaporation
Hawaii 1077, 1112
Panamint Butte Quadrangle,
California
Economic Geology 490
Panamint Range, California

Petrology 540
Stratigraphy 540
Panhandle
Texas 3624
Panola Mountain State Park, Georgia
Geology 918, 919
Panola Plains, Michigan
Geology 2351
Magnetic Data 2351
Panther Creek, Idaho
Geochemistry 1187
Geology 1187
Papaloapan River Basin 3002
Paragenesis
Kentucky
--Central Kentucky Mineral Dis-
trict 1982
Paris Quadrangle, Arkansas
Geology 388
Park County, Wyoming
Gypsum 4856
Mineralization 4849
Park River, North Dakota
Geology 3419
Parke County, Indiana
Geology 1528
Parkston, South Dakota
Groundwater 4100
Parnassus Quadrangle, Virginia
Geology 4573
Particulates 1300
Pasco County, Florida
Hydrogeology 821
Suwannee Limestone 748, 749
Passaic County, New Jersey 3087
Passaic River, New Jersey
Aeration 2966
Engineering Economy 2974
Pollution 2566
Passaic Valley, New Jersey
Water Quality 2973
Passerine Birds
Florida
--Reddick 738
Patch Reefs 4746
Patrognathus
Kansas
--Western 1756
Nebraska
--Southwestern 1756
Patuxent River Basin, Maryland
Trace Elements 2279
Water 2279
Pawlet Quadrangle, Vermont
Bedrock Geology 4483
Pawnee County, Oklahoma
Petroleum Geology 3659
Pawpaw Creek Watershed, Ohio
Coal 3520

Paxson, Alaska
Geochemistry 248
Geology 248
Peace River Basin, Florida
Fluoride 833
Hydrology 836
Peat
California 424
Illinois
--Illinois Basin 1376
--Northeastern 1359
Indiana 1516
Pennsylvania
--Northeastern 3837
Pecos Country, New Mexico
Geology 3256
Pecos River, New Mexico
Geology 3276
Pecten Perplanus Stock
Southeastern United States 39
Pectinidae
Georgia 953
Peever, South Dakota
Groundwater 4104
Pegmatite 3124
California
--Ramona Pegmatite District
504
--San Diego County 504
Colorado
--South Platte District 634
New Mexico 3159
Pelekunu Valley, Hawaii
Water Resources 1093
Peloncillo Mountains, New Mexico
Geology 3212
Pembina County, North Dakota
Cement Rock 3478
Geology 3369
Glaciation 3369
Groundwater 3370, 3371
Magnetic Anomalies 3479
Springs 3370
Water 3370
Wells 3370
Pend Oreille Lake Region, Idaho
Geology 1154
Pendleton County, West Virginia
Nittany Anticlinorium 4698
Trenton Group 4698
Peninsular Florida
Cretaceous 776
Geology 776
Sand 871
Peninsular Ranges, California
Geology 424
Pennington County, South Dakota
Radiocarbon Dating 4061
Terraces 4061

Pennington Formation
Alabama 87
Pennsylvanian 3510, 3885
Alabama
--Black Warrior Basin 153
--Warrior Basin 88
Arkansas 3696
--Northwestern 3695
Illinois 1285, 1319, 1348, 1371
--Clark County 1236
--Dewitt County 1329
--Edgar County 1236
--Illinois Basin 1192, 1195,
1240, 1267, 1314, 1376, 1477,
1478
--La Salle County 1472
--McLean County 1329
--Piatt County 1329
--West Central 1356
Indiana 1591
--Greene County 1489
Iowa 1664
Kansas 1710, 1716, 1743, 1865,
4704
--Eastern 1740, 1744, 1821
--Rawlins County 1700
--South-Central 1727
--Southeastern 1695, 1766
--Stafford County 1700
Kentucky 2137
--Edmonson County 1967
--Hart County 1967
--Henderson County 2132
--Webster County 2132
--Western 1952, 2069
--Western Coal Field Region
1961
Mid-Continent 2613
Missouri 2611
--Central 2612
New Mexico 3258
--Chloride Flat 3185
--Grant County 3185
--Manzanita Mountains 3189
--Sangre De Cristo Mountains
3259
--Southeast 3241
Ohio 3506
--Northeastern 3525
Oklahoma 3626, 3654, 3671, 3672,
3673, 3696
--Ardmore Basin 3637
--Carter County 3637
--Love County 3637
--Northeastern 3695
Pennsylvania
--Clearfield County 3833
--Southeastern 3882
Tennessee

--Campbell County 4159
--Interstate 75 4159
--Jellico 4159
Texas 3258, 3654, 4236
--Dalhart Basin 4231
--Marathon Basin 4292
--North-Central 4175, 4192,
 4248, 4299, 4310, 4317
--Palo Duro Basin 4231
--Texas Panhandle 4231
Virginia
--Southwestern 4511
Pensacola, Florida
Monitor Wells 791
Water 812
Penstock Alinement
Hawaii 1010
Peoria County, Illinois
Gravel 1237
Sand 1237
Peoria Lake, Illinois
Sediments 1432
Trace Elements 1432
Percussion Tools 2102
Peridotite
Michigan
--Northern Peninsula 2365
Perlite
Arizona 375
California 424
Idaho
--Oneida 1145
Permeability 2068, 4689
Southwestern States 49
Permian
Kansas 1681, 1699, 1853
--Southern 1782
--Western 1781
Oklahoma 1853, 3654, 3669,
 3675
--Anadarko Basin 3624
--Jackson County 3663
--Northern 1782
--Western 3624
Texas 3654
--Anadarko Basin 3624
--North 4315
--North-Central 4175
--Palo Duro Basin 4337
--Panhandle 3624, 4232
Perry County, Alabama
Phosphate 144
Perry Quadrangle, Georgia
Economic Geology 900
Paleontology 900
Stratigraphy 900
Pershing County, Nevada
Geology 2887, 2890
Minerals 2887, 2890

Petersburg Formation
Indiana
--Sullivan County 1600
Petrified Forests
Alaska
--Unga Island 329
Petrochemistry
Missouri
--St. Francois Mountains 2631
Petrogenesis
New Mexico
--Sacramento Mountains 3195
Petrography 1397, 1414, 1426
California
--Mariposa County 512
--Yosemite Valley 512
Colorado 627, 630
--Raton Mesa 630
--San Juan 630
--Uinta 630
Florida
--Hernando County 740
--Pasco County 749
Hawaii
--Lanai Island 981
--Kahoolawe Island 981
--Niihau Island 987
--Oahu Island 980
Illinois 1222, 1285, 1301
--Centura 1234
Missouri
--Western 2616
New Mexico
--Eastern 3219
--Luna County 3186
--Sacramento Mountains 3195
North Carolina
--Carolina Slate Belt 3307
--Union County 3307
Oregon
--Central 3767
Pennsylvania
--Allegheny County 3861
--Central 3803
--Clearfield County 3874
--Westmoreland County 3861
Utah
--Uinta Basin 4428
West Virginia 4739
Wyoming
--Lincoln County 4864
--Salt River Range 4864
Petroleum 736, 2061
Alaska
--Alaska Peninsula 297
Appalachians 4700, 4701
California 424
Georgia 944
--Coastal Plain 906

Petroleum (Cont.)
  Indiana
    --Southwestern 1574
  Kentucky 2060
    --Butler County 2110
    --Cincinnati Arch Province
      2092
    --Jackson Purchase Region
      1956
  New Mexico 3126, 3150
    --Catron County 3096
    --Hidalgo County 3206
  North Dakota 3410
    --Northwestern 3473
    --Southwestern 3472
  Oklahoma
    --Love County 3662
    --McIntosh County 3633,
      3635
    --Pawnee County 3659
  Oregon 3754
  Pennsylvania 3884
  Texas
    --Palo Duro Basin 4233,
      4337
  Utah 4445
    --San Juan County 4415
Petroleum Drilling 736
Petroleum Exploration 736
  Florida 765
  Georgia 958
  Indiana 1586
Petroleum Industry 1882, 2070
  Alabama 103, 106, 111, 199
  Indiana 1515
Petroleum Production 736
  Florida 758, 761, 765
Petroleum Prospects
  Florida 765
Petrology 2934, 4747
  Alaska
    --Alaska Range 280
    --Gulkana Glacier 280
    --Healy 287
    --Rainbow Mountain 280
  California
    --Inyo County 540
    --Lost Burro Formation
      540
    --Panamint Range 540
  Connecticut
    --Central 697
  Maine
    --Western 2219
  Maryland
    --Piedmont 2303
  Middle Atlantic Coastal Plain
    2289
  Minnesota

    --Duluth 2418
    --Itasca County 2432
  Montana
    --Bitterroot Range 2788
  Oklahoma
    --Nowata County 3629
    --Washington County 3629
  Texas
    --Marathon Basin 4292
    --South 4298
  Utah
    --Dinosaur Quarry Quadrangle
      4440
    --Uinta Basin 4373
  West Virginia 4741
    --Wayne County 4699
  Wisconsin
    --Madison 4788
Phenol
  Wisconsin
    --Walworth County 4830
Philadelphia, Pennsylvania
  Geology 3791
Philmont Ranch Region, New Mexico
  Geochemical Anomalies 3146
Philpott Reservoir Quadrangle, Vir-
  ginia
  Geology 4576
Phosphate 2815
  Alabama
    --Perry County 144
  California 424
  Florida 789, 885
  Georgia
    --Atkinson County 967
    --Ben Hill County 967
    --Berrien County 967
    --Brooks County 965
    --Camden County 965
    --Chatham County 901, 971
    --Clinch County 965
    --Cook County 967
    --Echols County 971
    --Effingham County 965
    --Irwin County 967
    --Jeff Davis County 971
    --Lanier County 965, 971
    --Liberty County 971
    --Long County 971
    --Lowndes County 965, 971
    --Montgomery County 971
    --North-Central 967
    --Peninsular Georgia 969
    --Savannah 972
    --Thomas County 965
    --Ware County 965
    --Wayne County 971
  Idaho
    --Southeastern 1134

--Western Phosphate Field
   1175
Kansas 1728
Pennsylvania
--Juniata County 3836
Phosphate Sand
Georgia
--Effingham County 968
Phosphatic Facies
California
--Coast Ranges 419
Phosphoria Formation
Wyoming 4859
Phosphorite 973
Georgia
--Charlton County 966
--Clinch County 966
--Echols County 966
--Lowndes County 966
Photogeology 45
Alabama
--Alabama River 142
Physical Resources
Maine
--Knox County 2221
Physiography
Colorado
--Crested Butte 613
--Gunnison County 613
Mississippi
--Adams County 2540
Tennessee
--Cumberland Plateau 4143
--Sequatchie Valley 4143
Phytoplankton
New Jersey
--Hackensack Meadowlands
   3031
Piatt County, Illinois
Coal 1329
Geology 1329
Pennsylvanian 1329
Pickaway County, Ohio
Groundwater 3597
Pickens County, Alabama
Groundwater 30
Piedmont Area, Alabama
Electrical Resistivity Surveys
   115
Geophysical Methods 123
Geophysical Prospecting 100
Groundwater 100, 123
Test Drilling 123
Piedmont Delaware
Geophysics 719
Groundwater 730
Highway Design 719
Piedmont Georgia 959
Piedmont Maryland 2244, 2276

Bear Island 2246
Geology 2286
Groundwater 2288
Potomac River 2246
Rocks 2303, 2305
Piedmont North Carolina
Diabase Dikes 3301
Piedmont Pennsylvania
Geology 3796
Piedmont Province, Alabama
Residual Clay 74
Piedmont Upland, Maryland
Barite 2234
Cobalt 2234
Copper 2234
Iron 2234
Lead 2234
Zinc 2234
Piedmont Virginia
Magnetic Data 4519
Radiometric Data 4519
Pierce County, North Dakota
Aquifers 3364
Geology 3362
Glaciation 3362, 3364
Groundwater 3363, 3364
Water 3363, 3364
Wells 3363
Pierce Field, Pennsylvania
Medina Production 3879
Oriskany Production 3879
Piercement Structure
Washington
--Coast 4645
Pierre, South Dakota
Groundwater 4097
Pierre Formation
North Dakota
--Knife River Basin 3494
Pierson Formation
Missouri
--Southwestern 2619
Pike County, Alabama
Iron Ore 170
Pike County, Illinois
Coal 1230
Pike County, Kentucky
Sand 1953, 2030
Pike County, Ohio
Bedrock Geology 3596
Pima County, Arizona
Mining Properties 360
Pinchouts
Maryland
--Salisbury Embayment 2307
Pine Mountain, Kentucky
Geology 2125
Piney Point Aquifer, Delaware
Digital Model 732

Piney Point Aquifer, Delaware
(Cont.)
Hydraulics 729
Piney Point Aquifer, Maryland
Water Levels 2309
Pioneer Mountains, Idaho
Bedrock Geology 1162
Piper Diagrams 1883
Pitkin County, Colorado
Geology 604
Groundwater 604
Mineral Resources 604
Pits 3545
Pitt County, North Carolina
Upper Miocene Foraminifera
3306
Pittsburgh (No. 8) Coal
Ohio
--Belmont Field 3606, 3607
Pittsburgh, Pennsylvania
Clay 3890
Construction Aggregate 3886
Gas 3852
Geology 3810
Groundwater 3921
Oil 3852, 3889
Shale 3890
Surface Structure 3852
Piute County, Utah
Minerals 4375
Place Names
Ohio
--Northeast 3540
Tennessee 4117
Virginia 4527
Placer Deposits 466
Placer Gold 961
Placer Mining 300, 473
Plains
Idaho
--Snake River 1168
Texas
--Panhandle 4234
--Southern High Plains
4286
Plant Microfossils
Mississippi
--Hinds County 2489
Plantation Canal, Florida
Organic Contamination 857
Plants
Illinois 1371
Oklahoma
--Southern 3657
West Virginia 4708, 4709,
4710
Wisconsin
--Baraboo District 4801
Plaquemine, Louisiana
Groundwater 2176

Plaquemines Parish, Louisiana
Paleontology 2146
Plastic Clay
Hawaii
--Kauai 1006
Platinum
California 424
Wyoming
--Albany County 4854
Platte County, Wyoming 4852
Platteville Group
Illinois
--Northern 1358
Plattsmouth Limestone
Kansas 1716
Playas
Nevada 2888
Pleasanton Group
Kansas 1710
Pleasanton Shale
Kansas 1728
Plecoptera
Connecticut 663
Pleistocene
Delaware 706, 708, 714
--New Castle County 713
Georgia
--Coastal 937
Illinois 1196
--Central Northern 1212
--East Central 1471
--Lake Michigan 1436
--Lake Saline 1327, 1436
--Southeastern 1327
Iowa 1621, 1622, 1623
Indiana 1485
Kansas
--Meade County 1820
--Missouri River Valley 1891
--Neosho River Valley 1729
--Northeastern 2590
--Southeastern 1729
--Upper Tuttle Creek Reservoir
1774
Louisiana 2458
Maryland
--Eastern Shore 2280
--Salisbury 2280
Mississippi
--Mississippi Sound 2480
--Southern 2458
Missouri 2622
--Missouri River Valley 1891,
2590
Montana
--Bitterroot Range 2787
--Lake Missola 2787
Nebraska 2849
New Mexico
--East-Central 3263

North Dakota 3420, 3474
--Logan County 3422
--Sheridan County 3431
--Southeastern 3423
Ohio 3502, 3503, 3504, 3505
Pennsylvania
--Delaware Valley 3811
--Luzerne County 3939
--Northwestern 3805, 3815
--Wyoming Valley 3939
Texas
--Red River Basin 4284
--San Patricio County 4312
Virginia
--Coastal Plain 4510
--Southeastern 4563
Vermont 4484
West Virginia
--Morgantown 4688
Wisconsin
--Southern 4802
Pliocene
California
--Los Angeles Basin 523
--Los Angeles County 523
--Orange County 523
New Mexico
--East-Central 3263
South Dakota 4053
Plum River Fault Zone, Illinois
1347
Plum River Fault Zone, Iowa
1649
Plumas Copper Belt, California
Trace Elements 544
Plumas County, California
Trace Elements 544
Plume Fields 3028
Plummer Field, Indiana 1608
Plutonic Rocks
California
--Ben Lomond Mountain 514
--Santa Cruz County 514
Pluvial
Oregon
--Lake County 3776
Pluvial Climate
Nevada 2895
Pluvial Lakes
Nevada 2895
Pocahontas Quadrangle, Mississippi
Environmental Geology 2537
Pocatalico River Basin, West Vir-
ginia
Salt Groundwater 4696
Point Bars 4198
Point Grenville, Washington
Quinault Formation 4614
Point Loma Quadrangle, California

Geology 438
Point Reyes Peninsula, California
Geology 423, 440
Pointe Coupee Parish, Louisiana
Water Resources 2171, 2196
Pokagon State Park, North Dakota
Geology 1611
Pokakuloa, Hawaii
Drilling Logs 1020
Polk County, Florida
Groundwater 831
Hydrology 851
Water Resources 754
Wells 754
Polk County, Georgia
Geology 945
Groundwater 945
Pollen
Iowa
--Sumner Bog 1653
Maryland 2233
Western United States 3224
Pollutants 1454, 1455
New Jersey
--Hackensack Meadowlands 3031
Pollution 138, 2949, 2955, 2958,
2970, 2977, 2979, 2981, 2983,
2988, 2991, 2992, 2996, 2997,
3001, 3003, 3011, 3016, 3021,
3025, 3040, 3057, 3059, 3067,
3068
Connecticut 703
Illinois
--Greater Chicago Area 1428
Japan 3009
New Jersey
--Passaic River 2966
New York
--New York City 2986
North Dakota
--Divide County 3492
--McKenzie County 3492
--Williams County 3492
Rhine River 2984
Ruhr 2984
Polypora M'Coy
Indiana
--Southeastern 1530
Pomfret Dome, Vermont
Geochemistry 4492
Pompano Beach, Florida
Biscayne Aquifer 823
Hydrology 823
Ponds
Kentucky
--Inner Blue Grass Region 2127
Pools
Kentucky
--Southeastern 2106

Pools (Cont.)
  Tennessee
    --Northeastern 2106
Pope County, Arkansas
  Geology 391
Pope County, Illinois
  Fluorspar 1473
  Geology 1473
Poplar Interval
  North Dakota 3482
Population 1381
  Alabama 193
  Hawaii
    --Hawaii Island 1066
Poquoson Quadrangle, Virginia
  Geology 4584
Porosity
  Kentucky
    --Northeastern 2040
Portage Creek, Alaska
  Geology 238
Porter County, Indiana
  Geology 1602
Portsmouth, Ohio
  Geology 3516
Portuguese Bend, California
  Clay 537
Post-Glacial Stratigraphy
  Connecticut
    --Coasts 665
Post-Illinoian
  Iowa 1622
Post-Knox Unconformity
  Indiana 1586
Post-Miocene
  Virginia
    --Southeastern 4506
Post-Woodfordian
  Illinois
    --Northeastern 1343
Potash
  California 424
  Michigan
    --Michigan Basin 2366
  Montana
    --Northeastern 3498
  New Mexico
    --Carlsbad Potash District
      3104
  North Dakota 3436
    --Williston Basin 3498
  Utah
    --San Juan County 4415
Potassium Permanganate 3160
Potassium Salts
  Michigan
    --Michigan Basin 2366
Potholes
  Pennsylvania

    --Archbald Pothole State Park
      3900
Potomac Formation
  Delaware 715
Potomac Group
  Maryland 2233
Potomac River, Maryland
  Rocks 2246
Potomac River Basin, West Virginia
  Springs 4751
  Streams 4751
  Water 4756
  Wells 4751
Potomac Sediments
  Middle Atlantic Coastal Plain 2289
Potrillo Basalt Field, New Mexico
  3167
  Geology 3203
Pottsville Group
  Pennsylvania
    --Clearfield 3833
Poway Quadrangle, California
  Geology 438
Powder River, Montana
  Water 2776
Powder River Coal Basin, Wyoming
  Coal Geology 4878
Powder River County, Montana
  Coal 2727
Powder River Valley, Montana
  Groundwater 2722
Powell County, Montana
  Metallic Minerals 2763
Power 1416, 2960
Prairie Du Chien Group
  Wisconsin
    --Western 4796
Pratt County, Kansas
  Geology 1764
  Groundwater 1764
Precambrian 899
  California
    --Death Valley 521
    --Inyo County 521
    --Salt Spring Hills 521
  Illinois 1238
    --Northeastern 1482
  Kansas 1686, 1698, 1846, 1863,
    1904
    --Northwestern 1696
  Michigan
    --Upper Peninsula 2353
  Minnesota
    --Cook County 2395
    --East-Central 2424, 2434
    --Granite Falls 2441
    --Minnesota River Valley 2394
    --Montevideo 2441
    --North Shore 2392

--Northeastern 2425, 2443,
2449
--Southwestern 2441
--Vermilion District 2391,
2425
Missouri 2581, 2582, 2636
--St. Francois Mountains
2631, 2641
--Southeast 2624, 2626,
2627
Nebraska 2851, 2852, 2857
New Mexico 3114
--Central 3268
--La Madera Quadrangle
3091
--Rio Arriba County 3091
--South-central 3268
Ohio 3565
Pennsylvania
--Erie County 3851
--Northwestern 3834
Texas
--Llano Region 4222
Wisconsin
--Baraboo District 4799
--South-Central 4787
Wyoming 4871
--Medicine Bow Mountains
4847
--Southeastern 4847
Pre-Chappel
Texas
--Llano 4303
Precious Metals
Alabama 113
Nevada
--Central 2921
--Western 2921
Virginia 4530
Precipitation 2957, 3065
Hawaii 1060, 1072, 1073
--Kona 1068, 1069
--Molokai 1050
Idaho
--Caribou County 1140
--Dry Valley 1140
--Little Long Valley 1140
New Jersey 3062
Pre-Illinoian
Iowa 1621, 1622
Pre-Knox Stratigraphy
Indiana 1531
Pre-Olentangy
Ohio
--Erie County 3571
--Logan County 3571
Pre-Pennsylvanian
Illinois
--Illinois Basin 1240

New Mexico
--Roosevelt County 3184
Presidio Bolson
Texas 4311
Presidio County, Texas
Minerals 4181
Structure 4235
Pre-Silurian
West Virginia 4692
Presque Isle Quadrangle, Maine
Bedrock Geology 2210
Pressure Chambers 4689
Price Quadrangle, Virginia
Geology 4587
Price River, Utah
Hydrogeology 4451
Pride Quadrangle, Alabama
Geology 205
Mineral Resources 205
Prince George's County, Maryland
Faulting 2260
Groundwater 2235
Prince of Wales Island, Alaska
Geochemistry 252, 283
Geology 252, 283
Mineral Deposits 262
Rock 232
Stream Sediments 232, 235
Prince William Sound, Alaska
Geology 251
Mineral Deposits 251
Propane Gas
Nebraska
--Omaha 2856
Prospect-Darby Fault, Wyoming
Geometry 4888
Prospecting 961, 2108, 2251, 4811
Alabama
--Piedmont Area 100
Alaska 301, 303, 312, 315
--Bokan Mountain 276
--Southeastern 276
Missouri 2573
Washington 4651
Wisconsin
--Southwest 4793
Prospectors
Alaska 304, 312
California 472
Prospectors' Samples 371
Protozoic
Wyoming
--Medicine Bow Mountain 4883
Prudhoe Bay Oil Field, Alaska
Hydrology 342
Surficial Geology 342
Pryor Mountain Quadrangle, Montana
Geology 2832

Ptychaspid Biomere
  Texas
    --Central 4301
Public Health 3020
Public Utilities
  California
    --San Fernando Valley 432
  Mississippi
    --Hinds County 2537
Public Water
  Kentucky 1547
Puget Lowland, Washington
  Geology 4635
Puget Sound, Washington
  Geology 4659
  Groundwater 4659
  Land Use 4655
  Minerals 4659
  Shore Bluffs 4655
Pukoo, Hawaii
  Water 1078
Pulama, Hawaii
  Drilling Logs 1017
  Pumping Tests 1017
Pulaski County, Arkansas
  Cookeite 386
  Quartz 386
  Rectorite 386
Pulaski County, Kentucky
  Devonian 2128
  Geology 1966
  Silurian 2128
  Stratigraphy 2128
Pulmonate Gastropods
  Mississippi 2515
Pumice
  California 424
Pumicite
  California 424
Pumping
  Georgia
    --Liberty County 952
    --McIntosh County 952
  Hawaii
    --Alaeloa 1026
    --Haena 1032
    --Hamakuapoko 1025
    --Hanalei 998
    --Hanapepe 996, 1033
    --Hilo 1055
    --Kalaheo 1016
    --Kamalo 1002
    --Kamiloloa 1013
    --Kauai 999, 1016, 1032,
      1033, 1035
    --Kawaihae 1015
    --Keahole 1039
    --Keei 1014
    --Kepaniwai 1056
    --Kohala 994, 997, 1015
    --Kona 1014, 1039, 1054
    --Lahaina 1000, 1028
    --Lihue 999
    --Maalaea 1034
    --Maka Ridge 998
    --Makena 1019
    --Malama-Ki 1003
    --Maui 995, 1000, 1019, 1025,
      1026, 1028, 1034, 1056, 1057
    --Molokai 1002, 1013
    --Pulama 1017
    --Puna 1003, 1017
    --Waikapu 995, 1057
    --Waimea 1035
  Kentucky 1945
    --Mayfield 1942
  Maryland
    --Anne Arundel County 2232
    --Harford County 2231
    --Worcester County 2230
  South Dakota
    --Wall 4037
Pumpkinvine Creek Formation
  Georgia
    --Northern 959
Puna, Hawaii
  Drilling Logs 1003, 1017
  Pumping Tests 1003, 1017
  Water Resources 1036
Pungo River Formation
  South Carolina
    --Beaufort County 3285
Putnam County, Florida
  Ceramic Clay 762
  Geology 319
  Groundwater 753
  Surface Water 755
Putnam County, Illinois
  Coal 1275
Putnam County, Indiana
  Geology 1528
Putnam County, Missouri
  Minerals 2609
  Water 2648
Putnam County, West Virginia
  Groundwater 4680
Pyrite
  Alabama
    --Clay County 85
    --Pyriton 85
  Indiana
    --Greene County 1575
    --Latta 1575
    --Sullivan County 1600
Pyriton, Alabama
  Copper-Bearing Pyrite Ore 85
Pyroclastics
  Missouri

--Washington County 2623
Pyrophyllite
California 424
Nevada 2885
North Carolina 3286
Utah
--North-Central 4426
--Utah County 4413
Pyroxenite
Kansas
--Riley County 1748
Pyrrhotite
Vermont
--Cuttingsville 4488

Quarries 3545
Hawaii
--Oahu 1058
Illinois 1193
Quarry Aggregate
Colorado
--Front Range 639, 640
Quarry Mountain Formation
Oklahoma
--Eastern 3652
Quartz 370, 3127
Arkansas
--Jeffery Quarry 386
--North Little Rock 386
--Pulaski County 386
California 424
Montana 2708
West Virginia 4685
Quartzite
California 410, 424
Quaternary
Arkansas
--Northeastern 405
California
--Coast Ranges 419
--Sierra Nevada 417
Gulf Basin 4186
Idaho
--Snake River Plain 1180
Iowa 1658
Minnesota 2426
--Red River Valley 3462
--Western 2396
Nevada 2903
--Truckee Meadows 2924
--Washoe County 2924
North Dakota 3464
--Red River Valley 3462
South Dakota
--Eastern 2396
Texas
--East 4224
--Gulf 4245

Quay County, New Mexico
Geology 3229
Groundwater 3229
Queen City Formation
Texas 4184
--East 4227
Queen of Sheba Lead Mine, California
Geology 508
Queenston Formation
Pennsylvania
--Western 3892
Questa Mine, New Mexico
Molybdenum 3135
Quicksilver 372
Oregon 3697, 3760
Quinault Formation
Washington
--Coast 4614
--Point Grenville 4614
--Raft River 4614

Rabun County, Georgia
Geology 902
Radiation
Alabama
--Gilbertown 167
--Marion County 89
Hawaii 1118
Radioactive Minerals
Colorado 594
Nevada 2882
South Carolina 3953
Radioactive Wastes
Kansas 1898
New Mexico
--Southeastern 3197
Radioactivity 3930
Colorado 614
--Crested Butte 613
--Gunnison County 613
Utah 4354
Radiocarbon Dating
Illinois 1404
North Dakota 3463, 3474
South Dakota
--Bear Creek 4061
--Pennington County 4061
Radiography 1299
Radiometric Data
Virginia
--Piedmont 4519
Radiometric Dates
Alaska 336
Raft River, Washington
Quinault Formation 4614
Railroads
New Mexico 3281

404 / SUBJECT INDEX

Rainbow Mountain, Alaska
  Bedrock Geology 237
  Geology 280
  Paleozoic 280
  Petrology 280
  Rocks 280
  Sedimentology 280
  Tertiary 280
Rainy Creek, Alaska
  Geochemistry 255
  Geology 249, 255
  Mineral Deposits 249
Ramona Pegmatite District, Cal-
    ifornia
  Geology 504
Rampart District, Alaska
  Geochemistry 220
Ramsey County, North Dakota
  Geophysics 3393
  Groundwater 3393
  Water 3393
  Wells 3393
Randalia Magnetic Anomaly, Indiana
  Fayette County 1648
Randolph County, Alabama
  Sulfide Mineralization 108
Randolph County, Missouri
  Minerals 2620
Randolph Quadrangle, Vermont
  Bedrock Geology 4474
Rankin County, Mississippi
  Clay 2521
  Geology 2520
  Mineral Industries 2522
  Mineral Resources 2520
  Subsurface Stratigraphy 2523
  Water Resources 2524
Ransom County, North Dakota
  Geology 3388
  Glaciation 3388
  Groundwater 3389
  Water 3389
  Wells 3389
Rapides Parish, Louisiana
  Water Resources 2168
Rare Earth
  California 424
  Idaho
    --Lemhi Pass 1147
Rare Species
  Connecticut 700
Raton, New Mexico
  Geology 3277
Raton Mesa, Colorado
  Coking Coals 630
Rattlesnake Formation
  Oregon
    --Central 3767
Rattlesnake Gulch, Montana

  Limestone 2807
Ravalli County, Montana
  Geology 2789
  Painted Rocks Lake 2810
Ravensburg State Park, Pennsylvania
  Guide 3908
Rawlins County, Kansas
  Calcareous Rock 1700
  Upper Pennsylvanian 1700
  Wells 1700
Raymond Fault, California
  Geology 542
Raymond Quadrangle, Mississippi
  Environmental Geology 2537
Reactors
  California
    --Alameda County 488
  Reclamation 2358
  California 483
  Kansas
    --Southeast 1834
Reconnaissance Geology
  California
    --Helena Quadrangle 522
    --Trinity County 522
  New Mexico
    --Lincoln County 3129
    --Little Black Peak Quadrangle
      3129
    --Socorro County 3129
Recovery 2037, 2083
  Kansas 1844, 1854, 1858, 1868,
    1870, 1874, 1884, 1887, 1895,
    1900
  Kentucky 1910, 1923
Recreation
  Michigan
    --Upper Peninsula 2370
  Mississippi
    --Adams County 2540
    --Hinds County 2538
Rectorite
  Arkansas
    --Jeffery Quarry 386
    --North Little Rock 386
    --Pulaski County 386
  Utah
    --North-Central 4426
Red Beds
  Connecticut
    --Central 668
  Kansas
    --Western 1781
  Texas
    --North 4315
Red Bluff Clay
  Alabama
    --Southwestern 27
Red Branch Member

Oklahoma
--Bryan County 3636
Red Cave Formation
Texas
--Panhandle 4232
Red Dog Loess
South Dakota
--Southwestern 4051
Red Eagle Cyclothem
Kansas 1681
Red Lake County, North Dakota
Molluscan Fossils 3430
Red River, New Mexico
Geology 3272
Red River Basin, Texas
Pleistocene Geology 4284
Red River Formation
North Dakota 3496
Red River Valley, Minnesota
Glacial Drift 3462
Late Quaternary 3462
Stratigraphy 3462
Red River Valley, North Dakota
Glacial Drift 3462
Late Quaternary 3462
Stratigraphy 3462
Red Valley Sandstone
Pennsylvania
--Forest County 3876
--Venango County 3876
Reddick, Florida
Birds 738
Redfield, South Dakota
Water 4069
Reduction Roasting
Alabama
--Crenshaw County 95
Redwood County, Minnesota
Kaolin Clay 2414
Reeds Spring Formation
Missouri
--Southwestern 2618, 2619
Reedy Creek Improvement Dis-
trict, Florida
Hydrology 866
Reefs 4746
Great Lakes 1606
Illinois
--Chicago 1463
--Dewitt County 1205, 1233
--McLean County 1233
--Southwestern 1338
--Thornton 1466
--Wapella East Oil Pool
1205
Indiana
--Southwestern 1574
Ohio
--Lake Erie 3583

Texas
--Central 4249
Refuse Disposal
Idaho 1165
Illinois
--Northeastern 1393
Remote Sensing 1629
Renville County, Minnesota
Kaolin Clay 2414
Renville County, North Dakota
Geology 3341, 3342
Glacial Drift 3342
Groundwater 3341, 3342
Water 3341, 3342
Repetto Hills, California
Geology 542
Reptiles
Oklahoma 3654
Texas 3654
Republican County, Kansas
Groundwater 1730
Republican River, Kansas
Groundwater 1730
Research and Development 1894
Reservoirs 1389, 2093
Alabama 177
--North-Central 29
California
--Van Norman Reservoir 432
Illinois
--McHenry County 1401
Iowa 1656
Kentucky
--Western 2051
Missouri 2671
Residual Clay
Alabama
--Piedmont Province 74
Resistivity 4219
Alabama
--Mobile 61
--Piedmont Area 115
Delaware 714
Illinois 1232
Vermont
--Monkton Kaolin Deposit 4486
Resources
Missouri
--St. Charles County 2595
Rex Lake Quadrangle, Wyoming
Structural Geology 4858
Reynolds Creek Experimental Water-
shed, Idaho
Cenozoic 1171
Geology 1171
Rhenium 3162
California 424
Rhine River
Pollution 2984

Rhodium
  California 424
Rice County, Kansas
  Geology 1768
  Hydrology 1768
Rice Lake, Wisconsin
  Groundwater 4819
  Irrigation 4819
Rich County, Utah
  Environmental Geology 4369
Rich Mountain, Oklahoma and
    Arkansas
  Stratigraphy 3623
  Structure 3623
Richardson Area, Alaska
  Geochemistry 211
Richland County, North Dakota
  Geology 3328, 3329, 3330
  Glaciation 3328
  Groundwater 3328, 3329, 3330
  Water 3329
Richland County, Ohio
  Glacial Geology 3589
Richmond, California
  Geology 458
  Geophysics 458
Richmond Conodonts
  Indiana 1996
  Kentucky 1996
  Ohio 1996
Richmond Quadrangle, Virginia
  Geology 4592
Ricketts Glen State Park, Pennsyl-
    vania
  Guide 3910
Ridgeland Quadrangle, Mississippi
  Environmental Geology 2538
Ridges
  Pennsylvania
    --Northwestern 3815
Rifle River Basin, Michigan
  Water 2367
Rifts
  Colorado
    --Rio Grande Rift 3217
  New Mexico
    --Rio Grande Rift 3174,
      3207, 3217
Riley County, Kansas
  Geochronology 1747
  Pyroxenite 1748
  Stockdale Kimberlite 1747,
    1748, 1749, 1751
  Xenoliths 1747
Rincon Quadrangle, New Mexico
  Geology 3112
Ringgold Quadrangle, Virginia
  Geology 4538
Rio Arriba County, New Mexico

  Coal 3154
  Geology 3091, 3102, 3103
  Iron Bearing Rocks 3175
  Minerals 3102
  Precambrian 3091
Rio Grande Rift
  Guidebooks 3207, 3217
  Seismicity 3174
Rio Grande River, Texas
  Geology 4331
River Basins 2968
  Idaho
    --Coeur D'Alene 1169
  Papaloapan 3002
Rivers 2972
  Florida
    --Dade County 832
    --St. Johns River 799
  Michigan
    --Upper Peninsula 2370
  Missouri 2568
  New Jersey
    --Passaic River 2966
Riverside County, California
  Faulting 572
  Field Sampling 519
  Geology 524
  Mineralization 496
Roach River, Maine
  Geology 2216
Road Logs
  Idaho 1143, 1144
  Missouri 2610, 2611, 2641
    --St. Charles County 2614
    --St. Louis County 2614
  Nevada 2929
    --Black Mountain Volcanic
      Center 2917
  North Dakota 3450
    --Bismarck 3414
    --Burleigh County 3452
    --Dickinson 3415
    --Eddy County 3441
    --Foster County 3441
    --Grand Forks 3419
    --Highway 10 3418
    --Interstate 3 398
    --McKenzie County 3416
    --Mandan 3414
    --Minot 3412
    --North-Central 3404
    --Northeastern 3399
    --Northwest 3405
    --Park River 3419
    --South-Central 3403
    --Southeastern 3400
    --Southwest 3406
    --Theodore Roosevelt National
      Memorial Park 3401

--Valley City 3411
--Williams County 3416
--Williston 3416
Virginia
--Daleville Quadrangle 4598
Washington
--Chinook Highway 4648
--Ellensburg Highway 4648
--White Pass Highway 4648
Roads and Highways
Alaska
--East-Central 217
--Elliott Highway 219
--Nenana Highway 218
--South-Central 215
--Taylor Highway 217
--Valdez to Chitina 223
Roaring Fork Valley, Colorado
Geology 604
Groundwater 604
Mineral Resources 604
Robeline, Louisiana
Water Resources 2200
Rochester, Minnesota
Geology 2390
Rock Asphalt
Kentucky 1965
Rock-Fox River Basin, Wisconsin
Hydrogeology 4805
Rock Hounds
Alaska 306
Rock Island, Illinois
Geology 1484
Groundwater 1484
Test Wells 1250
Rock Island County, Illinois
Coal 1295
Rock Pressures 3123
Rock River, Illinois 1367
Gravel 1270
Sand 1270
Rock Salt
Ohio 3591
Rockdale County, Georgia
Geology 939
Groundwater 939
Rockfalls
Colorado
--Crested Butte 613
--Gunnison County 613
--Larimer County 606
Rockford Limestone
Indiana 1490
Rockingham County, Virginia
Igneous Rocks 4523
Silica 4534
Rocks 465, 1397, 1414, 1426,
1431, 1444, 1576, 2399, 2761,
2899, 2900, 3088, 3546, 3549,
3550, 4188, 4686, 4689
Alabama 60
--Central 43, 47
Alaska 306
--Alaska Peninsula 297
--Alaska Range 280
--Anchorage Quadrangle 253
--Arctic 233
--Craig Quadrangle 232
--Eklutna 253
--Gulkana Glacier 280
--Hetta Inlet 232
--Mount Hayes A-6 Quadrangle
234
--Prince of Wales Island 232
--Rainbow Mountain 280
--Teklanika River 285
--Toklat River 285
--Wild Lake 233
--Wiseman Quadrangle 233
Appalachians 2310
Arizona
--East-Central 382
California 424, 538
--Ben Lomond Mountain 514
--Coast Ranges 419
--Death Valley 521
--Geysers Geothermal Area 583
--Inyo County 521
--Salt Spring Hills 521
--Santa Cruz County 514
--Western 406
Central Europe 3012
Colorado
--Front Range Urban Corridor
603
Florida 739, 750
Georgia 895
--Chattooga 933
--Clay County 904
--Dawson County 936
--Lake Chatuge 904
--Towns County 904
Great Plains 1811
Idaho 1122
--Lewiston 1146
--Snake River 1146
Illinois 1368, 1370, 1479
--Northeastern 1335
--Northwestern 1335
Indiana 1497, 1517, 1585
--Cincinnati Arch 1524
--Northern 1498
Iowa 1640
--North-Central 1647
--Southwestern 1624
Kansas 1686, 1698, 1787, 1790,
1791, 1984, 1905
--Cherokee County 1810

Rocks (Cont.)
Kansas (Cont.)
 --Rawlins County 1700
 --South-Central 1735
 --Stafford County 1700
Kentucky 2032
 --Boone County 1949
Maryland 2304
 --Baltimore 2305
 --Bear Island 2246
 --Frederick Valley 2297
 --Hagerstown Valley 2297
 --Piedmont 2303, 2305
 --Potomac River 2246
Michigan 2312, 2313, 2340,
 2343
Minnesota
 --Gabbro Lake Quadrangle
  2449
 --Granite Falls 2441
 --Montevideo 2441
 --North Shore 2392
 --Northeastern 2425, 2449
 --Southeastern 2393
 --Southwestern 2441
 --Vermilion District 2391,
  2425
Missouri 2582, 2592, 2629,
 2636, 2637
 --Bates County 2639
 --Interstate 55 2642
 --St. Francois Mountains
  2631, 2635, 2641
 --Southeast 2624, 2627
 --Southwestern 2619
 --Springfield 2678
Montana 2718
Nebraska 2851, 2852, 2857
Nevada 2904, 2941
 --Central 2921
 --Eastern 2905
 --Reno 2925
 --Storey County 2870
 --Washoe County 2870
 --Western 2921
New Mexico 3278
 --Central 3268
 --Delaware Basin 3213
 --Hidalgo County 3206
 --Rio Arriba County 3175
 --Roosevelt County 3184
 --South-Central 3268
 --Southeast 3241
 --Todilto Park 3238
North Carolina
 --Coastal Plain 3308
North Dakota 3427
 --Burke County 3353
 --Eastern 3312, 3471, 3478

 --Mountrail County 3353
 --Western 3444
Northeast 3012
Ohio 3507, 3528, 3550
 --Erie County 3571
 --Fayette County 3553
 --Logan County 3571
 --Northwestern 3601
Oklahoma
 --Ellis County 3668
 --Osage County 3677
 --Southern 3619
Oregon 3744, 3774
 --Benton County 3769
 --Clackamas County 3772
 --Columbia County 3772
 --Curry County 3735
 --Multowah County 3772
 --Washington County 3772
Pennsylvania 3781
 --Central 3799, 3803
 --Chester County 3936
 --French Creek State Park 3903
 --Gifford Pinchot State Park
  3907
 --Lebanon Valley 3929
 --McConnells Mill State Park
  3906
 --South-Central 3937
 --Valley Forge State Park 3905
 --Western 3799
South Dakota 4039, 4041
Southwestern States 49
Tennessee 4110, 4118
Texas 4239, 4291, 4293
 --Big Bend 4240
 --East 4289
 --Llano Region 4222
 --North-Central 4310
 --West 4305
Utah 4354
 --Needles Range 4421
 --San Juan County 4416
 --Western 4348, 4421
Virginia
 --Augusta County 4523
 --Bath County 4523
 --Highland County 4523
 --Rockingham County 4523
 --Southwestern 4511
Washington 4633
 --Stevens County 4622
West Virginia 4713
 --Allegheny Front 4672
 --Wayne County 4699
Wisconsin 4767
 --Baraboo District 4799
 --Madison 4788
 --Western 4795

Wyoming 4833, 4838, 4871
--Lincoln County 4864
--Medicine Bow Mountains
  4847, 4883
--Northern 4834
--Salt River Range 4864
--Southeastern 4847
--Southwestern 4890
--Teton County 4879
Rocky Mountains
  Basement Tectonics 4877
Rolette County, North Dakota
  Geology 3361
  Glaciation 3361
Rolla, Missouri
  Geology 2584
Rome Trough 2104
Roosevelt County, New Mexico
  Rocks 3184
Rosebud County, Montana
  Bentonite 2814
Rosebud Formation
  South Dakota 4050
Rostroconchia 3510
Roswell, New Mexico
  Geology 3273
Roswell Artesian Basin, New
  Mexico
  Hydrogeology 3147
  Stratigraphy 3147
  Structure 3147
Rough Creek Tectonic Element
  Kentucky 2060
Rove Formation
  Minnesota
  --Northeastern 2443
Rowland Quadrangle, Nevada
  Geology 2866
Rozel Point, Utah
  Oil Seeps 4395
Rubber Industry 3590
Rubidium
  California 424
Ruby Ridge, Alaska
  Geology 293
Rugose Corals
  Oklahoma 3626, 3631
Ruhr
  Pollution 2984
Ruidoso, New Mexico
  Geology 3273
Rule River, Michigan
  Geology 2351
  Magnetic Data 2351
Rush County, Kansas
  Geology 1769
  Groundwater 1769
Rush Springs Sandstone
  Oklahoma

--Caddo County 3658
Rush Valley, Utah
  Geology 4446
  Seismic Risk 4446
  Water 4458
Russell County, Alabama
  Groundwater 22
Russellville Area, Alabama
  Geology 24
  Groundwater 24
Russian River, California
  Geology 455, 459
Ruthenium
  California 424
Rutland, Vermont
  Geology 4494

Sabine Parish, Louisiana
  Water Resources 2163
Sabine Pass, Texas
  Shoreline 4195
Sabkha Complex
  Texas
  --Panhandle 4232
Sacramento, California
  Gravel 562
  Sand 562
Sacramento County, California
  Geophysics 558
Sacramento (High Rolls) Mining Dis-
  trict, New Mexico
  Geology 3097
  Ore 3097
Sacramento Mountains, New Mexico
  Camptonites 3195
  Diorites 3195
  Petrogenesis 3195
  Petrography 3195
  Tertiary 3195
Sacramento Valley, California
  Geology 423
St. Albans, Vermont
  Geology 4497
St. Charles County, Missouri
  Bedrock 2595
  Geology 2595
  Hydrology 2595
  Land 2595
  Middle Ordovician 2614
  Minerals 2595
  Mississippian 2614
  Topography 2595
  Water 2595, 2674
St. Clair County, Alabama
  Geology 20
  Groundwater 20
St. Clair County, Illinois
  Geology 1321

St. Clair County, Illinois (Cont.)
    Planning 1321
St. Croix County, Wisconsin
    Geology 4820
    Groundwater 4820
St. Francois County, Missouri
    Floods 2663
St. Francois Mountains, Missouri
    Ash Flow Tuffs 2626
    Cambrian 2621, 2623, 2635
    Carbonate Mud Facies 2621
    Geology 2641
    Igneous 2631
    Lapilli Tuffs 2623
    Petrochemistry 2631
    Precambrian 2641
    Rocks 2627, 2631
    Sediments 2623
    Stromatulite 2621
    Traverse 2635
St. Helens Mining District, Wash-
    ington 4658
St. James Pool
    Illinois
        --Fayette County 1218
St. Joe National Forest, Idaho
    Streams 1133
St. Johns County, Florida
    Geology 819
    Groundwater 753
    Surface Water 755
St. Johns River, Florida
    Chemical Characteristics 799
    Flow 799
St. Joseph Island, Texas
    Hurricane Washover Fan 4302
St. Landry Parish, Louisiana
    Water Resources 2198
St. Louis, Missouri
    Paleozoic 2577
    Water 2674
St. Louis County, Minnesota
    Geology 2444
St. Louis County, Missouri
    Engineering Geology 2552,
        2553, 2555
    Middle Ordovician 2614
    Mississippian 2614
    Paleozoic 2577
    Water 2674
St. Louis Formation
    Illinois 1211
    Indiana 1211
    Missouri 1211
St. Louis Limestone
    Illinois 1330
    Kentucky
        --Northwestern 1940
St. Lucie County, Florida

Canals 849
Sediments 849
Water 849
St. Maries River, Idaho
    Fossils 1143
    Guidebooks 1143
    Road Logs 1143
St. Mary's Formation
    Maryland
        --Southern 2282
St. Paul, Minnesota
    Artesian Basin 2397
    Environment 2389
    Geology 2389
    Hydrogeology 2397
St. Paul Pool
    Illinois
        --Fayette County 1218
St. Peter Sandstone
    Kentucky
        --Clark County 2029
        --Estill County 2029
St. Regis, Montana
    Water 2771
Ste. Genevieve County, Missouri
    Floods 2663
Ste. Genevieve Limestone
    Illinois
        --Effingham 1310
        --Exchange 1292
        --Marion County 1292
    Illinois Basin 1522
Salado Formation
    New Mexico
        --Carlsbad Potash District 3104
Salamonie Dolomite
    Indiana
        --Northeastern 1532
        --Southeastern 1500
    Kentucky 1500
Salem Formation
    Illinois 1244
Salem Limestone
    Illinois 1330
    Indiana 1585
Salem Quadrangle, Virginia
    Geology 4591
Salina Formation
    West Virginia 4744
Salina Group
    Pennsylvania 3877
Saline Bayou Member
    Louisiana 2150
        --Winn Parish 2150
Saline County, Missouri
    Groundwater 2670
Saline Formation 1249, 1253
Saline Water 3036
    Florida

--Lee County 856
--Sanibel Island 856
Kansas
--Little Arkansas River Basin 1785
Saline Water Intrusion
Florida
--Lee County 792
--McGregor Isles 792
Salinity 3022, 3037
Alaska
--Cook Inlet Basin 289
Delaware 716
--Delaware Estuary 704
Florida
--Dade County 853
--East Glades 853
Kansas 1885
Salinity Changes
Alabama
--Mobile Bay 140
Salisbury, Maryland
Channel Fill 2237
Geohydrology 2237
Pleistocene 2280
Stratigraphy 2280
Water Resources 2281
Salisbury Embayment, Maryland
Basement 2307
Danian 2307
Paleocene 2307
Pinchouts 2307
Senonian 2307
Upper Cretaceous 2307
Salisbury Quadrangle, Virginia
Geology 4578
Sallisaw Formation
Oklahoma 3617
Salmon River Drainage Basin,
Idaho
Mineral Resources 1167
Salona Formation
Pennsylvania
--Central 3788
Salt
California 424
--San Francisco Bay 529
Kansas
--Central 1780
Michigan
--Michigan Basin 2366
Montana
--Northeastern 3498
North Dakota
--Williston Basin 3498
Ohio 3591
Oklahoma
--Anadarko Basin 3624
Pennsylvania 3877

Texas
--Anadarko Basin 3624
--Panhandle 3624
Utah
--Great Salt Lake 4467, 4470
--Millard County 4422
--Sevier Lake 4422
Salt Creek Valley, Indiana
Glacial Lake Sediments 1556
Salt Domes
Louisiana 2159
Texas
--East 4228
Salt Flats
Utah
--Bonneville Salt Flats 4468
Salt Intrusion
Florida 816
Salt Lake City, Utah
Geology 4401, 4430
Sediments 4400
Salt Lake County, Utah
Geology 4344, 4387
History 4387
Ore 4387
Sediments 4400
Salt Lake Mining Review 4364
Salt River Range, Wyoming
Petrography 4864
Rocks 4864
Salt Sand 2107
Salt Spring Hills, California
Cambrian 521
Precambrian 521
Sedimentary Rock 521
Salt Water
Appalachians 4676
Florida
--Dade County 832
--Miami River 832
Mississippi 2545
West Virginia
--Pocatalico River Basin 4696
Salt Water Contamination
Florida
--Caldosahatchee River 779
Salt Water Encroachment 2190
Louisiana
--Baton Rouge 2173, 2192
--Southwestern 2170
Salt Water Fishes
Connecticut 661
Salt Water Intrusion
Florida
--Ft. Myers 778
Salt Water Leakage
Florida
--Venice 825

Salton Trough, California
  Geology 424
Saluda River, South Carolina
  Drainage Basin 3952
  Fluvial Monazite 3992
Sampling
  Mississippi 2487
San Andreas Fault, California
  421, 423, 474, 498, 559
San Andreas Quadrangle, California
  Geology 428
San Angelo Formation
  Texas
    --North 4315
San Bernardino, California
  Geology 451
San Bernardino County, California
  Colemanite 482
  Geological Hazards 554
  Geology 411, 494, 500, 501,
    526
  Landsliding 577
  Mineral Deposits 494, 501
  Mudflows 577
  Talc 525
  Vulcanism 493
San Clemente, California
  Geology 534
  Slope Stability 534
San Diego, California
  Faulting 564
  Geology 438
San Diego County, California
  Coccoliths 539
  Cretaceous 539
  Eocene 539
  Geology 503, 534, 539
  Mineral Resources 442
  Mines 442
  Slope Stability 534
San Diego Mountains, New Mexico
  Geology 3108
San Fernando, California
  Earthquakes 429, 450
San Fernando Valley, California
  Bridges 432
  Buildings 432
  Faults 430
  Geology 430
  Groundwater 430
  Highways 431
  Landslides 430
  Magnetic Anomalies 430
  Seismology 431
San Francisco Bay, California
  Engineering 527
  Fill 527, 531, 532
  Geology 527, 528
  Mud 530

Salt 529
Sand 529
Seismology 531, 532
Shells 529
San Francisco Mountains, Utah
  Aeromagnetic Data 4371
  Gravity Data 4371
San Francisco Peninsula, California
  Geology 423
San Gabriel Mountains, California
  Faulting 546
  Geology 430, 546
  Landslides 546
  Tectonics 430
San Gregorio-Hosgri Fault Zone,
  California 578
San Jacinto Fault, California 474
San Joaquin County, California
  Cretaceous 545
San Jose, California
  Shale 507
San Jose Island, Texas
  Shoreline 4202
San Juan, Colorado
  Coking Coals 630
San Juan Basin, Colorado
  Coal 3257
San Juan Basin, New Mexico
  Coal 3209, 3257
  Geology 3209
  Morrison Formation 3183
  Niobium 3122
  Sanostee Heavy Minerals 3122
San Juan Capistrano Quadrangle, Cal-
  ifornia
  Geology 553
San Juan County, Utah
  Groundwater 4415
  Minerals 4414
  Petroleum 4415
  Potash 4415
  Uranium 4416
San Luis Obispo County, California
  Geology 437
  K-Feldspar 569
  Mesozoic 569
  Sandstone 569
San Mateo County, California
  Franciscan Formation 495
San Pablo, California
  Geology 458
  Geophysics 458
San Patricio County, Texas
  Fossil Vertebrates 4312
  Ingleside Fauna 4312
  Late Pleistocene 4312
San Rafael Group
  Utah
    --Southcentral 4360

--Southwestern 4360
San Rafael Mountain, California
  Geology 409
San Rafael River Mining Area,
  Utah
  Geology 4386
  Uranium 4386
  Vanadium 4386
San Simeon, California
  Geology 467
Sanborn County, South Dakota
  Geology 3998, 3999
  Groundwater 3998, 4000
Sand 1300, 2107, 3645
  Alabama 102, 147
  California 424
  --Sacramento 562
  --San Francisco Bay 529
  Colorado
  --Front Range 639, 640
  Florida 877
  --Northern Peninsula 871
  Georgia
  --Brooks County 965
  --Camden County 965
  --Coastal 970
  --Echols County 964
  --Effingham County 965,
    968
  --Lanier County 965
  --Lowndes County 965
  --Thomas County 965
  --Ware County 965
  Illinois 1247
  --Boone County 1273
  --Dekalb County 1223
  --Macon County 1302
  --Mason County 1320
  --Northeastern 1215, 1359
  --Peoria County 1237
  --Rock River 1270
  --Sangamon County 1308
  --Southern 1284
  --Tazewell County 1255
  Indiana 1505, 1507
  --Indiana Dunes 1599
  --Johnson County 1595
  --Shelby County 1595
  Kentucky 1965
  --Pike County 1953, 2030
  Louisiana
  --Southeast 2155
  Michigan 2328, 2339, 2341,
    2361, 2362
  --Lake Michigan 2364
  Mississippi 2542
  Missouri
  --Barton County 2599
  --Cedar County 2599

  --Dade County 2599
  --Interstate 55 2642
  --Macon County 2620
  --Putnam County 2609
  --Randolph County 2620
  --Southwestern 2599
  --Vernon County 2599
  North Dakota 3476
  Ohio
  --Madison County 3518
  --Maumee River 3577
  Pennsylvania
  --Pittsburgh 3886
  South Dakota
  --Bon Homme County 4033
  --Charles Mix County 4036
  --Douglas County 4036
  Tennessee
  --West 4152
  Texas 4217, 4294, 4340
  --Gulf Coast 4295
  --Llano Region 4288
  Wabash River 1258
  Washington
  --Grays Harbor 4671
  West Virginia 2082, 4692
  --Barbour County 4679
  --Upshur County 4679
Sand Bodies
  Illinois
  --Southern 1469
Sand Creek Facies
  Wisconsin
  --Washakie County 4848
Sand Plains
  Wisconsin
  --Central 4824
Sanders Group
  Indiana 1525, 1589
  Kentucky 1525
Sandia Mountains, New Mexico
  Geology 3262
Sandoval County, New Mexico
  Clay 3164
  Minerals 3092
Sandstone 1299, 1707, 3645, 4694
  California
  --Atascadero 569
  --San Luis Obispo County 569
  --Santa Lucia Range 569
  Illinois 1259, 1281
  --Illinois Basin 1480
  Illinois Basin 1924
  Indiana 1516
  --Greene County 1575
  --Latta 1575
  --South-Central 1957, 2094
  Kansas
  --South-Central 1727

Sandstone (Cont.)
Kentucky
--Clark County 2029
--Edmonson County 1967
--Estill County 2029
--Hart County 1967
--South-Central 1959
--Western 1957, 2094
--Western Coal Field Region
1961
Michigan 2362
New Mexico
--Chaves County 3176
--Eastern 3219
--Northeastern 3204
--Northwestern 3218
--Tucumcari Basin 3211
Oklahoma
--Caddo County 3658
Pennsylvania
--Forest County 3876
--Venango County 3876
South Dakota
--Butte County 4059
--Harding County 4059
--Wall 4037
Texas
--Coast 4210, 4223, 4225,
4326
--Llano Region 4288
--North-Central 4175,
4199
--West 4307
Utah 4431, 4445
--Garfield County 4362
--Grand County 4423, 4429
--Uinta Basin 4425, 4428
--Uintah County 4409, 4423,
4429
West Virginia 4678, 4683,
4685, 4691, 4692, 4739
--Brooke County 4681
--Eastern 4693
--Hancock County 4681
--Ohio County 4681
Wisconsin 4806
--Dane County 4816
Sandwich Fault Zone, Illinois
1361
Sandy Hook, New Jersey
Beaches 3018, 3030
Sediments 3018, 3030
Spit-Bar Development 2956
Sandy Hook Bay, New Jersey
Sediments 3032
Shoals 3032
Sangamon Arch 1239
Sangamon County, Illinois
Gravel 1308

Sand 1308
Sangamonian
Illinois
--Central 1475
Sangre De Cristo Mountains, New
Mexico 3235
Biostratigraphy 3259
Brachiopods 3259
Pennsylvanian 3259
Sanibel Island, Florida
Freshwater 856
Saline Water 856
Sanitary Landfills 4719
Alabama 129
Florida
--Old Tampa Bay 855
Mississippi
--Hinds County 2538
Ohio 3536
Tennessee 4125
Texas
--Coast 4183
West Virginia
--Central West 4723
--Eastern Panhandle 4720
--Northern 4725
--Northwestern 4724
--Southeastern 4721
--Southwestern 4722
Wisconsin 4823
Sanostee Heavy Mineral
New Mexico
--San Juan Basin 3122
Sanpete County, Utah
Land 4358
Minerals 4358
Santa Barbara California
Earthquakes 461, 462, 585
Santa Barbara County, California
Geology 409
Santa Clara County, California
Environmental Geology 456, 457
Seismic Hazards 548
Shale 507
Santa Clara Valley, California
Geology 408, 423
Hydrogeology 423
Santa Cruz County, California
Earthquakes 515
Mineralogy 518
Rock 514
Santa Fe, New Mexico
Geology 3271
Santa Fe County, New Mexico
Clay 3164
Minerals 3092
Santa Lucia Range, California
K-Feldspar 569
Mesozoic 569

Sandstone 569
Santa Margarita, California
  Geology 437
Santa Monica Fault, California
  Geology 542
Santa Rita, New Mexico
  Geology 3275
Santa Rosa County, Florida
  Geology 740
  Water Resources 766, 827
Santa Susana Mountains, California
  Earthquakes 430
  Geology 430
Santa Ynez Mountains, California
  Geology 409
Sargent County, North Dakota
  Geology 3388
  Glaciation 3388
  Groundwater 3389
  Water 3389
  Wells 3389
Satellite Imagery 1177
  West Virginia 4732
Satsop River, Washington
  Foraminifera 4605
  Stratigraphy 4605
Saugus, California
  Earthquakes 430
  Geology 430
Sauk Sequence
  Ohio 3528
Savannah, Georgia
  Phosphate 972
Savannah River, South Carolina
  Drainage Basin 3992
  Fluvial Monazite 3992
Sawtooth Range, Idaho
  Geology 1149
Scallion Subinterval
  North Dakota
  --Central 3497
Scatopsidae
  Connecticut 649
Scenery
  Idaho
  --Snake River 1138
  New Mexico 3278
Schoolcraft County, Michigan
  Drill Cores 2359
  Fiborn Limestone 2359
Schuyler County, Illinois
  Coal 1230
  Limestone 1226
  Mississippian 1226
Science 2998
Scotland, South Dakota
  Water 4065
Scott County, Illinois
  Limestone 1328

Scott County, Kansas
  Groundwater 1814, 1823
Scott County, Tennessee
  Drillers' Logs 4142
  Well Samples 4142
Scranton Quadrangle, Arkansas
  Geology 393
Sea Level Stands
  Florida 746
Sea Water Intrusion
  Florida
  --Broward County 814
Seas 3034
Seattle, Washington
  Landslides 4646
Seaweeds
  Connecticut 664
Sebago Lake State Park, Maine
  Geology 2206
Sebastian County, Arkansas
  Geology 389, 390
Secondary Faulting
  California
  --Transverse Ranges 441
Secondary Recovery 2037
  Kansas 1795, 1844, 1854, 1858,
    1868, 1870, 1874, 1884, 1887,
    1895, 1900
  Kentucky 1910, 1923
  --Union County 2034
Section-Township-Range Conversion
  2811
Sedgwick County, Kansas
  Geohydrology 1711
  Test Holes 1860
  Well Logs 1860
Sediment Fossils
  Alabama
  --Black Warrior Basin 153
Sedimentary Environments
  Connecticut
  --Norwalk Islands 650
Sedimentary History
  California
  --Los Angeles Basin 523
  --Los Angeles County 523
  --Orange County 523
Sedimentary Rocks
  California
  --Death Valley 521
  --Inyo County 521
  --Salt Spring Hills 521
  Kansas
  --South-Central 1735
  Minnesota
  --Vermilion District 2391
Sedimentation 1693
  California
  --Los Angeles County 520

Sedimentation (Cont.)
  Hawaii 1117
  Illinois
    --Illinois Basin 1240
    --Lake Douglas 1322
    --Lake Michigan 1352
    --Southern 1469
  Indiana
    --Lake Lemon 1564
    --Monroe County 1564
    --Southwestern 1579
  Kansas
    --Neosho River Valley 1729
    --Southeastern 1729
  Maine
    --Montsweag Bay 2220
  Maryland 2247
    --Coastal Plains 2299
  Utah
    --Great Salt Lake 4427
    --Millard County 4422
    --Salt Lake Metro Area
      4400
    --San Juan County 4416
    --Sevier Lake 4422
    --Western 4348
  Vermont
    --West Danville 4500
Sedimentology 1439
  Alaska
    --Alaska Range 280
    --Gulkana Glacier 280
    --Rainbow Mountain 280
  Illinois
    --Chicago 1474
    --Coast 1474
    --Northshore 1474
  Kentucky
    --Edmonson County 1967
    --Hart County 1967
    --South-Central 1959
  Maryland
    --Frederick Valley 2301
  Minnesota
    --East-Central 2424
  New Mexico
    --Northeastern 3204
    --Todilto Park 3238
    --Tucumcari Basin 3211
  Utah
    --Uinta Basin 4425
  Wisconsin
    --Baraboo District 4799
    --Madison 4788
Sediments 1341, 1443, 1866,
  2972, 4188
  Alabama
    --Black Warrior Basin 153
    --Mobile River 137

    --Tennessee River 137
    --Tombigbee River 137
    --Warrior River 137
  Alaska
    --Anton Harsen Bay 266
    --Arctic 233
    --Craig Quadrangle 232, 235
    --Elliott Highway 219
    --Hetta Inlet 232
    --Kodiak Island 266
    --Maclaren River 324
    --Matanuska Valley 332
    --Mt. Hayes A-6 Quadrangle
      234
    --Prince of Wales Island 232,
      235
    --South-Central 324
    --Southeastern 235
    --Wild Lake 233
    --Wiseman Quadrangle 233
  Delaware 706, 708
    --Middletown 707
    --Odessa 707
  Florida
    --St. Lucie County 849
  Georgia
    --Coastal Plain 946
    --Lincoln County 950
    --Metasville Quadrangle 950
    --Wilkes County 950
  Hawaii
    --Oahu 1024
  Illinois
    --Illinois River 1432
    --Johnston City Lake 1458
    --Lake Du Quoin 1458
    --Lake Michigan 1331, 1406,
      1408, 1411, 1413, 1415,
      1417, 1420, 1423, 1430,
      1434, 1436, 1445, 1450,
      1460
    --Lake Saline 1436
    --Little Grassy Lake 1458
    --Peoria Lake 1432
    --Southern 1257
  Indiana
    --Bedford 1556
    --Salt Creek Valley 1556
  Iowa 1656
  Lake Erie 3603
  Maryland
    --Severn River 2270
    --Southern 2285
  Middle Atlantic Coastal Plain 2289
  Mississippi 2498, 2508
  Missouri
    --Washington County 2623
  Montana
    --Flathead County 2700

--Lincoln County 2700
New Jersey
--Brigantine Inlet 3024, 3045
--Sandy Hook 3018, 3030
--Sandy Hook Bay 3032
New Mexico
--Magdalena Mining District 3139
--Northeastern 3214
--Southeastern 3199
New York
--Long Island 2980
North Dakota
--Grand Forks 3454
--Lake Agassiz 3490
--Red Lake County 3430
--Sheridan County 3431
--Western 3453
--Williston Basin 3438, 3453
Ohio
--Lake Erie 3535
Pennsylvania
--Bradford County 3804
--Luzerne County 3939
--Newark-Gettysburg Basin 3793
--Northeastern 3814
--Wyoming Valley 3939
Tennessee
--East 4153
--Northwest 4153
Texas
--Central 4249
--Marathon Basin 4292
--Matagorda Bay 4221
--Southeast 4244
Washington
--Okanogan County 4617
West Virginia
--Little Kanawha River Basin 4750
Wyoming
--Carbon County 4863
--Teton County 4879
Seepage
Florida
--Hoover Dike 845
--Lake Okeechobee 775, 845
Georgia 923
Kansas 1885
Missouri 2666
Sego Coal Field, Utah 4391
Seismic Events
North Carolina 3309
Seismic Evidence
California

--Northern 422
Seismic Hazards
California
--Santa Clara County 548
Seismic Investigations
Tennessee
--Eastern 4122
Seismic Refraction
Delaware 714
Illinois
--Illinois Valley 1363
--Meredosia Channel 1344
--Northwestern 1344
Seismic Risks 4656
Utah
--Rush Valley 4446
--Tooele County 4446
--Tooele Valley 4446
Seismic Safety
California
--Auburn Damsite 486
Seismic Shotholes
Montana
--Eastern 2830
Seismicity
California 576
Colorado
--Crested Butte 613
--Gunnison County 613
Indiana 1571
New Mexico
--Rio Grande Rift 3174
--Southeastern 3197
Seismograms 1568
Seismology
California
--San Fernando Valley 431
--San Francisco Bay 531, 532
Michigan 2331
Minnesota 2422
Wisconsin 2422
Selden, New Mexico
Tertiary Geology 3187
Selenium
California 424
Illinois 1441
Selway-Bitterroot Wilderness Area, Idaho
Geology 1127, 1174
Semianthracite Coal
Arkansas
--Arkvalley Coal Field 397
Semiarid Regions 60
Seminole County, Florida
Hydrology 868
Seminole Dam Quadrangle, Wyoming
Gravity 4853
Semivariograms 1838

Senonian
Maryland
--Salisbury Embayment
2307
Sentinel Butte Interval
North Dakota 3483
--Western 3475
--Williston Basin 3453,
3488
Septic Tank Effluents
Wisconsin 4808
Sequatchie Valley, Tennessee
Middle Ordovician 4151
Physiography 4143
Stratigraphy 4151
Serial Micropeel Technique 1688
Serpent Mount Crytoexplosion,
Ohio
Bedrock Geology 3596
Settling Ponds 1184
Seven Pines Quadrangle, Virginia
Geology 4592
Severn River, Maryland
Sediments 2270
Sevier County, Utah
Coal 4405
Sevier Lake, Utah
Brines 4422
Salts 4422
Sediments 4422
Sevier-Sanpete Coal Field, Utah
4392
Sewage
Developing Countries 2978
Hawaii
--Kapaa 1085
--Kauai 1085
--Oahu 1045
--Waimanalo 1045
Missouri 2554, 2556
New Jersey
--Hackensack River 3013
Seward Peninsula, Alaska
Geochemistry 226, 230, 258,
259, 264, 268, 271
Geology 226, 230, 245, 246,
247, 258, 259, 264, 268,
271
Shades Creek Basin, Alabama
Flooding 191
Urbanization 191
Shades State Park, Indiana
Geology 1614
Shadow Mountains, California
Geology 451
Shady Valley Quadrangle, Virginia
Geology 4553
Shale 1299, 4437, 4697
Alabama 84

--Northeastern 76
--Northwestern 75
--Southeastern 77
Arkansas
--North-Central 399
California 424
--Gilroy 507
--San Jose 507
--Santa Clara County 507
Central Appalachian Area 1909
Colorado
--Mesa County 625
Indiana 1490, 1491, 1513, 1517
--Delphi 1562
--Northwestern 1587
--Southern 1501
Kansas 1687, 1728
--Central 1713
Kentucky 1958
--Eastern 1912, 2020, 2099
Michigan
--Michigan Basin 2363
Minnesota 2442
--Southeastern 2421
Missouri
--Macon County 2620
--Putnam County 2609
--Randolph County 2620
Montana 2695, 2709, 2728, 2740,
2752
New Mexico 3099, 3141
North Dakota
--Central 3497
Ohio 3519
Oklahoma
--Jackson County 3663
--Osage County 3677
--Southern 3657
Pennsylvania 3866, 3867, 3870
--Greene County 3839
--Pittsburgh 3890
--Southeastern 3882
--Southwestern 3896
South Carolina 3970
Tennessee
--East 4146
Utah
--North-Central 4426
--Uinta Basin 4418
Virginia
--Southern 4535
--Southwestern 4529
--West-Central 4528
Shale Micropore Systems 1249
Shale Stimulation 2063
Shale Wells 2100
Shaly Chalk
Kansas 1755
Sharon Quadrangle, Connecticut

Bedrock Geology 694
Shasta County, California
Economic Geology 502
Geology 425
Mineral Resources 445
Mines 445
Shauntie Hills, Utah
Geology 4396
Hydrothermal Alteration 4396
Shelby County, Alabama
Sinkholes 159
Shelby County, Indiana
Gravel 1595
Sand 1595
Shelbyville, Illinois
Glacial Drift 1315
Hydrogeology 1391
Water 1391
Shelbyville Moraine
Illinois
--Shelbyville 1315
Shell Rock
Iowa
--North-Central 1647
Shellfish 3020
Shells 143, 3237
California
--Coast Ranges 435
--San Francisco Bay 529
North Carolina
--Albemarle Sound 3291
Shenandoah County, Virginia
Silica 4534
Shenandoah National Park, Virginia
Geology 4513
Groundwater 4533
Sherando Quadrangle, Virginia
Geology 4541
Sheridan County, North Dakota
Fish 3431
Late Pleistocene 3431
Sediments 3431
Sheridan District, Montana
Bedrock Geology 2786
Sherman County, Oregon
Geologic Hazards 3733
Shoals
New Jersey
--Sandy Hook Bay 3032
Shores 2239
Alabama 186
Illinois
--Lake Michigan 1452
Massachusetts
--Cape Cod Bay 3041
Michigan
--Lake Michigan 2364
New Mexico

--Northwestern 3218
Ohio
--Lake County 3600
--Lucas County 3608
Texas 4208, 4211
--Bolivar Roads 4195
--Brazos Island 4191
--Brazos River Delta 4193
--Central Padre Island 4207
--Galveston Island 4187
--Matagorda Island 4202
--Matagorda Peninsula 4204
--Mustang Island 4206
--North Padre Island 4206
--Sabine Pass 4195
--San Jose Island 4202
--South Padre Island 4191
Washington
--Puget Sound 4655
Wisconsin 4831
Shotholes
Montana
--Eastern 2830
Shubuta Member
Alabama
--Southwestern 27
Shungnak D-2 Quadrangle, Alaska
Geochemistry 272, 274
Geology 272, 274
Sicily Island, Louisiana
Indian Village Site 2139
Sierra Alta Quadrangle, New Mexico
Geology 3113
Sierra Nevada, California
Cenozoic 417
Geology 417, 423, 424
Minerals 417
Quaternary 417
Tertiary 417
Vulcanism 417
Silica
South Carolina 3972
Virginia
--Clarke County 4534
--Frederick County 4534
--Page County 4534
--Rockingham County 4534
--Shenandoah County 4534
--Warren County 4534
Wisconsin 4806
Silica Sand 1300
Silicates 3136, 3143, 3144, 3151
Sills
Minnesota 2433
Silt
Missouri 2622
North Dakota 3464
Siltation
Iowa 1656

Siltstone
  Illinois 1281
  --Southwestern 1242
Silurian 4747
  Alabama
    --Northwestern 50
  Appalachian Basin
    --Central 4745
  Great Lakes 1606
  Illinois 1339
    --Dewitt County 1205
    --Northeastern 1335
    --Northwestern 1335
    --Southwestern 3646
    --Wapella East Oil Pool
      1205
  Indiana 1497, 1570
    --Cincinnati Arch 1496,
      1524
    --Clark County 1607
    --Northeastern 1532
    --Northern 1498, 1492
    --Southeastern 1500
    --Southwestern 1574, 1579
  Iowa
    --Cedar Rapids 1677
    --Eastern 1643
  Kentucky 1500, 1907
    --East-Central 1927
    --Jefferson County 1607
    --Pulaski County 2128
  Michigan 2343
  Missouri
    --Eastern 3646
  Ohio 1927, 3537, 3576, 3591
    --Allen County 3521
    --Erie County 3571
    --Logan County 3571
    --Moreland Oil Pool 3547
    --Northwestern 3601
    --Southwestern 3609
    --Wayne County 3547
  Oklahoma 3622, 3630, 3631,
    3642, 3646
    --Anadarko Basin 3648
    --Arkoma Basin 3656
    --Eastern 3652
    --Northeastern 3627
  Tennessee
    --West 4157
  West Virginia 4692, 4741
    --Wayne County 4699
Silver 1121, 2715
  Alaska
    --Chugach Mountains 326
    --Nelchina Glacier 326
  California 424
    --Inyo County 490
    --Modoc District 490

  --Panamint Butte Quadrangle 490
  --Plumas County 544
  Missouri
    --Southeast 2638
  Oregon 3703
  Texas 4270
  Washington 4621
Silver Bow Creek Drainage Area,
  Montana
  Hydrogeology 2734
Silver City, Idaho
  Geochemistry 1182
  Geology 1158, 1181, 1182
  Mineral Resources 1158
  Ore Deposits 1181
Silver City, New Mexico
  Geology 3275, 3280
Silver Crown Mining District, Wyo-
  ming
  Geology 4861
  Minerals 4861
Simpson, Louisiana
  Water Resources 2199
Simpson County, Kentucky
  Economic Geology 1933
Simulation Models 2267, 2306, 2309,
  3010, 3064
Simuliidae
  Connecticut 653
Sink Valley, Utah
  Water 4458
Sinkholes
  Alabama
    --Birmingham 126
    --Greenwood 141
    --Shelby County 159
  Florida
    --Seminole County 868
  Missouri 2556
  Tennessee
    --Montgomery County 4130
Sinks
  Florida
    --West-Central 826
Sinuk Area, Alaska
  Geochemistry 264, 271
  Geology 264, 271
Sinuk River, Alaska
  Geochemistry 259
  Geology 259
Sioux County, North Dakota
  Groundwater 3383, 3384
  Water 3383
  Wells 3383
Siphonodella
  Kansas
    --Western 1756
  Nebraska

--Southwestern 1756
Sithylemenkat Lake, Alaska
  Geochemistry 270
  Geology 270
Skamania County, Washington
  Heat Flow 4660
Skin Diving
  Alaska 307
Skull Valley, Utah
  Water 4458
Slagle, Louisiana
  Water Resources 2199
Slana District, Alaska
  Geochemistry 210
  Geology 256, 265
Slate Range, California
  Geology 525
Sloans Valley, Kentucky
  Geology 1966
Slope Formation
  North Dakota 3489
Slope Stability 4654
  California 574
    --Geysers Geothermal Area
      583
    --Orange County 534
    --San Clemente 534
    --San Diego County 534
  Colorado
    --Crested Butte 613
    --Gunnison County 613
    --Larimer County 606
Sludge 3044
Slurries 2074
Smelting
  Mississippi
    --Winborn 2477
Smith Basin Limestone
  New Mexico 3251
Smith County, Mississippi
  Clay 2526
  Geology 2525
  Mineral Resources 2525
  Water Resources 2527
Smithfield Quadrangle, Virginia
  Geology 4571
Smithland Quadrangle, Illinois
  Geology 1216
Snails
  Louisiana 2458
  Mississippi 2457
    --Southern 2458
Snake River, Idaho
  Carbonate Rocks 1146
  Geology 1138, 1146, 1156
  Scenery 1138
Snake River Basin, Idaho
  Drain Wells 1168
  Groundwater 1168

Quaternary 1180
Volcanoes 1180
Snakes
  Connecticut 659
Snoqualmie Highway, Washington
  Geology 4626
Snow
  Colorado 592
Snow Creek Quadrangle, Virginia
  Geology 4587
Snowfall
  New Jersey 3062
Soap 1388
Soapstone
  California 424
Socorro, New Mexico
  Springs 3169
  Wells 3169
Socorro County, New Mexico
  Beryllium Mineralization 3157
  Fusulinids 3255
  Geology 3129, 3157, 3222
  Gravity Surveys 3145
  Groundwater 3178
  Jornada Del Muerto Coal Field
    3222
  Springs 3169
  Wells 3169
Socorro Peak Area, Arizona
  Geology 383
Sodium Carbonate
  California 424
Sodium Hydroxide Leaching
  Alabama
    --Crenshaw County 95
Sodium Sulfate
  California 424
Soils 1341, 1342, 1405, 1778, 2761,
  4653, 4815
  Alabama 40, 41
  Colorado 646
    --Crested Butte 613
    --Denver 597
    --Front Range Urban Corridor
      603
    --Gunnison County 613
    --Jefferson County 590
  Hawaii
    --Kauai 1102
    --Lanai 1105
    --Maui 1106
    --Molokai 1104
  Idaho 1124
  Illinois
    --Lake County 1433
  Iowa 1657
  Kansas 1770
  Missouri 2592

Soils (Cont.)
Tennessee
--Knox County 4114
Texas
--South 4336
Wisconsin 4763, 4770, 4776,
4808
--Dodge County 4759
--Jefferson County 4762
--Menominee County 4761
Solar Radiation
Hawaii 1118
Solemint, California
Earthquakes 430
Geology 430
Solid Wastes 3066
Iowa 1630
Maryland 2296
Ohio
--Summit County 3599
Solomon C-5 Quadrangle, Alaska
Geochemistry 268
Geology 268
Solomon D-6 Quadrangle, Alaska
Geochemistry 226
Geology 226, 245
Soluble Rock
Missouri 2556
Somerset, Kentucky
Limestone 1954
Somerset County, Maryland
Hydrogeology 2226
Somervell County, Texas
Glen Rose Formation 4171
Sonar Investigations
Illinois
--Lake Michigan 1450
Sonoma County, California
Engineering Geology 563
Erosion 575
Geology 455, 459, 561
Sound Velocity
Illinois
--Lake Michigan 1331
Souse Springs Quadrangle, New
Mexico
Geology 3111
South Birmingham Pool Area,
Ohio
Rocks 3571
South Bloomingville Quadrangle,
Ohio
Bedrock Geology 3564
South Boston Quadrangle, Vir-
ginia
Geology 4542
South Canaan Quadrangle, Con-
necticut
Bedrock Geology 688

South County Study Area, California
Environmental Geology 457
South Dakota Geological Survey 4040
South Mountain, Idaho
Geochemistry 1182, 1186
Geology 1182, 1186
South Mountain Anticlinorium, Mary-
land
Erosion 2266
South Platte District, Colorado
Pegmatites 634
South Westhope Oil Field, North
Dakota 3457
Southbury Quadrangle, Connecticut
Bedrock Geology 686
Space Stations 2967
Space Views
Georgia 916
Illinois 1449
Sparta Formation
Texas
--Gulf Coast Basin 4212
Spearfish Formation
North Dakota
--Northern 3346
--Williston Basin 3346
Spectograph 371
Spencer, South Dakota
Groundwater 4099
Spirit Mountains, Alaska
Copper 276
Geology 276
Nickel 276
Spit-Bar Development
New Jersey 2956
Spoon Formation
Illinois
--Illinois Basin 1195
Spores
Illinois 1319
--Illinois Basin 1192
Maryland 2233
Spray Quadrangle, Virginia
Geology 4587
Spring Grove Pool, Kentucky
Inert Gas Injection 2034
Secondary Recovery 2034
Spring Hill Quadrangle, Connecticut
Surficial Geology 682
Spring Mill State Park, Indiana
Geology 1617
Springdale Mine, Pennsylvania
Petrography 3861
Springfield, Illinois
Geology 1353
Springfield, Missouri
Geology 2678
Hydrology 2678
Water 2678

Springfield Coal Member (V)
Indiana
--Sullivan County 1600
Springs
Colorado 593, 612, 614, 636
Florida
--Sun Coast 818
--West-Central 826
Maryland
--Baltimore County 2224
--Charles County 2225
Missouri 2568, 2673
New Mexico
--Magdalena 3169
--Socorro 3169
--Socorro County 3169
North Dakota
--Burke County 3352
--Cavalier County 3370
--Dickey County 3391
--Grand Forks County 3348
--Lamoore County 3391
--Mercer County 3356
--Mountrail County 3352
--Oliver County 3356
--Pembina County 3370
--Wells County 3344
Oregon 3759
South Dakota
--Black Hills 4060
Utah 4465
West Virginia
--Little Kanawha River Ba-
sin 4750
--Monongahela River Basin
4749
--New River Basin 4752
--Potomac River Basin 4751
Wisconsin
--Southwestern 4793
Wyoming 4060, 4842
Staats Mine, Utah
Hydrothermal Alteration 4402
Mineralization 4402
Stafford County, Kansas
Calcareous Rock 1700
Upper Pennsylvanian 1700
Wells 1700
Stanislaus County, California
Cretaceous 545
Stanton County, Kansas
Geohydrology 1692
Water Levels 1848, 1856,
1875
Stanton Formation
Kansas
--Southeastern 1772
Stark County, Ohio
Bedrock Geology 3566

Geology 3501
Starr District, Utah
Geology 4438
Mineralogy 4438
State Parks
Oklahoma 3689
Texas 4243
Staunton Quadrangle, Virginia
Copper 4585
Geology 4506
Lead 4585
Zinc 4585
Steamboat Mountain, Washington
Heat Flow 4660
Steele County, North Dakota
Geology 3373
Glaciation 3373
Groundwater 3374, 3375
Water 3374
Wells 3374
Stemple Pass, Montana
Geochemistry 2827
Stephenson Quadrangle, Virginia
Geology 4588
Stephensport Group
Indiana 1557
Stepovak Bay, Alaska
Mineralization 330
Stevens County, Washington
Carbonate Rocks 4622
Lead 4622
Zinc 4622
Stevens Highway, Washington
Geology 4626
Stevenson, Alabama
Groundwater Contamination 134
Limestone Aquifers 134
Stewart County, Tennessee
Faulting 4137
Stillwater River Valley, Montana
Geology 2741
Water 2741
Stockdale Kimberlite
Kansas
--Riley County 1747, 1748
Stockton Formation
Pennsylvania
--Southeastern 3925
Stokesville Quadrangle, Virginia
Geology 4573
Stone 1561
Indiana
--Fort Wayne 1559
Montana 2758
Pennsylvania 3785
Washington 4607
Wisconsin 4764

Stone Creek Section
  Indiana
    --West-Central 1566
Stone Mountain Park, Georgia
  Geology 920
Stoneflies
  Connecticut 663
Storage
  Kansas 1898
  Nebraska
    --Omaha 2856
  Pennsylvania 3860
Storey County, Nevada
  Geology 2869
  Industrial Rock 2870
  Minerals 2869
Storms
  Hawaii 1117
  Ohio
    --Lake Erie 3534
Strasburg Quadrangle, Virginia
  Geology 4602
Strata
  Connecticut
    --South-Central 667
  Illinois
    --Illinois Basin 1192,
      1267
    --Northeastern 1481
    --Southern 1297
    --Southwestern 1338
  Michigan Basin 1519
  New Mexico
    --South-Central 3090
    --Southwestern 3090
Stratigraphic Data
  California
    --Cheyenne Basin 620
Stratigraphic Geology
  Alabama 195
    --Northwestern 50
    --Warrior Basin 88
  Alaska
    --Alaska Range 280
    --Cook Inlet 331
    --Gulkana Glacier 280
    --Rainbow Mountain 280
  California
    --Church Creek 505
    --Monterey County 505
  Colorado 638
    --Cheyenne Basin 628
    --Denver Basin 628
  Connecticut
    --Central 697
    --Coast 665
    --New Haven Quadrangle
      698
  Gulf Coast

    --East 196
Stratigraphy 1569, 1581
  Appalachian Piedmont 2284
  Appalachians
    --Central 4742
  Arkansas
    --Northeastern 3627
    --Northwestern 3695
    --Rich Mountain 3623
  Border 3158
  California
    --Inyo County 540
    --Lost Burro Formation 540
    --Panamint Range 540
    --San Joaquin County 545
    --Stanislaus County 545
  Georgia 953
    --Chatham County 901
    --Cochran Quadrangle 900
    --Perry Quadrangle 900
  Great Basin 2872
  Gulf Basin 4186
  Idaho
    --Benewah County 1160
    --Weiser 1137
  Illinois 1196, 1197
    --Batavia 1312
    --Central 1234, 1475
    --Central Northern 1212
    --Central Western 1467
    --East Central 1471
    --Northeastern 1335
    --Northwestern 1335
    --Western 1465
  Indiana 1517, 1518, 1521, 1525,
    1531, 1585
    --Cincinnati Arch 1496
    --Northeastern 1532
    --Northern 1527
    --Southeastern 1500
    --West-Central 1566
  Iowa 1664
    --North-Central 1647
    --Southeastern 1638
    --Southern 1665
  Kansas 1731, 1736, 1771, 1772
    --Central 1713
    --Missouri River Valley 1891
    --Northeastern 2590
    --South-Central 1727
    --Southern 1782
    --Western 1822
  Kentucky 1500, 1525, 1907
    --Pulaski County 2128
    --Southeastern 2106
  Maine
    --Munsungun Anticlinorium 2217
    --Vassalboro 2215
    --Waterville 2215

--Western 2219
Maryland
--Eastern Shore 2280
--Frederick Valley 2301
--Salisbury 2280
Mexico
--Coahuila 4300
Michigan
--Barry County 2356
--Crystal Falls 2349, 2350,
  2352
--Michigan Basin 2360, 2363
--Upper Peninsula 2353
Minnesota
--Duluth 2418
--East-Central 2434
--Hollandale 2423
--Northeastern 2425
--Red River Valley 3462
--Southeastern 2427
--Vermilion District 2425
--Western 2396
Mississippi 2479, 2506, 2532
--Central 2456, 2507
--Claibore County 2500
--Copiah County 2511
--George County 2505
--Jasper County 2452
--Missouri River Valley
  1891
--Rankin County 2523
--Southern 2456, 2507
Missouri 2587, 2608, 2611,
  2622, 2637
--Highway 65 2617
--Missouri River Valley
  2590
--Southwestern 2625
Montana 2794, 2801, 2817,
  2818
--Central 2710
--Dawson County 2824
--Great Falls-Lewistown
  Coal Field 2710
Nevada
--Eastern 2905
--Yerington District 2933
New Mexico
--Big Hatchet Mountains
  3240
--Catron County 3096
--Chaves County 3147
--Roswell Artesian Basin
  3147
North Carolina
--Carolina Slate Belt 3307
--Union County 3307
North Dakota 3350, 3368,
  3464, 3474

--Eastern 3312
--Lake Agassiz 2490
--Red River Valley 3462
--Riverdale 3481
--Southwestern 3486
Ohio 3507
--Erie County 3571
--Logan County 3571
--Northeastern 3552
Oklahoma
--Custer County 3639
--Northeastern 3695
--Northern 1782
--Rich Mountain 3623
Oregon
--Southwestern 3725
Pennsylvania 3789
--Bradford County 3804
--Central 3797, 3799, 3802,
  3821
--Dauphin County 3794
--Eastern 3801
--Northeastern 3814
--Northwestern 3805
--South-Central 3798
--Western 3799
South Dakota 3996
--Eastern 2396
Tennessee 4145
--East 4144
--Hamilton County 4123
--Knox County 4114
--Lauderdale County 4147
--Northeastern 2106
--Sequatchie Valley 4151
--Wells Creek Structure 4112
--Western 4119
Texas 4294
--Central 4306
--South-Central 4287
Utah 4343
--South-Central 4360
--Southeastern 4377
--Southwest 4360
--Uinta Basin 4370, 4373, 4419
Vermont
--Castleton Quadrangle 4478
Virginia
--Coastal Plain 4508
--Southeastern 4506
--Southwestern 4511
Washington
--Coast 4614
--Olympic Peninsula 4605
--Point Grenville 4614
--Raft River 4614
--Satsop River 4605
West Virginia 4739, 4741, 4742
--Allegheny Front 4672

Stratigraphy (Cont.)
  West Virginia (Cont.)
    --Charleston 4682
    --Eastern 4698
    --Pendleton County 4698
  Wisconsin 4791
    --Southwest 4789
    --Western 4790
  Wyoming
    --Medicine Bow Mountain
      4883
    --Southwestern 4890
Streamflow
  Alabama 34, 68, 72, 90
    --Jefferson County 16
  Florida 786
    --Myakka River Basin 840
  Georgia 892
  Hawaii 1088
    --Haiku 1012
    --Kahaluu 1012
    --Oahu 1012
  Maryland 2294
  Michigan 2371
  Minnesota 2417
  Missouri 2664, 2666, 2669,
    2672, 2676
    --Jefferson County 2674
    --Joplin 2668
    --St. Charles County 2674
    --St. Louis 2674
    --St. Louis County 2674
  West Virginia
    --Little Kanawha River
      Basin 4750
    --New River Basin 4752
Streams 3010
  Alaska
    --Anton Larsen Bay 266
    --Arctic 233
    --Craig Quadrangle 232,
      235
    --Elliott Highway 219
    --Hetta Inlet 232
    --Kodiak Island 266
    --Maclaren River 324
    --Mt. Hayes A-6 Quadrangle
      234
    --Prince of Wales Island
      232, 235
    --South-Central 324
    --Southeastern 235
    --Wild Lake 233
    --Wiseman Quadrangle 233
  Arkansas 401, 402
  Australia 2993
  Connecticut
    --Eastern 662
  Delaware 737

  Florida 786
  Georgia
    --Lincoln County 950
    --Metasville Quadrangle 950
    --Wilkes County 950
  Hawaii 1088, 1097
    --Molokai 1050
    --Oahu 1024
  Idaho
    --St. Joe National Forest 1133
  Maryland 2283, 2294
  Michigan 2371
  Missouri 2664, 2666, 2669, 2671,
    2672, 2676
    --Jefferson County 2674
    --St. Charles County 2674
    --St. Louis 2674
    --St. Louis County 2674
  Montana
    --Flathead County 2700
    --Lincoln County 2700
  New Mexico
    --Magdalena Mining District
      3139
  Tennessee
    --East 4153
    --Northwest 4153
  West Virginia
    --Potomac River Basin 4751
  Wyoming
    --Teton County 4879
Stress Accumulation
  California
    --Transverse Ranges 441
Strip Mining
  Indiana 1597
  Kansas
    --Southeast 1834
  Montana
    --Decker 2757
    --Southeastern 2757
  North Dakota 3402, 3434
    --Dunn Center 3491
  Texas
    --East 4200
Strom Quadrangle, Virginia
  Geology 4578
Stromatolite
  Missouri
    --St. Francois Mountain 2621
Strong Motion Instrumentation 480,
  487, 549
Strong Motion Records
  California
    --Santa Barbara 585
Strontium
  California 424
Strontium Sulfate 3170
Stroudsburg Quadrangle, Pennsylvania

Surficial Geology 3808
Strouss Hill Quadrangle, Wyoming
Structural Geology 4860
Structural Analysis
Louisiana
--South 2153
Structural Clay 1970
Structural Geology
Alabama
--Birmingham Red Iron District 79
Arkansas
--Rich Mountain 3623
California
--Northern 422
Colorado
--Cheyenne Basin 628
--Denver Basin 628
Connecticut
--New Haven Quadrangle 698
Georgia
--Southwestern 947
Illinois
--Southern 1207
Kansas
--Barton County 1775
--Cheyenne Bottoms 1775
--Manhattan 1763
Maine
--Vassalboro 2215
--Waterville 2215
--Western 2219
Mississippi
--Claiborne County 2501
--Hinds County 2494
Missouri 2629
--St. Charles County 2614
--St. Louis County 2614
New Mexico
--Animas Mountains 3179
--Big Hatchet Peak Quadrangle 3200
--Chaves County 3147
--Hidalgo County 3179, 3200
--Roswell Artesian Basin 3147
North Dakota
--Eastern 3312
Oklahoma
--Rich Mountain 3623
--Southern 3619
Tennessee
--Dumplin Valley 4150
--East 4150
Utah 4343
Vermont
--Castleton Quadrangle 4478

Washington
--Coast 4645
West Virginia
--Allegheny Front 4672
--Allegheny Plateau 4743
--Charleston 4682
--Western Valley and Ridge 4743
Wyoming
--Albany County 4860
--Arlington 4862
--Carbon County 4860
--James Lake Quadrangle 4860
--Laramie Basin 4858
--Lincoln County 4864
--Morgan Quadrangle 4860
--Rex Lake Quadrangle 4858
--Salt River Range 4864
--Strouss Hill Quadrangle 4860
--Wagonhound Creek 4862
Stuart City Trend
Texas
--South 4313
Stuarts Draft Quadrangle, Virginia
Geology 4566
Studley Quadrangle, Virginia
Geology 4592
Stutsman County, North Dakota
Aquifers 3315
Geology 3313, 3314, 3315
Glaciation 3313
Groundwater 3313, 3314, 3315
Water 3314
Style Manuals
Missouri 2586
Sub-Pennsylvanian
Illinois
--Illinois Basin 1314
Sublette County, Wyoming
Faults 4888
Talconite 4857
Subsidence
Alabama
--Columbiana 114
California
--Great Valley Province 418
Colorado
--Boulder-Weld Coal Field 605
--Crested Butte 613
--Gunnison County 613
Kansas
--Central 1780
Subsurface Beds
Kansas
--Cherokee County 1776
Subsurface Geology
Alabama
--Gilbertown Oil Field 93
--Southeastern 35

428 / SUBJECT INDEX

Subsurface Geology (Cont.)
Alabama (Cont.)
  --Southwest 54
Czechoslovakia
  --East 2963
Georgia
  --Coastal Plain 931, 937
Illinois
  --Clark County 1236
  --Dewitt County 1329
  --Edgar County 1236
  --McLean County 1329
  --Piatt County 1329
Iowa
  --Laporte City 1641
Kansas
  --Western 1683
Kentucky
  --Monroe County 2065
  --Sulphur Lick 2065
  --Tompkinsville 2065
Louisiana
  --Central 2152
  --South 2153
Maryland
  --Coastal Plain 2299
  --Eastern Shore 2280
  --Salisbury 2280
Michigan
  --Barry County 2356
  --Michigan Basin 2360
Minnesota
  --Southeastern 2427
Mississippi
  --Attala County 2464
Nebraska 2852, 2857
North Dakota
  --Grand Forks 3454
Ohio 3530, 3538, 3541, 3576
  --Northwestern 3579, 3601
Pennsylvania 3783
  --Northwestern 3834
  --Western 3892
Tennessee 4120
Utah
  --Millard County 4422
  --Sevier Lake 4422
West Virginia
  --Wayne County 4699
Subsurface Investigations
Minnesota
  --Northwestern 2430
Subsurface Movements
California 430
Subsurface Pressures 2321
Subsurface Storage
Maryland 2292
Subsurface Stratigraphy
Illinois

  --Central Northern 1212
Indiana 1521
Kentucky 1907
Mississippi
  --Claiborne County 2500
  --Copiah County 2511
  --George County 2505
  --Jasper County 2452, 2453
Rankin County 2523
Subsurface Temperatures
New Mexico
  --Jemez Mountains 3205
Sub-Trenton Formation 3549, 3550,
  3553
Ohio 3550
  --Fayette County 3553
Sugarloaf Mountain Quadrangle, Vir-
  ginia
Geology 4567
Sulfate 4718
Sulfide 2705
Arizona 379, 380
Minnesota
  --Itasca County 2433
Ohio 3605
Sulfide Mineralization
Alabama
  --Clay County 108
  --Cragford District 108
  --Randolph County 108
Minnesota
  --Birchdale 2428
  --Indus 2428
  --Koochiching County 2428
Sulfonate Flotation 3119, 3127
Sulfur 1252, 1442, 2705
California 424
Colorado 615
Illinois 1288, 1318, 1340, 1410
Indiana 1529
Texas 4256
  --West 4173
Sulfur Dioxide 1397, 1414, 1426,
  1431, 3586
Sulfur Isotopic Studies
Utah
  --Uinta Basin 4433
Sulfuric Acid 3166
Tennessee 4136
Sullivan County, Indiana
  Petersburg Formation 1600
  Pyrite 1600
  Springfield Coal Member (V) 1600
Sullivan County, Missouri
  Water 2654
Sullivan County, Tennessee
  Geology 4156
Sulphate
  West Virginia 4673

Sulphur Lick, Kentucky
  Subsurface Geology 2065
Summit County, Ohio
  Surficial Materials 3610
  Waste Disposal 3599
Sumner Bog, Iowa
  Pollen 1653
Sunbury Shale
  Michigan
  --Michigan Basin 2363
Suplee, Oregon
  Geology 3700
Surface Geology
  Alaska
  --Prudhoe Bay Oil Field
    342
  California
  --San Fernando 450
  Colorado
  --Denver 597
  Connecticut
  --Ansonia Quadrangle 679
  --Branford Quadrangle 670
  --Clinton Quadrangle 684
  --Essex Quadrangle 687
  --Guilford Quadrangle 684
  --Haddam Quadrangle 672
  --Hartford Quadrangle 676
  --Milford Quadrangle 679
  --Naugatuck Quadrangle 691
  --New Haven Quadrangle 674
  --Old Lyme Quadrangle 687
  --Spring Hill Quadrangle
    682
  --Woodmont Quadrangle 674
  Illinois
  --McHenry County 1395
  Montana
  --Stillwater River Valley
    2741
  --Tobacco Valley 2741
  Nebraska 2851
  Pennsylvania
  --Stroudsburg Quadrangle
    3808
  Utah 4343
  Vermont 4484
  --Brandon Quadrangle 4501
  --Ticonderoga Quadrangle
    4501
Surface Mining
  California 483
  New Mexico 3168
Surface Movements
  California 430
Surface Structure
  Pennsylvania
  --Pittsburgh 3852
Surface Temperatures 1651
Surface Water

Alabama 94
  --Calhoun County 91
  --Southwestern 31
Colorado
  --Boulder County 596
  --Jefferson County 590
Delaware
  --Coastal Plain 727
Florida
  --Flagler County 755
  --Putnam County 755
  --St. Johns County 755
Georgia
  --Gwinnett County 928
  --Yellow River Basin 928
Iowa 1672
Louisiana
  --Southwestern 2202
Missouri 2550
  --Joplin 2668
New Mexico 3216
South Carolina 3964, 3965, 3966,
  3967, 3968
Surveying 3051
  Illinois 1232
  New Jersey 3080
Surveys
  Aeromagnetic
  --Alaska 317
  --Illinois 1351
  --Maryland 3824
  --Ohio 3824
  --Pennsylvania 3824, 3849
  --West Virginia 3824
  Electrical Resistivity
  --Alabama 115
  Gamma Rays
  --Louisiana 2157
  Geodetic
  --New Jersey 2944, 3049
  Geophysical
  --California 502
  Gravity
  --Alabama 109, 112, 117, 118,
    119, 120, 121, 132, 145,
    146
  --Alaska 284
  --Georgia Piedmont 947
  --Illinois 1306
  --Iowa 1648
  --Kansas 1761
  --New Mexico 3145
  --South Dakota 4047
  --Tennessee 4114
  Land 3075
  Magnetic
  --Kansas 1761
  Marine Boundaries 3052
  Radiation
  --Alabama 89, 167

Surveys (Cont.)
Seepage
--Kansas 1885
Seismic Refraction
--Illinois 1344, 1363
Soil
--Hawaii 1102, 1104, 1105,
1106
Thermal
--New York 2969
Susitna River, Alaska
Geology 238
Susquehanna Aqueduct, Maryland
Geology 2286
Sussex County, New Jersey
Geology 3014
Groundwater 3014
Ogdensburg-Culvers Gap Moraine
3083
Suwannee Limestone
Florida
--Hernando County 748, 749
--Pasco County 748, 749
Swauk Highway, Washington
Geology 4626
Sweetwater Creek State Park,
Georgia
Geology 917
Swope Limestone
Kansas
--Southeast 1766
Sylmer, California
Earthquakes 430
Geology 430
Symbionts 3022
Synclines
Georgia
--Tate Quadrangle 894
Oklahoma
--Le Flore County 3676
Synclinoria
Maine
--Northwestern 2216
Systems Analysis 2968, 2989

Tabby Mountain Coal Field, Utah
4391
Tailings 2358
Missouri
--Washington County 2633
Utah
--Kennecott Copper Company
4420
Talbot County, Maryland
Water 2295
Talc
Alabama
--Chambers County 38

--Tallapoosa County 38
California 424
--Death Valley 525
--Inyo County 525
--Kingston Range 525
--San Bernardino County 525
Montana 2790
Nevada 2885
--Grantham Talc Mine 2932
Talconite
Wyoming
--Sublette County 4857
--Wind River Mountains 4857
Talcose Minerals
Nevada 2885
Talkeetna Mountains, Alaska
Copper 325
Geochemistry 227, 269
Geology 227, 267, 269
Gold 325
Talkeetna River, Alaska
Geology 267
Talladega County, Alabama
Groundwater 28
Tallahassee, Florida
Environmental Geology 883
Hydrology 883
Tallahatchie County, Mississippi
Geology 2467
Tallapoosa County, Alabama
Anthophyllite 38, 107
Gold 125
Talc 38
Tampa, Florida
Environmental Geology 886
Groundwater 861
Hydrology 861, 866
Lake Levels 861
Water Levels 861
Tampa Bypass Canal System, Florida
Hydrology 869
Tampa Formation
Florida 751
Tanana, Alaska
Geochemistry 221
Tanana Region, Alaska
Mineral Resources 328
Tanks
Hawaii 1073
Tantalum
California 424
Taos, New Mexico
Geology 3272
Tar
Utah
--Grand County 4429
--Uintah County 4429
Tar Sand
Alabama 147

Kentucky 1965
Missouri
--Barton County 2599
--Cedar County 2599
--Dade County 2599
--Southwestern 2599
--Vernon County 2599
Tarkio Limestone
Nebraska
--Southeastern 2858
Tate Quadrangle, Georgia
Murphy Syncline 894
Tax Laws 2042
Taylor County, Kentucky
Oil 2003
Taylor Highway, Alaska
Geochemistry 217
Taylorsville, California
Geology 417
Tazewell County, Illinois
Gravel 1255
Sand 1255
Technology 2998
Tectonics
Appalachians
--Central 4742
California 430
--San Fernando Valley 431
--San Gabriel Mountains 430
Illinois 1350
--Northeastern 1482
Mexico 4331
North Dakota
--Williston Basin 3438
Rocky Mountains 4877
Texas
--Southern Edwards Plateau
4215
--Trans-Pecos 4331
West Virginia 4742
Western United States 2868
Tehama County, California
Geology 425
Tuscan Buttes 513
Tuscan Formation 513
Teklanika Formation
Alaska
--Alaska Range 282
Teklanika River, Alaska
Metamorphic Rock 285
Temperature Prospecting
Illinois 1289
Temperatures 1651
Alabama 171
--North Central 29
Arkansas
--Coastal Plain 404
Montana 2842
New Mexico

--Jemez Mountains 3205
Tennessee River, Alabama
Mercury 137
Terraces
South Dakota
--Bear Creek 4061
--Pennington County 4061
Terranes 43, 44, 45, 46, 47, 48, 49
Alabama 60
--Columbiana 114
Kentucky
--Bowling Green 1963
--Warren County 1963
Minnesota
--Minnesota River Valley 2394
Southwestern States 49
Terrestrial Heat Flow 1584
Tertiary
Alaska
--Alaska Peninsula 297
--Alaska Range 280, 294
--Gulkana Glacier 280
--Mount Galen 294
--Rainbow Mountain 280
California
--Church Creek 505
--Coast Ranges 419
--Monterey County 505
--Sierra Nevada 417
Delaware
--Greenwood 726
Idaho
--Weiser 1137
Illinois
--Southern 1284
Nevada
--Central 2916, 2921
--Eastern 2905
--Reno 2925
--Western 2921
--Yerington District 2933
New Mexico
--Cedar Hills 3187
--Dona Ana County 3187
--Sacramento Mountains 3195
--Selden 3187
North Carolina 3310
North Dakota
--Burke County 3353
--Mercer County 3357
--Mountrail County 3353
--Oliver County 3357
--Williams County 3334
Oregon
--Western 4670
Texas
--Gulf Coast 4203, 4210, 4225
--West 4305

Tertiary (Cont.)
  Utah
    --Needles Range 4421
    --Western 4348, 4421
  Washington
    --Western 4670
  Wyoming
    --Carbon County 4863
Test Borings
  West Virginia
    --Monongahela River Basin
      4749
Test Drilling
  Alabama
    --Piedmont Area 123
  Georgia
    --Brunswick 929
    --Glynn County 929
Test Holes
  California
    --Gobernadora Quadrangle
      449
    --Orange County 449
  Kansas
    --Sedgwick County 1860
  Mississippi 2541
  Nebraska
    --Council Bluffs 2853
    --Omaha 2853
Test Wells
  Delaware
    --Greenwood 726
  Florida
    --Myakka River Basin 772
  Illinois
    --Rock Island 1250
  Minnesota
    --Hollandale 2423
  West Virginia 4692
Teton County, Wyoming
  Rock and Stream Sediments
    4879
Thallium
  California 424
Thaumaleidae
  Connecticut 653
Theodore Roosevelt National Me-
  memorial Park, North Dakota
  Geology 3401
Thermal Energy 3071
Thermal Expansion
  Illinois 1271
Thermal Metamorphism
  New Mexico
    --Hidalgo County 3206
Thermal Models 1651
Thermal Plume Fields 3028
Thermal Recovery 2027
Thermal Springs

Colorado 612, 614, 636
  Oregon 3759
  Wyoming 4842
Thermal Surveys
  New York 2969
Thermal Water
  Mexico 4331
  Nevada 2892
  New Mexico 3137, 3234
  Texas
    --Trans-Pecos 4331
Thermal Wells
  Colorado 612, 614, 636
  Oregon 3759
Thermoluminescence 1684
Thin Elk Formation
  South Dakota 4053
Thomas County, Georgia
  Geology 940
  Groundwater 940
  Phosphate 965
  Sand 965
Thomasville, Georgia
  Aquifers 942
Thomson Formation
  Minnesota
    --East-Central 2424
Thorium 376
  Idaho
    --Lemhi Pass 1147
  Montana
    --Beaverhead County 2804
    --Lemhi Pass District 2804
Thornton, Illinois
  Niagaran Reef 1466
Thrust Belts
  Idaho 4874
  Utah 4874
  Wyoming 4874
Ticonderoga Quadrangle, Vermont
  Surface Geology 4501
Tidal Datum 3052
Tidal Boundary 3051
Tidal Flats
  Kentucky
    --Central 1964
Tidal Rivers
  Maryland 2291
Tidal Surveying
  New Jersey 3080
Tidelands
  New Jersey 3078
Tides 1577, 3055
Tierra Amarilla Coal Field, New
  Mexico 3154
Tift County, Georgia
  Geohydrology 925
Tikchik Lake, Alaska
  Geochemistry 225

Till Sheets
Ohio
--Northeastern 3522
Tillamook County, Oregon
Environmental Geology 3716,
3721
Tills
Illinois 1203, 1256
--Lake Michigan 1445
--Northwestern 1293
Indiana
--Eastern 1523
Timbered Hills Group
Oklahoma
--Arbuckle Mountains 3632
--Murray County 3632
Time-Trend Curves 1850
Tin
California 424
--Plumas County 544
Tinsley Field, Mississippi 2533
Tisza Valley, Hungary
Water 2987
Titanium
California 424
Nevada 2896
North Carolina 3297
Tobacco Valley, Montana
Geology 2741
Water 2741
Tocks Island, New Jersey
Geology 3014
Groundwater 3014
Todilto Park, New Mexico
Sedimentology 3238
Upper Cretaceous Rocks 3238
Toklat River, Alaska
Metamorphic Rock 285
Tombigbee River, Alabama
Mercury 137
Tombigbee River, Mississippi
Landslides 2546
Tompkinsville, Kentucky
Subsurface Geology 2065
Toms Brook Quadrangle, Virginia
Geology 4602
Tonalite
California
--Lakeview Mountains 519
--Riverside County 519
Toney Quadrangle, Alabama
Geology 203
Mineral Resources 203
Tongue River Interval
North Dakota 3483
--Western 3475, 3495
--Williston Basin 3453,
3488
Tonoloway Limestone 4747

West Virginia 4744
Tonopah Mining District, Nevada 2919
Tonopah Quadrangle, Nevada
Geology 2893
Tooele County, Utah
Geology 4750, 4446
Mineralization 4356
Minerals 4372
Seismic Risk 4446
Tooele Valley, Utah
Geology 4446
Seismic Risk 4446
Toole County, Montana
Water 2716
Topeka, Kansas
Groundwater 1867
Hydrology 1867
Topography
California 424
--Los Angeles County 520
--San Fernando Valley 431
Kentucky 2126
Mississippi
--Hinds County 2537, 2538
Missouri
--St. Charles County 2595
Oklahoma 3685
South Carolina 3990
West Virginia
--Charleston 4682
Toronto Limestone
Kansas 1743
Torrance County, New Mexico
Geology 3106
Minerals 3106
Tower-Soudan State Park, Minnesota
Geology 2387
Towns
Kansas
--Southeastern 1890
Towns County, Georgia
Rock 904
Toxic Wastes
California
--Southern 476
Tornadoes
Alabama 182
Torrington Quadrangle, Connecticut
Bedrock Geology 681
Trace Base Metals
New Mexico
--Luna County 3186
Trace Elements 1355, 1437, 1448
Illinois
--Illinois River 1432
--Johnston City Lake 1458
--Lake Du Quoin 1458
--Lake Michigan 1436
--Lake Saline 1436

Trace Elements (Cont.)
Illinois (Cont.)
--Little Grassy Lake 1458
--Peoria Lake 1432
Iowa 1663
Maryland
--Patuxent River Basin
2279
Ohio 3604
Southeast 4138
Texas
--South 4336
Trace Fossils
Nevada 2891
Trail Guides
New Mexico
--Pecos 3276
Trail Mountain, Utah
Coal 4385
Trail Ridge, Florida
Sand 871
Traill County, North Dakota
Geology 3337, 3338, 3339
Glaciation 3337, 3339
Groundwater 3337, 3338, 3339
Water 3338, 3339
Trans-Pecos, Texas
Cenozoic Geology 4253
Presidio Bolson 4311
Volcanic Fields 4253
Transmissivity Zones
Florida
--South 887
Transportation
Mississippi
--Adams County 2540
--Hinds County 2537, 2538
Transverse Ranges, California
424, 498
Secondary Faulting 441
Stress Accumulation 441
Vertical Deformation 441
Traverse
Missouri
--St. Francois Mountains
2635
--Southern 2635
Oklahoma
--Delaware County 2635
Travertine
Montana 2694
Treasure County, Montana
Bentonite 2814
Tree Fern
Illinois 1348
Trego County, Kansas
Geology 1705
Groundwater 1705
Trempealeau Remnants 2055

Trend-Surface 1852, 1864
Trend-Surface Analysis
Kansas 1849
Trend-Surface Mapping 1841
Trenton Formation 3549, 3550, 3553
Ohio 3550
--Fayette County 3553
Trenton Group
West Virginia
--Eastern 4698
--Pendleton County 4698
Trepostomatous Bryozoan Fauna
Indiana
--Eastern 1493
Ohio
--Western 1493
Tres Hermanas Stock
New Mexico
--Luna County 3186
Triana Area, Alabama
Environmental Geology 4
Hydrology 4
Triassic 4539
Connecticut
--South-Central 667
--Southern 696
Maryland 2304
North Carolina
--Deep River Basin 3305
Pennsylvania 3840
--Cornwall 3875
--Newark-Gettysburg Basin 3793
Texas
--Panhandle 4332
Virginia
--Danville Basin 4560
Trilobites
Oklahoma 3642
--Arbuckle Mountains 3632,
3650
--Murray County 3632
--Wichita Mountains 3651
Texas
--Central 4301
Trimmers Rock
Pennsylvania
--Eastern 3801
Trinity County, California
Geology 502, 522
Magnetic Study 506
Mineral Resources 443
Mines 443
Trinity Deposits
Texas
--Central 4306
Trip Guides
Nevada 2917
Trough Creek State Park, Pennsylvania
Guide 3898

Troughs
  Georgia
    --Southwestern 947
Truckee Meadows, Nevada
  Geology 2924
  Geothermal Activity 2927
  Quaternary 2924
Trumbull County, Ohio
  Glacial Geology 3581
Tualatin Valley, Oregon
  Engineering Geology 3702
Tucumcari Basin, New Mexico
  Mesa Rica Sandstone 3211
  Sedimentology 3211
Tuff
  Missouri
    --Southeast 2626
    --Washington County 2623
  Nevada
    --Northumberland 2918
    --Yerington District 2933
Tully Formation
  Pennsylvania
    --Western 3892
Tully Limestone
  New York 3832
  Pennsylvania 3832
Tungsten
  California 424
Tunnels 4694
  Hawaii
    --Haiku 1012
    --Kahaluu 1012
    --Oahu 1012, 1089
    --Waihee Valley 1089
  Illinois
    --Greater Chicago Area
      1428
Tununk Member 4437
Turbidites
  Illinois 1281
Turkey Heaven Mountain, Alabama
  Kyanite 99
Turkey Run State Park, Indiana
  Geology 1615
Turner County, Georgia
  Geohydrology 925
Turquoise
  Nevada 2914
Tuscaloosa County, Alabama
  Groundwater 83
Tuscan Buttes, California 513
Tuscan Formation
  California
    --Tehama County 513
Tuscarora Sandstone
  West Virginia 4692
Tuscola, Illinois
  Geology 1280

Oil Production 1280
Tuscumbia Quadrangle, Alabama
  Geology 207
  Mineral Resources 207
Tustin Quadrangle, California
  Geology 567
Tuttle Creek Reservoir, Kansas
  Pleistocene Drainage 1774
Twelve-Mile Creek, Alaska
  Geochemistry 252
  Geology 252
Two Creeks Forest Bed
  Wisconsin 4798
Tyger River, South Carolina
  Drainage Basin 3991
  Fluvial Monazite 3991
Tymochtee Dolomite
  Ohio
    --Allen County 3521

Uinta, Colorado
  Coking Coals 630
Uinta Basin, Utah
  Blackshale Facies 4373
  Currant Creek Formation 4419
  Duchesne River Formation 4370
  Green River Formation 4373, 4418
  Hydrocarbons 4433
  Oil Sandstone 4425, 4428
  Oil Shale 4418
  Sandstone 4428
  Stratigraphy 4370
Uintah County, Utah
  Bituminous Sandstone 4409
  Geology 4345
  Oil 4423, 4429
  Sandstone 4423, 4429
Ulm, Montana
  Geology 2706
  Groundwater 2706
Ultramafic Bodies
  Vermont
    --Dover 4485
    --Lowell 4490
    --Ludlow 4485
    --Windham 4485
Ultramafic Rock
  Alaska
    --Anchorage Quadrangle 253
    --Eklutna 253
  Georgia
    --Clay County 904
    --Lake Chatuge 904
    --Towns County 904
Unconformaties
  Indiana 1586
Underclay 3829
  Alabama

Underclay (Cont.)
  Alabama (Cont.)
    --Warrior Basin 86
    Illinois 1285
    Missouri
    --Western 2616
Underground Storage
  Indiana 1592
  Kentucky
  --Western 2035
  Pennsylvania 3860
Unga Island, Alaska
  Petrified Forests 329
Union County, Florida
  Water Resources 759, 822
Union County, Illinois
  Geology 1461
  Groundwater 1461
Union County, Kentucky
  Inert Gas Injection 2034
  Secondary Recovery 2034
Union County, New Mexico
  Groundwater 3228
Union County, North Carolina
  Petrography 3307
  Stratigraphy 3307
Union County, Ohio
  Glacial Geology 3567, 3570
Union Parish, Louisiana
  Water Resources 2177
Unionport Gas-Storage Project,
  Indiana 1586
Unit Regional-Value Concept
  Kansas 1876
United Verde Open Pit, Arizona
  349
Upland Silt
  North Dakota 3464
Uplifts
  Chihauhua
    --Cerro De Cristo Rey Up-
      lift 3264
  New Mexico
    --Central 3254
    --Cerro De Cristo Rey Up-
      lift 3264
Upper Cambrian
  Missouri
    --Washington County 2623
  Texas
    --Central 4301
Upper Chesterian
  Arkansas
    --Northwestern 3695
  Indiana
    --Southern 2580
  Oklahoma
    --Northeastern 3695
Upper Chetco Drainage, Oregon

  Geology 3730
  Minerals 3730
Upper Cretaceous
  Delaware 710
  Florida
    --Peninsular 776
  Gulf Coast 4281
  Kansas 1755, 1771
    --Central 1713
    --Western 1822
  Maryland
    --Salisbury Embayment 2307
  Montana
    --Southeastern 2823
  New Mexico
    --Northwestern 3218
    --Todilto Park 3238
  North Dakota 3368, 3466
  South Dakota 3466
  Tennessee
    --Western 4119
  Texas
    --Central 4164
    --Northeast 4308
  Utah
    --Garfield County 4362
  Western United States 3270
Upper Devonian 1642
  Indiana
    --Delphi 1562
  Iowa 1646
    --North-Central 1647
  Pennsylvania 3789
    --Bradford County 3804
    --Northeastern 3814
    --Southern 3813
Upper Dry Valley, Idaho
  Groundwater 1179
Upper East Fork Drainage Basin,
  Indiana
  Geology 1594
Upper Eocene
  Georgia 953
Upper Freeport
  Pennsylvania
    --Allegheny County 3861
    --Greene County 3895
    --Westmoreland County 3861,
      3894
Upper Hamakua Ditch, Hawaii 1009
Upper Miocene
  North Carolina
    --Grimesland 3306
    --Pitt County 3306
Upper Mississippian
  Illinois
    --Illinois Basin 1277, 1314
  Kansas 1736
    --Western 1683

Virginia
--Southwestern 4511
Upper Niagaran
Ohio
--Northeastern 3552
Upper Oligocene
Louisiana 2158
Upper Ordovician
Indiana 1996
Kentucky 1996
Minnesota 2440
Ohio 1996
Pennsylvania
--Western 3892
Upper Patoka Drainage Basin,
Indiana
Geology 1593
Upper Pennsylvanian
Kansas 1710
--Eastern 1740, 1821
--Rawlins County 1700
--Southeast 1766
--Stafford County 1700
Texas
--North-Central 4192, 4310
Upper Silurian
Ohio
--Allen County 3521
Upshur County, West Virginia
Oil and Gas 4679
Uranium 376, 2815
Alaska 273, 301, 338
--Bokan Mountain 276
--Southeastern 276, 279
Arizona 353
--East-Central 382
Arkansas
--North-Central 399
California 424
Colorado 607
--Cheyenne Basin 620
--Mesa County 625
Florida 889
Great Plains 1811
New Mexico 3172
--Grants 3116, 3239
North Dakota
--Southwestern 3486
--Williston Basin 3485,
3487, 3488
Texas 4167, 4179
--Coastal Plain 4322,
4335
--South 4246, 4252, 4336
Utah
--Emery County 4386
--Garfield County 4412
--San Juan County 4416
--San Rafael River Mining

Area 4386
Washington 4643
Wyoming 4871
--Medicine Bow Mountains 4883
Urban Development
Missouri 2592
Urban Geology
California 436
Urban Growth
Colorado 635
Urbanization 1392
Alabama
--Jefferson County 191
--Shades Creek Basin 191
Florida
--Dade County 834
Utah County, Utah
Pyrophyllite Clay 4413
Utah Lake, Utah
Hydrogeology 4472
Ute Indian Reservation
Coal 3188

Valderan Type Localities
Wisconsin 4798
Valdez-Chitina Highway, Alaska
Geochemistry 223
Valencia, California
Earthquakes 430
Geology 430
Valencia County, New Mexico
Geology 3227
Groundwater 3227
Valley and Ridge Province, Alabama
Geology 10
Valley and Ridge Province, Pennsyl-
vania
Caves 3819
Valley City, North Dakota
Geology 3411, 3418
Valley County, Idaho
Geology 1139
Valley Forge State Park, Pennsylvania
Guide 3905
Valley Systems
Illinois
--Northern 1210
Valmeyeran
Illinois 1211
Indiana 1211
Missouri 1211
Van Buren Quadrangle, Arkansas
Geology 395
Van Buren Quadrangle, Oklahoma
Geology 395
Van Horn Sandstone
Texas
--West 4307

Van Norman Reservoir, California
Earthquakes 432
Vanadium
California 424
Illinois 1339
Utah
--Emery County 4386
--San Rafael River Mining
Area 4386
Vananda, Montana
Bentonite 2814
Vandalia, Illinois
Glacial Geology 1298
Vascular Flora
Connecticut 702
Vassalboro, Maine
Geology 2215
Metamorphism 2215
Vegetation 2358
Minnesota
--Northeastern 2447
Venango County, Pennsylvania
Geology 3876
Red Valley Sandstone 3876
Venice, Florida
Salt Water Leakage 825
Venomous Snakes
Connecticut 659
Ventura County, California
Geology 453, 498
Mineral Resources 453
Vermiculite 367
California 424
Nevada
--Clark County 2913
Vermilion District, Minnesota
Geology 2438
Iron Ore 2438
Lower Precambrian 2425
Precambrian 2391
Rocks 2391, 2425
Sedimentation 2391
Stratigraphy 2425
Volcanism 2391
Vermilion West Quadrangle, Ohio
Geology 3561
Vermillion, South Dakota
Water 4066
Vernal Coal Field, Utah 4391
Vernon County, Missouri
Tar Sand 2599
Vernon Parish, Louisiana 2166
Verrazano Narrows Bridge 3056
Versailles State Park, Indiana
Geology 1616
Versene (EDTA) Compounds 1689
Versene (EDTA) Solutions 1709
Vertebrates
Florida 879

Oklahoma 3669, 3675
Tennessee 4121
Texas
--San Patricio County 4312
Vertical Deformation
California
--Transverse Ranges 441
Vestal, F.E. 2470
Vesuvius Quadrangle, Virginia
Geology 4561
Viborg, South Dakota
Groundwater 4088
Vicksburg Formation
Texas
--Gulf Coast 4223
Vienna District, Idaho
Geology 1166
Geochemistry 1166
Vinton County, Ohio
Bed Rock Geology 3564
Viola Formation
Oklahoma
--Arbuckle Mountains 3644
Virburnum Trend
Missouri
--Southeast 2638
Virgil Environment
Texas
--North-Central 4174
Virgilian
Kansas 1865
Virgilina Quadrangle, Virginia
Geology 4542
Virgin Mountains, Arizona
Geology 357
Volcanic Ash
Kansas 1857
Volcanic Cinders
California 424
Volcanic Construction Materials
Idaho 1155
Volcanic Fields
Texas
--Trans-Pecos 4253
Volcanic Formations
Alaska
--Alaska Range 282, 294
--Mount Galen 294
Volcanic Rock
California 538
Central Europe 3012
Minnesota
--North Shore 2392
--Vermilion District 2391
Northeast 3012
Utah
--Needles Range 4421
--Western 4421
Volcanism

Alaska
--Alaska Range 294
--Mount Galen 294
California
--Amboy Crater 493
--San Bernardino County
493
--Sierra Nevada 417
Idaho
--Snake River Plain 1180
Nevada
--Eastern 2905
Oregon 3755
Tennessee 4118
Utah
--West-Central 4447
Washington
--Cascade Mountains 4612
Volcanoes
Nevada
--Black Mountain Volcanic
Center 2917
--Central 2916, 2921
--Northumberland 2918
--Reno 2925
--Western 2921
New Mexico
--Albuquerque 3279
--Albuquerque Basin 3210
Oregon
--Mt. Hood 3777
Washington
--Mt. St. Helens 4669
Volga, South Dakota
Groundwater 4096
Volusia County, Florida
Water Resources 844

W. L. Newman Phosphate Mine,
Pennsylvania 3836
W. P. Franklin Dam, Florida
Saltwater Intrusion 778
Wabash Formation
Indiana
--Clark County 1607
Kentucky
--Jefferson County 1607
Wabash River
Heavy Minerals 1258
Sand 1258
Wabash Valley, Indiana
Geology 1582
Wabash Valley Fault System,
Illinois 1365
Wabasha County, Minnesota
Clay 2413
Geology 2413
Wadley North Quadrangle, Alabama

Geology 64
Wadsworth Quadrangle, Nevada
Geology 2871
Wagonhound Creek, Wyoming
Structural Geology 4862
Wah Wah Pass, Utah
Igneous Complex 4407
Wah Wah Range, Utah
Hydrothermal Alteration 4402
Mineralization 4402
Wahkiakum County, Washington
Geology 4606
Minerals 4606
Waialeale, Hawaii
Hydropower 1120
Waianae, Hawaii
Groundwater 1007
Waihee Valley, Hawaii
Water 1089
Waikapu, Hawaii
Drilling Logs 995, 1057
Pumping Tests 995, 1057
Waikolu Valley, Hawaii
Water Resources 1093
Wailuku, Hawaii
Water Resources 1052
Waimanalo, Hawaii
Deep Well Disposal 1045, 1092
Geohydrology 1045
Irrigation 1067
Ocean Outfall 1084
Sewage 1045
Waimanalo Irrigation System, Hawaii
Costs 1067
Waimea, Hawaii
Ditches 1009
Drilling Logs 1027, 1035
Pumping Tests 1035
Wake County, North Carolina
Geology 3292
Minerals 3292
Walker County, Alabama
Drainage 122
Walker County, Georgia
Geology 935
Groundwater 935
Wall, South Dakota
Dakota Sandstone 4037
Pump Test 4037
Wallace County, Kansas
Geology 1678
Groundwater 1678
Wallace Quadrangle, Virginia
Geology 4579
Walnut Springs Quadrangle, New
Mexico
Geology 3095
Walsh County, North Dakota
Aquifers 3360

Walsh County, North Dakota (Cont.)
  Esker 3442
  Geology 3358, 3442
  Glaciation 3358, 3360
  Groundwater 3359, 3360
  Water 3359, 3360
  Wells 3359
Walt Disney World Development,
    Florida
  Hydrology 866
Waltersburg Anomaly, Kentucky
  Daviess County 2025
Waltersburg Crudes
  Illinois
    --Illinois Basin 1277
Walton County, Florida
  Hydrology 795
  Mineral Resources 744
  Water Resources 863
Walworth County, Wisconsin
  Geology 4822, 4830
  Groundwater 4822, 4830
  Hydrology 4830
Wapanucka Formation
  Oklahoma 3626, 3672
Wapella East Oil Pool, Illinois
    1205
Ward County, North Dakota
  Geology 3341, 3342
  Glaciation 3342
  Groundwater 3341, 3342
  Water 3341, 3342
Ware County, Georgia
  Phosphate 965
  Sand 965
Warm Springs Valley, Nevada
  Gravity Study 2912
Warren County, Illinois
  Coal 1204, 1230, 1295
Warren County, Kentucky
  Coal 1938
  Economic Geology 1936
  Limestone Terrane 1963
  Water 1963
Warren County, New Jersey
  Geology 3014
  Groundwater 3014
Warren County, Pennsylvania
  Oil and Gas 3881
Warren County, Virginia
  Geology 4552
  Land Modification 4552
  Silica 4534
Warren Quadrangle, Pennsylvania
  Oil and Gas 3871
Warrior Basin, Alabama
  Pennsylvanian 88
  Underclay 86
Warrior River, Alabama

Mercury 137
Warsaw Formation
  Illinois 1244
Wasatch Plateau, Utah
  Coal 4385, 4392
Wasatch Range, Nevada
  Quaternary 2903
Wasco County, Oregon
  Geologic Hazards 3733
Washakie County, Wyoming
  Geology 4848
  Lower Eocene 4848
  Lower Willwood Formation 4848
  Mammalian Paleontology 4848
  San Creek Facies 4848
Washington County, Florida
  Mineral Resources 744
Washington County, Indiana
  Geology 1499
  Mineral Resources 1499
Washington County, Mississippi
  Geology 2467
Washington County, Missouri
  Barite 2633
  Lapalli Tuffs 2627
  Sediments 2623
  Upper Cambrian 2623
Washington County, Ohio
  Coal 3520
  Geology 3509
  Minerals 3509
Washington County, Oklahoma
  Hogshooter Formation 3629
  Missourian 3629
  Petrology 3629
Washington County, Oregon
  Rocks 3772
Washington County, Pennsylvania
  Geology 3806, 3807
  Groundwater 3949
  Land Use 3806, 3807
Washington County, Wisconsin
  Geology 4826
  Groundwater 4826
Washita Group
  Texas
    --North-Central 4329
Washoe County, Nevada
  Geology 2869, 2924
  Geothermal Activity 2927
  Gravity Study 2912
  Industrial Rock 2870
  Minerals 2869, 2870
  Quaternary 2924
Washougal Mining District, Washing-
    ton 4658
Washover Fan
  Texas
    --Coast 4302

Waste Disposal 1396, 2071,
3066
California
--Southern 476
Colorado 622
Florida 787
Idaho 1165
Illinois 1282
--Northeastern 1421, 1456
Indiana 1596
Kansas 1770
Maryland 2296
Mississippi
--Adams County 2540
New Mexico
--Southeastern 3197
Ohio
--Summit County 3599
Oklahoma 3681, 3682, 3683,
3684
Pennsylvania 3783
Texas 4177
--East 4228
--Palo Duro Basin 4220,
4230
Waste Effluents 1399
Waste Injection
Ohio 3538
Waste Injection Monitor Wells
Florida
--Pensacola 791
Waste Storage
Iowa 1630
Kansas 1898
Waste Water 1184
Arizona 380
Idaho 1163
Wastes
Maryland 2292
New Jersey
--Hackensack River 3013
Water 1249, 1253, 2949, 2950,
2951, 2952, 2954, 2955, 2958,
2960, 2962, 2965, 2967, 2970,
2971, 2981, 2982, 2983, 2985,
2989, 2991, 2992, 2999, 3001,
3003, 3006, 3016, 3022, 3036,
3039, 3040, 3057, 3058, 3068,
3075, 3085, 4815
Alabama 40, 41, 81, 91, 95,
133, 154, 165, 178, 180,
184, 194
--Athens Area 18
--Barbour County 173
--Birmingham Valley 161
--Black Warrior Basin 12
--Bryce State Hospital 83
--Butler County 172
--Cahaba Valley 161

--Calhoun County 91
--Cane Creek Basin 122
--Cherokee County 26
--Colbert County 163
--Escambia County 21
--Franklin County 19
--Geneva County 175
--Greene County 33
--Houston County 174
--Huntsville Area 82
--Jefferson County 11, 101
--Lauderdale County 162
--Lawrence County 25
--Limestone County 164
--Marshall County 32
--Mobile 150
--Montgomery Area 17
--Montgomery County 17
--Morgan County 23
--Negro Colony 83
--Oakman 122
--Pickens County 36
--Piedmont Area 100, 123
--Russell County 22
--Russellville Area 24
--St. Clair County 20
--South 201
--Southwestern 31
--Stevenson Area 134
--Talladega County 28
--Tuscaloosa County 83
--Walker County 122
--Wilcox County 176
--Winston County 181
Alaska 322
Appalachians 4676
Arizona 351, 380
Arkansas 401
Colorado 586, 638
--Boulder County 596
--Denver Basin 621
--Jefferson County 590
Delaware 708, 711, 718, 723
--Coastal Plain 727
Developing Countries 2978
Florida 764, 768, 777, 785, 787,
790, 798, 800, 802, 810, 811,
878, 880
--Alachua County 759, 822
--Alafia Basin 833
--Bradford County 759, 822
--Broward County 814, 838,
852
--Charlotte County 865
--Clay County 759, 822
--East-Central 848
--Econfina Creek Basin 773,
828
--Escambia County 766, 827

Water (Cont.)
  Florida (Cont.)
    --Flagler County 755
    --Hillsborough County 760
    --Indian River County 867
    --Jacksonville 774, 815
    --Lake Jackson 835
    --Lee County 856
    --Miami 862
    --North-Central 860
    --North-East 841
    --Okaloosa County 784
    --Orange County 757, 817,
      837
    --Palm Beach County 854
    --Peace River Basin 833
    --Pensacola 812
    --Polk County 754
    --Putnam County 755
    --Sanibel Island 856
    --Santa Rosa County 766,
      827
    --St. Johns County 755
    --St. Lucie County 849
    --Southeastern 847
    --Tampa 861
    --Union County 759, 822
    --Volusia County 844
    --Walton County 863
  Georgia 927
    --Atlanta 974
    --Brunswick 929
    --Coastal 924
    --Glynn County 929
    --Northwest 910
  Germany
    --Saxony 2961
  Hawaii 989, 990, 991, 1029,
    1079, 1095, 1116, 1119
    --Hamakua 1004, 1008,
      1086, 1090
    --Hawaii Island 1044, 1074,
      1082, 1090, 1113
    --Hilo 1021, 1036
    --Hoolehua 1061
    --Kahuku 1030
    --Kahakuloa 1115
    --Kau District 1018
    --Kauai 1080, 1082, 1083,
      1101
    --Kaunakakai 1078
    --Kehena Ditch 1044
    --Kihei-Makena 1099
    --Kohala 1004, 1008, 1086,
      1096, 1113
    --Hohala Mountain 1005
    --Kona 1037, 1071
    --Lahaina District 1042,
      1081, 1094
    --Maui 1042, 1049, 1051, 1052,
      1053, 1081, 1094, 1099
    --Mauna Kea 1005
    --Molokai 1061, 1075, 1078,
      1093
    --Ninole 1021
    --Oahu Island 978, 979, 1001,
      1030, 1089
    --Pelekunu Valley 1093
    --Pukoo 1078
    --Puna 1036
    --Wainee Valley 1089
    --Waikolu Valley 1093
    --Wailuku 1052
  Hungary
    --Tisza Valley 2987
  Idaho 1124, 1163, 1188
    --Latah County 1132, 1157
    --Moscow Basin 1157
  Illinois
    --Shelbyville 1391
  Iowa 1672
    --Central 1666
    --East-Central 1671
    --South-Central 1670
    --Southeast 1669
  Kansas 1840, 1886
    --Cedar Bluff Irrigation District
      1783
    --Finney County 1786
    --Ford County 1789, 1827
    --Grant County 1848, 1856,
      1875
    --Gray County 1786
    --Great Bend Prairie 1789
    --Hamilton County 1786
    --Hodgeman County 1827
    --Kearny County 1786
    --Little Arkansas River Basin
      1785
    --Ness County 1825
    --Southwestern 1788
    --Stanton County 1848, 1856,
      1875
    --West-Central 1784
  Kentucky 1944, 1947, 1948, 2077
    --Bowling Green 1963
    --Eastern 1951
    --Jackson Purchase 2097
    --Middlesboro 1955
    --Muhlenberg County 1939
    --New Cypress Pool 1939
    --Warren County 1963
  Louisiana
    --Alexandria 2186
    --Batchelor 2196
    --Belmont 2200
    --Bossier Parish 2165
    --Caddo Parish 2165

--Flatwoods 2199
--Lake Pontchartrain 2172
--Lettsworth 2196
--Marthaville 2200
--Natchitoches Parish 2164
--Ouachita Parish 2174
--Pointe Coupee Parish
  2171, 2196
--Rapides Parish 2168
--Robeline 2200
--Sabine Parish 2163
--Saint Landry Parish 2198
--Simpson 2199
--Slagle 2199
--Southeastern 2197
--Southwestern 2170, 2181,
  2182
--Union Parish 2177
--Vernon Parish 2166
Maryland 2240, 2283, 2287
--Dorchester County 2295
--Patuxent River Basin
  2279
--Piney Point Aquifer 2309
--Salisbury 2281
--Southern 2227, 2306
--Talbot County 2295
--Tidal Rivers 2291
Mexico 3038, 4331
Michigan
--Branch County 2377
--Rifle River Basin 2367
--Van Buren County 2374
Mississippi 254, 2518
--Adams County 2540
--Alcorn College 2544
--Attala County 2465
--Copiah County 2510
--Hinds County 2495, 2537,
  2538
--Mississippi Valley 1200
--Rankin County 2524
--Smith County 2527
--Wayne County 2529
Missouri 2550, 2568, 2592,
  2596, 2607, 2666, 2671,
  2677
--Andrew County 2659
--Buchanan County 2658
--Carroll County 2657
--Chariton County 2656
--Davies County 2653
--Dekalb County 2652
--Gentry County 2651
--Grundy County 2645
--Harrison County 2647
--Jefferson County 2674
--Joplin 2668
--Linn County 2655

--Livingston County 2650
--Mercer County 2646
--North 2580
--Putnam County 2648
--St. Charles County 2595,
  2674
--St. Louis 2674
--St. Louis County 2674
--Springfield 2678
--Sullivan County 2654
--Worth County 2649
Montana 2711, 2721, 2729, 2735,
  2791, 2829, 2842
--Clark Fork 2771
--Glacier County 2716
--Judith Basin 2749, 2767
--Kalispell Valley 2707
--Missoula Valley 2685
--Powder River 2776
--St. Regis 2771
--Southeastern 2776
--Stillwater River Valley 2741
--Tobacco Valley 2741
--Toole County 2716
Nevada 2864, 2892
New Jersey 3073
--Bergen County 3087
--Hackensack River 3013
--Morris County 3087
--Passaic County 3087
--Passaic Valley 2973
New Mexico 3098, 3137, 3192,
  3216, 3234
--De Baca County 3230
--Grant County 3232
--Guadalupe County 3233
New York 2986
--Thermal Surveys 2969
North Dakota 3372, 3426
--Adams County 3377
--Barnes County 3320
--Benson County 3363, 3364
--Bowman County 3377
--Burke County 3353
--Burleigh County 3317
--Cass County 3332, 3333
--Cavalier County 3370
--Dickey County 3391
--Divide County 3326, 3327
--Dunn County 3386
--Eddy County 3323, 3324
--Emmons County 3380
--Foster County 3323, 3324
--Grand Forks County 3348,
  3349
--Grant County 3383
--Griggs County 3374
--Lamoure County 3391
--McIntosh County 3397

Water (Cont.)
  North Dakota (Cont.)
    --McLean County   3366
    --Mercer County   3357
    --Montrail County   3357
    --Nelson County   3359, 3360
    --Oliver County   3357
    --Pembina County   3370
    --Pierce County   3363, 3364
    --Ramsey County   3393
    --Ransom County   3389
    --Richland County   3329
    --Sargent County   3389
    --Sioux County   3383
    --Steele County   3374
    --Stutsman County   3314
    --Traill County   3338, 3339
    --Walsh County   3359, 3360
    --Ward County   3341, 3342
    --Wells County   3345
    --Williams County   3335,
      3336
  Ohio
    --Lake Erie   3559, 3575
  Oklahoma   3685
  Oregon   3706
  Pennsylvania
    --Lehigh County   3942
  South Carolina   3964, 3965,
    3966, 3967, 3968
  South Dakota   3997, 4030
    --Beadle County   4001, 4003
    --Beresford   4067
    --Bon Homme County   4012,
      4014
    --Brown County   4021, 4023
    --Campbell County   4008,
      4010
    --Charles Mix County   4015,
      4017
    --Claremont   4070
    --Clay County   4006
    --Day County   4020
    --Douglas County   4015,
      4017
    --Edmunds County   4024
    --Faulk County   4024
    --Lesterville   2068
    --McPherson County   4024
    --Marshall County   4018,
      4019
    --Redfield   4069
    --Scotland   4065
    --Vermilion   4066
  Southeast   3000
  Southwestern States   49
  Tennessee
    --Knox County   4114
  Texas
    --Corpus Christi   4330
    --Coast   4180, 4205
    --East   4226
    --Mustang Island   4327
    --Padre Island   4327
    --Trans-Pecos   4331
  Utah   4346
    --Dugway Valley   4458
    --Escalante   4460
    --Garfield County   4460
    --Government Creek   4458
    --Kane County   4452, 4457
    --Old River Bed   4458
    --Rush Valley   4458
    --Sink Valley   4458
    --Skull Creek   4458
    --Western   4458
  West Virginia   4717
    --Little Kanawha River Basin
      4750, 4755
    --New River Basin   4752
    --Potomac River Basin   4756
  Wisconsin   4769, 4777
    --Central   4824
Water Analysis   2576
Water Desalting
  Hawaii   1111
Water Development Tunnels
  Hawaii
    --Haiku   1012
    --Kahaluu   1012
    --Oahu   1012
Water Laws   3149
  Alabama   36, 37
Water Mobility   2037
Water Pollution   2997
Water Pumpage   2185
  Louisiana   2195, 2201
    --Southwestern   2202
Water Quality   1877, 1881, 2194,
  3010, 3059, 3060, 3070
  Idaho
    --Coeur D'Alene River Basin
      1169
  Kansas
    --Kansas River Basin   1762
  Louisiana
    --Southwestern   2191
  Maryland
    --Patuxent River Basin   2279
  New Jersey   3008
    --Passaic Valley   2973
Water Reservoirs
  Alabama   177
    --North-Central   29
Water Softener   1388
Water Towers   3061
Waterbury Quadrangle, Connecticut
  Bedrock Geology   678

Waterflooding
Kentucky
--Lee County 2039
Watersheds
Czechoslovakia
--East 2963
Idaho
--Owyhee County 1171
Oregon
--Bull Run 3724
--Clackamas County 3724
--Multnomah County 3724
Watertown, South Dakota
Groundwater 4073
Waterville, Maine
Geology 2215
Metamorphism 2215
Waterways
Florida
--Venice 825
Illinois 1334
Waubay, South Dakota
Groundwater 4085
Waukesha County, Wisconsin
Groundwater 4817
Waves 3029
Wayne County, Georgia
Phosphate 971
Wayne County, Iowa
Groundwater 1668
Wayne County, Michigan
Geology 2344
Groundwater 2344
Land 2344
Wayne County, Mississippi
Clay 2530
Geology 2528
Mineral Resources 2528
Water Resources 2529
Wayne County, Missouri
Brown Iron Ore 2578
Wayne County, Ohio
Bedrock Geology 3562
Geology 3547
Glacial Geology 3563
Wayne County, West Virginia
Gas 4741
McKenzie Formation 4699
Middle Silurian 4699
Petrology 4699
Waynesboro Quadrangle, Virginia
Geology 4540
Weather
Nevada 2942
Weather Modification 2951
Weathering 1203, 1256
Maryland
--Piedmont 2303
Webster, South Dakota

Groundwater 4102
Webster County, Kentucky
Pennsylvanian Channel 2132
Weiser, Idaho
Later Tertiary 1137
Paleobotany 1137
Stratigraphy 1137
Weld County, Colorado
Geology 602
Wellborn Formation
Texas
--Burleson County 4290
--Lee County 4290
Wells 1627, 1637, 1871, 2049, 2107,
2327, 3538, 3923, 3930
Alabama 59
--Baldwin County 116
--Escambia County 116
--Mobile County 116
Arizona 356
Colorado 593, 612, 614, 636
--Denver Basin 621
Delaware
--Newark 731
Florida
--Jacksonville 781
--Polk County 754
--Tampa 861
--Venice 825
Hawaii
--Oahu Island 979
Idaho
--Latah County 1132
--Snake River Plain 1168
Kansas 1686, 1691, 1698, 1704,
1712, 1721, 1814, 1817, 1863,
1904, 3690
--Northwestern 1696
--Rawlins County 1700
--Sedgwick County 1860
--Stafford County 1700
Kentucky 1946, 2021
--Western 2035
Maryland 2229, 2263
--Anne Arundel County 2232
--Baltimore County 2224
--Charles County 2225
--Harford County 2231
--Southern 2227
--Worcester County 2230
Michigan 2368
Minnesota
--Hollandale 2423
Mississippi 2491, 2541, 2545
Missouri
--North 2580
--Northwest 2557
Nebraska 2846, 2848, 2850, 2851
New Mexico 3142, 3163

Wells (Cont.)
  New Mexico (Cont.)
    --Magdalena 3169
    --Socorro 3169
    --Socorro County 3169
  North Dakota 3426, 3482
    --Adams County 3377
    --Benson County 3363
    --Bowman County 3377
    --Burke County 3352
    --Cavalier County 3370
    --Dickey County 3391
    --Dunn County 3386
    --Emmons County 3380
    --Grand Forks County
      3348
    --Grant County 3383
    --Griggs County 3374
    --Lamoure County 3391
    --McIntosh County 3397
    --McLean County 3366
    --Mercer County 3356
    --Morton County 3395
    --Mountrail County 3352
    --Nelson County 3359
    --Oliver County 3356
    --Pembina County 3370
    --Pierce County 3363
    --Ramsey County 3393
    --Ransom County 3389
    --Sioux County 3383
    --Steele County 3374
    --Walsh County 3359
    --Wells County 3344
    --Western 3444
  Ohio 3537, 3565, 3598
  Oklahoma 3690
  Oregon 3753, 3759
  Pennsylvania 3931
  South Dakota 4026, 4030
  Southeastern United States
    3284
  Tennessee 4120
    --Morgan County 4141
    --Scott County 4142
  West Virginia 4692
    --Little Kanawha River
      Basin 4750
    --Monongahela River Ba-
      sin 4749
    --New River Basin 4752
    --Potomac River Basin
      4751
Wells County, North Dakota
  Aquifers 3345
  Geology 3343, 3344, 3345
  Glaciation 3343, 3345
  Groundwater 3343, 3344,
    3345

  Springs 3344
  Water 3345
  Wells 3344
  Williston Basin 3477
Wells Creek Basin, Tennessee
  Faulting 4137
Wells Creek Structure, Tennessee
  Geology 4112
Wessington Springs, South Dakota
  Groundwater 4083
West Baden Group
  Indiana 1521
West Danville, Vermont
  Morphometry 4500
  Sedimentation 4500
West Fork Glacier, Alaska
  Mineral Occurrence 340
West Mansfield Quadrangle, Ohio
  Glacial Geology 3570
West Point Quadrangle, Mississippi
  Bentonite 2481
  Geology 2481
West Torrington Quadrangle, Connec-
  ticut
  Bedrock Geology 673
Western Coal Field Region, Kentucky
  Aquifers 1961
  Geology 1961
  Hydrology 1961
  Pennsylvanian 1961
  Sandstone 1961
Western Phosphate Field, Idaho 1175
Westmoreland County, Pennsylvania
  Geology 3894
  Groundwater 3948
  Petrography 3861
  Upper Freeport Coal 3894
Weston County, Wyoming
  Bentonite 4851
Wetlands
  Illinois
    --Alexander County 1461
    --Union County 1461
Whatcom County, Washington
  Minerals 4609
  Mines 4609
Wheeler County, Oregon
  Bedrock Geology 3714
White Castle, Louisiana
  Groundwater 2176
White Mountains, Arizona
  Cenozoic 384
  Geology 384
White Pass Highway, Washington
  Geology 4648
White Pine County, Nevada
  Geology 2886
  Minerals 2886
White River, Alaska

Geochemistry 229
Geology 229
White River Formation
South Dakota
--Big Badlands 4052
Whiterock 3261
Whiterock Cephalopods 3244
Whitewater Formation
Indiana
--Eastern 1493
Ohio
--Western 1493
Whitfield County, Georgia
Geology 954
Groundwater 954
Wichita County, Kansas
Groundwater 1814, 1824
Wichita Mountains, Oklahoma
Late Cambrian 3651
Ordovician 3651
Trilobites 3651
Wilberns Formation
Texas
--Central 4297, 4301
Wilberton Pool
Illinois
--Fayette County 1218
Wilcox County, Alabama
Energy Resources 176
Gravity Surveys 120
Mineral Resources 124, 176
Water Resources 176
Wilcox Group
Mississippi 2479
Texas 4170
--East 4226, 4333
Wild Cow Formation
New Mexico
--Manzanita Mountains 3189
Wild Lake, Alaska
Rock 233
Stream Sediments 233
Wilderness Areas
Idaho
--Selway-Bitterroot Wilder-
ness Area 1127, 1174
Wilkes County, Georgia
Copper 950
Lead 950
Stream Sediments 950
Zinc 950
Will County, Illinois
Coal 1275
Drilling 1386
Williams County, North Dakota
Geology 3334, 3335, 3336,
3416
Glaciation 3334
Groundwater 3334, 3335,

3336, 3492
Land Use 3492
Minerals 3492
Pollution 3492
Tertiary 3334
Water 3335, 3336
Williamsburg Quadrangle, Virginia
Geology 4572
Williamsport Sandstone
West Virginia 4683, 4739
Williston, North Dakota
Geology 3416
Williston Basin, North Dakota
Cannonball Formation 3487
Fort Union Group 3488
Fox Hills Formation 3485
Geology 3485, 3487, 3488
Newcastle Formation 3477
Oil and Gas 3467
Paleocene 3453
Potash Salts 3498
Sediments 3438, 3453
Spearfish Formation 3346
Tectonics 3438
Uranium 3485, 3487, 3488
Willow Creek, Alaska
Geochemistry 222
Wills Mountain Anticline, West Vir-
ginia 4748
Willwood Formation
Wyoming
--Washakie County 4848
Wilson Canyon, California
Earthquakes 430
Geology 430
Wilson County, Tennessee
Geology 4161
Winborn, Mississippi
Smelting 2477
Wind River Mountains, Wyoming
Talconite 4857
Windbreaks 2957
Windham, Vermont
Magnetic Data 4485
Ultramafic Bodies 4485
Winding Stair Range, Oklahoma
Geology 3625, 3664
Windsor, Vermont
Geology 4498
Windsor Quadrangle, Virginia
Geology 4571
Windsor Study Area, Colorado
Geology 602
Winn Parish, Louisiana
Ostracoda 2150
Winner, South Dakota
Groundwater 4081
Winston County, Alabama
Energy Resources 181

Winston County, Alabama (Cont.)
  Geology 52
  Gravity Surveys 121
  Mineral Resources 181
  Water 52
  Water Resources 181
Winston Mining District, Montana
  Economic Geology 2691
  Geochemistry 2691
Wisconsinan 1341
  Illinois
  --Central 1475
  --Central Western 1467
  Indiana
  --Eastern 1523
  North Dakota
  --Logan County 3422
  Pennsylvania
  --Northeastern 3822
Wiseman Quadrangle, Alaska
  Rock 233
  Stream Sediments 233
Wolf Gap Quadrangle, Virginia
  Geology 4589
Wolfcamp Environment
  Texas
  --North-Central 4174
Wolfcampian
  New Mexico
  --Central 3254
  --Southeast 3241
Wollastonite
  California 424
Wolsey, South Dakota
  Groundwater 4108
Wood River, Alaska
  Mineral Occurrences 340
Wood River Lake, Alaska
  Geochemistry 225
Woodbine Formation
  Oklahoma
  --Bryan County 3636
  Texas
  --Northeast 4308
Woodfordian
  Illinois 1283
  --Northeastern 1343
Woodfordian (Jules) 1342
Woodmont Quadrangle, Connecticut
  Surficial Geology 674
Woods County, Oklahoma
  Geology 3628
  Minerals 3628
Woodstock Quadrangle, Virginia
  Geology 4589
Woodstock Quadrangle, Vermont
  Bedrock Geology 4482
Woodward County, Oklahoma

  Guide 3692
Worcester County, Maryland
  Groundwater 2230, 2302
  Wells 2230
Worlds End State Park, Pennsylvania
  Guide 3909
Worth County, Georgia
  Geohydrology 925
Worth County, Missouri
  Water 2649
Wounded Knee, South Dakota
  Early Miocene 4058
  Microfossils 4058
  Monroe Creek 4058
Wreford Megacyclothem
  Kansas
  --Southern 1782
  Oklahoma
  --Northern 1782
Wrightwood, California
  Landsliding 577
  Mudflows 577
Wutinoceratidae 3260
Wyandotte Limestone
  Kansas
  --Eastern 1744
Wyndale Quadrangle, Virginia
  Geology 4553
Wynoochee Valley Quadrangle, Washington
  Geology 4608
Wyoming Valley, Pennsylvania
  Hydrology 3939
  Pleistocene Sediments 3939

X-Ray Diffraction 4697, 4740
X-Ray Powder Diffraction 1576
Xenoliths
  Kansas
  --Riley County 1747
Xenotime Mineralization
  California
  --Music Valley 496
  --Riverside County 496

Yazoo Clay
  Alabama
  --Southwestern 27
Yazoo County, Mississippi
  Geology 2467
Yellow Dog Peridotite
  Michigan
  --Northern Peninsula 2365
Yellow River Basin, Georgia
  Surface Water 928
Yellow Spring Group
  Iowa 1646

Yellow Tavern Quadrangle, Virginia
  Geology 4592
Yerington District, Nevada
  Ash 2933
  Stratigraphy 2933
  Tertiary 2933
  Tuff 2933
York County, Maine
  Geology 2209
York County, Pennsylvania
  Groundwater 3953
  Hydrology 3932
  New Oxford Formation 3932
York County, South Carolina
  Geology 3982
  Minerals 3982
Yorktown Quadrangle, Virginia
  Geology 4584
Yosemite Valley, California
  Geology 423
  Granitic Intrusive Units 512
Youngsville Quadrangle, Pennsylvania
  Oil and Gas 3872
Yukon, Alaska
  Geochemistry 221
  Mineral Resources 328

Zenith Mine, Minnesota
  Geology 2438
  Iron Ore 2438
Zeolite
  Alaska
    --Horn Mountains 335
    --Matanuska Valley 332
    --South-Central 335
  Nevada 2880
Zinc
  California 424
    --Inyo County 490
    --Modoc District 490
    --Panamint Butte Quadrangle 490
    --Plumas County 544
  Georgia
    --Lincoln County 950
    --Metasville Quadrangle 950
    --Wilkes County 950
  Kansas
    --Linn County 1761
  Maryland
    --Piedmont Upland 2234
  Missouri
    --Southeast 2638
  Pennsylvania 3891
  Tennessee 4134

  --East 4153
  --Henry County 4148
  --Northwest 4153
  Utah 4436
  Virginia
    --Staunton Quadrangle 4585
  Washington
    --Kootenay Arc 4613
    --Northeastern 4613
    --Stevens County 4622
  Wisconsin 4781
    --Southwest 4793
Zircon 1130
Zirconium
  California 424
Zonation
  Illinois 1283
Zoning
  Kentucky
    --Central Kentucky Mineral District 1982
  Wisconsin 4832
Zuni Mountains, New Mexico
  Geology 3274